Rectangle

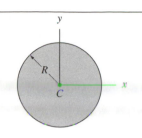

$$\bar{I}_x = \frac{bh^3}{12} \qquad \bar{I}_y = \frac{b^3h}{12} \qquad \bar{I}_{xy} = 0$$

$$I_x = \frac{bh^3}{3} \qquad I_y = \frac{b^3h}{3} \qquad I_{xy} = \frac{b^2h^2}{4}$$

Circle

$$I_x = I_y = \frac{\pi R^4}{4} \qquad I_{xy} = 0$$

Half parabolic complement

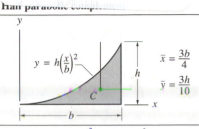

$$\bar{x} = \frac{3b}{4} \qquad \bar{y} = \frac{3h}{10}$$

$$\bar{I}_x = \frac{37bh^3}{2100} \qquad I_x = \frac{bh^3}{21}$$

$$\bar{I}_y = \frac{b^3h}{80} \qquad I_y = \frac{b^3h}{5}$$

$$\bar{I}_{xy} = \frac{b^2h^2}{120} \qquad I_{xy} = \frac{b^2h^2}{12}$$

Right triangle

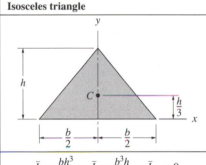

$$\bar{I}_x = \frac{bh^3}{36} \qquad \bar{I}_y = \frac{b^3h}{36} \qquad \bar{I}_{xy} = -\frac{b^2h^2}{72}$$

$$I_x = \frac{bh^3}{12} \qquad I_y = \frac{b^3h}{12} \qquad I_{xy} = \frac{b^2h^2}{24}$$

Semicircle

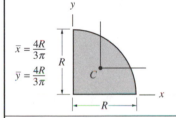

$$\bar{I}_x = 0.1098R^4 \qquad \bar{I}_{xy} = 0$$

$$I_x = I_y = \frac{\pi R^4}{8} \qquad I_{xy} = 0$$

Half parabola

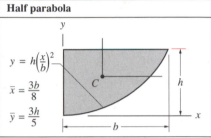

$$y = h\left(\frac{x}{b}\right)^2$$

$$\bar{x} = \frac{3b}{8} \qquad \bar{y} = \frac{3h}{5}$$

$$\bar{I}_x = \frac{8bh^3}{175} \qquad I_x = \frac{2bh^3}{7}$$

$$\bar{I}_y = \frac{19b^3h}{480} \qquad I_y = \frac{2b^3h}{15}$$

$$\bar{I}_{xy} = \frac{b^2h^2}{60} \qquad I_{xy} = \frac{b^2h^2}{6}$$

Isosceles triangle

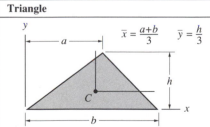

$$\bar{I}_x = \frac{bh^3}{36} \qquad \bar{I}_y = \frac{b^3h}{48} \qquad \bar{I}_{xy} = 0$$

$$I_x = \frac{bh^3}{12} \qquad I_{xy} = 0$$

Quarter circle

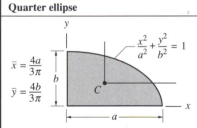

$$\bar{x} = \frac{4R}{3\pi}$$

$$\bar{y} = \frac{4R}{3\pi}$$

$$\bar{I}_x = \bar{I}_y = 0.054\,88R^4 \qquad I_x = I_y = \frac{\pi R^4}{16}$$

$$\bar{I}_{xy} = -0.016\,47R^4 \qquad I_{xy} = \frac{R^4}{8}$$

Circular sector

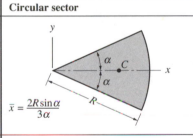

$$\bar{x} = \frac{2R\sin\alpha}{3\alpha}$$

$$I_x = \frac{R^4}{8}(2\alpha - \sin 2\alpha)$$

$$I_y = \frac{R^4}{8}(2\alpha + \sin 2\alpha)$$

$$I_{xy} = 0$$

Triangle

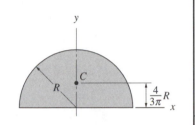

$$\bar{x} = \frac{a+b}{3} \qquad \bar{y} = \frac{h}{3}$$

$$\bar{I}_x = \frac{bh^3}{36} \qquad\qquad I_x = \frac{bh^3}{12}$$

$$\bar{I}_y = \frac{bh}{36}(a^2 - ab + b^2) \qquad I_y = \frac{bh}{12}(a^2 + ab + b^2)$$

$$\bar{I}_{xy} = \frac{bh^2}{72}(2a - b) \qquad I_{xy} = \frac{bh^2}{24}(2a + b)$$

Quarter ellipse

$$\frac{x^2}{a^2} + \frac{y^2}{b^2} = 1$$

$$\bar{x} = \frac{4a}{3\pi}$$

$$\bar{y} = \frac{4b}{3\pi}$$

$$\bar{I}_x = 0.054\,88ab^3 \qquad I_x = \frac{\pi ab^3}{16}$$

$$\bar{I}_y = 0.054\,88a^3b \qquad I_y = \frac{\pi a^3b}{16}$$

$$\bar{I}_{xy} = -0.016\,47a^2b^2 \qquad I_{xy} = \frac{a^2b^2}{8}$$

Engineering Mechanics

STATICS

Second Edition

Engineering Mechanics

STATICS

Second Edition

Andrew Pytel
The Pennsylvania State University

Jaan Kiusalaas
The Pennsylvania State University

Brooks/Cole Publishing Company

I⟨T⟩P® *An International Thomson Publishing Company*

Pacific Grove · Albany · Belmont · Boston · Cincinnati · Detroit · Johannesburg · London · Madrid
Melbourne · Mexico City · New York · Scottsdale · Singapore · Tokyo · Toronto

Sponsoring Editor: *Bill Stenquist*
Project Development Editor: *Suzanne Jeans*
Marketing Team: *Nathan Wilbur, Aaron
    Eden, Jean Vevers Thompson*
Editorial Assistant: *Meg Weist*
Production Coordinator: *Laurel Jackson*
Production Service: *Helen Walden*
Manuscript Editor: *Lynn Richardson*

Interior Design: *Helen Walden*
Cover Design: *Michael Rogondino*
Cover Photo: *Tom Tracy*
Typesetting: *Publication Services*
Cover Printing: *Phoenix Color Corporation*
Printing and Binding: *R. R. Donnelley &
    Sons/Crawfordsville Mfg. Division*

For more information, contact:

BROOKS/COLE PUBLISHING COMPANY
511 Forest Lodge Road
Pacific Grove, CA 93950
USA

International Thomson Editores
Seneca 53
Col. Polanco
11560 México, D. F., México

International Thomson Publishing Europe
Berkshire House 168-173
High Holborn
London WC1V 7AA
England

International Thomson Publishing GmbH
Königswinterer Strasse 418
53227 Bonn
Germany

Thomas Nelson Australia
102 Dodds Street
South Melbourne, 3205
Victoria, Australia

International Thomson Publishing Asia
60 Albert St.
#15-01 Albert Complex
Singapore 189969

Nelson Canada
1120 Birchmount Road
Scarborough, Ontario
Canada M1K 5G4

International Thomson Publishing Japan
Hirakawacho Kyowa Building, 3F
2-2-1 Hirakawacho
Chiyoda-ku, Tokyo 102
Japan

Printed in the United States of America

10 9 8 7 6 5 4 3 2 1

Library of Congress Cataloging-in-Publication Data
Pytel, Andrew.
    Engineering mechanics. Statics / Andrew Pytel, Jaan Kiusalaas.
    — 2nd ed.
        p.   cm.
    ISBN 0-534-95741-2
    1. Statics.   2. Mechanics, Applied.   I. Kiusalaas, Jaan.
II. Title.
TA351.P97   1999
620.1′03—dc21
                                                            98–53174
                                                              CIP

To Jean, Leslie, Lori, John, Nicholas

and

To Judy, Nicholas, Jennifer, Timothy

Contents

*Indicates optional articles.

Preface

Statics and dynamics form the foundation of many engineering disciplines and are, therefore, essential to the training of an engineer. Mastery of these subjects requires a clear understanding of the principles, and experience in applying the principles to a wide range of situations. Because the applications require reasoning rather than memorization, statics and dynamics are perceived by students as difficult courses, making their teaching particularly challenging.

This textbook on statics and its companion volume on dynamics were developed over many years of teaching. The salient features of *Statics* are as follows:

- The selection of homework problems strives for a balanced presentation. It includes many "textbook" problems that illustrate the principles in a straightforward manner. In addition, there are numerous problems of direct engineering relevance, which are more interesting and challenging.
- The problems are evenly divided between SI and U.S. Customary units.
- The analysis of equilibrium is presented in three articles. The first teaches how to draw a free-body diagram, and the second shows how to write the equilibrium equations from a given free-body diagram. The third article combines the methods just learned to arrive at a logical scheme for the complete analysis of a problem.
- The Sample Problems separate the method of analysis from the mathematical details of the solution. This approach to equilibrium analysis teaches the student to "think before calculating."
- The text continually emphasizes the importance of comparing the number of independent equations with the number of unknowns.

The book contains several optional topics, which are marked by an asterisk (*). This material can be omitted without jeopardizing the presentation of other parts of *Statics*. An asterisk is also used to mark problems that require advanced reasoning.

New to the Second Edition   The considerable feedback we received from the users of the first edition was very helpful in the preparation of the second edition. As the result of their comments and of suggestions made by the reviewers, we made the following changes:

- The number of homework problems has been increased.
- Approximately one-third of the homework problems are new or have changed numerical values.
- A new problem set involving units and dimensions has been added to the first chapter.

- Several chapter introductions have been rewritten to place the topics in a better perspective.
- A number of improvements in notation have been made.

Ancillary   *Study Guide to Accompany Pytel and Kiusalaas, Engineering Mechanics, Statics, Second Edition,* J. L. Pytel, 1999. The goal of this study guide is to help students master the problem-solving skills required in their study of *Statics.* Students are prompted to interact with the material as they work through "guided" problems. The study guide contains additional problems that are accompanied by complete solutions.

Acknowledgments   Statics and dynamics is a mature field of study that has been built over many generations. Thus any new textbook pays tacit homage to the books that preceded it and to their authors. We are also grateful to the following reviewers for their valuable suggestions: Duane Castaneda, University of Alabama–Birmingham; Scott G. Danielson, North Dakota State University; Richard N. Downer, University of Vermont; Howard Epstein, University of Connecticut; Ralph E. Flori, University of Missouri–Rolla; Li-Sheng Fu, Ohio State University; Susan L. Gerth, Kansas State University; Edward E. Hornsey, University of Missouri–Rolla; Cecil O. Huey, Clemson University; Thomas J. Kosic, Texas A&M University; Dahsin Liu, Michigan State University; Mark Mear, University of Texas–Austin; Satish Nair, University of Missouri–Columbia; Hamid Nayeb-Hashemi, Northeastern University, Boston; Robert Price, Louisiana Tech University; Robert Schmidt, University of Detroit–Mercy; Robert Seabloom, University of Washington–Seattle; Kassim M. Tarhini, Valparaiso University; Dennis Vandenbrink, Western Michigan University; Carl Vilmann, Michigan Tech University.

We especially recognize our indebtedness to Dr. Christine Masters, who checked the solutions to the homework problems.

Andrew Pytel
Jaan Kiusalaas

Engineering Mechanics

STATICS

Second Edition

1

Introduction to Statics

1.1 Introduction

a. What is engineering mechanics?

Statics and dynamics are introductory engineering mechanics courses, and they are among the first engineering courses encountered by most students. Therefore, it is appropriate that we begin with a brief exposition on the meaning of the term *engineering mechanics* and on the role that these courses play in engineering education. Before defining engineering mechanics, we must first consider the similarities and differences between physics and engineering.

In general terms, *physics* is the science that relates the properties of matter and energy, excluding biological and chemical effects. Physics includes the study of mechanics,* thermodynamics, electricity and magnetism, and nuclear physics. On the other hand, *engineering* is the application of the mathematical and physical sciences (physics, chemistry, and biology) to the design and manufacture of items that benefit humanity. *Design* is the key concept that distinguishes engineers from scientists. According to the Accreditation Board for Engineering and Technology (ABET), engineering design is the process of devising a system, component, or process to meet desired needs.

Mechanics is the branch of physics that considers the action of forces on bodies or fluids that are *at rest* or *in motion.* Correspondingly, the primary topics of mechanics are statics and dynamics. The first topic that you studied in your initial physics course, in either high school or college, was undoubtedly mechanics. Thus, *engineering mechanics* is the branch of engineering that applies the principles of mechanics to mechanical design (i.e., any design that must take into account the effect of forces). The primary goal of engineering mechanics courses is to introduce the student to the engineering applications of mechanics. Statics and Dynamics are generally followed by one or more courses that introduce material properties and deformation, usually called Strength of Materials or Mechanics of Materials. This sequence of courses is then followed by formal training in mechanical design.

Of course, engineering mechanics is an integral component of the education of engineers whose disciplines are related to the mechanical sciences,

---

*When discussing the topics included in physics, the term *mechanics* is used without a modifier. Quite naturally, this often leads to confusion between "mechanics" and "engineering mechanics."

such as aerospace engineering, architectural engineering, civil engineering, and mechanical engineering. However, a knowledge of engineering mechanics is also useful in most other engineering disciplines, because there, too, the mechanical behavior of a body or fluid must often be considered. Because mechanics was the first physical science to be applied to everyday life, it follows that engineering mechanics is the oldest branch of engineering. Given the interdisciplinary character of many engineering applications (e.g., robotics and manufacturing), a sound training in engineering mechanics continues to be one of the most important aspects of engineering education.

b. *Problem formulation and the accuracy of solutions*

As you will soon discover, your mastery of the principles of engineering mechanics will be reflected in your ability to formulate and solve problems. Unfortunately, there is no simple method for teaching problem-solving skills. Nearly all individuals require a considerable amount of practice in solving problems before they begin to develop the analytical skills that are so necessary for success in engineering. For this reason, a relatively large number of sample problems and homework problems are placed at strategic points throughout the text.

To help you develop an "engineering approach" to problem analysis, we think that you will find it instructive to divide your solution for each homework problem into the following parts:

1. GIVEN: After carefully reading the problem statement, list all the data provided. If a figure is required, sketch it neatly and approximately to scale.
2. FIND: State precisely the information that is to be determined.
3. SOLUTION: Solve the problem, showing all the steps that you used in the analysis. Work neatly so that your work can be easily followed by others.
4. VALIDATE: Many times, an invalid solution can be uncovered by simply asking yourself, "Does the answer make sense?"

When reporting your answers, use only as many digits as the least accurate value in the given data. For example, suppose that we are required to convert 12 500 ft (assumed to be accurate to three significant digits) to miles. Using a calculator, we would divide 12 500 ft by 5280 ft/mi and report the answer as 2.37 mi (three significant digits), although the quotient displayed on the calculator would be 2.367 424 2. Reporting the answer as 2.367 424 2 implies that all eight digits are significant, which is, of course, untrue. It is your responsibility to round off the answer to the correct number of digits. *In this text,* you should assume that given data are accurate to three significant digits unless stated otherwise. For example, a length that is given as 3 ft should be interpreted as 3.00 ft.

When performing intermediate calculations, a good rule of thumb is to carry one more digit than will be reported in the final answer; for example, use four-digit intermediate values if the answer is to be significant to three digits. Furthermore, it is common practice to report an additional digit if the first digit in an answer is 1; for example, use 1.392 rather than 1.39.

1.2 *Newtonian Mechanics*

a.   *Scope of Newtonian mechanics*

In 1687 Sir Isaac Newton (1642–1727) published his celebrated laws of motion in *Principia (Mathematical Principles of Natural Philosophy)*. Without a doubt, this work ranks among the most influential scientific books ever published. We should not think, however, that its publication immediately established classical mechanics. Newton's work on mechanics dealt primarily with celestial mechanics and was thus limited to particle motion. Another two hundred or so years elapsed before rigid-body dynamics, fluid mechanics, and the mechanics of deformable bodies were developed. Each of these areas required new axioms before it could assume a usable form.

Nevertheless, Newton's work is the foundation of classical, or Newtonian, mechanics. His efforts have even influenced two other branches of mechanics, born at the beginning of the twentieth century: relativistic and quantum mechanics. *Relativistic mechanics* addresses phenomena that occur on a cosmic scale (velocities approaching the speed of light, strong gravitational fields, etc.). It removes two of the most objectionable postulates of Newtonian mechanics: the existence of a fixed or inertial reference frame and the assumption that time is an absolute variable, "running" at the same rate in all parts of the universe. (There is evidence that Newton himself was bothered by these two postulates.) *Quantum mechanics* is concerned with particles on the atomic or subatomic scale. It also removes two cherished concepts of classical mechanics: determinism and continuity. Quantum mechanics is essentially a probabilistic theory; instead of predicting an event, it determines the likelihood that an event will occur. Moreover, according to this theory, the events occur in discrete steps (called *quanta*) rather than in a continuous manner.

Relativistic and quantum mechanics, however, have by no means invalidated the principles of Newtonian mechanics. In the analysis of the motion of bodies encountered in our everyday experience, both theories converge on the equations of Newtonian mechanics. Thus the more esoteric theories actually reinforce the validity of Newton's laws of motion.

b.   *Newton's laws for particle motion*

Using modern terminology, Newton's laws of particle motion may be stated as follows:

1. If a particle is at rest (or moving with constant velocity in a straight line), it will remain at rest (or continue to move with constant velocity in a straight line) unless acted on by a force.
2. A particle acted on by a force will accelerate in the direction of the force. The magnitude of the acceleration is proportional to the magnitude of the force and inversely proportional to the mass of the particle.
3. For every action, there is an equal and opposite reaction; that is, the forces of interaction between two particles are equal in magnitude and opposite in direction.

Although the first law is simply a special case of the second law, it is customary to state the first law separately because of its importance to the subject of statics.

c.   Inertial reference frames

When applying Newton's second law, attention must be paid to the coordinate system in which the accelerations are measured. An *inertial reference frame* (also known as a Newtonian or Galilean reference frame) is defined to be any rigid coordinate system in which Newton's laws of particle motion relative to that frame are valid with an acceptable degree of accuracy. In most design applications used on the surface of the earth, an inertial frame can be approximated with sufficient accuracy by attaching the coordinate system to the earth. In the study of earth satellites, a coordinate system attached to the sun usually suffices. For interplanetary travel, it is necessary to use coordinate systems attached to the so-called fixed stars.

It can be shown that any frame that is translating with constant velocity relative to an inertial frame is itself an inertial frame. It is a common practice to omit the word *inertial* when referring to frames for which Newton's laws obviously apply.

d.   Units and dimensions

The standards of measurement are called *units*. The term *dimension* refers to the type of measurement, regardless of the units used. For example, kilogram and foot/second are units, whereas mass and length/time are dimensions. Throughout this text we use two standards of measurement: U.S. Customary system and SI system (from *Système internationale d'unités*). In the *U.S. Customary system* the base (fundamental) dimensions are force $[F]$, length $[L]$, and time $[T]$. The corresponding base units are pound (lb), foot (ft), and second (s). The base dimensions in the *SI system* are mass $[M]$, length $[L]$, and time $[T]$, and the base units are kilogram (kg), meter (m), and second (s). All other dimensions or units are combinations of the base quantities. For example, the dimension of velocity is $[L/T]$, the units being ft/s, m/s, and so on.

A system with the base dimensions $[FLT]$ (such as the U.S. Customary system) is called a *gravitational system*. If the base dimensions are $[MLT]$ (as in the SI system), the system is known as an *absolute system*. In each system of measurement, the base units are defined by physically reproducible phenomena or physical objects. For example, the second is defined by the duration of a specified number of radiation cycles in a certain isotope, the kilogram is defined as the mass of a certain block of metal kept near Paris, France, and so on.

All equations representing physical phenomena must be *dimensionally homogeneous;* that is, each term of an equation must have the same dimension. Otherwise, the equation will not make physical sense (it would be meaningless, for example, to add a force to a length). Checking equations for dimensional homogeneity is a good habit to learn, as it can reveal mistakes made during algebraic manipulations.

e. Mass, force, and weight

If a force $\mathbf{F}$ acts on a particle of mass m, Newton's second law states that

$$\mathbf{F} = m\mathbf{a} \tag{1.1}$$

where $\mathbf{a}$ is the acceleration vector of the particle. For a gravitational $[FLT]$ system, dimensional homogeneity of Eq. (1.1) requires the dimension of mass to be

$$[M] = \left[\frac{FT^2}{L}\right] \tag{1.2a}$$

In the U.S. Customary system, the derived unit of mass is called a *slug*. A slug is defined as the mass that is accelerated at the rate of 1.0 ft/s^2 by a force of 1.0 lb. Substituting units for dimensions in Eq. (1.2a), we get for the unit of a slug

$$1.0 \, \text{slug} = 1.0 \, \text{lb} \cdot \text{s}^2/\text{ft}$$

For an absolute $[MLT]$ system of units, dimensional homogeneity of Eq. (1.1) yields for the dimension of force

$$[F] = \left[\frac{ML}{T^2}\right] \tag{1.2b}$$

The derived unit of force in the SI system is a *newton* (N), defined as the force that accelerates a 1.0-kg mass at the rate of 1.0 m/s^2. From Eq. (1.2b), we obtain

$$1.0 \, \text{N} = 1.0 \, \text{kg} \cdot \text{m/s}^2$$

Weight is the force of gravitation acting on a body. Denoting gravitational acceleration (free-fall acceleration of the body) by g, the weight W of a body of mass m is given by Newton's second law as

$$W = mg \tag{1.3}$$

Note that mass is a constant property of a body, whereas weight is a variable that depends on the local value of g. The gravitational acceleration on the surface of the earth is approximately 32.2 ft/s^2, or 9.81 m/s^2. Thus the mass of a body that weighs 1.0 lb on earth is (1.0 lb)/(32.2 ft/s^2) = 1/32.2 slug. Similarly, if the mass of a body is 1.0 kg, its weight on earth is (9.81 m/s^2)(1.0 kg) = 9.81 N.

At one time, the pound was also used as a unit of mass. The *pound mass* (lbm) was defined as the mass of a body that weighs 1.0 lb on the surface of the earth. Although pound mass is an obsolete unit, it is still used occasionally, giving rise to confusion between mass and weight. In this text, we use the pound exclusively as a unit of force.

f. Conversion of units

A convenient method for converting a measurement from one set of units to another is to multiply the measurement by appropriate conversion factors. For example, to convert 240 mi/h into ft/s, we proceed as follows:

$$240\,\text{mi/h} = 240\frac{\cancel{\text{mi}}}{\cancel{\text{h}}} \times \frac{1.0\,\cancel{\text{h}}}{3600\,\text{s}} \times \frac{5280\,\text{ft}}{1.0\,\cancel{\text{mi}}} = 352\,\text{ft/s}$$

where the multipliers 1.0 h/3600 s and 5280 ft/1.0 mi are conversion factors. Because 1.0 h = 3600 s and 5280 ft = 1.0 mi, we see that each conversion factor is dimensionless and of magnitude 1. Therefore, a measurement is unchanged when it is multiplied by conversion factors—only its units are altered. Note that it is permissible to cancel units during the conversion as if they were algebraic quantities.

Conversion factors applicable to mechanics are listed inside the front cover of the book.

g. Law of gravitation

In addition to his many other accomplishments, Newton also proposed the law of universal gravitation. Consider two particles of mass m_A and m_B that are separated by a distance R, as shown in Fig. 1.1. The law of gravitation states that the two particles are attracted to each other by forces of magnitude F that act along the line connecting the particles, where

$$F = G\frac{m_A m_B}{R^2} \tag{1.4}$$

The universal gravitational constant G is equal to 3.44×10^{-8} ft^4/(lb · s^4), or 6.67×10^{-11} m^3/(kg · s^2). Although this law is valid for particles, Newton showed that it is also applicable to spherical bodies, provided that their masses are distributed uniformly. (When attempting to derive this result, Newton was forced to develop calculus.)

If we let $m_A = M_e$ (the mass of the earth), $m_B = m$ (the mass of a body), and $R = R_e$ (the mean radius of the earth), then F in Eq. (1.4) will be the weight W of the body. Comparing $W = GM_e m/R_e^2$ with $W = mg$, we find that $g = GM_e/R_e^2$. Of course, adjustments may be necessary in the value of g for some applications in order to account for local variation of the gravitational attraction.

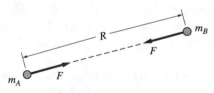

Fig. 1.1

Sample Problem 1.1

Convert 5000 lb/in.^{2} to Pa (1 Pa = 1 N/m^2).

Solution

Using the conversion factors listed inside the front cover, we obtain

$$5000 \text{ lb/in.}^2 = 5000 \frac{\cancel{lb}}{\cancel{in.}^2} \times \frac{4.448 \text{ N}}{1.0\cancel{lb}} \times \left(\frac{39.37 \cancel{in.}}{1.0 \text{ m}}\right)^2$$

$$= 34.5 \times 10^6 \text{ N/m}^2 = 34.5 \text{ MPa} \qquad \textit{Answer}$$

Sample Problem 1.2

The acceleration a of a particle is related to its velocity v, its position coordinate x, and time t by the equation

$$a = Ax^3t + Bvt^2 \qquad \text{(a)}$$

where A and B are constants. The dimension of the acceleration is length per unit time squared; that is, $[a] = [L/T^2]$. The dimensions of the other variables are $[v] = [L/T]$, $[x] = [L]$, and $[t] = [T]$. Derive the dimensions of A and B if Eq. (a) is to be dimensionally homogeneous.

Solution

For Eq. (a) to be dimensionally homogeneous, the dimension of each term on the right-hand side of the equation must be $[L/T^2]$, the same as the dimension for a. Therefore, the dimension of the first term on the right-hand side of Eq. (a) becomes

$$[Ax^3t] = [A][x^3][t] = [A][L^3][T] = \left[\frac{L}{T^2}\right] \qquad \text{(b)}$$

Solving Eq.(b) for the dimension of A, we find

$$[A] = \frac{1}{[L^3][T]}\left[\frac{L}{T^2}\right] = \frac{1}{[L^2T^3]} \qquad \textit{Answer}$$

Performing a similar dimensional analysis on the second term on the right-hand side of Eq. (a) gives

$$[Bvt^2] = [B][v][t^2] = [B]\left[\frac{L}{T}\right][T^2] = \left[\frac{L}{T^2}\right] \qquad \text{(c)}$$

Solving Eq. (c) for the dimension of B, we find

$$[B] = \left[\frac{L}{T^2}\right]\left[\frac{T}{L}\right]\left[\frac{1}{T^2}\right] = \left[\frac{1}{T^3}\right] \qquad \textit{Answer}$$

Sample Problem 1.3

Find the gravitational force exerted by the earth on a 70-kg man whose elevation above the surface of the earth equals the radius of the earth. The mass and radius of the earth are $M_e = 5.9742 \times 10^{24}$ kg and $R_e = 6378$ km, respectively.

Solution

Consider a body of mass m located at the distance $2R_e$ from the center of the earth (of mass M_e). The law of universal gravitation, from Eq. (11.4), states that the body is attracted to the earth by the force F given by

$$F = G \frac{mM_e}{(2R_e)^2}$$

where $G = 6.67 \times 10^{-11}$ m^3/(kg·s^2) is the universal gravitational constant. Substituting the values for G and the given parameters, the earth's gravitational force acting on the 70-kg man is

$$F = (6.67 \times 10^{-11}) \frac{(70)(5.9742 \times 10^{24})}{[2(6378 \times 10^3)]^2} = 171.4 \text{ N} \qquad \textit{Answer}$$

Problems

1.1   A person weighs 30 lb on the moon, where $g = 5.32$ ft/s^2. Determine (a) the mass of the person; and (b) the weight of the person on earth.

1.2   The radius and length of a steel cylinder are 60 mm and 120 mm, respectively. If the mass density of steel is 7850 kg/m^3, determine the weight of the cylinder in pounds.

1.3   Convert the following: (a) 600 lb/ft^2 to kN/m^2; (b) 60 mi/h to m/s; (c) 10 Mg to slugs; and (d) 14.7 lb/in.^{2} to kN/m^2.

1.4   The mass moment of inertia of a certain body is $I = 20$ kg $\cdot$ m^2. Express I in terms of the base units of the U.S. Customary system.

1.5   When a rigid body of mass m undergoes plane motion, its kinetic energy (KE) is

$$ KE = \frac{1}{2}mv^2 + \frac{1}{2}mk^2\omega^2 $$

where v is the velocity of its mass center, k is a constant, and ω is the angular velocity of the body in rad/s. Express the units of KE and k in terms of the base units of (a) the SI system and (b) the U.S. Customary system.

1.6   In a certain application, the acceleration a and the position coordinate x of a particle are related by

$$ a = \frac{gkx}{W} $$

where g is the gravitational acceleration, k is a constant, and W is the weight of the particle. Show that this equation is dimensionally consistent if the dimension of k is $[F/L]$.

1.7   When a force F acts on a linear spring, the elongation x of the spring is given by $F = kx$, where k is called the stiffness of the spring. Determine the dimension of k in terms of the base dimensions of an absolute $[MLT]$ system of units.

1.8   Determine the dimensions of the following in terms of the base dimensions of a gravitational $[FLT]$ system of units: (a) mv^2; (b) mv; and (c) ma.

1.9   A geometry textbook gives the equation of a parabola as $y = x^2$, where x and y are measured in inches. How can this equation be dimensionally correct?

1.10   The mass moment of inertia I of a homogeneous sphere about its diameter is $I = (2/5)mR^2$, where m and R are its mass and radius, respectively. Find the dimension of I in terms of the base dimensions of (a) a gravitational $[FLT]$ system; and (b) an absolute $[MLT]$ system.

1.11   Determine the dimensions of constants A and B in the following equations, assuming each equation to be dimensionally correct: (a) $v^2 = Ax^3 + Bvt$; and (b) $x^3 = Ate^{Bt}$. The dimensions of the variables are $x[L]$, $t[T]$ and $v[L/T]$.

*1.12   In a certain vibration problem the differential equation describing the motion of a particle of mass m is

$$ m\frac{d^2x}{dt^2} + c\frac{dx}{dt} + kx = P_0 \sin \omega t $$

where x is the displacement of the particle and t is time. What are the dimensions of the constants c, k, P_0, and ω in terms of the base dimensions of a gravitational $[FLT]$ system?

1.13 Using Eq. (1.4), derive the dimensions of the universal gravitational constant G in terms of the base dimensions of (a) a gravitational $[FLT]$ system; and (b) an absolute $[MLT]$ system.

1.14 A famous equation of Einstein states that $E = mc^2$, where E is energy, m is mass, and c is the speed of light. Determine the dimension of energy in terms of the base dimensions of (a) a gravitational $[FLT]$ system; and (b) an absolute $[MLT]$ system.

1.15 Two 10-kg particles are placed 500 mm apart. Express the gravitational attraction acting on one of the particles as a percentage of its weight on earth.

1.16 Two identical spheres of radius 8 in. and weighing 2 lb on the surface of the earth are placed in contact. Find the gravitational attraction between them.

Use the following data for Problems 1.17–1.21: mass of earth = 5.9742×10^{24} kg, radius of earth = 6378 km, mass of moon = $0.073\,483 \times 10^{24}$ kg, radius of moon = 1738 km.

1.17 Find the mass of an object (in kg) that weighs 2 kN at a height of 1800 km above the earth's surface.

1.18 Use Eq. (1.4) to show that the weight of an object on the moon is approximately 1/6 its weight on earth.

1.19 Plot the earth's gravitational acceleration g (m/s^2) against the height h (km) above the surface of the earth.

1.20 Find the height h (km) above earth's surface where the gravitational attraction is one-half its value on earth.

1.21 Calculate the gravitational force between the earth and the moon in newtons. The distance between the earth and the moon is 384×10^3 km.

1.3 *Fundamental Properties of Vectors*

A knowledge of vectors is a prerequisite for the study of statics. In this article, we describe the fundamental properties of vectors, with subsequent articles discussing some of the more important elements of vector algebra. (The calculus of vectors will be introduced as needed in *Dynamics*.) We assume that you are already familiar with vector algebra—our discussion is intended only to be a review of the basic concepts.

The differences between scalar and vector quantities must be understood:

> A *scalar* is a quantity that has magnitude only. A *vector* is a quantity that possesses magnitude and direction and obeys the parallelogram law for addition.

Because scalars possess only magnitudes, they are real numbers that can be positive, negative, or zero. Physical quantities that are scalars include temperature, time, and speed. As shown later, force, velocity, and displacement are examples of physical quantities that are vectors. The magnitude of a vector is always taken to be a nonnegative number. When a vector represents a physical quantity, the units of the vector are taken to be the same as the units of its magnitude (pounds, meters per second, feet, etc.).

The algebraic notation used for a scalar quantity must, of course, be different from that used for a vector quantity. In this text, we adopt the following conventions: (1) Scalars are written as italicized English or Greek letters—for example, *t* for time and *θ* for angle; (2) vectors are written as boldface letters—for example, **F** for force; and (3) the magnitude of a vector **A** is denoted as |**A**| or simply as *A* (italic).

There is no universal method for indicating vector quantities when writing by hand. The more common notations are $\vec{A}$, $\underset{\rightarrow}{A}$, $\overline{A}$, and $\underline{A}$. Unless instructed otherwise, you are free to use the convention that you find most comfortable. However, it is imperative that you take care to always distinguish between scalars and vectors when you write.

The following summarizes several important properties of vectors.

Vectors as Directed Line Segments   Any vector **A** can be represented geometrically as a directed line segment (an arrow), as shown in Fig. 1.2(a). The magnitude of **A** is denoted by *A,* and the direction of **A** is specified by the sense of the arrow and the angle *θ* that it makes with a fixed reference line. When using graphical methods, the length of the arrow is drawn proportional to the magnitude of the vector. Observe that the representation shown in Fig. 1.2(a) is complete because both the magnitude and direction of the vector are indicated. In some instances, it is also convenient to use the representation shown in Fig. 1.2(b), where the vector character of **A** is given additional emphasis by using boldface. Both of these representations for vectors are used in this text.

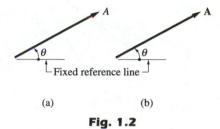

(a)   (b)

Fig. 1.2

We see that a vector does not possess a unique line of action, because moving a vector to a parallel line of action changes neither its magnitude nor its direction. In some engineering applications, the definition of a vector is more restrictive to include a line of action or even a point of application—see Art. 2.2.

Equality of Vectors   Two vectors **A** and **B** are said to be equal, written as **A** = **B**, if (1) their magnitudes are equal—that is, *A* = *B*, and (2) they have the same direction.

Scalar-Vector Multiplication   The multiplication of a scalar *m* and a vector **A**, written as *m***A** or as **A***m*, is defined as follows.

1. If *m* is positive, *m***A** is the vector of magnitude *mA* that has the same direction as **A**.
2. If *m* is negative, *m***A** is the vector of magnitude |*m*|*A* that is oppositely directed to **A**.
3. If *m* = 0, *m***A** (called the null or zero vector) is a vector of zero magnitude and arbitrary direction.

For *m* = −1, we see that (−1)**A** is the vector that has the same magnitude as **A** but is oppositely directed to **A**. The vector (−1)**A**, usually written as −**A**, is called the *negative of* **A**.

Unit Vectors   A unit vector is a dimensionless vector with magnitude 1. Therefore, if **λ** represents a unit vector (|**λ**| = 1) with the same direction as **A**, we can write **A** = *A***λ**.

The Parallelogram Law for Addition and the Triangle Law   The addition of two vectors **A** and **B** is defined to be the vector **C** that is determined by the

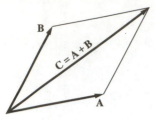

(a) Parallelogram law

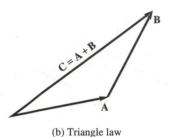

(b) Triangle law

Fig. 1.3

geometric construction shown in Fig. 1.3(a). Observe that **C** is the diagonal of the parallelogram formed by **A** and **B**. The operation depicted in Fig. 1.3(a), written as **A** + **B** = **C**, is called the *parallelogram law for addition*. The vectors **A** and **B** are referred to as *components* of **C**, and **C** is called the *resultant* of **A** and **B**. The process of replacing a resultant with its components is called *resolution*. For example, **C** in Fig. 1.3(a) is resolved into its components **A** and **B**.

An equivalent statement of the parallelogram law is the *triangle law*, which is shown in Fig. 1.3(b). Here the tail of **B** is placed at the tip of **A**, and **C** is the vector that completes the triangle, drawn from the tail of **A** to the tip of **B**. (Observe that the result is identical if the tail of **A** is placed at the tip of **B** and **C** is drawn from the tail of **B** to the tip of **A**.)

Letting **E**, **F**, and **G** represent any three vectors, we have the following two important properties (each follows directly from the parallelogram law):

- Addition is commutative: **E** + **F** = **F** + **E**
- Addition is associative: **E** + (**F** + **G**) = (**E** + **F**) + **G**

It is often convenient to find the sum **E** + **F** + **G** (no parentheses are needed) by adding the vectors from tip to tail, as shown in Fig. 1.4. The sum of the three vectors is seen to be the vector drawn from the tail of the first vector (**E**) to the tip of the last vector (**G**). This method, called the *polygon rule for addition,* can easily be extended to any number of vectors.

The subtraction of two vectors **A** and **B**, written as **A** − **B**, is defined as **A** − **B** = **A** + (−**B**), as shown in Fig. 1.5.

Because of the geometric nature of the parallelogram law and the triangle law, vector addition can be accomplished graphically. A second technique is to analytically determine the relationships between the various magnitudes and angles by applying the laws of sines and cosines to a sketch of the parallelogram (or the triangle). Both the graphical and the analytical methods are illustrated in the sample problems that follow.

Some words of caution: It is unfortunate that the symbols +, −, and = are commonly used in both scalar algebra and vector algebra, because they have completely different meanings in the two systems. For example, note the different meanings for + and = in the following two equations: **A** + **B** = **C** and 1 + 2 = 3. Unless you are extremely careful, this double meaning for symbols can easily lead to invalid expressions—for example, **A** + 5 (a vector cannot be added to a scalar!) and **A** = 1 (a vector cannot equal a scalar!).

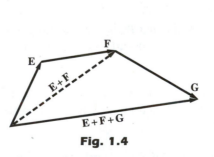

Fig. 1.4

Fig. 1.5

Sample Problem 1.4

Figure (a) shows two position vectors, the magnitudes of which are $A = 60$ ft and $B = 100$ ft. (A position vector is a vector drawn between two points in space.) Determine the resultant $\mathbf{R} = \mathbf{A} + \mathbf{B}$ using the following methods: (1) analytically, using the triangle law; and (2) graphically, using the triangle law.

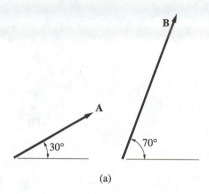

(a)

Solution

Part 1

The first step in the analytical solution is to draw a sketch (approximately to scale) of the triangle law. The magnitude and direction of the resultant are then found by applying the laws of sines and cosines to the triangle.

In this problem, the triangle law for the vector addition of $\mathbf{A}$ and $\mathbf{B}$ is shown in Fig. (b). The magnitude R of the resultant and the angle α are the unknowns to be determined. Applying the law of cosines, we obtain

$$R^2 = 60^2 + 100^2 - 2(60)(100)\cos 140°$$

which yields $R = 151.0$ ft.

The angle α can now be found from the law of sines:

$$\frac{100}{\sin \alpha} = \frac{R}{\sin 140°}$$

Substituting $R = 151.0$ ft and solving for α, we get $\alpha = 25.2°$. Referring to Fig. (b), we see that the angle that $\mathbf{R}$ makes with the horizontal is $30° + \alpha = 30° + 25.2° = 55.2°$. Therefore, the resultant of $\mathbf{A}$ and $\mathbf{B}$ is

$R = 151.0$ ft

55.2°

Answer

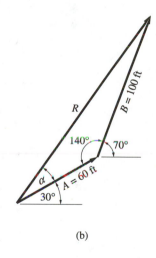

(b)

Part 2

In a graphical solution, Fig. (b) is drawn to scale with the aid of a ruler and a protractor. We first draw the vector $\mathbf{A}$ at $30°$ to the horizontal and then append vector $\mathbf{B}$ at $70°$ to the horizontal. The resultant $\mathbf{R}$ is then obtained by drawing a line from the tail of $\mathbf{A}$ to the head of $\mathbf{B}$. The magnitude of $\mathbf{R}$ and the angle it makes with the horizontal can now be measured directly from the figure.

Of course, the results would not be as accurate as those obtained in the analytical solution. If care is taken in making the drawing, two-digit accuracy is the best we can hope for. In this problem we should get $R \approx 150$ ft, inclined at $55°$ to the horizontal.

Sample Problem 1.5

The vertical force **P** of magnitude 100 kN is applied to the frame shown in Fig. (a). Resolve **P** into components that are parallel to the members AB and AC of the truss.

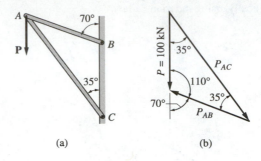

(a)　　　　　　　(b)

Solution

The force triangle in Fig. (b) represents the vector addition $\mathbf{P} = \mathbf{P}_{AC} + \mathbf{P}_{AB}$. The angles in the figure were derived from the inclinations of AC and AB with the vertical: $\mathbf{P}_{AC}$ is inclined at 35° (parallel to AC), and $\mathbf{P}_{AB}$ is inclined at 70° (parallel to AB). Applying the law of sines to the triangle, we obtain

$$\frac{100}{\sin 35°} = \frac{P_{AB}}{\sin 35°} = \frac{P_{AC}}{\sin 110°}$$

which yields for the magnitudes of the components

$$P_{AB} = 100.0 \, \text{kN} \qquad P_{BC} = 163.8 \, \text{kN} \qquad \qquad \textit{Answer}$$

Problems

Solve the problems in this set analytically, unless a graphical solution is specified.

1.22 The magnitudes of the three velocity vectors are $A = 50$ mi/h, $B = 40$ mi/h, and $C = 30$ mi/h. Determine (a) $\mathbf{A} + \mathbf{B}$; (b) $\mathbf{A} + \mathbf{B} + \mathbf{C}$; and (c) $\mathbf{A} - \mathbf{B}$. Solve graphically.

1.23 Using the vectors given in Prob. 1.22, find the scalars a and b such that $a\mathbf{A} + b\mathbf{B} = \mathbf{C}$. Solve graphically.

1.24 The total aerodynamic force $\mathbf{F}$ acting on the airplane has a magnitude of 6250 lb. Resolve this force into vertical and horizontal components (called the lift and the drag, respectively).

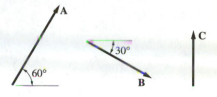

Fig. P1.22, P1.23

Fig. P1.24

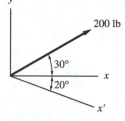

Fig. P1.25, P1.26

1.25 Resolve the 200-lb force into components along (a) the x- and y-axes; and (b) the x'- and y-axes.

1.26 Solve Prob. 1.25 graphically using the triangle law.

1.27 The two tugboats apply the forces $\mathbf{P}$ and $\mathbf{Q}$ to the barge, where $P = 76$ kN and $Q = 52$ kN. Determine the resultant of $\mathbf{P}$ and $\mathbf{Q}$.

1.28 Solve Prob. 1.27 graphically.

1.29 Determine the resultant of the position vectors $\mathbf{A}$ and $\mathbf{B}$.

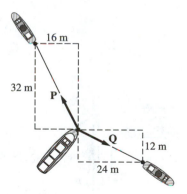

Fig. P1.27, P1.28

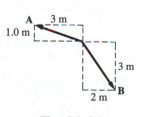

Fig. P1.29

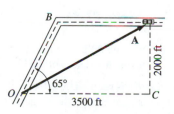

Fig. P1.30

1.30 Resolve the position vector $\mathbf{A}$ of the car (measured from fixed point O) into components along OB and OC.

1.31 Resolve the 360-lb force into components along the cables AB and AC. Use $\alpha = 55°$ and $\beta = 30°$.

1.32 The supporting cables AB and AC are oriented so that the components of the 360-lb force along AB and AC are 185 lb and 200 lb, respectively. Determine the angles α and β.

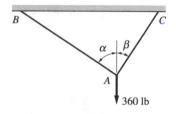

Fig. P1.31, P1.32

1.33   The two forces shown act on the structural member *AB*. Determine the magnitude of **P** such that the resultant of these forces is directed along *AB*.

1.34   The resultant of the two forces has a magnitude of 650 lb. Determine (a) the direction of the resultant; and (b) the magnitude of **P**.

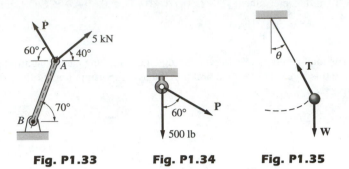

Fig. P1.33   **Fig. P1.34**   **Fig. P1.35**

1.35   The forces acting on the bob of the pendulum are its weight **W** ($W = 2$ lb) and the tension **T** in the cord. When the pendulum reaches the limit of its swing at $\theta = 30°$, it can be shown that the resultant of **W** and **T** is perpendicular to the cord. Determine the magnitude of **T** in this position.

1.36   The force **P** of magnitude 5 tons is to be resolved into components P_{OA} and P_{OB} acting along *OA* and *OB*, respectively. Determine (a) P_{OA} and P_{OB} if $\theta = 45°$; and (b) P_{OB} and θ if $P_{OA} = 6$ tons.

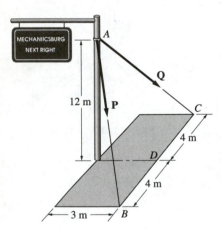

Fig. P1.38, P1.39

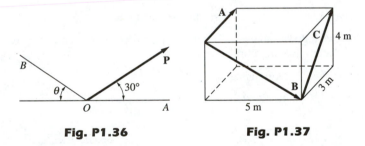

Fig. P1.36   **Fig. P1.37**

1.37   Determine the following resultants of the position vectors given in the figure, and show the results in a sketch of the "box": (a) **A** + **B**; and (b) **B** + **C**.

1.38   A signpost is secured by cables *AB* and *BC* that are pretensioned to $P = 800$ N and $Q = 1200$ N. Determine the magnitude of the resultant of **P** and **Q** and the angle between **P** and **Q**.

1.39   The forces **P** and **Q** are applied by the supporting cables to the signpost. If the magnitude of **P** + **Q** is 1265 N, determine *P* if (a) $P = Q$; and (b) $P = 2Q$.

1.4 *Representation of Vectors Using Rectangular Components*

The fundamental properties of vectors discussed in the preceding article are independent of coordinate systems. However, in engineering and other applied fields, it is usually convenient to describe the vectors relative to a coordinate system and then to perform mathematical operations, such as addition, using these descriptions. The rectangular coordinate system discussed here is the only reference frame we use in *Statics*.*

a. Rectangular components and direction cosines

The rectangular coordinate system is the *xyz* reference frame seen in Fig. 1.6, with which you are undoubtedly familiar. As shown in Fig. 1.6(a), the unit vectors that act in the positive *x*-, *y*-, and *z*-coordinate directions—called the base vectors—are labeled **i, j,** and **k,** respectively. Using the properties of vector addition, the vector **A** shown in Fig. 1.6(b) can be written as the sum of its three components:

$$\mathbf{A} = A_x\mathbf{i} + A_y\mathbf{j} + A_z\mathbf{k} \tag{1.5}$$

where

$$A_x = A\cos\theta_x \qquad A_y = A\cos\theta_y \qquad A_z = A\cos\theta_z \tag{1.6}$$

Note that θ_x, θ_y, and θ_z are the angles between the positive direction of **A** and the positive *x*-, *y*-, and *z*-axes, respectively. The cosines of these angles are called the *direction cosines* of **A** and we denote them by

$$\lambda_x = \cos\theta_x \qquad \lambda_y = \cos\theta_y \qquad \lambda_z = \cos\theta_z \tag{1.7}$$

In terms of the direction cosines, the rectangular form of **A** is

$$\begin{aligned}\mathbf{A} &= A\cos\theta_x\mathbf{i} + A\cos\theta_y\mathbf{j} + A\cos\theta_z\mathbf{k} \\ &= A(\lambda_x\mathbf{i} + \lambda_y\mathbf{j} + \lambda_z\mathbf{k})\end{aligned} \tag{1.8}$$

where the magnitude of **A** is

$$A = \sqrt{A_x^2 + A_y^2 + A_z^2} \tag{1.9}$$

Knowing that **A** can be written as $\mathbf{A} = A\boldsymbol{\lambda}$, where $\boldsymbol{\lambda}$ is the unit vector in the direction of **A**, as shown in Fig. 1.6(b), we see from Eq. (1.8) that

$$\boldsymbol{\lambda} = \lambda_x\mathbf{i} + \lambda_y\mathbf{j} + \lambda_z\mathbf{k} \tag{1.10}$$

The fact that the magnitude of $\boldsymbol{\lambda}$ is unity yields

$$\lambda_x^2 + \lambda_y^2 + \lambda_z^2 = 1 \tag{1.11}$$

which is an identity that must be satisfied by the direction cosines.

The terms A_x, A_y, and A_z are called the *scalar components* of **A,** whereas $A_x\mathbf{i}$, $A_y\mathbf{j}$, and $A_z\mathbf{k}$ are the *vector components* of **A.** For the sake of brevity, we will use the term *component,* omitting the word *scalar* or *vector* when the type of component is clear from the context. When a scalar component is positive (negative), the corresponding vector component is directed along the positive (negative) coordinate axis.

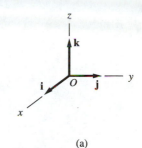

(a)

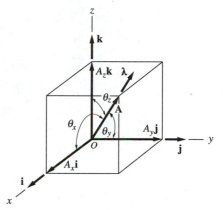

(b)

Fig. 1.6

---

*This is a special case of the Cartesian coordinate system, in which the coordinate axes are perpendicular.

b. *Vector addition using rectangular components*

Assume that two vectors **A** and **B** are expressed in their rectangular forms as follows: $\mathbf{A} = A_x\mathbf{i} + A_y\mathbf{j} + A_z\mathbf{k}$ and $\mathbf{B} = B_x\mathbf{i} + B_y\mathbf{j} + B_z\mathbf{k}$. Letting **C** be the sum of **A** and **B**, we have

$$\mathbf{C} = \mathbf{A} + \mathbf{B} = (A_x\mathbf{i} + A_y\mathbf{j} + A_z\mathbf{k}) + (B_x\mathbf{i} + B_y\mathbf{j} + B_z\mathbf{k})$$

which, using the properties of vector addition stated in the preceding article, can be written as

$$\mathbf{C} = C_x\mathbf{i} + C_y\mathbf{j} + C_z\mathbf{k}$$
$$= (A_x + B_x)\mathbf{i} + (A_y + B_y)\mathbf{j} + (A_z + B_z)\mathbf{k} \qquad (1.12)$$

Equating like components, the rectangular components of **C** are

$$\boxed{C_x = A_x + B_x \qquad C_y = A_y + B_y \qquad C_z = A_z + B_z} \qquad (1.13)$$

From Eq. (1.13), we see that each component of the sum equals the sum of the components. This result is depicted in Fig. 1.7, where, for simplicity's sake, the *xy*-plane has been chosen as a plane that contains the vectors **A** and **B**. Equations (1.12) and (1.13) can, of course, be extended to include the sum of any number of vectors.

c. *Position and relative position vectors and unit vectors*

The vector drawn from the origin O of a coordinate system to point B, denoted by $\overrightarrow{OB}$, is called the *position vector of B*. The vector $\overrightarrow{AB}$, drawn from point A to point B, is called the *position vector of B relative to A*. (Note that the position vector of B relative to A is the negative of the position vector of A relative to B; that is, $\overrightarrow{AB} = -\overrightarrow{BA}$.)

Figure 1.8 shows the relative position vector $\overrightarrow{AB}$: the vector drawn from $A(x_A, y_A, z_A)$ to $B(x_B, y_B, z_B)$. The rectangular representation of this vector is

$$\overrightarrow{AB} = (x_B - x_A)\mathbf{i} + (y_B - y_A)\mathbf{j} + (z_B - z_A)\mathbf{k} \qquad (1.14)$$

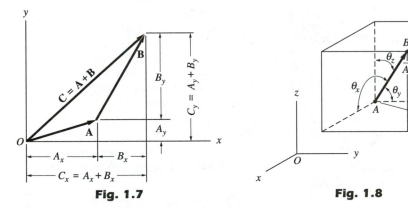

Fig. 1.7   **Fig. 1.8**

The magnitude of $\overrightarrow{AB}$ (the distance d in Fig. 1.8) is

$$|\overrightarrow{AB}| = d = \sqrt{(x_B - x_A)^2 + (y_B - y_A)^2 + (z_B - z_A)^2} \qquad (1.15)$$

The unit vector directed from point A toward B, which we indicate as $\boldsymbol{\lambda}$ in Fig. 1.8, can be found by dividing the vector $\overrightarrow{AB}$ by its magnitude:

$$\boldsymbol{\lambda} = \frac{\overrightarrow{AB}}{|\overrightarrow{AB}|} = \frac{(x_B - x_A)\mathbf{i} + (y_B - y_A)\mathbf{j} + (z_B - z_A)\mathbf{k}}{d} \qquad (1.16)$$

Referring to Fig. 1.8, we see that the direction cosines of $\overrightarrow{AB}$, or, equivalently, the components of $\boldsymbol{\lambda}$, are

$$\lambda_x = \cos\theta_x = \frac{x_B - x_A}{d}$$

$$\lambda_y = \cos\theta_y = \frac{y_B - y_A}{d} \qquad (1.17)$$

$$\lambda_z = \cos\theta_z = \frac{z_B - z_A}{d}$$

d. Techniques for writing a vector in rectangular form

Figure 1.9 shows a vector $\mathbf{F}$ that acts along the line AB. Here we summarize two useful methods for representing $\mathbf{F}$ in rectangular form—important skills that must be mastered if rectangular components are to be used effectively in problem analysis.

Method 1: Using Direction Cosines   If the angles θ_x, θ_y, and θ_z in Fig. 1.9 are known or can be easily determined, it is straightforward to use Eq. (1.8) to write $\mathbf{F}$ in rectangular form as follows:

$$\mathbf{F} = F\boldsymbol{\lambda} = F\cos\theta_x \mathbf{i} + F\cos\theta_y \mathbf{j} + F\cos\theta_z \mathbf{k}$$
$$= F(\lambda_x \mathbf{i} + \lambda_y \mathbf{j} + \lambda_z \mathbf{k}) \qquad (1.18)$$

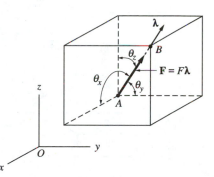

Fig. 1.9

Method 2: Using the Relative Position Vector   If the coordinates of points A and B in Fig. 1.9 are known or can easily be determined, the following steps lead to the description of $\mathbf{F}$ in rectangular form:

Step 1: Write the relative position vector $\overrightarrow{AB}$ in rectangular form. This step can be accomplished by substituting the coordinates of points A and B into Eq. (1.14) or by inspection of a carefully drawn sketch similar to Fig. 1.9.

Step 2: Evaluate the unit vector $\boldsymbol{\lambda} = \overrightarrow{AB} / |\overrightarrow{AB}|$.

Step 3: Write $\mathbf{F} = F\boldsymbol{\lambda}$.

Because the above two methods are equivalent, either may be used to determine the expression for $\mathbf{F}$. You should, of course, select the technique that is more convenient for the problem at hand. Regardless of the method employed, you will find that a carefully drawn sketch aids your understanding of the mathematical processes and reduces the opportunities for error.

Sample Problem 1.6

Determine the rectangular representation of the force **F** shown in Fig. (a), given that its magnitude is 500 lb.

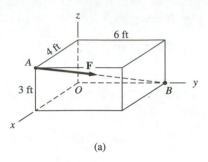

(a)

Solution

Because the coordinates of points A and B on the line of action of **F** are known, the following is a convenient method for obtaining the rectangular representation of **F**.

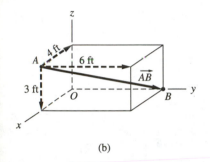

(b)

1. Write $\overrightarrow{AB}$, the vector from A to B, in rectangular form.

The vector $\overrightarrow{AB}$ and its rectangular components are shown in Fig. (b). Two common errors made by students at this point are choosing the wrong signs and mixing up the scalar components. You can avoid both of these difficulties by taking the time to show the vector on a carefully drawn sketch of the appropriate parallelepiped. From Fig. (b) we see that $\overrightarrow{AB}$ is

$$\overrightarrow{AB} = -4\mathbf{i} + 6\mathbf{j} - 3\mathbf{k} \text{ ft}$$

2. Evaluate $\boldsymbol{\lambda}$, the unit vector from A toward B:

$$\boldsymbol{\lambda} = \frac{\overrightarrow{AB}}{|\overrightarrow{AB}|} = \frac{-4\mathbf{i} + 6\mathbf{j} - 3\mathbf{k}}{\sqrt{(-4)^2 + 6^2 + (-3)^2}}$$

$$= -0.5122\mathbf{i} + 0.7682\mathbf{j} - 0.3841\mathbf{k}$$

3. Write $\mathbf{F} = F\boldsymbol{\lambda}$:

$$\mathbf{F} = 500(-0.5122\mathbf{i} + 0.7682\mathbf{j} - 0.3841\mathbf{k})$$

$$= -256\mathbf{i} + 384\mathbf{j} - 192\mathbf{k} \text{ lb} \qquad \textit{Answer}$$

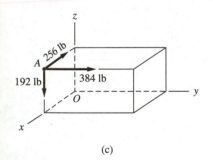

(c)

The rectangular components of **F** are shown in Fig. (c).

Sample Problem 1.7

Referring to Fig. (a), determine (1) the rectangular representation of the position vector **A**; and (2) the angles between **A** and each of the positive coordinate axes.

(a)

Solution

Part 1

We first resolve **A** into two components as shown in Fig. (b): A_z along the z-axis and A_{xy} in the xy-plane. (Once again we see that a carefully drawn sketch is an essential aid in performing vector resolution.) Because **A**, A_z, and A_{xy} lie in the same plane (a diagonal plane of the parallelepiped), we obtain by trigonometry

$$A_z = A\cos 30° = 12\cos 30° = 10.392\,\text{m}$$

$$A_{xy} = A\sin 30° = 12\sin 30° = 6\,\text{m}$$

The next step, illustrated in Fig. (c), is to resolve A_{xy} into the components:

$$A_x = A_{xy}\cos 40° = 6\cos 40° = 4.596\,\text{m}$$

$$A_y = A_{xy}\sin 40° = 6\sin 40° = 3.857\,\text{m}$$

Therefore, the rectangular representation of **A** is

$$\mathbf{A} = A_x\mathbf{i} + A_y\mathbf{j} + A_z\mathbf{k} = 4.60\mathbf{i} + 3.86\mathbf{j} + 10.39\mathbf{k}\,\text{m} \qquad \textit{Answer}$$

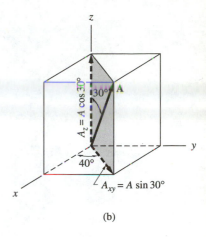

(b)

Part 2

The angles between **A** and the coordinate axes can be computed from Eq. (1.6):

$$\theta_x = \cos^{-1}\frac{A_x}{A} = \cos^{-1}\frac{4.596}{12} = 67.5°$$

$$\theta_y = \cos^{-1}\frac{A_y}{A} = \cos^{-1}\frac{3.857}{12} = 71.3° \qquad \textit{Answer}$$

$$\theta_z = \cos^{-1}\frac{A_z}{A} = \cos^{-1}\frac{10.392}{12} = 30.0°$$

These angles are shown in Fig. (d). Note that it was not necessary to compute θ_z, because it was already given in Fig (a).

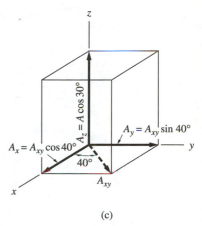

(c)

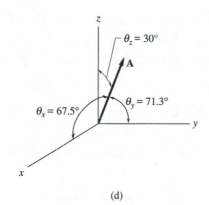

(d)

Sample Problem 1.8

Using rectangular components, find the resultant **R** of the vectors **P** and **Q** shown in Fig. (a).

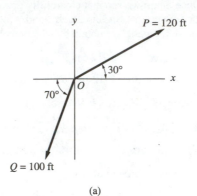

(a)

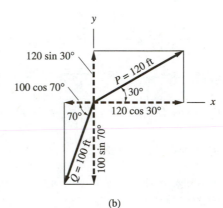

(b)

Solution

Referring to Fig. (b), the rectangular representations of **P** and **Q** are

$$\mathbf{P} = 120\cos 30°\mathbf{i} + 120\sin 30°\mathbf{j} = 103.9\mathbf{i} + 60.0\mathbf{j} \text{ ft}$$

$$\mathbf{Q} = -100\cos 70°\mathbf{i} - 100\sin 70°\mathbf{j} = -34.2\mathbf{i} - 94.0\mathbf{j} \text{ ft}$$

The resultant of **P** and **Q** is found by adding their components:

$$\mathbf{R} = \mathbf{P} + \mathbf{Q} = (103.9 - 34.2)\mathbf{i} + (60.0 - 94.0)\mathbf{j}$$

$$= 69.7\mathbf{i} - 34.0\mathbf{j} \text{ ft}$$ *Answer*

Calculating the magnitude and direction of **R**, we obtain

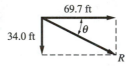

$$R = \sqrt{34.0^2 + 69.7^2} = 77.6 \text{ ft} \qquad \theta = \tan^{-1}\frac{34.0}{69.7} = 26.0°$$

Problems

1.40 Obtain the rectangular representation of the force **P**, given that its magnitude is 30 lb.

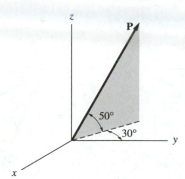

Fig. P1.40

1.41 The length of the position vector **r** is 240 mm. Determine the rectangular components of (a) **r**; and (b) the unit vector directed from O toward A.

1.42 A velocity vector **v** is directed from point $A\,(0, 10, -8)$ toward point $B\,(24, -18, 0)$, where the coordinates are in feet. Find the rectangular representation of **v**, given that its magnitude is 12 ft/s.

1.43 For points $A\,(-3, 0, 2)$ and $B\,(4, 1, 7)$, where the coordinates are in feet, determine (a) the distance between A and B; and (b) the rectangular representation of the unit vector directed from A toward B.

1.44 The velocity vector **v**, directed along the line AB, has a magnitude of 6 m/s. Determine the rectangular representations of (a) the unit vector directed from A toward B; and (b) **v**.

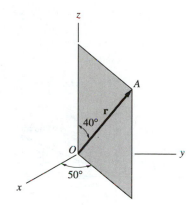

Fig. P1.41

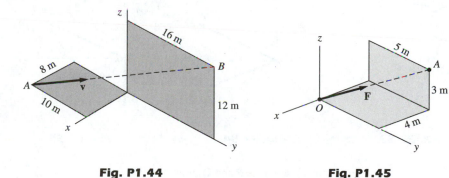

Fig. P1.44   **Fig. P1.45**

1.45 Find the rectangular representation of the force **F**, given that its magnitude is 240 N.

1.46 The magnitude of the force **F** is 120 lb. Find its rectangular representation.

1.47 The angles between a position vector **A** of magnitude 240 m and the x-, y-, and z-axes are $60°$, $45°$, and θ_z, respectively. Determine (a) θ_z; and (b) the rectangular representation of **A**.

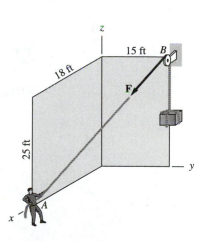

Fig. P1.46

1.48 Find the angles between the force $\mathbf{F} = 1000\mathbf{i} + 707\mathbf{j} - 1000\mathbf{k}$ lb and the *x*-, *y*-, and *z*-axes. Show your results on a sketch of the coordinate system.

1.49 Find the resultant of the two forces, each of which is of magnitude *P*.

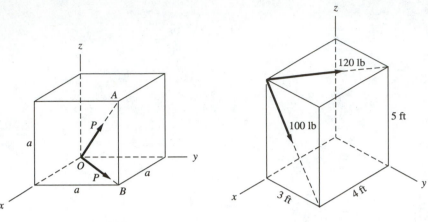

Fig. P1.49 **Fig. P1.50**

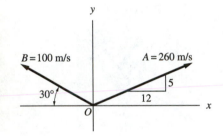

Fig. P1.51

1.50 Determine the resultant of the two forces shown.

1.51 Find the resultant of the two velocity vectors $\mathbf{A}$ and $\mathbf{B}$.

1.52 Given that $P = 120$ lb and $Q = 130$ lb, find the rectangular representation of $\mathbf{P} + \mathbf{Q}$.

1.53 Knowing that $P = 120$ lb and that the resultant of $\mathbf{P}$ and $\mathbf{Q}$ lies in the positive *x*-direction, determine Q and the magnitude of the resultant.

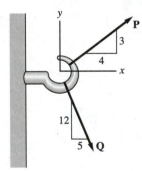

Fig. P1.52, P1.53

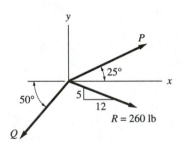

Fig. P1.54

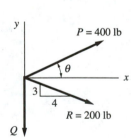

Fig. P1.55

1.54 If $\mathbf{R}$ is the resultant of the forces $\mathbf{P}$ and $\mathbf{Q}$, find P and Q.

1.55 The force $\mathbf{R}$ is the resultant of $\mathbf{P}$ and $\mathbf{Q}$. Determine Q and the angle θ.

1.56 Solve Prob. 1.27 using rectangular components of the forces.

1.5 *Vector Multiplication**

a. *Dot (scalar) product*

Figure 1.10 shows two vectors **A** and **B**, with θ being the angle between their positive directions. The *dot product* of **A** and **B** is defined as

$$\mathbf{A} \cdot \mathbf{B} = AB \cos\theta \qquad (0 \leq \theta \leq 180°) \tag{1.19}$$

Because the dot product is a scalar, it is also called the *scalar product*.

The dot product is positive if $\theta < 90°$ and negative if $\theta > 90°$. If **A** and **B** are parallel and have the same sense ($\theta = 0$), then $\mathbf{A} \cdot \mathbf{B} = AB$. For a vector dotted with itself, we see that $\mathbf{A} \cdot \mathbf{A} = A^2$. If **A** and **B** are parallel but have opposite sense ($\theta = 180°$), then $\mathbf{A} \cdot \mathbf{B} = -AB$.

The following two properties of the dot product follow from its definition in Eq. (1.19).

- The dot product is commutative: $\mathbf{A} \cdot \mathbf{B} = \mathbf{B} \cdot \mathbf{A}$
- The dot product is distributive: $\mathbf{A} \cdot (\mathbf{B} + \mathbf{C}) = \mathbf{A} \cdot \mathbf{B} + \mathbf{A} \cdot \mathbf{C}$

From the definition of the dot product, we also note that the base vectors of a rectangular coordinate system satisfy the following identities:

$$\begin{aligned} \mathbf{i} \cdot \mathbf{i} = \mathbf{j} \cdot \mathbf{j} = \mathbf{k} \cdot \mathbf{k} = 1 \\ \mathbf{i} \cdot \mathbf{j} = \mathbf{j} \cdot \mathbf{k} = \mathbf{k} \cdot \mathbf{i} = 0 \end{aligned} \tag{1.20}$$

When **A** and **B** are expressed in rectangular form, their dot product becomes

$$\mathbf{A} \cdot \mathbf{B} = (A_x\mathbf{i} + A_y\mathbf{j} + A_z\mathbf{k}) \cdot (B_x\mathbf{i} + B_y\mathbf{j} + B_z\mathbf{k})$$

which, using the distributive property of the dot product and Eqs. (1.20), reduces to

$$\mathbf{A} \cdot \mathbf{B} = A_x B_x + A_y B_y + A_z B_z \tag{1.21}$$

Equation (1.21) is a powerful and relatively simple method for computing the dot product of two vectors that are given in rectangular form.

The following are two of the more important applications of the dot product.

Finding the Angle Between Two Vectors The angle θ between the two vectors **A** and **B** in Fig. 1.11 can be found from the definition of the dot product in Eq. (1.19), which can be rewritten as

$$\cos\theta = \frac{\mathbf{A} \cdot \mathbf{B}}{AB} = \frac{\mathbf{A}}{A} \cdot \frac{\mathbf{B}}{B}$$

---

*Note that division by a vector, such as 1/**A** or **B**/**A**, is not defined.

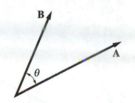

Fig. 1.10

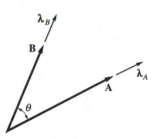

Fig. 1.11

Letting $\lambda_A = \mathbf{A}/A$ and $\lambda_B = \mathbf{B}/B$ be the unit vectors that have the same directions as $\mathbf{A}$ and $\mathbf{B}$, as shown in Fig. 1.11, the last equation becomes

$$\cos \theta = \lambda_A \cdot \lambda_B \tag{1.22}$$

If the unit vectors are written in rectangular form, this dot product is easily evaluated.

Determining the Orthogonal Component of a Vector in a Given Direction

If we project $\mathbf{B}$ onto $\mathbf{A}$ as in Fig. 1.12, the projected length $B \cos \theta$ is called the *orthogonal component of* $\mathbf{B}$ *in the direction of* $\mathbf{A}$. Because θ is the angle between $\mathbf{A}$ and $\mathbf{B}$, the definition of the dot product, $\mathbf{A} \cdot \mathbf{B} = AB \cos \theta$, yields

$$B \cos \theta = \frac{\mathbf{A} \cdot \mathbf{B}}{A} = \mathbf{B} \cdot \frac{\mathbf{A}}{A}$$

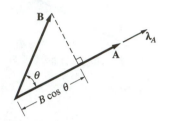

Fig. 1.12

Because $\mathbf{A}/A = \lambda_A$ (the unit vector in the direction of $\mathbf{A}$), as shown in Fig. 1.12, the last equation becomes

$$B \cos \theta = \mathbf{B} \cdot \lambda_A \tag{1.23}$$

Therefore,

> The orthogonal component of $\mathbf{B}$ in the direction of $\mathbf{A}$ equals $\mathbf{B} \cdot \lambda_A$. 　　(1.24)

The fact that the dot product is positive for $0 < \theta < 90°$ and negative for $90° < \theta < 180°$ means that Eq. (1.23) can be used to find any value of θ between $0°$ and $180°$.

b. Cross (vector) product

The *cross product* $\mathbf{C}$ of two vectors $\mathbf{A}$ and $\mathbf{B}$, denoted by

$$\mathbf{C} = \mathbf{A} \times \mathbf{B} \tag{1.25}$$

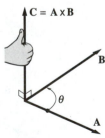

Fig. 1.13

has the following characteristics (see Fig. 1.13):

- $C = AB \sin \theta$, where θ ($0 \leq \theta \leq 180°$) is the angle between the positive directions of $\mathbf{A}$ and $\mathbf{B}$. (Note that C is always a positive number.)
- $\mathbf{C}$ is perpendicular to both $\mathbf{A}$ and $\mathbf{B}$.
- The sense of $\mathbf{C}$ is determined by the right-hand rule, which states that when the fingers of your right hand are curled in the direction of the angle θ (directed from $\mathbf{A}$ toward $\mathbf{B}$), your thumb points in the direction of $\mathbf{C}$.*

The cross product of two vectors is also called their *vector product.*

---

*An alternative statement of the right-hand rule is this: The direction of $\mathbf{C}$ is the direction in which a right-hand screw would advance when turned in the direction of θ (directed from $\mathbf{A}$ toward $\mathbf{B}$).

It can be shown that the cross product is distributive; that is,

$$\mathbf{A} \times (\mathbf{B} + \mathbf{C}) = (\mathbf{A} \times \mathbf{B}) + (\mathbf{A} \times \mathbf{C})$$

However, the cross product is neither associative nor commutative. In other words,

$$\mathbf{A} \times (\mathbf{B} \times \mathbf{C}) \neq (\mathbf{A} \times \mathbf{B}) \times \mathbf{C}$$
$$\mathbf{A} \times \mathbf{B} \neq \mathbf{B} \times \mathbf{A}$$

In fact, it can be deduced from the right-hand rule that $\mathbf{A} \times \mathbf{B} = -\mathbf{B} \times \mathbf{A}$.

From the definition of the cross product, we see that (1) if $\mathbf{A}$ and $\mathbf{B}$ are perpendicular ($\theta = 90°$), then $C = AB$; and (2) if $\mathbf{A}$ and $\mathbf{B}$ are parallel ($\theta = 0°$ or $180°$), then $C = 0$.

From the properties of the cross product, we deduce that the base vectors of a rectangular coordinate system satisfy the following identities:

$$\begin{array}{lll} \mathbf{i} \times \mathbf{i} = 0 & \mathbf{j} \times \mathbf{j} = 0 & \mathbf{k} \times \mathbf{k} = 0 \\ \mathbf{i} \times \mathbf{j} = \mathbf{k} & \mathbf{j} \times \mathbf{k} = \mathbf{i} & \mathbf{k} \times \mathbf{i} = \mathbf{j} \end{array} \qquad (1.26)$$

where the equations in the bottom row are valid in what is defined as a *right-handed* coordinate system. If the coordinate axes are labeled such that $\mathbf{i} \times \mathbf{j} = -\mathbf{k}, \mathbf{j} \times \mathbf{k} = -\mathbf{i}$, and $\mathbf{k} \times \mathbf{i} = -\mathbf{j}$, the system is said to be *left-handed*. Examples of both right- and left-handed coordinate systems are shown in Fig. 1.14.*

When $\mathbf{A}$ and $\mathbf{B}$ are expressed in rectangular form, their cross product becomes

$$\mathbf{A} \times \mathbf{B} = (A_x\mathbf{i} + A_y\mathbf{j} + A_z\mathbf{k}) \times (B_x\mathbf{i} + B_y\mathbf{j} + B_z\mathbf{k})$$

Using the distributive property of the cross product and Eqs. (1.26), this equation reduces to

$$\mathbf{A} \times \mathbf{B} = (A_y B_z - A_z B_y)\mathbf{i} - (A_x B_z - A_z B_x)\mathbf{j} + (A_x B_y - A_y B_x)\mathbf{k} \qquad (1.27)$$

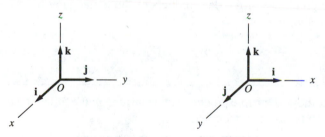

Right-handed coordinate system ($\mathbf{i} \times \mathbf{j} = \mathbf{k}$, etc.)   Left-handed coordinate system ($\mathbf{i} \times \mathbf{j} = -\mathbf{k}$, etc.)

Fig. 1.14

*In this text, we assume that all rectangular coordinate systems are right-handed.

The identical expression is obtained when the rules for expanding a 3×3 determinant are applied to the following array of nine terms (because the terms are not all scalars, the array is not a true determinant):

$$\mathbf{A} \times \mathbf{B} = \begin{vmatrix} \mathbf{i} & \mathbf{j} & \mathbf{k} \\ A_x & A_y & A_z \\ B_x & B_y & B_z \end{vmatrix} \tag{1.28}$$

You may use any method for determinant expansion, but you will find that the following technique, called *expansion by minors using the first row,* is very convenient.

$$\begin{vmatrix} a & b & c \\ d & e & f \\ g & h & i \end{vmatrix} = a \begin{vmatrix} e & f \\ h & i \end{vmatrix} - b \begin{vmatrix} d & f \\ g & i \end{vmatrix} + c \begin{vmatrix} d & e \\ g & h \end{vmatrix}$$

$$= a(ei - fh) - b(di - fg) + c(dh - eg)$$

Expanding Eq. (1.28) by this method, we find that the 2×2 determinants equal the $\mathbf{i}$, $\mathbf{j}$, and $\mathbf{k}$ components of the cross product.

c. Scalar triple product

Of the vector products that involve three or more vectors, the one that is most useful in statics is the scalar triple product. The *scalar triple product* arises when the cross product of two vectors is dotted with a third vector—for example, $\mathbf{A} \times \mathbf{B} \cdot \mathbf{C}$. When writing this product, it is not necessary to show the parenthe- ses, because $\mathbf{A} \times \mathbf{B} \cdot \mathbf{C}$ can be interpreted only in one way—the cross product must be done first; otherwise the expression is meaningless.

Assuming that $\mathbf{A}$, $\mathbf{B}$, and $\mathbf{C}$ are expressed in rectangular form and recall- ing Eqs. (1.21) and (1.27), the scalar triple product becomes

$$\mathbf{A} \times \mathbf{B} \cdot \mathbf{C} = \left[(A_y B_z - A_z B_y)\mathbf{i} - (A_x B_z - A_z B_x)\mathbf{j} \right.$$
$$\left. + (A_x B_y - A_y B_x)\mathbf{k} \right] \cdot (C_x \mathbf{i} + C_y \mathbf{j} + C_z \mathbf{k})$$

Using the properties of the dot products of the rectangular base vectors, this expression simplifies to

$$\mathbf{A} \times \mathbf{B} \cdot \mathbf{C} = (A_y B_z - A_z B_y)C_x - (A_x B_z - A_z B_x)C_y$$
$$+ (A_x B_y - A_y B_x)C_z \tag{1.29}$$

Therefore, the scalar triple product can be written in the following determinant form, which is easy to remember:

$$\mathbf{A} \times \mathbf{B} \cdot \mathbf{C} = \begin{vmatrix} A_x & A_y & A_z \\ B_x & B_y & B_z \\ C_x & C_y & C_z \end{vmatrix} \tag{1.30}$$

The following identities relating to the scalar triple product are useful:

$$\mathbf{A} \times \mathbf{B} \cdot \mathbf{C} = \mathbf{A} \cdot \mathbf{B} \times \mathbf{C} = \mathbf{B} \cdot \mathbf{C} \times \mathbf{A} = \mathbf{C} \cdot \mathbf{A} \times \mathbf{B} \tag{1.31}$$

Observe that the value of the scalar triple product is not altered if the locations of the dot and cross are interchanged or if the positions of $\mathbf{A}$, $\mathbf{B}$, and $\mathbf{C}$ are changed—provided that the cyclic order A–B–C is maintained.

Sample Problem 1.9

Given the vectors

$$\mathbf{A} = 8\mathbf{i} + 4\mathbf{j} - 2\mathbf{k} \text{ lb}$$

$$\mathbf{B} = 2\mathbf{j} + 6\mathbf{k} \text{ ft}$$

$$\mathbf{C} = 3\mathbf{i} - 2\mathbf{j} + 4\mathbf{k} \text{ ft}$$

calculate the following: (1) $\mathbf{A} \cdot \mathbf{B}$; (2) the orthogonal component of $\mathbf{B}$ in the direction of $\mathbf{C}$; (3) the angle between $\mathbf{A}$ and $\mathbf{C}$; (4) $\mathbf{A} \times \mathbf{B}$; (5) a unit vector $\boldsymbol{\lambda}$ that is perpendicular to both $\mathbf{A}$ and $\mathbf{B}$; and (6) $\mathbf{A} \times \mathbf{B} \cdot \mathbf{C}$.

Solution

Part 1

From Eq. (1.21), the dot product of $\mathbf{A}$ and $\mathbf{B}$ becomes

$$\mathbf{A} \cdot \mathbf{B} = A_x B_x + A_y B_y + A_z B_z = 8(0) + 4(2) + (-2)(6)$$

$$= -4 \text{ lb} \cdot \text{ft} \qquad \textit{Answer}$$

The negative sign indicates that the angle between $\mathbf{A}$ and $\mathbf{B}$ is greater than 90°.

Part 2

Letting θ be the angle between $\mathbf{B}$ and $\mathbf{C}$, we obtain from Eq. (1.23)

$$B \cos\theta = \mathbf{B} \cdot \boldsymbol{\lambda}_C = \mathbf{B} \cdot \frac{\mathbf{C}}{C} = (2\mathbf{j} + 6\mathbf{k}) \cdot \frac{3\mathbf{i} - 2\mathbf{j} + 4\mathbf{k}}{\sqrt{3^2 + (-2)^2 + 4^2}}$$

$$= \frac{(0)(3) + (2)(-2) + (6)(4)}{\sqrt{29}} = 3.71 \text{ ft} \qquad \textit{Answer}$$

Part 3

Letting α be the angle between $\mathbf{A}$ and $\mathbf{C}$, we find from Eq. (1.22)

$$\cos\alpha = \boldsymbol{\lambda}_A \cdot \boldsymbol{\lambda}_C = \frac{\mathbf{A}}{A} \cdot \frac{\mathbf{C}}{C}$$

$$= \frac{8\mathbf{i} + 4\mathbf{j} - 2\mathbf{k}}{\sqrt{8^2 + 4^2 + (-2)^2}} \cdot \frac{3\mathbf{i} - 2\mathbf{j} + 4\mathbf{k}}{\sqrt{3^2 + (-2)^2 + 4^2}}$$

$$= \frac{(8)(3) + (4)(-2) + (-2)(4)}{\sqrt{84}\sqrt{29}} = 0.162\,09$$

from which the angle between $\mathbf{A}$ and $\mathbf{C}$ is found to be

$$\alpha = 80.7° \qquad \textit{Answer}$$

Part 4

Referring to Eq. (1.22), the cross product of $\mathbf{A}$ and $\mathbf{B}$ is

$$\mathbf{A} \times \mathbf{B} = \begin{vmatrix} \mathbf{i} & \mathbf{j} & \mathbf{k} \\ A_x & A_y & A_z \\ B_x & B_y & B_z \end{vmatrix} = \begin{vmatrix} \mathbf{i} & \mathbf{j} & \mathbf{k} \\ 8 & 4 & -2 \\ 0 & 2 & 6 \end{vmatrix}$$

$$= \mathbf{i} \begin{vmatrix} 4 & -2 \\ 2 & 6 \end{vmatrix} - \mathbf{j} \begin{vmatrix} 8 & -2 \\ 0 & 6 \end{vmatrix} + \mathbf{k} \begin{vmatrix} 8 & 4 \\ 0 & 2 \end{vmatrix}$$

$$= 28\mathbf{i} - 48\mathbf{j} + 16\mathbf{k} \text{ lb} \cdot \text{ft} \qquad \textit{Answer}$$

Part 5

The cross product $\mathbf{A} \times \mathbf{B}$ is perpendicular to both $\mathbf{A}$ and $\mathbf{B}$. Therefore, a unit vector in that direction is obtained by dividing $\mathbf{A} \times \mathbf{B}$, which was evaluated above, by its magnitude

$$\frac{\mathbf{A} \times \mathbf{B}}{|\mathbf{A} \times \mathbf{B}|} = \frac{28\mathbf{i} - 48\mathbf{j} + 16\mathbf{k}}{\sqrt{28^2 + (-48)^2 + 16^2}}$$

$$= 0.484\mathbf{i} - 0.830\mathbf{j} + 0.277\mathbf{k}$$

Because the negative of this vector is also a unit vector that is perpendicular to both $\mathbf{A}$ and $\mathbf{B}$, we obtain

$$\boldsymbol{\lambda} = \pm(0.484\mathbf{i} - 0.830\mathbf{j} + 0.277\mathbf{k}) \qquad \textit{Answer}$$

Part 6

The scalar triple product $\mathbf{A} \times \mathbf{B} \cdot \mathbf{C}$ is evaluated using Eq. (1.30).

$$\mathbf{A} \times \mathbf{B} \cdot \mathbf{C} = \begin{vmatrix} A_x & A_y & A_z \\ B_x & B_y & B_z \\ C_x & C_y & C_z \end{vmatrix} = \begin{vmatrix} 8 & 4 & -2 \\ 0 & 2 & 6 \\ 3 & -2 & 4 \end{vmatrix}$$

$$= 8\begin{vmatrix} 2 & 6 \\ -2 & 4 \end{vmatrix} - 4\begin{vmatrix} 0 & 6 \\ 3 & 4 \end{vmatrix} + (-2)\begin{vmatrix} 0 & 2 \\ 3 & -2 \end{vmatrix}$$

$$= 160 + 72 + 12 = 244 \text{ lb} \cdot \text{ft}^2 \qquad \textit{Answer}$$

Problems

1.57 Compute the dot product $\mathbf{A} \cdot \mathbf{B}$ for each of the following cases. Identify the units of each product.

(a) $\mathbf{A} = 6\mathbf{j} + 9\mathbf{k}$ ft $\mathbf{B} = 7\mathbf{i} - 3\mathbf{j} + 2\mathbf{k}$ ft
(b) $\mathbf{A} = 2\mathbf{i} - 3\mathbf{j}$ m $\mathbf{B} = 6\mathbf{i} - 13\mathbf{k}$ N
(c) $\mathbf{A} = 5\mathbf{i} - 6\mathbf{j} - \mathbf{k}$ m $\mathbf{B} = -5\mathbf{i} + 8\mathbf{j} + 6\mathbf{k}$ m

1.58 Compute the cross product $\mathbf{C} = \mathbf{A} \times \mathbf{B}$ for each of the cases given in Prob. 1.57. Identify the units of each product.

1.59 Given

$$\mathbf{r} = 5\mathbf{i} + 4\mathbf{j} + 3\mathbf{k} \text{ m (position vector)}$$
$$\mathbf{F} = 30\mathbf{i} - 20\mathbf{j} - 10\mathbf{k} \text{ N (force vector)}$$
$$\boldsymbol{\lambda} = 0.6\mathbf{j} + 0.8\mathbf{k} \text{ (dimensionless unit vector)}$$

compute (a) $\mathbf{r} \times \mathbf{F} \cdot \boldsymbol{\lambda}$; and (b) $\boldsymbol{\lambda} \times \mathbf{r} \cdot \mathbf{F}$.

1.60 Compute $\overrightarrow{BA} \times \overrightarrow{OA}$ and $\overrightarrow{BO} \times \overrightarrow{OA}$ for the vectors shown. Explain why the results are identical.

1.61 Use the dot product to find the angle between the vectors $\overrightarrow{BA}$ and $\overrightarrow{OA}$. Check your results by trigonometry.

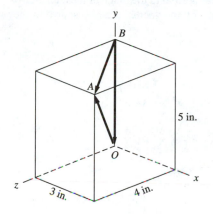

Fig. P1.60, P1.61

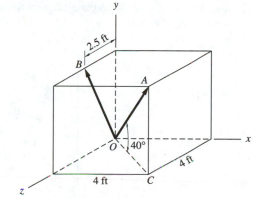

Fig. P1.62

1.62 Use the dot product to find the angle between the lines OA and OB.

1.63 Let $\mathbf{A}$ and $\mathbf{B}$ be two nonparallel vectors that lie in a common plane S. If $\mathbf{C} = \mathbf{A} \times (\mathbf{A} \times \mathbf{B})$, which of the following statements are true: (i) $\mathbf{C} = \mathbf{0}$; (ii) $\mathbf{C}$ lies in plane S; (iii) $\mathbf{C}$ is perpendicular to plane S?

1.64 Determine which of the following position vectors $\mathbf{B}$ is perpendicular to $\mathbf{A} = 3\mathbf{i} - 5\mathbf{j} + 2\mathbf{k}$ m:

(a) $\mathbf{B} = 5\mathbf{i} + 3\mathbf{j} - 2\mathbf{k}$ m
(b) $\mathbf{B} = 2\mathbf{i} + 3\mathbf{j} + 4\mathbf{k}$ m
(c) $\mathbf{B} = \mathbf{i} + \mathbf{j} + \mathbf{k}$ m
(d) $\mathbf{B} = 3\mathbf{i} + \mathbf{j} - 2\mathbf{k}$ m

1.65 Find a unit vector that is perpendicular to both $\mathbf{A} = 8\mathbf{i} - 3\mathbf{j} + 2\mathbf{k}$ ft and $\mathbf{B} = -6\mathbf{i} + 4\mathbf{j} + 3\mathbf{k}$ ft.

1.66 The three points $A(0, 0, 4)$, $B(-1, 4, 1)$, and $C(3, 2, 0)$ define a plane. The coordinates are in inches. Find a unit vector that is perpendicular to this plane.

1.67 Show that the position vectors $\mathbf{A} = \mathbf{i} + 3\mathbf{j} - 5\mathbf{k}$ m, $\mathbf{B} = \mathbf{i} + 3\mathbf{j} + 2\mathbf{k}$ m, and $\mathbf{C} = 21\mathbf{i} - 7\mathbf{j}$ m are mutually perpendicular.

1.68 Compute the magnitude of the orthogonal projection of $\mathbf{F} = 6\mathbf{i} + 20\mathbf{j} - 12\mathbf{k}$ lb in the direction of the vector $\mathbf{A} = 2\mathbf{i} - 3\mathbf{j} + 5\mathbf{k}$ ft.

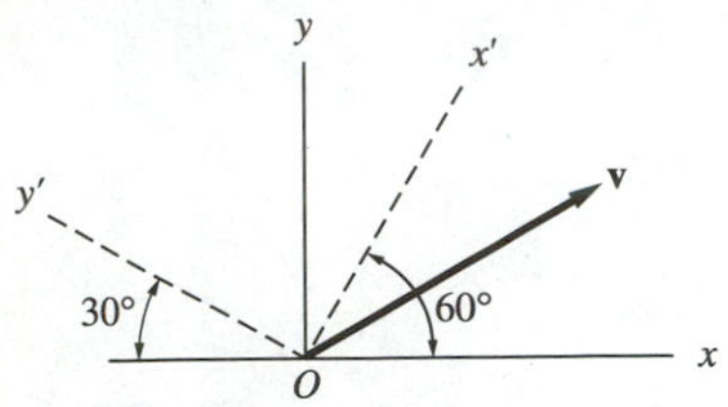

1.69 Using the dot product, find the components of the velocity vector $\mathbf{v} = 2\mathbf{i} + \mathbf{j}$ km/h in the directions of the x'- and y'-axes.

*1.70 Resolve $\mathbf{A} = 3\mathbf{i} + 5\mathbf{j} - 4\mathbf{k}$ in. into two vector components—one parallel to and the other perpendicular to $\mathbf{B} = 6\mathbf{i} + 2\mathbf{k}$ in. Express each of your answers as a magnitude multiplied by a unit vector.

1.71 Vectors $\mathbf{a}$, $\mathbf{b}$, and $\mathbf{c}$ form an orthogonal set of unit vectors, where $\mathbf{b} = 0.745\mathbf{i} + 0.596\mathbf{j} + 0.298\mathbf{k}$ and $\mathbf{c} = -0.371\mathbf{i} + 0.743\mathbf{j} - 0.557\mathbf{k}$. (a) Determine $\mathbf{a}$; and (b) express the velocity vector $\mathbf{v} = 16\mathbf{i} - 24\mathbf{j} - 8\mathbf{k}$ ft/s in terms of $\mathbf{a}$, $\mathbf{b}$, and $\mathbf{c}$.

1.72 Determine the value of the scalar a if the following three vectors are to lie in the same plane: $\mathbf{A} = 2\mathbf{i} - \mathbf{j} + 2\mathbf{k}$ m, $\mathbf{B} = 6\mathbf{i} + 3\mathbf{j} + a\mathbf{k}$ m, and $\mathbf{C} = 16\mathbf{i} + 46\mathbf{j} + 7\mathbf{k}$ m.

*1.73 Resolve the force $\mathbf{F} = 20\mathbf{i} + 30\mathbf{j} + 50\mathbf{k}$ lb into two components—one perpendicular to plane ABC and the other lying in plane ABC.

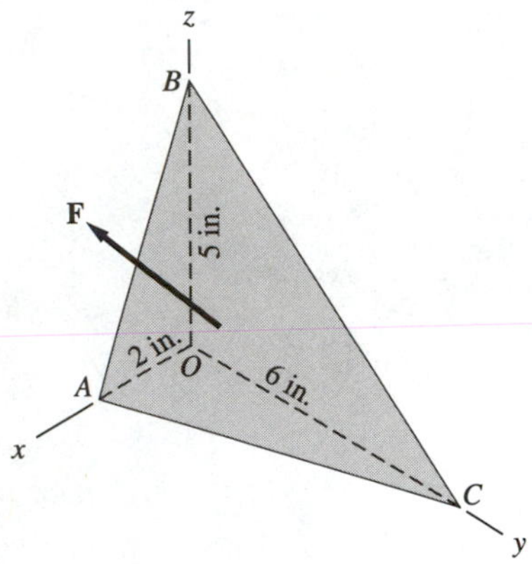

1.74 It can be shown that a plane area may be represented by a vector $\mathbf{A} = A\boldsymbol{\lambda}$, where A is the area and $\boldsymbol{\lambda}$ represents a unit vector normal to the plane of the area. Show that the area vector of the parallelogram formed by the vectors $\mathbf{a}$ and $\mathbf{b}$ shown in the figure is $\mathbf{A} = \mathbf{a} \times \mathbf{b}$.

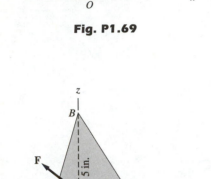

Fig. P1.69

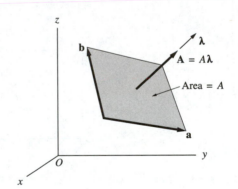

Fig. P1.74

Fig. P1.73

1.75 The coordinates of the corners of a triangle ABC are $A(3, -1, 0)$, $B(-2, 2, 3)$, and $C(0, 0, 4)$. The units are inches. Calculate the area of triangle ABC. (*Hint:* See Prob. 1.74.)

1.76 Show that $|\mathbf{a} \times \mathbf{b} \cdot \mathbf{c}|$ equals the volume of a parallelepiped that has $\mathbf{a}$, $\mathbf{b}$, and $\mathbf{c}$ as its edges. (*Hint:* See Prob. 1.74.)

2

Basic Operations with Force Systems

2.1 *Introduction*

The usefulness of vector algebra in real-world problems stems from the fact that several commonly encountered physical quantities possess the properties of vectors. One such quantity is force, which was shown to obey the parallelogram law of addition by Stevinus (1548–1620).

In this chapter we begin to study the effects of forces on particles and rigid bodies. In particular, we learn how to use vector algebra to reduce a system of forces to a simpler, equivalent system. If the forces are concurrent (all forces intersect at the same point), we show that the equivalent system is a single force. The reduction of a nonconcurrent force system requires two additional vector concepts: the moment of a force and the couple. Both of these concepts are introduced in this chapter.

2.2 *Equivalence of Vectors*

We recall that vectors are quantities that have magnitude and direction, and combine according to the parallelogram law for addition. Two vectors that have the same magnitude and direction are said to be *equal*.

In mechanics, the term *equivalence* implies interchangeability; two vectors are considered to be equivalent if they can be interchanged without changing the outcome of the problem. Equality does not always result in equivalence. For example, a force applied to a certain point in a body does not necessarily produce the same effect on the body as an equal force acting at a different point.

From the viewpoint of equivalence, vectors representing physical quantities are classified into the following three types:

- *Fixed vectors:* Equivalent vectors have the same magnitude, direction, and point of application.
- *Sliding vectors:* Equivalent vectors have the same magnitude, direction, and line of action.
- *Free vectors:* Equivalent vectors have the same magnitude and direction.

It is possible for a physical quantity to be one type of vector—say, fixed—in one application and another type of vector, such as sliding, in another application. In vector algebra, reviewed in Chapter 1, all vectors were treated as free vectors.

2.3 *Force*

Force is the name assigned to mechanical interaction between bodies. A force can affect both the motion and the deformation of the body on which it acts. Forces may arise from direct contact between bodies, or they may be applied at a distance (such as gravitational attraction). Contact forces are distributed over a surface area of the body, whereas forces acting at a distance are distributed over the volume of the body.

Sometimes the area over which a contact force is applied is so small that it may be approximated by a point, in which case the force is said to be *concentrated* at the point of contact. The contact point is also called the *point of application* of the force. The *line of action* of a concentrated force is the line that passes through the point of application and is parallel to the force. In this chapter we consider only concentrated forces; the discussion of distributed forces begins in the next chapter.

Force is a *fixed vector,* because one of its characteristics (in addition to its magnitude and direction) is its point of application. As an informal proof, consider the three identical bars in Fig. 2.1, each loaded by two equal but opposite forces of magnitude *P*. If the forces are applied as shown Fig. 2.1(a), the bar is under tension, and its deformation is an elongation. By interchanging the forces, as seen in Fig. 2.1(b), the bar is placed in compression, resulting in its shortening. The loading in Fig. 2.1(c), where both forces are acting at point *A*, produces no deformation. Note that the forces in all three cases have the same line of action and the same zero resultant; only the points of application are different. Therefore, we conclude that the point of application is a characteristic of a force, as far as deformation is concerned.

If the bar is rigid, however (meaning that the deformation is negligible), there will be no observable differences in the behavior of the three bars in Fig. 2.1. In other words, the *external effects** of the three loadings are identical. It follows that if we are interested only in the external effects, a force can be

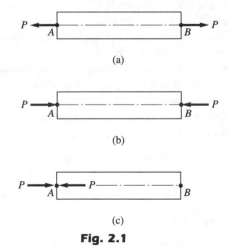

(a)

(b)

(c)

Fig. 2.1

---

*The external effects that concern us most are the motion (or state of rest) of the body, and the support reactions.

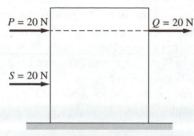

Fig. 2.2

treated as a *sliding vector.* The above conclusion is summarized by the *principle of transmissibility:*

> A force may be moved anywhere along its line of action without changing its external effects on a rigid body.

Two force systems that produce the same external effects on a rigid body are said to be *equivalent.* (Sometimes the term *rigid-body equivalent* is used.)

In summary, a force is a fixed vector tied to a point of application, but if one is interested only in its external effect on a rigid body, a force may be treated as a sliding vector.

As a further illustration of the principle of transmissibility, consider the rigid block shown in Fig. 2.2. The block is subjected to three forces **P**, **Q**, and **S** acting in the plane of the paper, each with magnitude 20 N. The three forces are equal in the mathematical sense: $\mathbf{P} = \mathbf{Q} = \mathbf{S}$. However, only **P** and **Q** would produce identical external effects because they have the same line of action. Because **S** has a different line of action, its external effect would be different.

2.4 *Reduction of Concurrent Force Systems*

In this article, we discuss the method for replacing a system of concurrent forces with a single equivalent force.

Consider the forces $\mathbf{F}_1, \mathbf{F}_2, \mathbf{F}_3, \ldots$ acting on the rigid body in Fig. 2.3(a) (for convenience, only three of the forces are shown). All the forces are concurrent at point O. (Their lines of action intersect at O.) These forces can be reduced to a single, equivalent force by the following two steps.

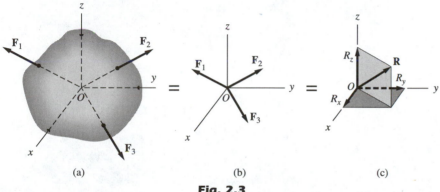

(a)   (b)   (c)

Fig. 2.3

1. Move the forces along their lines of action to the point of concurrence O, as indicated in Fig. 2.3(b). According to the principle of transmissibility, this operation does not change the external effects on the body. Therefore, the force systems in Figs. 2.3(a) and (b) are equivalent, which is indicated by the equal sign between the figures.

2. With the forces now at the common point O, compute their resultant **R** from the vector sum

$$\mathbf{R} = \Sigma\mathbf{F} = \mathbf{F}_1 + \mathbf{F}_2 + \mathbf{F}_3 + \cdots \qquad (2.1)$$

This resultant, which is also equivalent to the original force system, is shown in Fig. 2.3(c) together with its rectangular components. Note that Eq. (2.1) determines only the magnitude and direction of the resultant. The line of action of **R** must pass through the point of concurrency O in order for the equivalence to be valid.

When evaluating Eq. (2.1), any of the graphical or analytical methods for vector addition discussed in Chapter 1 may be used. If rectangular components are chosen, the equivalent scalar equations for determining the resultant force **R** are

$$R_x = \Sigma F_x \qquad R_y = \Sigma F_y \qquad R_z = \Sigma F_z \qquad (2.2)$$

Thus we see that three scalar equations are required to determine the resultant force for a concurrent system of forces. If the original forces lie in a common plane—say, the xy-plane—the equation $R_z = \Sigma F_z$ yields no independent information and only the following two equations are necessary to determine the resultant force.

$$R_x = \Sigma F_x \qquad R_y = \Sigma F_y \qquad (2.3)$$

We emphasize that the method described here for determining the resultant force is valid only for forces that are concurrent. Because a force is tied to its line of action, the reduction of nonconcurrent force systems will require additional concepts, which are discussed later.

Sample Problem 2.1

Determine the resultant of the three concurrent forces shown in Fig. (a).

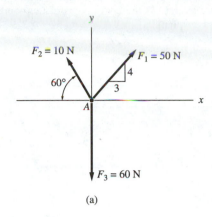

(a)

Solution

Because the three forces are concurrent at point A, they may be added immediately to obtain the resultant force $\mathbf{R}$.

The rectangular components of each of the three forces are shown in Fig. (b). Using the two scalar equations to determine the components of the resultant, we have

$$R_x = \Sigma F_x \quad \xrightarrow{+} \quad R_x = 30 - 5 = 25 \text{ N}$$

and

$$R_y = \Sigma F_y \quad +\uparrow \quad R_y = 40 + 8.66 - 60 = -11.34 \text{ N}$$

The signs in these equations indicate that R_x acts to the right and R_y acts downward. The resultant force $\mathbf{R}$ is as shown in Fig. (c). Note that the magnitude of the resultant is 27.5 N and that it acts through point A (the original point of concurrency) at the 24.4° angle shown.

The foregoing solution could also have been accomplished using vector notation. The forces would first be written in vector form as follows,

$$\mathbf{F}_1 = 30\mathbf{i} + 40\mathbf{j} \text{ N}$$
$$\mathbf{F}_2 = -5\mathbf{i} + 8.66\mathbf{j} \text{ N}$$
$$\mathbf{F}_3 = -60\mathbf{j} \text{ N}$$

and the resultant force $\mathbf{R}$ would then be determined from the vector equation

$$\mathbf{R} = \Sigma \mathbf{F} = \mathbf{F}_1 + \mathbf{F}_2 + \mathbf{F}_3$$
$$\mathbf{R} = (30\mathbf{i} + 40\mathbf{j}) + (-5\mathbf{i} + 8.66\mathbf{j}) + (-60\mathbf{j})$$
$$\mathbf{R} = 25\mathbf{i} - 11.34\mathbf{j} \text{ N} \qquad \textit{Answer}$$

Whether you use scalar or vector notation in the solution of this type of problem is a matter of personal preference.

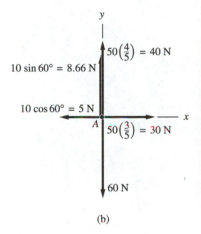

(b)

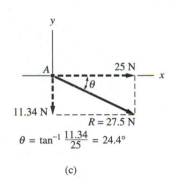

$$\theta = \tan^{-1} \frac{11.34}{25} = 24.4°$$

(c)

Sample Problem 2.2

Let **R** refer to the resultant of the three forces shown in Fig. (a). Given that $F_1 = 260$ lb, $F_2 = 75$ lb, and $F_3 = 60$ lb, determine (1) the magnitude of **R**; (2) the angles between **R** and the coordinate axes; and (3) the coordinates of the point at which the line of action of **R** intersects the yz-plane.

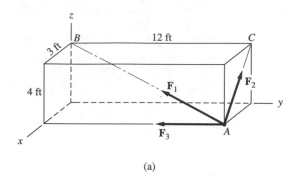

(a)

Solution

Part 1

The forces are concurrent at point A and thus may be added immediately. Because the forces do not lie in a coordinate plane, it is convenient to use vector notation.

One method for expressing each of the forces in vector notation is to use the form $\mathbf{F} = F\boldsymbol{\lambda}$, where $\boldsymbol{\lambda}$ is the unit vector in the direction of the force. Thus

$$\mathbf{F}_1 = 260\boldsymbol{\lambda}_{AB}$$

$$= 260\frac{\overrightarrow{AB}}{|\overrightarrow{AB}|} = 260\left(\frac{-3\mathbf{i} - 12\mathbf{j} + 4\mathbf{k}}{13}\right)$$

$$= -60\mathbf{i} - 240\mathbf{j} + 80\mathbf{k} \text{ lb}$$

$$\mathbf{F}_2 = 75\boldsymbol{\lambda}_{AC}$$

$$= 75\frac{\overrightarrow{AC}}{|\overrightarrow{AC}|} = 75\left(\frac{-3\mathbf{i} + 4\mathbf{k}}{5}\right)$$

$$= -45\mathbf{i} + 60\mathbf{k} \text{ lb}$$

$$\mathbf{F}_3 = -60\mathbf{j} \text{ lb}$$

The resultant force is given by

$$\mathbf{R} = \Sigma\mathbf{F} = \mathbf{F}_1 + \mathbf{F}_2 + \mathbf{F}_3$$

$$\mathbf{R} = (-60\mathbf{i} - 240\mathbf{j} + 80\mathbf{k}) + (-45\mathbf{i} + 60\mathbf{k}) + (-60\mathbf{j})$$

$$\mathbf{R} = -105\mathbf{i} - 300\mathbf{j} + 140\mathbf{k} \text{ lb}$$

The magnitude of **R** is

$$R = \sqrt{(-105)^2 + (-300)^2 + (140)^2} = 347.3 \text{ lb} \qquad \textit{Answer}$$

The resultant **R** is shown in Fig. (b). Note that the line of action of **R** must pass through point A, the point of concurrency for the original three forces.

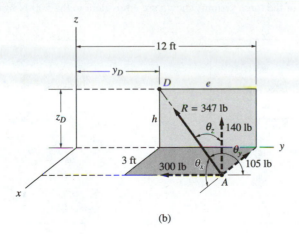

(b)

Part 2

The unit vector $\boldsymbol{\lambda}$ in the direction of **R** is

$$\boldsymbol{\lambda} = \frac{\mathbf{R}}{R} = \frac{-105\mathbf{i} - 300\mathbf{j} + 140\mathbf{k}}{347.3}$$

$$= -0.3023\mathbf{i} - 0.8638\mathbf{j} + 0.4031\mathbf{k}$$

The angles between **R** and the coordinate axes, shown in Fig. (b), are

$$\theta_x = \cos^{-1}(-0.3023) = 107.6°$$
$$\theta_y = \cos^{-1}(-0.8638) = 149.7°$$
$$\theta_z = \cos^{-1}(0.4031) \quad = 66.2°$$

Answer

Part 3

Let D be the point where the line of action of the resultant **R** intersects the yz-plane. The horizontal distance e and the vertical distance h, shown in Fig. (b), can be determined by proportions:

$$\frac{e}{300} = \frac{h}{140} = \frac{3}{105}$$

from which $e = 8.57$ ft and $h = 4.0$ ft.

From Fig. (b) the coordinates of point D can be seen to be

$$x_D = 0$$
$$y_D = 12 - e = 12 - 8.57 = 3.43 \text{ ft}$$
$$z_D = h = 4.0 \text{ ft}$$

Answer

Problems

2.1   Which of the force systems shown are equivalent to the 500-N force in (a)?

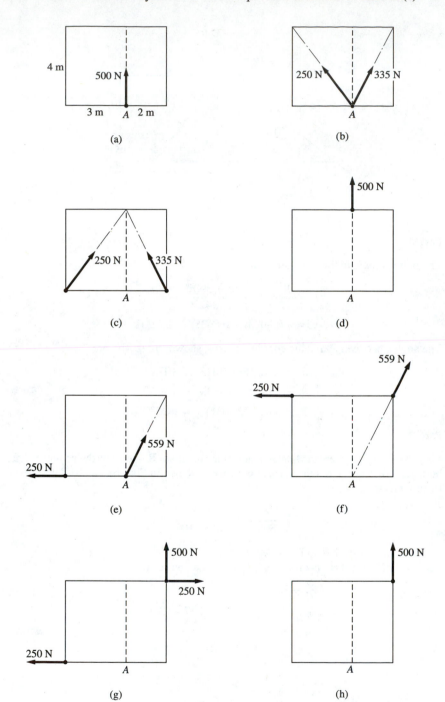

Fig. P2.1

2.2 Two people are trying to roll the boulder by applying the forces shown. Determine the magnitude and direction of the force that is equivalent to the two applied forces.

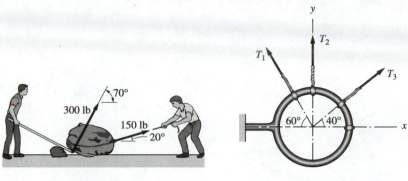

Fig. P2.2   **Fig. P2.3**

2.3 The resultant of the three forces acting on the ring bolt is $\mathbf{R} = 180\mathbf{j}$ lb. Given that $T_2 = 40$ lb, determine T_1 and T_3.

2.4 The hook is acted on by the three forces. Determine P and the angle θ, given that the resultant is $90\mathbf{i}$ kN.

2.5 Replace the three forces acting on the bracket by a single, equivalent force. Be sure to state the line of action of this force.

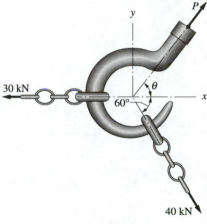

Fig. P2.4

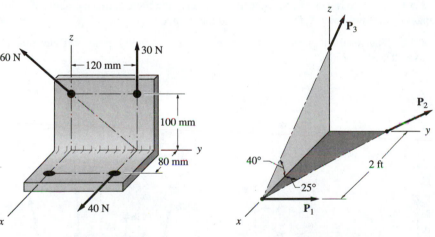

Fig. P2.5   **Fig. P2.6, P2.7**

2.6 Let $\mathbf{R}$ refer to the resultant of the three forces shown. Given that $P_1 = 110$ lb, $P_2 = 200$ lb, and $P_3 = 150$ lb, determine (a) the magnitude of $\mathbf{R}$; (b) the direction cosines of $\mathbf{R}$; and (c) the point at which the line of action of $\mathbf{R}$ intersects the yz-plane.

2.7 Determine the magnitudes of the three forces $\mathbf{P}_1$, $\mathbf{P}_2$, and $\mathbf{P}_3$, given that their resultant is $\mathbf{R} = -600\mathbf{i} + 500\mathbf{j} + 300\mathbf{k}$ lb.

2.8   Determine the resultant of the three forces shown, given that $P_1 = 50$ kN, $P_2 = 80$ kN, and $P_3 = 120$ kN and the dimension a is 4 m.

2.9   Determine the magnitudes of the three forces shown, given that their resultant is $\mathbf{R} = 200\mathbf{i} - 100\mathbf{j} + 50\mathbf{k}$ kN and the dimension a is 3 m.

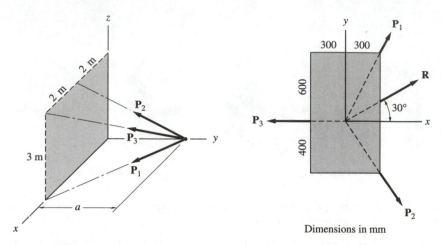

Fig. P2.8, P2.9

Dimensions in mm

Fig. P2.10

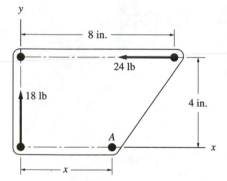

Fig. P2.11

2.10   The force $\mathbf{R}$ is the resultant of the forces $\mathbf{P}_1$, $\mathbf{P}_2$, and $\mathbf{P}_3$ acting on the rectangular plate. Find P_1 and P_2 if $R = 40$ kN and $P_3 = 20$ kN.

2.11   The two forces can be reduced to an equivalent force $\mathbf{R}$ that has a line of action passing through point A. Determine R and the distance x.

2.12   The resultant of the two forces shown passes through point A. Determine Q, given that $P = 25$ lb.

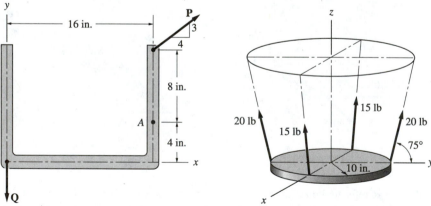

Fig. P2.12

Fig. P2.13

2.13   The four forces are to be replaced by a single, equivalent force. Determine the rectangular components of this force and the point of intersection of its line of action with the plate.

2.14   The three forces shown have the same magnitude T. Determine the magnitude of their resultant.

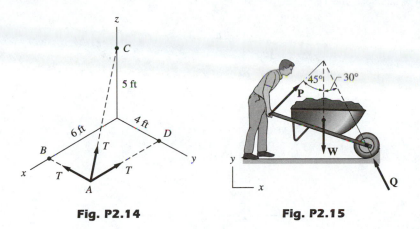

Fig. P2.14   **Fig. P2.15**

2.15   The man exerts a force **P** of magnitude 150 lb on the handles of the wheelbarrow. Knowing that the resultant of the forces **P**, **Q** (the reaction at the wheel), and **W** (the weight of the wheelbarrow) is zero, determine **W**.

2.16   The plate suspended from two wires is acted on by the three forces shown. Knowing that the resultant acting on the plate is zero, determine Q and the angle θ.

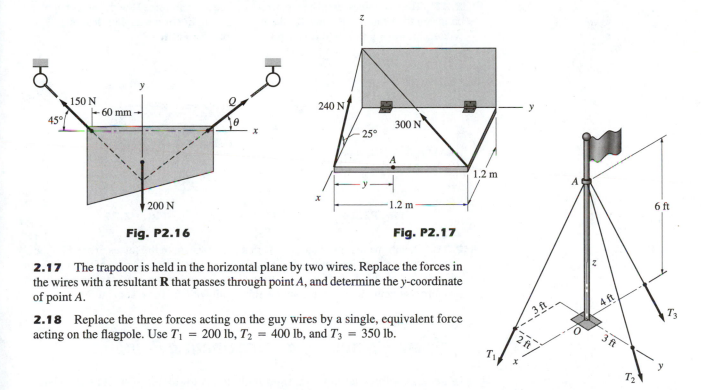

Fig. P2.16   **Fig. P2.17**

2.17   The trapdoor is held in the horizontal plane by two wires. Replace the forces in the wires with a resultant **R** that passes through point A, and determine the y-coordinate of point A.

2.18   Replace the three forces acting on the guy wires by a single, equivalent force acting on the flagpole. Use $T_1 = 200$ lb, $T_2 = 400$ lb, and $T_3 = 350$ lb.

Fig. P2.18

2.19 If **R** refers to the resultant of the three forces shown, determine (a) the magnitude of **R**; (b) the direction cosines of **R**; and (c) the point at which the line of action of **R** intersects the *xy*-plane.

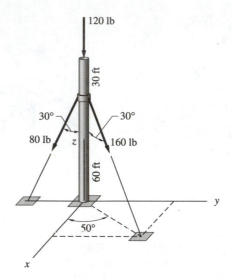

Fig. P2.19

2.20 The two forces are to be replaced by an equivalent force **R** with point of application on the beam. Determine (a) the magnitude of **R**; (b) the angle θ that **R** makes with the positive *x*-axis; and (c) the point of application of **R**.

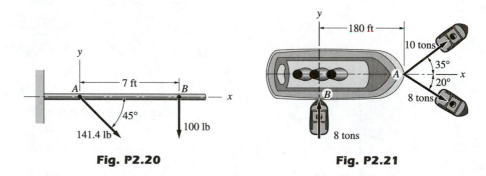

Fig. P2.20 **Fig. P2.21**

*2.21** Determine the resultant force **R** that is equivalent to the forces exerted by the three tugboats as they maneuver the ocean liner. Specify the coordinate of the point on the *x*-axis through which **R** passes. (*Hint:* First determine the resultant force for the two forces at point *A*, and then combine this result with the force at point *B*.)

2.5 *Moment of a Force about a Point*

In general, a force acting on a rigid body tends to rotate, as well as translate, the body. The force itself is the translational effect—the body tends to move in the direction of the force, and the magnitude of the force is proportional to its ability to translate the body. (The formal statement of this relationship is

Newton's second law: Force equals mass times acceleration.) Here we introduce the tendency of a force to rotate a body, called the *moment of a force about a point*. As you will see, this rotational effect depends on the magnitude of the force and the distance between the point and the line of action of the force. The tendency of a force to rotate a body about an axis, called the *moment of a force about an axis*, is discussed in the next article.

a. Definition

We begin by considering an arbitrary force $\mathbf{F}$ and an arbitrary point O, which is not on the line of action of $\mathbf{F}$, as shown in Fig. 2.4. Note that the force $\mathbf{F}$ and the point O determine a unique plane. We let A be any point on the line of action of $\mathbf{F}$ and define $\mathbf{r}$ to be the vector from point O to point A.

The moment of the force $\mathbf{F}$ about point O, called the *moment center*, is defined as

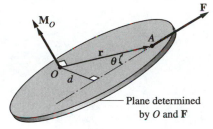

Fig. 2.4

$$\mathbf{M}_O = \mathbf{r} \times \mathbf{F} \tag{2.4}$$

Note that moment about a point has the dimension $[FL]$. In SI units, moment is measured in newton-meters (N · m). In U.S. Customary units, pound-inches (lb · in.) and pound-feet (lb · ft) are commonly used.

The moment of $\mathbf{F}$ about point O is a vector by definition. From the properties of the cross product of two vectors, $\mathbf{M}_O$ is perpendicular to both $\mathbf{r}$ and $\mathbf{F}$, with its sense determined by the right-hand rule, as shown in Fig. 2.4.*

b. Geometric interpretation

The moment of a force about a point can, of course, always be computed using the cross product in Eq. (2.4). However, a scalar computation of the magnitude of the moment can be obtained from the geometric interpretation of Eq. (2.4).

Observe that the magnitude of $\mathbf{M}_O$ is given by

$$M_O = |\mathbf{M}_O| = |\mathbf{r} \times \mathbf{F}| = rF \sin \theta \tag{2.5}$$

in which θ is the angle between $\mathbf{r}$ and $\mathbf{F}$. Returning to Fig. 2.4, we see that

$$r \sin \theta = d \tag{2.6}$$

where d is the perpendicular distance from the moment center to the line of action of the force $\mathbf{F}$. The perpendicular distance d, called the *moment arm* of the force, can therefore be used to calculate the magnitude of $\mathbf{M}_O$ using the scalar equation

$$M_O = Fd \tag{2.7}$$

Because the magnitude of $\mathbf{M}_O$ depends only on the magnitude of the force and the perpendicular distance d, a force may be moved anywhere along its line of action without changing its moment about a point. Therefore, in this application, a force may be treated as a sliding vector. This is the reason that any point A on the line of action of the force may be chosen when determining the vector $\mathbf{r}$ in Eq. (2.4).

---

*Moment vectors are drawn as double-headed arrows throughout this text.

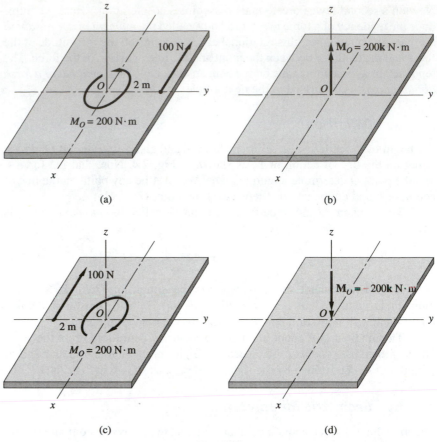

Fig. 2.5

Equation (2.7) is convenient only when the point O and the force $\mathbf{F}$ lie in a plane in which the moment arm can be easily determined. Furthermore, when using Eq. (2.7), the direction of $\mathbf{M}_O$ must be found by inspection. For example, the magnitude of the moment of the 100-N force about the point O in Fig. 2.5(a) is $(100)(2) = 200$ N · m, and its direction is counterclockwise, as viewed from the positive z-axis. Using the right-hand rule, the vector representation of this moment is $\mathbf{M}_O = 200\mathbf{k}$ N · m, as shown in Fig. 2.5(b). The magnitude of the moment about point O for the 100-N force in Fig. 2.5(c) is also 200 N · m, but in this case its direction is clockwise, as viewed from the positive z-axis. For this force, $\mathbf{M}_O = -200\mathbf{k}$ N · m, as shown in Fig. 2.5(d). Although the vector description for both forces is $-100\mathbf{i}$ N, their moments about point O are oppositely directed.

c. *Principle of moments*

When determining the moment of a force about a point, it is often convenient to use the *principle of moments*, also known as *Varignon's theorem:*

> The moment of a force about a point is equal to the sum of the moments of its components about that point.

Proof

To prove Varignon's theorem, consider the three forces $\mathbf{F}_1$, $\mathbf{F}_2$, and $\mathbf{F}_3$ concurrent at point A, as shown in Fig. 2.6, where $\mathbf{r}$ is the vector from point O to point A. The sum of the moments about point O for the three forces is

$$\mathbf{M}_O = \Sigma(\mathbf{r} \times \mathbf{F}) = (\mathbf{r} \times \mathbf{F}_1) + (\mathbf{r} \times \mathbf{F}_2) + (\mathbf{r} \times \mathbf{F}_3) \tag{a}$$

Using the properties of the cross product, Eq. (a) may be written as

$$\mathbf{M}_O = \mathbf{r} \times (\mathbf{F}_1 + \mathbf{F}_2 + \mathbf{F}_3) = \mathbf{r} \times \mathbf{R} \tag{b}$$

where $\mathbf{R} = \mathbf{F}_1 + \mathbf{F}_2 + \mathbf{F}_3$ is the resultant force for the three original forces. Equation (b) proves the principle of moments: The moment of $\mathbf{R}$ equals the moments of the components of $\mathbf{R}$. (Although the preceding proof has used only three components, it may obviously be extended to any number of components.)

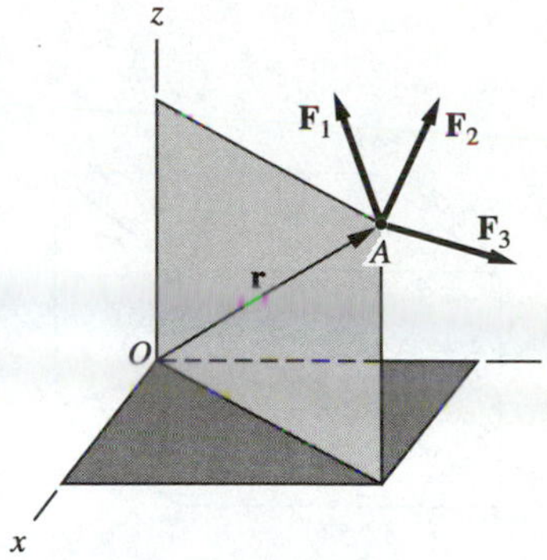

Fig. 2.6

d.   Vector and scalar methods

From the preceding discussion, we observe that the following are equivalent methods for computing the moment of a force $\mathbf{F}$ about a point O.

Vector Method   The vector method uses $\mathbf{M}_O = \mathbf{r} \times \mathbf{F}$, where $\mathbf{r}$ is a vector from point O to any point on the line of action of $\mathbf{F}$. The most efficient technique for using the vector method (with rectangular components) is the following: (1) Write $\mathbf{F}$ in vector form; (2) choose an $\mathbf{r}$, and write it in vector form; and (3) use the determinant form of $\mathbf{r} \times \mathbf{F}$ to evaluate $\mathbf{M}_O$:

$$\mathbf{M}_O = \mathbf{r} \times \mathbf{F} = \begin{vmatrix} \mathbf{i} & \mathbf{j} & \mathbf{k} \\ x & y & z \\ F_x & F_y & F_z \end{vmatrix} \tag{2.8}$$

where the second and third lines in the determinant are the rectangular components of $\mathbf{r}$ and $\mathbf{F}$, respectively. These components are shown in Fig. 2.7. Expansion of the determinant in Eq. (2.8) yields

$$\mathbf{M}_O = (yF_z - zF_y)\mathbf{i} + (zF_x - xF_z)\mathbf{j} + (xF_y - yF_x)\mathbf{k} \tag{2.9}$$

Scalar Method   In the scalar method, the magnitude of the moment of the force $\mathbf{F}$ about the point O is found from $M_O = Fd$, where d is the moment arm of the force. In this method, the sense of the moment must be determined by inspection. As mentioned previously, the scalar method is convenient only when the moment arm d can be easily determined.

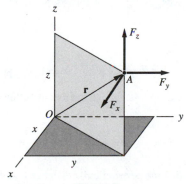

Fig. 2.7

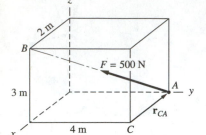

Sample Problem 2.3

Determine (1) the moment of the force $\mathbf{F}$ about point C; and (2) the perpendicular distance between C and the line of action of $\mathbf{F}$.

Solution

Part 1

The moment of a force about point C can be computed by either the scalar method ($M_C = Fd$), or the vector method ($\mathbf{M}_C = \mathbf{r} \times \mathbf{F}$). In this problem the scalar method would be inconvenient, because we have no easy means of determining d (the perpendicular distance between C and the line AB). Therefore, we use the vector method, which consists of the following three steps: (1) write $\mathbf{F}$ in vector form; (2) choose an $\mathbf{r}$, and write it in vector form; and (3) compute $\mathbf{M}_C = \mathbf{r} \times \mathbf{F}$.

Step 1: Write $\mathbf{F}$ in vector form.

Referring to the figure, we obtain

$$\mathbf{F} = 500\boldsymbol{\lambda}_{AB} = 500\frac{\overrightarrow{AB}}{|\overrightarrow{AB}|} = 500\left(\frac{2\mathbf{i} - 4\mathbf{j} + 3\mathbf{k}}{5.385}\right)$$

which yields

$$\mathbf{F} = 185.7\mathbf{i} - 371.4\mathbf{j} + 278.6\mathbf{k} \text{ N}$$

Step 2: Choose an $\mathbf{r}$, and write it in vector form.

The vector $\mathbf{r}$ is a vector from point C to any point on the line of action of $\mathbf{F}$. From the figure we see that there are two convenient choices for $\mathbf{r}$—the vector from point C to either point A or point B. As shown in the figure, let us choose $\mathbf{r}$ to be $\mathbf{r}_{CA}$. (As an exercise, you may wish to solve this problem by choosing $\mathbf{r}$ to be the vector from point C to point B.) Now we have

$$\mathbf{r} = \mathbf{r}_{CA} = -2\mathbf{i} \text{ m}$$

Step 3: Calculate $\mathbf{M}_C = \mathbf{r} \times \mathbf{F}$.

The easiest method for evaluating the cross product is to use the determinant expansion:

$$\mathbf{M}_C = \mathbf{r} \times \mathbf{F} = \mathbf{r}_{CA} \times \mathbf{F} = \begin{vmatrix} \mathbf{i} & \mathbf{j} & \mathbf{k} \\ -2 & 0 & 0 \\ 185.7 & -371.4 & 278.6 \end{vmatrix}$$

Expanding this determinant gives

$$\mathbf{M}_C = 557.2\mathbf{j} + 742.8\mathbf{k} \text{ N} \cdot \text{m} \qquad \textit{Answer}$$

Part 2

The magnitude of $\mathbf{M}_C$ is

$$M_C = \sqrt{(557.2)^2 + (742.8)^2} = 928.6 \text{ N} \cdot \text{m}$$

The perpendicular distance d from point C to the line of action of $\mathbf{F}$ may be determined by

$$d = \frac{M_C}{F} = \frac{928.6}{500} = 1.857 \text{ m} \qquad \textit{Answer}$$

Observe that, instead of using the perpendicular distance to determine the moment, we have used the moment to determine the perpendicular distance.

Caution A common mistake is choosing the wrong sense for **r** in Eq. (2.4). Note that **r** is directed from the moment center to the line of action of **F**. If the sense of **r** is reversed, **r** × **F** will yield the correct magnitude of the moment, but the wrong sense. To avoid this pitfall, it is strongly recommended that you draw **r** on your sketch before attempting to write it in vector form.

Sample Problem 2.4

Determine the moment of the force **F** in Fig. (a) about point A.

Solution

The force **F** and point A lie in the xy-plane. Problems of this type may be solved using either the vector method (**r** × **F**) or the scalar method (Fd). For illustrative purposes, we use the vector method and four equivalent scalar methods.

Vector Solution

Recall that the three steps in the vector method are to write **F** in vector form, choose **r** and write it in vector form, and then evaluate the cross product **r** × **F**.

Writing **F** in vector form, we get

$$\mathbf{F} = -\left(\frac{4}{5}\right)200\mathbf{i} + \left(\frac{3}{5}\right)200\mathbf{j}$$

$$= -160\mathbf{i} + 120\mathbf{j} \text{ lb}$$

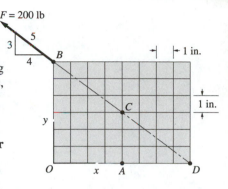

(a)

The components of **F** acting at points B, C, and D are shown in Fig. (b).

There are several good choices for **r** in this problem, three of which are the vector from A to B, the vector from A to C, and the vector from A to D. For reasons that will become apparent later, let us choose **r** to be the vector from A to B, as shown in Fig. (b), with the components of **F** located at point B. Therefore, we have

$$\mathbf{r} = \mathbf{r}_{AB} = -4\mathbf{i} + 6\mathbf{j} \text{ in.}$$

Then using the determinant form of the cross product, the moment about point A is

$$\mathbf{M}_A = \mathbf{r} \times \mathbf{F} = \mathbf{r}_{AB} \times \mathbf{F} = \begin{vmatrix} \mathbf{i} & \mathbf{j} & \mathbf{k} \\ -4 & 6 & 0 \\ -160 & 120 & 0 \end{vmatrix}$$

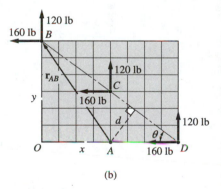

(b)

Expanding this determinant, we obtain

$$\mathbf{M}_A = \mathbf{k}[(120)(-4) + (160)(6)] = 480\mathbf{k} \text{ lb} \cdot \text{in.} \qquad \textit{Answer}$$

The magnitude of $\mathbf{M}_A$ is 480 lb · in. Note that the direction of $\mathbf{M}_A$ is the positive z direction, which by the right-hand rule means that the moment about point A is counterclockwise.

Scalar Solution: Components of F at Point B

In this problem, the scalar computation of the moment is as convenient as the vector method, because the perpendicular distances between A and each of the force components in Fig. (b) can be determined by inspection.

As the first scalar solution, let us locate the force **F** at point B and use the principle of moments. The moment of the 200-lb force about point A equals the sum of the moments of the 160-lb and 120-lb components about point A. The perpendicular distance from A to the line of action of the 120-lb component is 4 in., and for the 160-lb

49

component it is 6 in. Notice that the moment about point A for the 120-lb component is clockwise and for the 160-lb component it is counterclockwise. Considering the counterclockwise direction to be positive, we have the following scalar equation for the magnitude of the moment about point A.

$$\overset{\curvearrowleft}{+} \quad M_A = -(120)(4) + (160)(6) = +480 \text{ lb} \cdot \text{in.}$$

Because M_A is positive, the moment is counterclockwise. Using the right-hand rule, we can express the moment in vector form as

$$\mathbf{M}_A = +480\mathbf{k} \text{ lb} \cdot \text{in.} \qquad \qquad \textit{Answer}$$

Scalar Solution: Components of F at Point C

When calculating the moment of a force about a point, the force may be considered a sliding vector and thus may be moved to any point on its line of action. In this solution, the moment will be determined by placing the components of $\mathbf{F}$ at point C. Looking at Fig. (b) we see that the 120-lb component passes through point A ($d = 0$). The perpendicular distance for the 160-lb component is 3 in. and its moment about A is counterclockwise. Therefore, the magnitude of $\mathbf{M}_A$ is

$$\overset{\curvearrowleft}{+} \quad M_A = (0)(120) + (3)(160) = +480 \text{ lb} \cdot \text{in.}$$

which, of course, is the same value determined previously.

Scalar Solution: Components of F at Point D

Referring to Fig. (b) we see that, when the components of $\mathbf{F}$ are placed at point D, the magnitude of $\mathbf{M}_A$ becomes

$$\overset{\curvearrowleft}{+} \quad M_A = +(120)(4) + (160)(0) = +480 \text{ lb} \cdot \text{in.}$$

which is again the same value determined previously.

Scalar Solution: Using Force Times Perpendicular Distance

Referring to Fig. (b), we see that the perpendicular distance d from point A to the line of action of $\mathbf{F}$ is given by

$$d = 4 \sin \theta = 4\left(\frac{3}{5}\right) = 2.4 \text{ in.}$$

Therefore, the magnitude of $\mathbf{M}_A$ is

$$\overset{\curvearrowleft}{+} \quad M_A = Fd = +(200)(2.4) = +480 \text{ lb} \cdot \text{in.}$$

which once more agrees with the previous solutions.

Summary

This sample problem has been solved by five methods—the vector method and four equivalent scalar methods. Being able to work with both vector and scalar methods is one of the important skills to be learned in mechanics. If the plane determined by the force and the point is not a convenient plane, as in Sample Problem 2.3, there is no doubt that the vector solution ($\mathbf{r} \times \mathbf{F}$) is the most convenient. If the force and the point lie in a coordinate plane, such as in this sample problem, one of the scalar methods is usually easiest.

Problems

2.22 Calculate the moment of the 50-kN force about point A. Use the scalar method, placing the rectangular components of the force at (a) point B; (b) point C; and (c) point D.

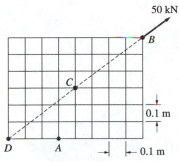

Fig. P2.22

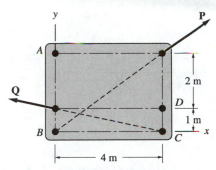

Fig. P2.23, P2.24

2.23 Determine the moments about points A, B, C, and D of the force $\mathbf{P}$, which has a magnitude of 250 N. Express your answers in vector form.

2.24 Determine the moments about points A, B, C, and D of the force $\mathbf{Q}$, which has a magnitude of 150 N. Express your answers in vector form.

2.25 A force $\mathbf{P}$ acts in the xy-plane. The moments of $\mathbf{P}$ about points O, A, and B are $M_O = 200$ N · m clockwise, $M_A = 0$, and $M_B = 0$. Determine $\mathbf{P}$.

2.26 A force $\mathbf{P}$ acts in the xy-plane. The moments of $\mathbf{P}$ about points O, A, and B are $M_O = 80$ N · m counterclockwise, $M_A = 200$ N · m clockwise, and $M_B = 0$. Determine $\mathbf{P}$.

2.27 Determine the moment of the force $\mathbf{F} = 9\mathbf{i} + 18\mathbf{j}$ lb about point O by each of the following methods: (a) vector method using $\mathbf{r} \times \mathbf{F}$; (b) scalar method using rectangular components of $\mathbf{F}$; and (c) scalar method using components of $\mathbf{F}$ that are parallel and perpendicular to the line OA.

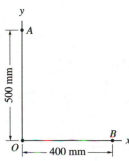

Fig. P2.25, P2.26

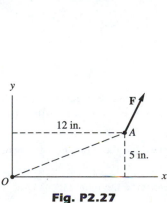

Fig. P2.27

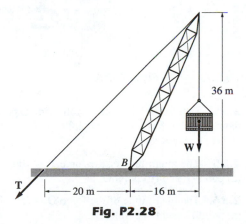

Fig. P2.28

2.28 Given that $T = 28.3$ kN and $W = 25$ kN, determine the magnitude and direction of the moments about point B of the following: (a) the force $\mathbf{T}$; (b) the force $\mathbf{W}$; and (c) forces $\mathbf{T}$ and $\mathbf{W}$ combined.

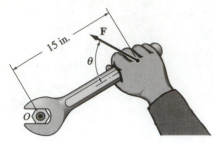

Fig. P2.29

2.29 A moment of 50 lb · ft about O is required to loosen the nut. Determine the smallest magnitude of the force **F** and the corresponding angle θ that will turn the nut.

2.30 Solve Prob. 2.12, recognizing that the moment of the resultant force about point A is zero.

2.31 The resultant of the two forces shown has a line of action that passes through point A. Recognizing that the moment of the resultant about A is zero, determine the distance x.

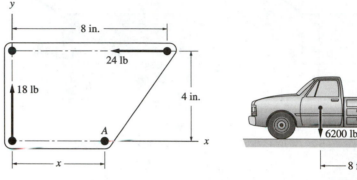

Fig. P2.31

Fig. P2.32

2.32 The tow truck's front wheels will be lifted off the ground if the moment of the load W about the rear axle exceeds the moment of the 6200-lb weight of the truck. Determine the smallest angle θ at which the boom may be safely positioned if $W = 9000$ lb.

2.33 The force **F** acts on the gripper of the robot arm. The moments of **F** about points A and B are 120 N · m and 60 N · m, respectively—both counterclockwise. Determine F and the angle θ.

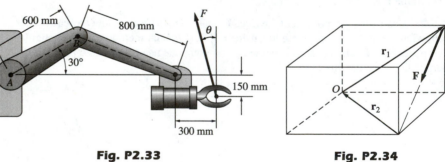

Fig. P2.33

Fig. P2.34

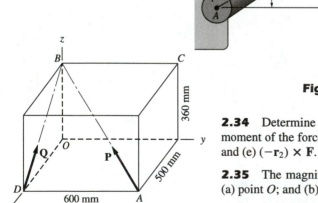

Fig. P2.35, P2.36

2.34 Determine which of the following expressions are valid representations for the moment of the force **F** about point O: (a) $\mathbf{r}_1 \times \mathbf{F}$; (b) $\mathbf{r}_2 \times \mathbf{F}$; (c) $\mathbf{F} \times \mathbf{r}_1$; (d) $\mathbf{F} \times \mathbf{r}_2$; and (e) $(-\mathbf{r}_2) \times \mathbf{F}$.

2.35 The magnitude of the force **P** is 100 N. Determine the moments of **P** about (a) point O; and (b) point C.

2.36 The magnitude of the force **Q** is 250 N. Determine the moments of **Q** about (a) point O; and (b) point C.

2.37 The magnitude of the moment of force **P** about point O is 200 kN · m. Determine the magnitude of **P**.

2.38 The magnitude of the force **P** is 50 kN. Determine the moment of **P** about (a) point A; and (b) point B.

2.39 Determine the moments of **Q** about (a) point O; and (b) point C. The magnitude of **Q** is 20 lb.

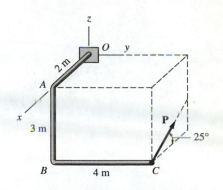

Fig. P2.37, P2.38

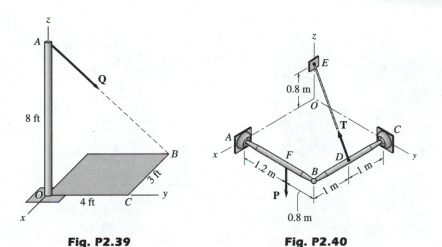

Fig. P2.39   **Fig. P2.40**

2.40 Given that $P = 200$ N and $T = 300$ N, compute the combined moment of **P** and **T** about point C. Express the answer in vector form.

2.41 The wrench is used to tighten a nut on the wheel. Determine the moment of the 120-lb force about point O. Express your answer in vector form.

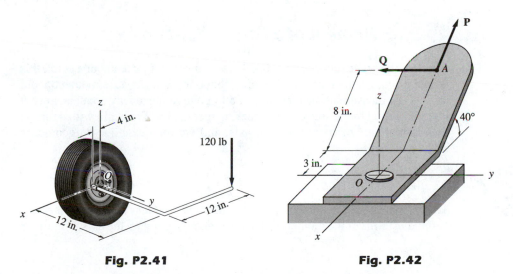

Fig. P2.41   **Fig. P2.42**

2.42 The magnitudes of the two forces shown are $P = 16$ lb and $Q = 22$ lb. Determine the magnitude of the combined moment of **P** and **Q** about point O and the direction cosines of this moment vector.

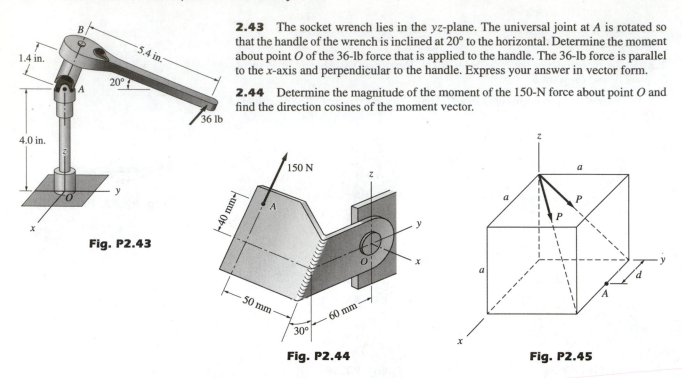

Fig. P2.43

Fig. P2.44

Fig. P2.45

2.43 The socket wrench lies in the *yz*-plane. The universal joint at *A* is rotated so that the handle of the wrench is inclined at 20° to the horizontal. Determine the moment about point *O* of the 36-lb force that is applied to the handle. The 36-lb force is parallel to the *x*-axis and perpendicular to the handle. Express your answer in vector form.

2.44 Determine the magnitude of the moment of the 150-N force about point *O* and find the direction cosines of the moment vector.

2.45 The combined moment of the two forces, each of magnitude *P*, about point *A* is zero. Determine the distance *d* that locates *A*.

2.46 A force $\mathbf{F} = 10\mathbf{i} + 30\mathbf{j} - 40\mathbf{k}$ N, acting at point *A* (3 m, 6 m, −2 m), intersects the *yz*-plane at point *D*. Determine the coordinates of point *D* using the concept of the moment of a force about a point.

2.6 *Moment of a Force about an Axis*

Whereas the preceding article defined the moment of a force about a point, this article discusses the moment of a force about an axis. Because moment about an axis is a measure of the tendency of a force to rotate a body about an axis, it is fundamental to the study of engineering mechanics. We begin with a formal definition of the moment about an axis, and we then examine its geometric interpretation.

a. Definition

The moment of a force about an axis, called the *moment axis,* is most easily defined in terms of the moment of the force about a point on the axis. Figure 2.8

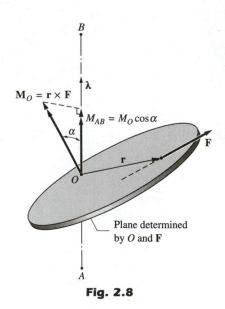

Fig. 2.8

shows the force $\mathbf{F}$ and its moment about point O, $\mathbf{M}_O = \mathbf{r} \times \mathbf{F}$, where O is any point on the axis AB. We define the moment about an axis as follows:

> The moment of $\mathbf{F}$ about the axis AB is the orthogonal component of $\mathbf{M}_O$ along the axis AB, where O is any point on AB.

Letting $\boldsymbol{\lambda}$ be a unit vector directed from A toward B, this definition gives for the moment of $\mathbf{F}$ about the axis AB

$$M_{AB} = M_O \cos \alpha \qquad (2.10)$$

where α is the angle between $\mathbf{M}_O$ and $\boldsymbol{\lambda}$, as shown in Fig. 2.8.

Because $M_O \cos \alpha = \mathbf{M}_O \cdot \boldsymbol{\lambda}$ (from the definition of the dot product), Eq. (2.10) can also be expressed in the form

$$M_{AB} = \mathbf{M}_O \cdot \boldsymbol{\lambda} = \mathbf{r} \times \mathbf{F} \cdot \boldsymbol{\lambda} \qquad (2.11)$$

Let us review each of the terms appearing in this equation:

- M_{AB} is the moment (actually, the magnitude of the moment) of the force $\mathbf{F}$ about the axis AB.
- $\mathbf{M}_O$ represents the moment of $\mathbf{F}$ about the point O, where O is any point on the axis AB.*
- $\boldsymbol{\lambda}$ is the unit vector directed from A toward B.
- $\mathbf{r}$ is the position vector drawn from O to any point on the line of action of $\mathbf{F}$.

Note that the direction of $\boldsymbol{\lambda}$ determines the positive sense of M_{AB} by the right-hand rule, as illustrated in Fig. 2.9. Paying heed to this sign convention will enable you to interpret the sign of M_{AB} in Eqs. (2.10) and (2.11).

Sometimes we wish to express the moment of $\mathbf{F}$ about the axis AB as a vector. We can do this by multiplying M_{AB} by the unit vector $\boldsymbol{\lambda}$ that specifies the direction of the moment axis, yielding

$$\mathbf{M}_{AB} = M_{AB}\boldsymbol{\lambda} = (\mathbf{r} \times \mathbf{F} \cdot \boldsymbol{\lambda})\boldsymbol{\lambda} \qquad (2.12)$$

Rectangular components of $\mathbf{M}_O$ Let $\mathbf{M}_O$ be the moment of a force $\mathbf{F}$ about O, where O is the origin of the xyz-coordinate system shown in Fig. 2.10. The moments of $\mathbf{F}$ about the three coordinate axes can be obtained from Eq. (2.11) by substituting $\mathbf{i}$, $\mathbf{j}$, and $\mathbf{k}$ in turn for $\boldsymbol{\lambda}$. The results are

$$M_x = \mathbf{M}_O \cdot \mathbf{i} \qquad M_y = \mathbf{M}_O \cdot \mathbf{j} \qquad M_z = \mathbf{M}_O \cdot \mathbf{k}$$

from which we draw the following conclusion:

> The rectangular components of the moment of a force about the origin O are equal to the moments of the force about the coordinate axes.

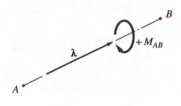

(a)

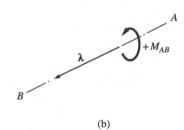

(b)

Fig. 2.9

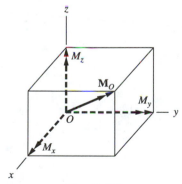

Fig. 2.10

---

*If we let O and C be two points on the moment axis AB, then $\mathbf{M}_O$ and $\mathbf{M}_C$ will, in general, be different. However, it can be shown that $\mathbf{M}_O \cdot \boldsymbol{\lambda} = \mathbf{M}_C \cdot \boldsymbol{\lambda}$, where $\boldsymbol{\lambda}$ is a unit vector parallel to AB. For this reason, O in Eq. (2.10) can be any point on AB.

In other words,

$$\mathbf{M}_O = M_x\mathbf{i} + M_y\mathbf{j} + M_z\mathbf{k} \qquad (2.13)$$

where M_x, M_y, and M_z, shown in Fig. 2.10, are equal to the moments of the force about the coordinate axes.

***Special Case: Moment Axis Perpendicular to* F**   Consider the case where the moment axis is perpendicular to the plane containing the force **F** and the point O, as shown in Fig. 2.11(a). Because the directions of $\mathbf{M}_O$ and $\mathbf{M}_{AB}$ now coincide, $\boldsymbol{\lambda}$ in Eq. (2.11) is in the direction of $\mathbf{M}_O$. Consequently, Eq. (2.11) yields

$$M_O = M_{AB} \qquad (2.14)$$

That is, the moment of **F** about point O equals the moment of **F** about the axis AB.

A two-dimensional representation of Fig. 2.11(a), viewed along the moment axis AB, is shown in Fig. 2.11(b). We will frequently use a similar figure in the solution of two-dimensional problems (problems where all forces lie in the same plane). In problems of this type, it is customary to use the term *moment about a point* (M_O), rather than *moment about an axis* (M_{AB}).

b.   *Geometric interpretation*

It is instructive to examine the geometric interpretation of the equation $M_{AB} = \mathbf{r} \times \mathbf{F} \cdot \boldsymbol{\lambda}$.

Suppose we are given an arbitrary force **F** and an arbitrary axis AB, as shown in Fig. 2.12. We construct a plane $\mathcal{P}$ that is perpendicular to the AB axis and let O and C be the points where the axis and the line of action of the force intersect $\mathcal{P}$, respectively. The vector from O to C is denoted by **r**, and $\boldsymbol{\lambda}$ is the unit vector along the axis AB. We then resolve **F** into two components: $\mathbf{F}_1$ and $\mathbf{F}_2$, which are parallel and perpendicular to the axis AB, respectively (observe that $\mathbf{F}_2$ lies in plane $\mathcal{P}$). In terms of these components, the moment of **F** about the

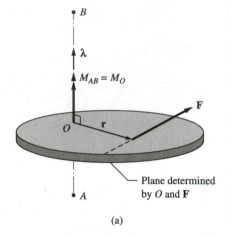

(a)

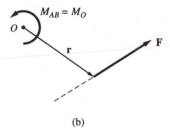

(b)

Fig. 2.11

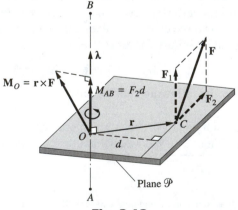

Fig. 2.12

axis AB is

$$M_{AB} = \mathbf{r} \times \mathbf{F} \cdot \boldsymbol{\lambda} = \mathbf{r} \times (\mathbf{F}_1 + \mathbf{F}_2) \cdot \boldsymbol{\lambda}$$
$$= \mathbf{r} \times \mathbf{F}_1 \cdot \boldsymbol{\lambda} + \mathbf{r} \times \mathbf{F}_2 \cdot \boldsymbol{\lambda}$$

Because $\mathbf{r} \times \mathbf{F}_1$ is perpendicular to $\boldsymbol{\lambda}$, $\mathbf{r} \times \mathbf{F}_1 \cdot \boldsymbol{\lambda} = 0$, and we get

$$M_{AB} = \mathbf{r} \times \mathbf{F}_2 \cdot \boldsymbol{\lambda}$$

Substitution of $\mathbf{r} \times \mathbf{F}_2 \cdot \boldsymbol{\lambda} = F_2 d$, where d is the perpendicular distance from O to the line of action of $\mathbf{F}_2$, yields

$$\boxed{M_{AB} = F_2 d} \tag{2.15}$$

We see that the moment of $\mathbf{F}$ about the axis AB equals the product of the component of $\mathbf{F}$ that is perpendicular to AB and the perpendicular distance of this component from AB. Observe that Eq. (2.15) gives only the magnitude of the moment about the axis; its sense must be determined by inspection.

Consideration of Eq. (2.15) reveals that the moment of a force about an axis, as defined in Eq. (2.10), possesses the following physical characteristics:

- A force that is parallel to the moment axis (such as F_1) has no moment about that axis.
- If the line of action of a force intersects the moment axis ($d = 0$), the force has no moment about that axis.
- The moment of a force is proportional to its component that is perpendicular to the moment axis (such as F_2), and the moment arm (d) of that component.
- The sense of the moment is consistent with the direction in which the force would tend to rotate a body.

To illustrate the above characteristics, consider opening the door in Fig. 2.13 by applying a force $\mathbf{P}$ to the handle. In the figure, $\mathbf{P}$ is resolved into the following rectangular components: P_x intersects the hinge axis, P_y is perpendicular to the door, and P_z is parallel to the hinge axis.

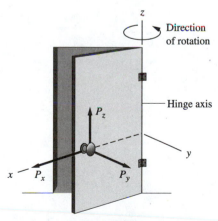

Fig. 2.13

To open the door, we must apply a moment about the z-axis (the hinge axis). Experience tells us that P_y is the only component of the force that would accomplish this task. The components P_x and P_z are ineffective, because their moments about the z-axis are zero. We also know that it is easier to open the door if we increase the distance between the handle and the hinge axis (the moment arm) or if the magnitude of P_y is increased. Finally, observe that P_y causes the door to rotate in the direction shown in the figure, which is also the sense of the moment about the z-axis.

c. Vector and scalar methods

From the preceding discussion we see that the moment of the force **F** about an axis AB can be computed by two methods.

Vector Method   The moment of **F** about AB is obtained from the triple scalar product $M_{AB} = \mathbf{r} \times \mathbf{F} \cdot \boldsymbol{\lambda}$, where **r** is a vector drawn from any point on the moment axis AB to any point on the line of action of **F** and $\boldsymbol{\lambda}$ represents a unit directed from A toward B. A convenient means of evaluating the scalar triple product is its determinant form

$$M_{AB} = \begin{vmatrix} x & y & z \\ F_x & F_y & F_z \\ \lambda_x & \lambda_y & \lambda_z \end{vmatrix} \tag{2.16}$$

where x, y, and z are the rectangular components of **r**.

Scalar Method   The moment of **F** about AB is obtained from the scalar expression $M_{AB} = F_2 d$. The sense of the moment must be determined by inspection. This method is convenient if AB is parallel to one of the coordinate axes (which is always the case in two-dimensional problems).

Sample Problem 2.5

The force **F** of magnitude 195 kN acts along the line AB. (1) Determine the moments of **F** about each of the coordinate axes by the scalar method; and (2) find the moment of **F** about point O by the vector method and verify that $\mathbf{M}_O = M_x\mathbf{i} + M_y\mathbf{j} + M_z\mathbf{k}$.

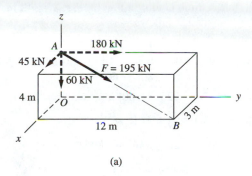

(a)

Solution

We start by computing the rectangular components of **F**:

$$\mathbf{F} = F\boldsymbol{\lambda}_{AB} = F\frac{\overrightarrow{AB}}{|\overrightarrow{AB}|} = 195\left(\frac{3\mathbf{i} + 12\mathbf{j} - 4\mathbf{k}}{\sqrt{3^2 + 12^2 + (-4)^2}}\right)$$

$$= 45\mathbf{i} + 180\mathbf{j} - 60\mathbf{k} \text{ kN}$$

When calculating the moment of a force, the force may be placed at any point on its line of action. As shown in Fig. (a), we chose to have the force acting at point A.

Part 1

The moment of **F** about a coordinate axis can be computed by summing the moments of the components of **F** about that axis (the principle of moments).

Moment about the x-Axis   Figure (b) represents a two-dimensional version of Fig. (a), showing the yz-plane. We see that the 45-kN and the 60-kN components of the force contribute nothing to the moment about the x-axis (the former is parallel to the axis, and the latter intersects the axis). The perpendicular distance (moment arm) between the 180-kN component and the x-axis is 4 m. Therefore, the moment of this component about the x-axis (which is also the moment of **F**) is $180(4) = 720$ kN · m, clockwise. According to the right-hand rule, the positive sense of M_x is counterclockwise, which means that M_x is negative; that is,

$$M_x = -720 \text{ kN} \cdot \text{m} \qquad \textit{Answer}$$

Moment about the y-Axis   To compute the moment about the y-axis, we refer to Fig. (c), which represents the xz-plane. We note that only the 45-kN force component has a moment about the y-axis, because the 180-kN component is parallel to the y-axis and the 60-kN component intersects the y-axis. Because the moment arm of the 45-kN component is 4 m, the moment of **F** about the y-axis is $45(4) = 180$ kN · m, counterclockwise. Therefore, we have

$$M_y = 45(4) = 180 \text{ kN} \cdot \text{m} \qquad \textit{Answer}$$

The sign of the moment is positive, because the right-hand rule determines positive M_y to be counterclockwise.

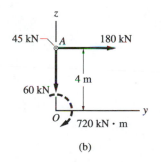

(b)

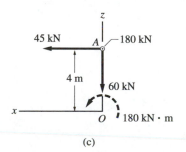

(c)

Moment about the z-Axis   The moment of **F** about the z-axis is zero, because **F** intersects that axis. Hence

$$M_z = 0 \qquad\qquad \textit{Answer}$$

Part 2

Recognizing that the vector from O to A in Fig. (a) is $\mathbf{r}_{OA} = 4\mathbf{k}$ m, the moment of **F** about point O can be computed as follows.

$$\mathbf{M}_O = \mathbf{r}_{OA} \times \mathbf{F} = \begin{vmatrix} \mathbf{i} & \mathbf{j} & \mathbf{k} \\ 0 & 0 & 4 \\ 45 & 180 & -60 \end{vmatrix} = -\mathbf{i}(4)(180) + \mathbf{j}(4)(45)$$

$$= -720\mathbf{i} + 180\mathbf{j} \text{ kN} \cdot \text{m} \qquad\qquad \textit{Answer}$$

Comparing with $\mathbf{M}_O = M_x\mathbf{i} + M_y\mathbf{j} + M_z\mathbf{k}$, we see that

$$M_x = -720 \text{ kN} \cdot \text{m} \qquad M_y = 180 \text{ kN} \cdot \text{m} \qquad M_z = 0$$

which agree with the results obtained in Part 1.

Sample Problem 2.6

The force **F** of Sample Problem 2.5 is shown again in Fig. (a). (1) Determine the moment of **F** about the axis CE; and (2) express the moment found in Part 1 in vector form.

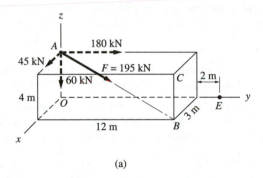

(a)

Solution

Part 1

Referring to Fig. (a), we see that it is not practical to calculate the moment about the axis CE by the scalar method. Because the axis CE is not parallel to a coordinate axis, the task of determining the perpendicular distance between **F** and CE would be tedious. However, if the vector method is used, the calculation of the moment is straightforward.

To employ the vector method we first express the force **F** in vector form. This was already done in the solution to Sample Problem 2.5:

$$\mathbf{F} = 45\mathbf{i} + 180\mathbf{j} - 60\mathbf{k} \text{ kN}$$

Next we calculate the moment of $\mathbf{F}$ about any convenient point on the axis CE. Inspection of Fig. (a) reveals that there are only two convenient points from which to choose—points C and E. Let us choose point C. Because we will use the cross product $\mathbf{r} \times \mathbf{F}$ to compute the moment about C, our next step is to choose the vector $\mathbf{r}$ and to write it in vector form (remember that $\mathbf{r}$ must be a vector from point C to any point on the line of action of $\mathbf{F}$). From Fig. (a) we see that there are two convenient choices for $\mathbf{r}$: either the vector from C to A or the vector from C to B. Choosing the latter, we have

$$\mathbf{r} = \mathbf{r}_{CB} = -4\mathbf{k} \text{ m}$$

The moment of $\mathbf{F}$ about point C then becomes

$$\mathbf{M}_C = \mathbf{r}_{CB} \times \mathbf{F} = \begin{vmatrix} \mathbf{i} & \mathbf{j} & \mathbf{k} \\ 0 & 0 & -4 \\ 45 & 180 & -60 \end{vmatrix}$$

$$= 720\mathbf{i} - 180\mathbf{j} \text{ kN} \cdot \text{m}$$

Note that the z-component of $\mathbf{M}_C$ is zero. To understand this result, recall that the z-component of $\mathbf{M}_C$ equals the moment of $\mathbf{F}$ about the axis BC (the line parallel to the z-axis passing through C). Because $\mathbf{F}$ intersects BC, its moment about BC is expected to be zero.

Next, we calculate the unit vector $\boldsymbol{\lambda}_{CE}$ directed from point C toward point E:

$$\boldsymbol{\lambda}_{CE} = \frac{\overrightarrow{CE}}{|\overrightarrow{CE}|} = \frac{-3\mathbf{i} + 2\mathbf{j} - 4\mathbf{k}}{\sqrt{(-3)^2 + 2^2 + (-4)^2}} = -0.5571\mathbf{i} + 0.3714\mathbf{j} - 0.7428\mathbf{k}$$

The moment of $\mathbf{M}_C$ about the axis CE can now be obtained from Eq. (2.11):

$$M_{CE} = \mathbf{M}_C \cdot \boldsymbol{\lambda}_{CE}$$

$$= (720\mathbf{i} - 180\mathbf{j}) \cdot (-0.5571\mathbf{i} + 0.3714\mathbf{j} - 0.7428\mathbf{k})$$

$$= -468 \text{ kN} \cdot \text{m} \qquad\qquad\qquad \textit{Answer}$$

The negative sign indicates that the sense of the moment is as shown in Fig. (b)—that is, opposite to the sense associated with $\boldsymbol{\lambda}_{CE}$.

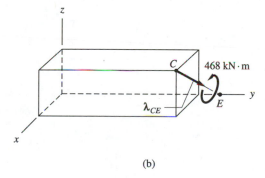

(b)

We could also compute M_{CE} without first determining $\mathbf{M}_C$ by using the scalar triple product:

$$M_{CE} = \mathbf{r}_{BC} \times \mathbf{F} \cdot \boldsymbol{\lambda}_{CE} = \begin{vmatrix} 0 & 0 & -4 \\ 45 & 180 & -60 \\ -0.5571 & 0.3714 & -0.7428 \end{vmatrix}$$

$$= -468 \text{ kN} \cdot \text{m}$$

This agrees, of course, with the result determined previously.

Part 2

To express the moment of $\mathbf{F}$ about the axis CE in vector form, we multiply M_{CE} by the unit vector $\boldsymbol{\lambda}_{CE}$, which gives

$$\mathbf{M}_{CE} = M_{CE}\boldsymbol{\lambda}_{CE} = -468(-0.5571\mathbf{i} + 0.3714\mathbf{j} - 0.7428\mathbf{k})$$
$$= 261\mathbf{i} - 174\mathbf{j} + 348\mathbf{k} \text{ kN} \cdot \text{m} \qquad \textit{Answer}$$

There is no doubt that using vector analysis is convenient when one wishes to calculate the moment about an axis such as CE, which is skewed relative to the coordinate system. However, there is a drawback to vector formalism: You can easily lose appreciation for the physical nature of the problem.

Problems

2.47 Solve Sample Problem 2.5 with the components of the force **F** located at point *B*.

2.48 Determine the moment of the 40-kN force about each of the following axes: (a) *AB*; (b) *CD*; (c) *CG*; (d) *CH*; and (e) *EG*.

2.49 Determine the moment of the 400-lb force about each of the following axes: (a) *AB*; (b) *CD*; (c) *BF*; (d) *DH*; and (e) *BD*.

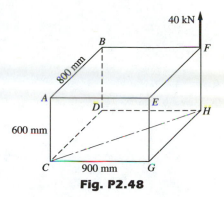

Fig. P2.48

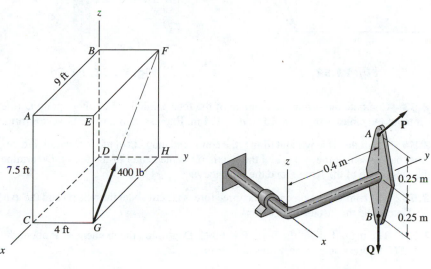

Fig. P2.49   **Fig. P2.50**

2.50 Determine the combined moment of **P** and **Q** about the *x*-axis, given that **P** = 60**i** + 120**k** N and **Q** = −30**k** N.

2.51 The force **F** = 12**i** − 8**j** + 6**k** N is applied to the gripper of the holding device shown. Determine the moment of **F** about (a) the *a*-axis; and (b) the *z*-axis.

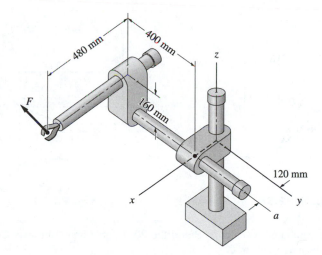

Fig. P2.51

2.52 The moment of the force **F** about the *x*-axis is 1080 N · m. Determine the moment of **F** about the axis *AB*.

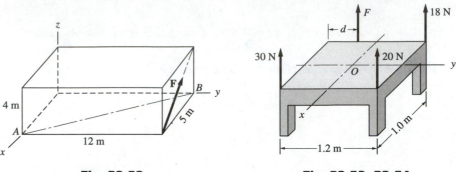

Fig. P2.52   Fig. P2.53, P2.54

2.53 Compute the combined moment of the four parallel forces about point *O* (the center of the table) using $F = 40$ N, $d = 0.4$ m. Express your answer in vector form.

2.54 To lift the table without tilting, the combined moment of the four parallel forces must be zero about the *x*-axis and the *y*-axis (*O* is the center of the table). Determine the magnitude of the force **F** and the distance *d*.

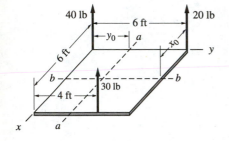

Fig. P2.55

2.55 The combined moment of the three forces is zero about the axis *aa* and the axis *bb*. Determine the distances x_0 and y_0.

2.56 The magnitude of the force **P** is 60 N. Determine the moment of **P** about the axis *EG*. Express your answer in vector form.

2.57 If the magnitude of the force **P** is 60 N, calculate its moment about the axis *FD*. Express your answer in vector form.

2.58 The magnitude of the force **Q** is 120 N. Determine the moment of **Q** about the axis *EB*. Express your answer in vector form.

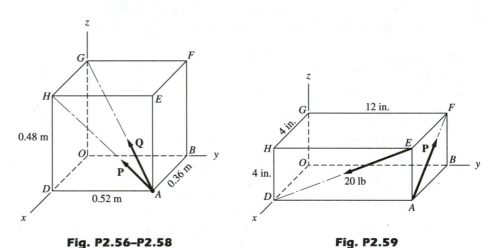

Fig. P2.56–P2.58   Fig. P2.59

2.59 The combined moment of **P** and the 20-lb force about the axis *GB* is zero. Determine the magnitude of **P**.

2.60 Determine the magnitude of the force **F** given that its moment about the axis BC is 137.3 lb · ft.

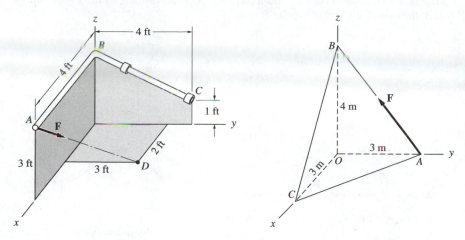

Fig. P2.60   **Fig. P2.61**

*2.61** Given that $F = 250$ N, determine the moment of **F** about the axis that is perpendicular to the plane ABC and passes through point O. Express your answer in vector form.

2.62 Calculate the moment of the force **P** about the axis AD using (a) A as the moment center; and (2) D as the moment center.

2.63 Calculate the combined moment of the two forces about the axis OD, using (a) the vector method ($\mathbf{r} \times \mathbf{F} \cdot \boldsymbol{\lambda}$); and (b) the scalar method (use trigonometry to find the moment arm of each force about the axis).

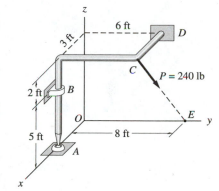

Fig. P2.62

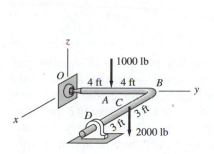

Fig. P2.63

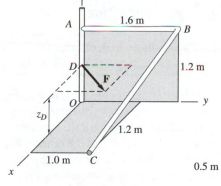

Fig. P2.64

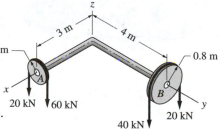

Fig. P2.65

2.64 The force $\mathbf{F} = F(0.6\mathbf{i} + 0.8\mathbf{j})$ is applied to the frame at the point D $(0, 0, z_D)$. If the moment of **F** about the axis BC is zero, determine the coordinate z_D.

2.65 Determine the combined moment of the four forces acting on the pulleys about the axis AB (A and B are the centers of the pulleys).

2.66 The flexible shaft *AB* of the wrench is bent into a horizontal arc with a radius of 24 in. The two 20-lb forces, which are parallel to the *z*-axis, are applied to the handle *CD*, as shown. Determine the combined moment of the two 20-lb forces about the *x*-axis, the axis of the socket at point *B*.

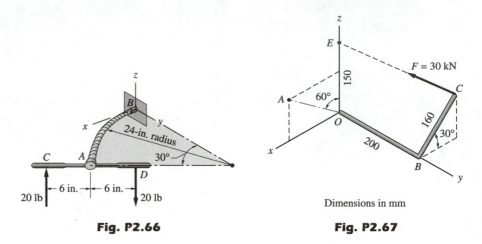

Fig. P2.66

Fig. P2.67

Dimensions in mm

2.67 Determine the moment of the force **F** about the axis *OA*.

2.7 Couples

As pointed out before, a force has two effects on a rigid body: translation due to the force itself and rotation due to the moment of the force. A couple, on the other hand, is a purely rotational effect—it has a moment but no resultant force. Couples play an important role in the analysis of force systems.

a. Definition

> Two parallel, noncollinear forces that are equal in magnitude and opposite in direction are known as a *couple*.

A typical couple is shown in Fig. 2.14. The two forces of equal magnitude *F* are oppositely directed along lines of action that are separated by the perpendicular distance *d*. (In a vector description of the forces, one of the forces would be labeled **F** and the other −**F**.) The lines of action of the two forces determine a plane that we call the *plane of the couple*. The two forces that form a couple have some interesting properties, which will become apparent when we calculate their combined moment about a point.

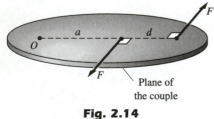

Plane of the couple

Fig. 2.14

b. Moment of a couple about a point

The moment of a couple about a point is the sum of the moments of the two forces that form the couple. When calculating the moment of a couple about a point, either the scalar method (force times perpendicular distance) or the vector method (**r** × **F**) may be used. For illustrative purposes, we will calculate the moment of a couple using both methods. Using two methods of analysis to determine the same quantity may appear redundant, but it is instructive because each method emphasizes different characteristics of couples.

Scalar Calculation   Let us calculate the moment of the couple shown in Fig. 2.14 about the point O. Note that O is an arbitrary point in the plane of the couple and that it is located a distance a from the force on the left. The sum of the moments about point O for the two forces is

$$\overset{+}{\curvearrowleft} \quad M_O = F(a + d) - F(a) = Fd \qquad (2.17)$$

Note that the moment of the couple about point O is *independent* of the location of O, because the result is independent of the distance a.

From the foregoing discussion, we see that a couple possesses two important characteristics: (1) A couple has no resultant force ($\Sigma \mathbf{F} = \mathbf{0}$), and (2) the moment of a couple is the same about any point in the plane of the couple.

Vector Calculation   When the two forces that form the couple are expressed as vectors, they can be denoted by $\mathbf{F}$ and $-\mathbf{F}$, as shown in Fig. 2.15. The points labeled in the figure are A, any point on the line of action of $\mathbf{F}$; B, any point on the line of action of $-\mathbf{F}$; and O, an arbitrary point in space (not necessarily lying in the plane of the couple). The vectors $\mathbf{r}_{OA}$ and $\mathbf{r}_{OB}$ are drawn from point O to points A and B, respectively. The vector $\mathbf{r}_{BA}$ connects points B and A. Using the cross product to evaluate the moment of the couple about point O, we get

$$\mathbf{M}_O = [\mathbf{r}_{OA} \times \mathbf{F}] + [\mathbf{r}_{OB} \times (-\mathbf{F})] = (\mathbf{r}_{OA} - \mathbf{r}_{OB}) \times \mathbf{F}$$

Noting from Fig. (2.15) that $\mathbf{r}_{OA} - \mathbf{r}_{OB} = \mathbf{r}_{BA}$, the moment of the couple about point O reduces to

$$\mathbf{M}_O = \mathbf{r}_{BA} \times \mathbf{F} \qquad (2.18)$$

Note again that the moment of the couple about point O is independent of the location of O. Although the choice of point O determines $\mathbf{r}_{OA}$ and $\mathbf{r}_{OB}$, neither of these vectors appear in Eq. (2.18). Thus we conclude the following:

> The moment of a couple is the same about every point.

In other words, the moment of a couple is a *free vector.* (Recall that, in the scalar calculation, point O was restricted to points in the plane of the couple. We see

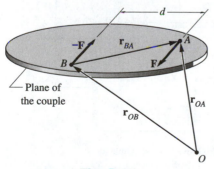

Fig. 2.15

now that this restriction is unnecessary; that is, *O* can be any point in space.) In contrast, the moment of a force about a point (the moment center) is a fixed vector, because the moment depends on the location of the moment center.

c.   Equivalent couples

Because a couple has no resultant force, its only effect on a rigid body is its moment. For this reason, two couples that have the same moment are said to be *equivalent* (have the same effect on a rigid body). Figure 2.16 illustrates the four operations that may be performed on a couple without changing its moment; all couples shown in the figure are equivalent. The operations are

1. Changing the magnitude *F* of each force and the perpendicular distance *d* while keeping the product *Fd* constant
2. Rotating the couple in its plane
3. Moving the couple to a parallel position in its plane
4. Moving the couple to a parallel plane

d.   Notation and terminology

Because the only rigid-body effect of a couple is its moment, we will use the terms *couple* and *moment of couple* synonymously. For example, consider

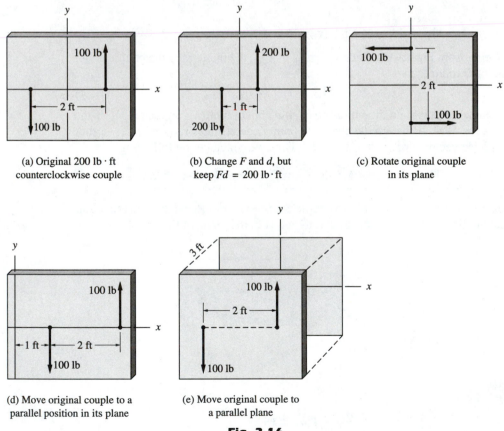

(a) Original 200 lb · ft
counterclockwise couple

(b) Change *F* and *d*, but
keep *Fd* = 200 lb · ft

(c) Rotate original couple
in its plane

(d) Move original couple to a
parallel position in its plane

(e) Move original couple to
a parallel plane

Fig. 2.16

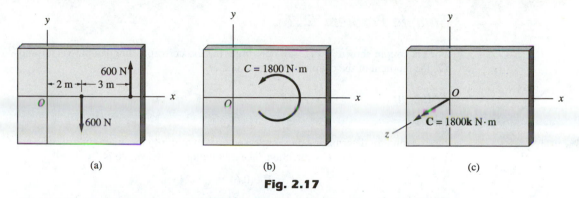

Fig. 2.17

the couple shown in Fig. 2.17(a). The moment of the couple has a magnitude $C = 1800$ N · m and is directed counterclockwise in the xy-plane, as shown in Fig. 2.17(b). Note that the moment can be placed anywhere in the figure (including moving it to a parallel plane) because the moment of a couple is, after all, a free vector. Rather than referring to Fig. 2.17(b) as the *moment of the couple,* we will from now on call it simply the *couple.* Analogously, we can replace a couple by its moment; for example, we can use the notation in Fig. 2.17(b) to represent the original couple.

Figure 2.17(c) shows the same couple as a vector, called the *couple-vector.* The couple-vector is perpendicular to the plane of the couple, and its sense is determined by the right-hand rule. The choice of point O for the location of the couple-vector was arbitrary; being a free vector, it could be shown at any point. We will employ the notations in Figs. 2.17(b) and (c) interchangeably, choosing the one that is more convenient for the problem at hand.

e. *The addition and resolution of couples*

Because couples are vectors, they may be added by the usual rules of vector addition. Being free vectors, the requirement that the couples to be added must have a common point of application does not apply. This is in contrast to the addition of forces, which can be added only if they are concurrent. Concurrency is also required for the addition of moments of forces about points, because these are fixed to a moment center. It follows that we must be careful when representing moments of forces and couples as vectors—it is easy to confuse these two concepts. To minimize the possibility of confusion, we will use **M** to denote moments of forces and reserve **C** for couples.

The resolution of couples is no different than the resolution of moments of forces. For example, the moment of a couple **C** about an axis AB can be computed from Eq. (2.11) by replacing $\mathbf{M}_O$ with **C**:

$$M_{AB} = \mathbf{C} \cdot \boldsymbol{\lambda} \qquad (2.19)$$

where $\boldsymbol{\lambda}$ is the unit vector in the direction of the axis. Note that the subscript O, which indicated that the moment must be taken about point O lying on the axis AB, is no longer present in Eq. (2.19). The reason is, of course, that the moment of **C** is the same about every point. As in the case of moments of forces, M_{AB} is equal to the rectangular component of **C** in the direction of AB, and is a measure of the tendency of **C** to rotate a body about the axis AB.

Sample Problem 2.7

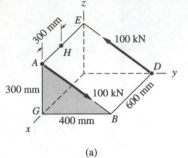

(a)

For the couple shown in Fig. (a), determine (1) the corresponding couple-vector and (2) the moment of the couple about the axis GH.

Solution

Part 1

One method for determining the couple-vector is to multiply the magnitude of the couple by the unit vector in its direction. The magnitude of the couple is

$$Fd = 100(0.6) = 60 \text{ kN} \cdot \text{m}$$

The sense of the couple is shown in Fig. (b)—counterclockwise looking down on the plane of the couple. Letting $\boldsymbol{\lambda}$ be the unit vector perpendicular to the plane of the couple, as shown in Fig. (c), the couple-vector $\mathbf{C}$ may be written as $\mathbf{C} = 60\boldsymbol{\lambda}$ kN · m. Because $\boldsymbol{\lambda}$ is perpendicular to the line AB, it can be seen that $\boldsymbol{\lambda} = (3\mathbf{j} + 4\mathbf{k})/5$ (recalling that perpendicular lines have negative reciprocal slopes). Therefore, the couple-vector is

$$\mathbf{C} = 60\boldsymbol{\lambda} = 60\left(\frac{3\mathbf{j} + 4\mathbf{k}}{5}\right) = 36\mathbf{j} + 48\mathbf{k} \text{ kN} \cdot \text{m} \qquad \textit{Answer}$$

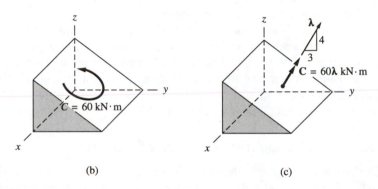

(b)   (c)

Alternative Solution   Because the couple-vector is equal to the moment of the couple about any point, it can also be determined by adding the moments of the two forces forming the couple about any convenient point, such as point B. Letting $\mathbf{F}$ be the 100-kN force that acts along the line DE, we have

$$\mathbf{F} = 100\boldsymbol{\lambda}_{DE} = 100\frac{\overrightarrow{DE}}{|\overrightarrow{DE}|} = 100\left(\frac{-0.4\mathbf{j} + 0.3\mathbf{k}}{0.5}\right)$$

$$= -80\mathbf{j} + 60\mathbf{k} \text{ kN}$$

Equating $\mathbf{C}$ to the moment of $\mathbf{F}$ about point B (the other force of the couple passes through B), we obtain

$$\mathbf{C} = \mathbf{r}_{BD} \times \mathbf{F} = \begin{vmatrix} \mathbf{i} & \mathbf{j} & \mathbf{k} \\ -0.6 & 0 & 0 \\ 0 & -80 & 60 \end{vmatrix}$$

$$= 36\mathbf{j} + 48\mathbf{k} \text{ kN} \cdot \text{m}$$

which agrees with the answer determined previously.

In the solution, the choice of point B as the moment center was arbitrary. Because the moment of a couple is the same about every point, the same result would have been obtained no matter which point had been chosen as the moment center.

Part 2

The most direct method for determining the moment of the couple about the axis GH is $M_{GH} = \mathbf{C} \cdot \boldsymbol{\lambda}_{GH}$. Because $\mathbf{C}$ has already been computed, all we need to do is compute the unit vector $\boldsymbol{\lambda}_{GH}$ and evaluate the dot product. Referring to Fig. (a), we have

$$\boldsymbol{\lambda}_{GH} = \frac{\overrightarrow{GH}}{|\overrightarrow{GH}|} = \frac{-0.3\mathbf{i} + 0.3\mathbf{k}}{0.3\sqrt{2}} = -0.7071\mathbf{i} + 0.7071\mathbf{k}$$

Hence the moment of the couple about axis GH is

$$M_{GH} = \mathbf{C} \cdot \boldsymbol{\lambda}_{GH} = (36\mathbf{j} + 48\mathbf{k}) \cdot (-0.7071\mathbf{i} + 0.7071\mathbf{k})$$

$$= +33.9 \text{ kN} \cdot \text{m} \qquad \textit{Answer}$$

The result is illustrated in Fig. (d). If you need help in interpreting the positive sign in the answer, you should refer back to Fig. 2.10.

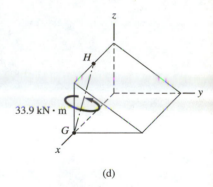

(d)

Sample Problem 2.8

The flat plate shown in Fig. (a) is acted on by the three couples. Replace the three couples with (1) a couple-vector; (2) two forces, one acting along the dashed line at point O and the other acting at point A; and (3) the smallest pair of forces, with one force acting at point O and the other at point A.

Solution

Part 1

The magnitudes (Fd) and senses of the couples, all of which lie in the xy-plane, are listed below.

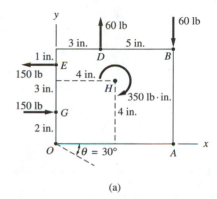

(a)

- Couple at H: 350 lb · in. clockwise.
- Couple acting on GE: $(150)(3) = 450$ lb · in. counterclockwise.
- Couple acting on DB: $(60)(5) = 300$ lb · in. clockwise.

Because all three couples lie in the same plane, they can be added, their sum being the resultant couple C^R. Choosing the counterclockwise sense as positive, we get

$$\overset{+}{\curvearrowleft} \quad C^R = -350 + 450 - 300 = -200 \text{ lb} \cdot \text{in.}$$

The negative sign shows that sense of the C^R is clockwise. Therefore, the corresponding couple-vector $\mathbf{C}^R$ is, according to the right-hand rule, in the negative z-direction. It follows that

$$\mathbf{C}^R = -200\mathbf{k} \text{ lb} \cdot \text{in.} \qquad \textit{Answer}$$

Note that more dimensions are given in Fig. (a) than are needed for the solution. The only relevant dimensions are the distances between the 60-lb forces (5 in.) and the 150-lb forces (3 in.).

Part 2

Two forces that are equivalent to the three couples shown in Fig. (a) must, of course, form a couple. The problem states that one of the forces acts along the dashed line at point O and the other acts at point A.

Because the two forces that form a couple must have parallel lines of action, the line of action of the force at point A must also be parallel to the dashed line at point O. From Fig. (b), we see that the perpendicular distance d between the lines of action of the two forces is $d = 8 \sin 30° = 4$ in. Having already determined that the magnitude of the resultant couple is 200 lb · in., the magnitudes of the forces that form the couple

71

are given by $C^R/d = 200/4 = 50$ lb. The sense of each force must be consistent with the clockwise sense of C^R. The final result is shown in Fig. (b).

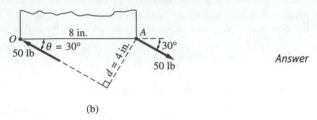

(b)

Answer

Part 3

Here we are to determine the smallest two forces acting at points O and A that are equivalent to the three couples shown in Fig. (a). Therefore, the two forces to be determined must form a couple that is equivalent to the resultant couple (200 lb · in., clockwise).

The magnitude of a couple (Fd) equals the product of the magnitude of the forces that form the couple (F) and the perpendicular distance (d) between the forces. For a couple of given magnitude, the smallest forces will be obtained when the perpendicular distance d is as large as possible. From Fig. (b) it can be seen that for forces acting at points O and A, the largest d will correspond to $\theta = 90°$, giving $d = 8$ in. Therefore, the magnitudes of the smallest forces are given by $C^R/d = 200/8 = 25$ lb. These results are shown in Fig. (c), where again note should be taken of the senses of the forces.

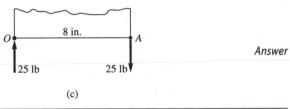

Answer

(c)

Sample Problem 2.9

A section of a piping system is acted on by the three couples shown in Fig. (a). Determine the magnitude of the resultant couple-vector and its direction cosines, given that the magnitudes of the applied couples are $C_1 = 50$ N · m, $C_2 = 90$ N · m, and $C_3 = 140$ N · m.

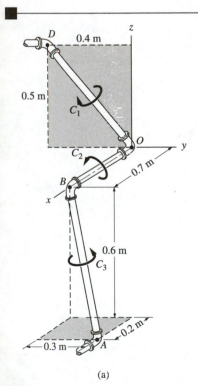

(a)

Solution

Applying the right-hand rule to each of the three couples in Fig. (a), we see that the corresponding couple-vectors will be directed as follows: $\mathbf{C}_1$, from point D toward point O; $\mathbf{C}_2$, from point O toward point B; and $\mathbf{C}_3$, from point A toward point B. Because these couple-vectors do not have the same directions, the most practical method of determining the resultant couple-vector $\mathbf{C}^R$ is to use the vector equation

$$\mathbf{C}^R = \mathbf{C}_1 + \mathbf{C}_2 + \mathbf{C}_3$$

Using the three unit vectors shown in Fig. (b), the couple-vectors $\mathbf{C}_1$, $\mathbf{C}_2$, and $\mathbf{C}_3$ can be written as

$$\mathbf{C}_1 = C_1 \, \boldsymbol{\lambda}_{DO} = 50 \frac{\overrightarrow{DO}}{|\overrightarrow{DO}|} = 50 \left(\frac{0.4\mathbf{j} - 0.5\mathbf{k}}{0.6403} \right)$$

$$= 31.24\mathbf{j} - 39.04\mathbf{k} \text{ N} \cdot \text{m}$$

$$\mathbf{C}_2 = C_2 \, \boldsymbol{\lambda}_{OB} = 90\mathbf{i} \text{ N} \cdot \text{m}$$

$$\mathbf{C}_3 = C_3 \, \boldsymbol{\lambda}_{AB} = 140 \frac{\overrightarrow{AB}}{|\overrightarrow{AB}|} = 140 \left(\frac{-0.2\mathbf{i} - 0.3\mathbf{j} + 0.6\mathbf{k}}{0.7000} \right)$$

$$= -40\mathbf{i} - 60\mathbf{j} + 120\mathbf{k} \text{ N} \cdot \text{m}$$

Adding these three couple-vectors gives

$$\mathbf{C}^R = 50\mathbf{i} - 28.76\mathbf{j} + 80.96\mathbf{k} \text{ N} \cdot \text{m}$$

The magnitude of $\mathbf{C}^R$ is

$$C^R = \sqrt{(50)^2 + (-28.76)^2 + (80.96)^2} = 99.41 \text{ N} \cdot \text{m} \qquad \textit{Answer}$$

and the direction cosines of $\mathbf{C}^R$ are the components of the unit vector $\boldsymbol{\lambda}$ directed along $\mathbf{C}^R$:

$$\lambda_x = \frac{50}{99.41} = 0.503$$

$$\lambda_y = -\frac{28.76}{99.41} = -0.289 \qquad \textit{Answer}$$

$$\lambda_z = \frac{80.96}{99.41} = 0.814$$

The resultant couple-vector is shown in Fig. (c). Although $\mathbf{C}^R$ is shown at point O, it must be remembered that couples are free vectors, so that $\mathbf{C}^R$ could be shown acting anywhere.

The couple-vector $\mathbf{C}^R$ can be represented as two equal and opposite parallel forces. However, because the two forces will lie in a plane perpendicular to the couple-vector, in this case a skewed plane, this representation is inconvenient here.

In general, given two forces that form a couple, the corresponding couple-vector is easily determined (e.g., by summing the moments of the two forces about any point). However, given a couple-vector, it is not always convenient (or even desirable) to determine two equivalent forces.

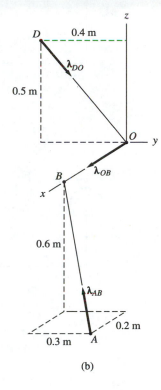

(b)

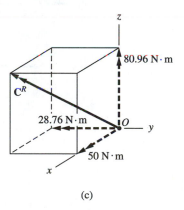

(c)

Problems

2.68 Which of the systems are equivalent to the couple in (a)?

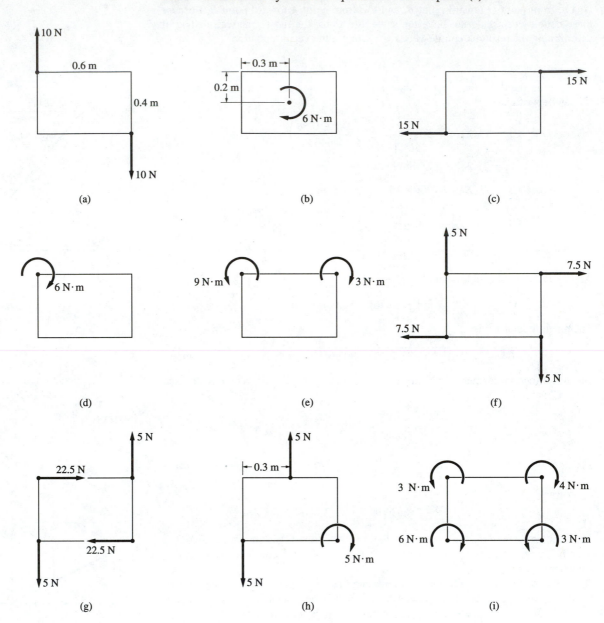

Fig. P2.68

2.69 Which of the systems are equivalent to the couple in (a)?

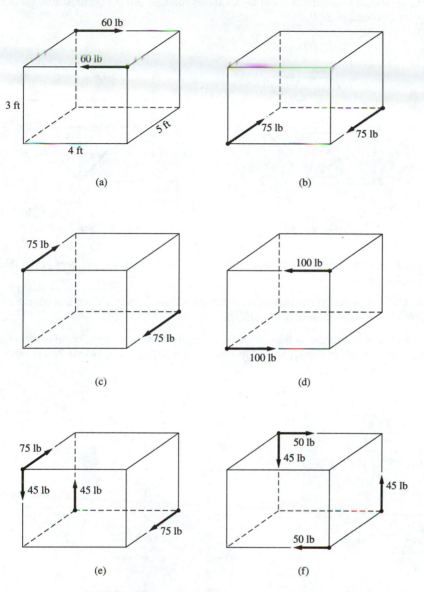

Fig. P2.69

2.70 A pair of parallel forces of magnitude F (not shown) that form the counter-clockwise 80-kN · m couple is acting on the plate. Determine (a) the smallest possible value of F; and (b) the value of F if the forces act at A and C, and are parallel to the dashed line shown at A.

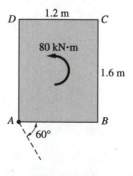

Fig. P2.70

2.71   Given that $C_1 = 200 \text{ N} \cdot \text{m}$ and $C_2 = 300 \text{ N} \cdot \text{m}$, determine the magnitude of the combined moment of $\mathbf{C}_1$ and $\mathbf{C}_2$ about (a) point G; and (b) the axis directed from point G toward point B.

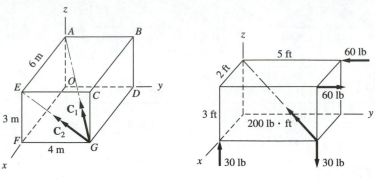

Fig. P2.71   **Fig. P2.72**

2.72   Determine the magnitude of the single couple that is equivalent to the three couples shown.

2.73   Calculate the combined moment of the couple $\mathbf{C}$ and the force $\mathbf{P}$ about the axis AB. Use $C = 80 \text{ N} \cdot \text{m}$ and $P = 400 \text{ N}$.

*2.74   Determine the couple-vector that is equivalent to the three couples acting on the gear box, given that $C_1 = 200 \text{ lb} \cdot \text{in.}$, $C_2 = 140 \text{ lb} \cdot \text{in.}$, and $C_3 = 220 \text{ lb} \cdot \text{in.}$

Fig. P2.73

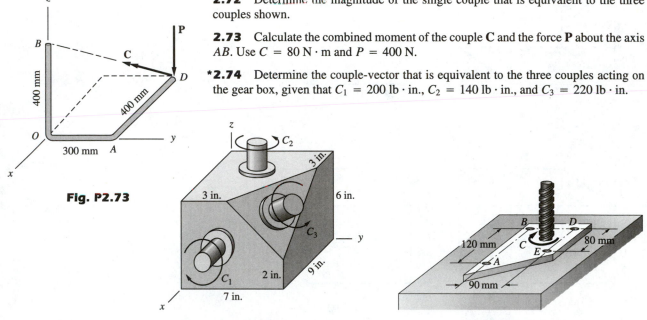

Fig. P2.74   **Fig. P2.75**

2.75   A drill applies a couple $\mathbf{C}$ to the quadrilateral plate. For the plate to remain stationary, an equal and opposite couple must be applied to it by the shearing forces in the bolts that attach the plate to the work table. (Shearing forces are forces inside the bolts that act parallel to the plate.) Although there are four bolt holes in the plate, assume that only two bolts are available. (a) Into which holes would you place the two bolts in order to minimize the shearing forces? (b) If the maximum allowable shearing force that can be carried by each bolt is 40 N, determine the magnitude of the largest couple $\mathbf{C}$ that can be applied by the drill.

2.76   Solve Prob. 2.66 by recognizing that the two 20-lb forces acting on the handle of the flexible wrench form a couple.

2.77 The two forces of magnitude P are equivalent to a 10-N · m counterclockwise couple. (a) Determine P as a function of θ. (b) Using the results of (a), sketch a graph of P versus θ for positive values of P. Label all significant values.

2.78 A couple of magnitude 360 lb · ft is applied about portion AB of the drive shaft (the drive shaft is connected by universal joints at points B and C). Compute the moment of the applied couple about the portion CD when the drive shaft is in the position shown.

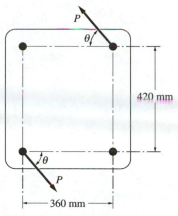

Fig. P2.77

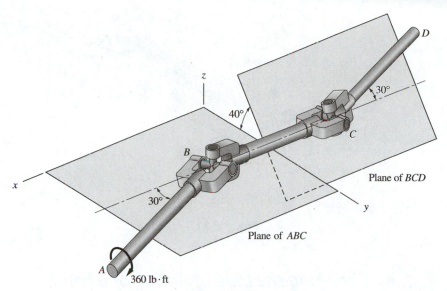

Fig. P2.78

2.79 The arm $ABCD$ of the industrial robot lies in a vertical plane that is inclined at 40° to the yz-plane. The arm CD makes an angle of 30° with the vertical. A socket wrench attached at point D applies a 52-lb · ft couple about the arm CD, directed as shown. (a) Find the couple-vector that represents the given couple. (b) Determine the moment of the couple about the z-axis.

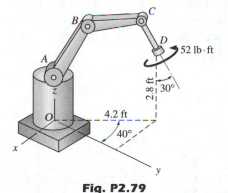

Fig. P2.79

2.80 The figure shows one-half of a universal coupling known as the *Hooke's joint*. The coupling is acted on by the three couples shown: (1) the input couple consisting of forces of magnitude P, (2) the output couple C_0, and (3) the couple formed by bearing reactions of magnitude R. If the resultant of these couples is zero, compute R and C_0 for $P = 600$ lb.

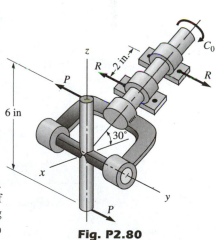

Fig. P2.80

2.81 The steering column of the rack-and-pinion steering mechanism lies in the *xz*-plane. The tube *AB* of the steering gear is attached to the automobile chassis at *A* and *B*. When the steering wheel is turned, the assembly is subjected to the four couples shown: the 3-N · m couple applied by the driver to the steering wheel, two 1.8-N · m couples (one at each wheel), and the couple formed by the two forces of magnitude *F* acting at *A* and *B*. If the resultant couple acting on the steering mechanism is zero, determine *F* and the angle θ (the magnitude and direction of the bearing reactions).

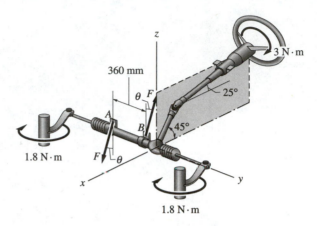

Fig. P2.81

2.8 *Changing the Line of Action of a Force*

In this article we show how to change the line of action of a force without affecting its external effect on a rigid body. This topic lays the foundation of the next chapter, in which we discuss the resultants of force systems.

Referring to Fig. 2.18(a), consider the problem of moving the force of magnitude *F* from point *B* to point *A*. We cannot simply move the force to *A*, because this would change its line of action, thereby altering the rotational effect (the moment) of the force. We can, however, counteract this change by introducing a couple that restores the rotational effect to its original state. The construction for determining the couple is illustrated in Fig. 2.18. It consists of the following two steps:

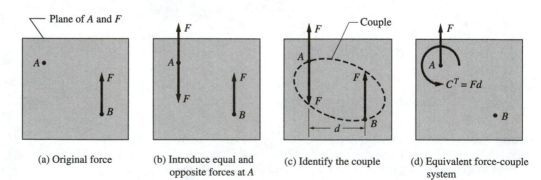

(a) Original force (b) Introduce equal and opposite forces at *A* (c) Identify the couple (d) Equivalent force-couple system

Fig. 2.18

- Introduce two equal and opposite forces of magnitude F at point A, as shown in Fig. 2.18(b). These forces are parallel to the original force at B. Because the forces at A have no net external effect on a rigid body, the force systems in Figs. 2.18(a) and (b) are equivalent.
- Identify the two forces that form a couple, as has been done in Fig. 2.18(c). The magnitude of the couple is $C^T = Fd$, where d is the distance between the lines of action of the forces at A and B. The third force and C^T thus constitute the *force-couple system* shown in Fig. 2.18(d), which is equivalent to the original force in Fig. 2.18(a).

We refer to the couple C^T as the *couple of transfer*, because it is the couple that must be introduced when a force is transferred from one line of action to another. From the construction in Fig. 2.18 we note the following:

> The couple of transfer is equal to the moment of the original force (acting at B) about the transfer point A.

In vector terminology, the line of action of a force $\mathbf{F}$ can be changed to a parallel line, provided that we introduce the couple of transfer

$$\mathbf{C}^T = \mathbf{r} \times \mathbf{F} \qquad (2.20)$$

where $\mathbf{r}$ is the vector drawn from the transfer point A to the point of application B of the original force, as illustrated in Fig. 2.19. It is conventional to show $\mathbf{C}^T$ acting at the transfer point, as in Fig. 2.19(b), but we must not forget that a couple is a free vector that could be placed anywhere.

According to the properties of the cross product in Eq. (2.20), the couple-vector $\mathbf{C}^T$ is perpendicular to $\mathbf{F}$. Thus a force at a given point can always be replaced by a force at a different point and a couple-vector that is perpendicular to the force. The converse is also true: A force and a couple-vector that are mutually perpendicular can always be reduced to a single, equivalent force by reversing the construction outlined in Fig. 2.18.

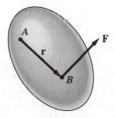

(a) Original force

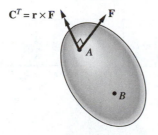

(b) Equivalent force-couple system

Fig. 2.19

Sample Problem 2.10

For the machine part shown in Fig. (a), replace the applied load of 150 kN acting at point A by (1) an equivalent force-couple system with the force acting at point B; and (2) two horizontal forces, one acting at point B and the other acting at point C.

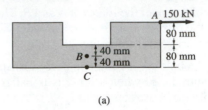

(a)

Solution

Part 1

First we move the 150-kN force to point B, and then we introduce the couple of transfer equal to the moment of the 150-kN force in Fig. (a) about point B, given by

$$\stackrel{+}{\curvearrowright} \quad C^T = M_B = -150(0.080 + 0.040) = -18 \text{ kN} \cdot \text{m}$$

The negative sign indicates that the sense of the couple is clockwise. The equivalent force-couple system is shown in Fig. (b).

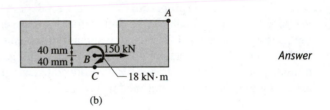

Answer

(b)

Part 2

The 18 kN · m clockwise couple in Fig. (b) can be replaced by two 450-kN forces, one acting at point B and the other at point C, as shown in Fig. (c). (The couple represented by these two forces is $450(0.040) = 18$ kN · m in the clockwise direction.) The two forces acting at point B can be added to get the system shown in Fig. (d). This is the answer, because we have replaced the original force with two horizontal forces, one at point B and the other at point C, as requested.

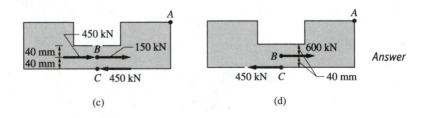

(c)   (d)   *Answer*

Sample Problem 2.11

Replace the system shown in Fig. (a) with an equivalent force-couple system, with the force acting at point A, given that $F = 100$ lb and $C = 120$ lb $\cdot$ in.

Solution

Moving the given force $\mathbf{F}$ from point B to point A requires the introduction of a couple of transfer $\mathbf{C}^T$. This couple is then added to the given couple-vector $\mathbf{C}$, thereby obtaining the resultant couple-vector, which we label $\mathbf{C}^R$. The couple-vector $\mathbf{C}^R$ and the force $\mathbf{F}$ located at point A will then be the requested force-couple system.

Owing to the three-dimensional nature of this problem, it is convenient to use vector methods in the solution. Writing $\mathbf{F}$ in vector form, we obtain

$$\mathbf{F} = 100\boldsymbol{\lambda}_{BE} = 100\frac{\overrightarrow{BE}}{|\overrightarrow{BE}|} = 100\left(\frac{-4\mathbf{i} + 2\mathbf{k}}{4.472}\right)$$

$$= -89.44\mathbf{i} + 44.72\mathbf{k}\ \text{lb}$$

The position vector from A to B is $\mathbf{r}_{AB} = 4\mathbf{j} - 2\mathbf{k}$ in. The couple of transfer is equal to the moment of the given force $\mathbf{F}$ about point A, so we have

$$\mathbf{C}^T = \mathbf{M}_A = \mathbf{r}_{AB} \times \mathbf{F} = \begin{vmatrix} \mathbf{i} & \mathbf{j} & \mathbf{k} \\ 0 & 4 & -2 \\ -89.44 & 0 & 44.72 \end{vmatrix}$$

$$= 178.9\mathbf{i} + 178.9\mathbf{j} + 357.8\mathbf{k}\ \text{lb} \cdot \text{in.}$$

Expressing the given couple-vector $\mathbf{C}$ shown in Fig. (a) in vector form,

$$\mathbf{C} = 120\boldsymbol{\lambda}_{DB} = 120\frac{\overrightarrow{DB}}{|\overrightarrow{DB}|} = 120\left(\frac{4\mathbf{i} + 4\mathbf{j} - 2\mathbf{k}}{6}\right)$$

$$= 80\mathbf{i} + 80\mathbf{j} - 40\mathbf{k}\ \text{lb} \cdot \text{in.}$$

Adding $\mathbf{C}^T$ and $\mathbf{C}$ (remember that couple-vectors are free vectors), the resultant couple-vector is

$$\mathbf{C}^R = \mathbf{C}^T + \mathbf{C} = 258.9\mathbf{i} + 258.9\mathbf{j} + 317.8\mathbf{k}\ \text{lb} \cdot \text{in.}$$

The magnitude of $\mathbf{C}^R$ is given by

$$C^R = \sqrt{(258.9)^2 + (258.9)^2 + (317.8)^2} = 485\ \text{lb} \cdot \text{in.}$$

The equivalent force-couple system is shown in Fig. (b). Note that the force acts at point A. For convenience of representation, $\mathbf{C}^R$ is shown at point O, but being a free vector, it could be placed anywhere.

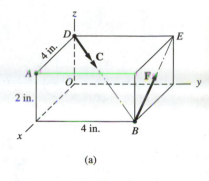

(a)

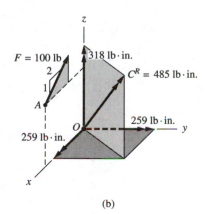

(b)

81

Problems

2.82 Which of the systems are equivalent to the force-couple system in (a)?

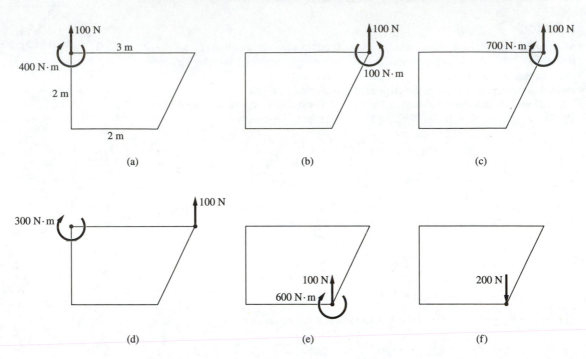

(a)   (b)   (c)

(d)   (e)   (f)

Fig. P2.82

2.83 A 15-lb force acts at point *A* on the high-pressure water cock. Replace this force with (a) a force-couple system, the force of which acts at point *B*; and (b) two horizontal forces, one acting at point *B* and the other acting at point *C*.

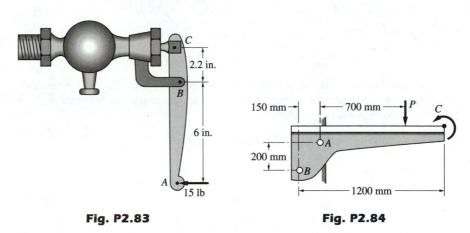

Fig. P2.83   **Fig. P2.84**

2.84 The bracket, which is fastened to the wall by anchor bolts at *A* and *B*, is loaded by the force *P* = 120 N and the couple *C* = 140 N · m. Replace *P* and *C* with (a) an equivalent force-couple system, the force of which acts at *A*; and (b) two vertical forces, one acting at *A* and the other at *B*.

2.85 The three forces shown are equivalent to a 50-kN upward force at A and a 170-kN·m counterclockwise couple. Determine P and F.

2.86 Replace the cable tensions $T_1 = 90$ lb and $T_2 = 50$ lb by a force-couple system with the force acting at O.

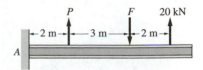

Fig. P2.85

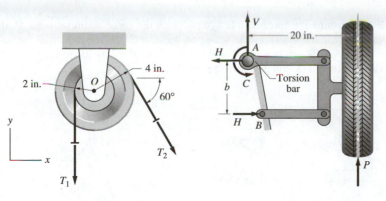

Fig. P2.86   **Fig. P2.87**

2.87 The figure shows the front view of an automobile torsion-bar suspension. (The torsion bar is shown in cross section at point A.) The vertical force P exerted by the road on the tire is known to be equivalent to the force system consisting of the vertical force V, the couple (torque) C, and the couple formed by the horizontal forces H. Given that $P = 1200$ lb and $C = 900$ lb · ft, determine the smallest possible value of the dimension b, if H is not to exceed 1400 lb.

2.88 The table can be lifted without tilting by applying the 100-N force at point O, the center of the table. Determine the force-couple system with the force acting at corner A that will produce the same result.

2.89 The magnitude of the tensile force $\mathbf{F}$ acting at point A is 160 kN. Determine the equivalent force-couple system with the force acting at point O.

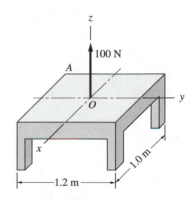

Fig. P2.88

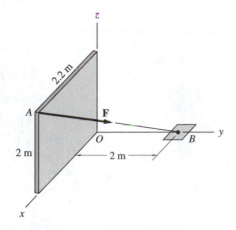

Fig. P2.89

2.90 Replace the force-couple system shown with another force-couple system in which the force acts at O. Use $P = 200$ lb and $C = 1200$ lb · ft.

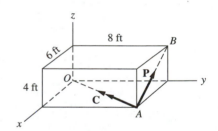

Fig. P2.90

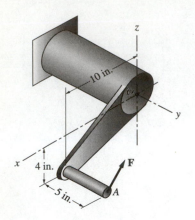

Fig. P2.91

*2.91   (a) Replace the force $\mathbf{F} = 2500\mathbf{i} + 4000\mathbf{j} + 3000\mathbf{k}$ lb acting at end A of the crank handle with a force $\mathbf{R}$ acting at O and a couple-vector $\mathbf{C}^R$. (b) Resolve $\mathbf{R}$ into the normal component P (normal to the cross section of the shaft) and the shear component V (in the plane of the cross section). (c) Resolve $\mathbf{C}^R$ into the twisting component T (normal to the cross section) and the bending component M (in the plane of the cross section).

2.92   Determine the force-couple system, with the force acting at point O, that is equivalent to the force and couple acting on the arm CD of the industrial robot. Note that the arm $ABCD$ lies in a vertical plane that is inclined at 40° to the yz-plane; the arm CD makes an angle of 30° with the vertical.

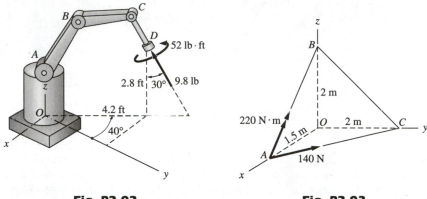

Fig. P2.92   **Fig. P2.93**

2.93   Replace the system shown with an equivalent force-couple system having the force at point O.

Review Problems

2.94   The moment of the force $\mathbf{P}$ about the axis AB is 600 lb · ft. Determine the magnitude of $\mathbf{P}$.

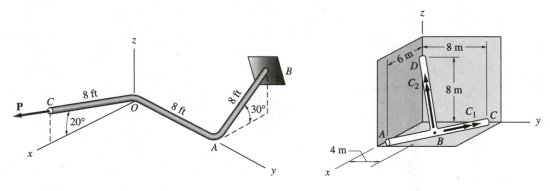

Fig. P2.94   **Fig. P2.95**

2.95   The magnitudes of the two couples that act on the T-bar are $C_1 = 220$ N · m and $C_2 = 180$ N · m. Calculate the total moment of the couples about the axis AD. Express your answer in vector form.

2.96 Three cable tensions T_1, T_2, and T_3 act at the top of the flagpole. Given that the resultant force for the three tensions is $R = -400k$ N, find the magnitudes of each of the cable tensions.

2.97 A force $F = 20i - 30j - 60k$ N acts at the point A $(x,$ 3 m, 5 m$)$. The moment of F about the y-axis is $M_y = 400j$ N · m. Determine the coordinate x and the moment of F about the x-axis and the y-axis, respectively.

2.98 The magnitude of the moment of the force P about the axis CD is 50 lb · in. Find the magnitude of P.

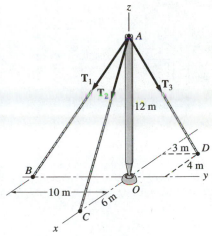

Fig. P2.96

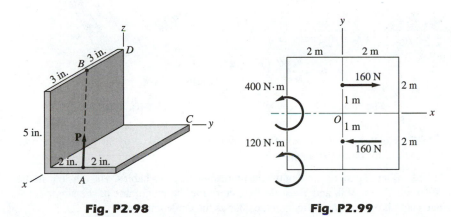

Fig. P2.98   **Fig. P2.99**

2.99 The three couples shown can be represented as two forces. If one of the forces is $F = 80i + 60j$ N, acting through the origin O, determine the coordinate of the point where the other force intersects the x-axis.

2.100 The magnitudes of the force P and couple C are 500 lb and 1200 lb · ft, respectively. Calculate the combined moment of P and C about (a) the origin O and (b) the axis OF.

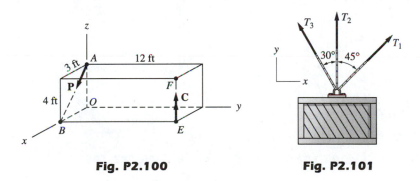

Fig. P2.100   **Fig. P2.101**

2.101 The resultant force of the three cable tensions that support the crate is $R = 500j$ lb. Find T_1 and T_3, given that $T_2 = 300$ lb.

2.102 A force system consists of the force $F = 200i + 100j + 250k$ lb, acting at the origin of a rectangular coordinate system, and a couple $C = -400i + 300j + 200k$ lb · in. (a) Show that F and C can be reduced to a single force. (b) Find the coordinates of the point in the xy-plane where the combined moment of F and C is zero.

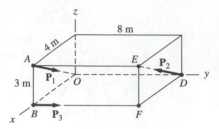

Fig. P2.103

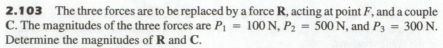

2.103 The three forces are to be replaced by a force **R**, acting at point *F*, and a couple **C**. The magnitudes of the three forces are $P_1 = 100$ N, $P_2 = 500$ N, and $P_3 = 300$ N. Determine the magnitudes of **R** and **C**.

2.104 The three forces of magnitude *P* can be replaced by a single, equivalent force **R** acting at point *A*. Determine the distance *x* and the magnitude and direction of **R**.

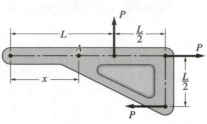

Fig. P2.104

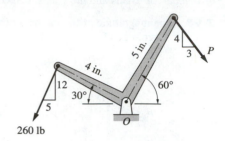

Fig. P2.105

2.105 Knowing that the two forces shown can be replaced by an equivalent force acting at *O* (no couple), determine *P*.

2.106 The trapdoor is held in the position shown by two cables. The tensions in the cables are $T_1 = 30$ lb and $T_2 = 90$ lb. Determine the magnitude of the single force that would have the same effect on the door as the cable tensions.

2.107 The force system consists of the force $\mathbf{P} = -300\mathbf{i} + 200\mathbf{j} + 150\mathbf{k}$ lb and the couple **C**. Determine the magnitude of **C** if the moment of this force system about the axis *DE* is 800 lb · ft.

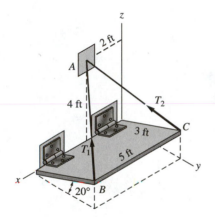

Fig. P2.106

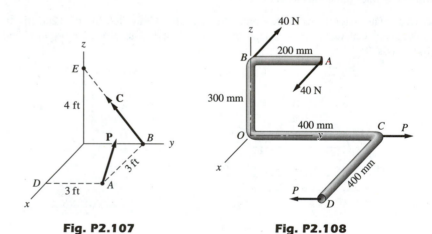

Fig. P2.107   **Fig. P2.108**

2.108 Given that the couple formed by the 40-N forces is equivalent to the couple formed by the forces of magnitude *P*, determine *P*.

3

Resultants of Force Systems

3.1 Introduction

In order to investigate the effects that a system of forces has on a body, it is often convenient to reduce the force system to its simplest equivalent representation. Some of these simplifications have been discussed in the preceding chapter. For example, you have learned that a system of concurrent forces can be replaced by a single force and that a system of couples can be replaced by a single couple.

The next article explains how an arbitrary force system can be reduced to a force and a couple. Subsequent articles discuss applications of the force-couple system to the determination of the resultants of coplanar and noncoplanar force systems.

3.2 Reduction of a Force System to a Force and a Couple

Here we show how a system of forces can be reduced to an equivalent system consisting of a force acting at an arbitrary point, plus a couple.

Consider the force system shown in Fig. 3.1(a), consisting of the forces $\mathbf{F}_1, \mathbf{F}_2, \mathbf{F}_3, \ldots$. The position vectors $\mathbf{r}_1, \mathbf{r}_2, \mathbf{r}_3, \ldots$ of the points where the forces act are measured from an arbitrarily chosen base point O. We can reduce this force system to an equivalent force-couple system, with the force acting at O, by the following procedure:

- Move each force to point O. As explained in Art. 2.8, the force $\mathbf{F}_1$ can be moved to O if we introduce the couple of transfer $\mathbf{C}_1^T = \mathbf{r}_1 \times \mathbf{F}_1$ (the moment of $\mathbf{F}_1$ about O). The forces $\mathbf{F}_2, \mathbf{F}_3, \ldots$ can be moved in the same manner, their couples of transfer being $\mathbf{C}_2^T = \mathbf{r}_2 \times \mathbf{F}_2, \mathbf{C}_3^T = \mathbf{r}_3 \times \mathbf{F}_3, \ldots$. After all the forces have been moved, we end up with the force system in Fig 3.1(b), which is equivalent to the original system. (The equal signs between the figures signify equivalence.)
- Because the forces are now concurrent at point O, they can be added to yield the resultant force $\mathbf{R}$:

$$\boxed{\mathbf{R} = \mathbf{F}_1 + \mathbf{F}_2 + \mathbf{F}_3 + \cdots = \Sigma\mathbf{F}} \qquad (3.1)$$

The couples of transfer can also be added, their sum being the resultant couple-vector $\mathbf{C}^R$:

$$\boxed{\mathbf{C}^R = \mathbf{r}_1 \times \mathbf{F}_1 + \mathbf{r}_2 \times \mathbf{F}_2 + \mathbf{r}_3 \times \mathbf{F}_3 + \cdots = \Sigma\mathbf{M}_O} \qquad (3.2)$$

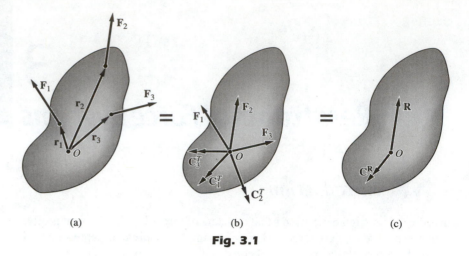

Fig. 3.1

The resultant force-couple system is displayed in Fig. 3.1(c), with both **R** and $\mathbf{C}^R$ shown acting at point O. It should be noted, however, that **R** is a sliding vector (its line of action must pass through O), whereas $\mathbf{C}^R$ is a free vector. Although $\mathbf{C}_1^T$ is perpendicular to $\mathbf{F}_1$, and so on, as pointed out in Art. 2.8, $\mathbf{C}^R$ is generally not perpendicular to **R**.

Using Eqs. (3.1) and (3.2), any force system can be reduced to an equivalent force-couple system, with the force acting at a reference point of your choosing. The resultant force **R** is simply the vector sum of all the forces—therefore, it is not affected by the location of the reference point. However, the resultant couple-vector $\mathbf{C}^R$, being the sum of the moments of all the forces about the reference point,* does depend on the choice of the reference point.

Two vector equations, Eqs. (3.1) and (3.2), are required to reduce a given force system to its equivalent force-couple representation. If we let O be the origin of a rectangular coordinate system, these two vector equations are equivalent to the following six scalar equations:

$$\begin{array}{ccc} R_x = \Sigma F_x & R_y = \Sigma F_y & R_z = \Sigma F_z \\ C_x^R = \Sigma M_x & C_y^R = \Sigma M_y & C_z^R = \Sigma M_z \end{array} \tag{3.3}$$

where R_x, R_y, and R_z are the rectangular components of the resultant force **R** and C_x^R, C_y^R, and C_z^R (the rectangular components of $\mathbf{C}^R$) are equal to the sums of the moments of the original force system about the coordinate axes.

If the forces of the original system lie in a plane—say, the xy-plane—the following three scalar equations are necessary to determine the force-couple system.

$$R_x = \Sigma F_x \qquad R_y = \Sigma F_y \qquad C^R = \Sigma M_O \tag{3.4}$$

The couple-vector $\mathbf{C}^R$ will always be in the z-direction, because the plane of the couple is the xy-plane. Because the resultant force **R** lies in the xy-plane, **R** and $\mathbf{C}^R$ will be mutually perpendicular. The last observation is significant—it implies that a coplanar force system can be further reduced to a single force or a single couple. This topic is discussed in more detail in the next article.

---

*If the original force system contains couples, their moments must be included in the sum.

Sample Problem 3.1

The force system shown in Fig. (a) consists of the couple-vector **C** and the forces F_1, F_2, and F_3. Determine the equivalent force-couple system with the force acting at (1) point O; and (2) point D. Use $C = 200$ lb · ft, $F_1 = 100$ lb, $F_2 = 90$ lb, and $F_3 = 120$ lb.

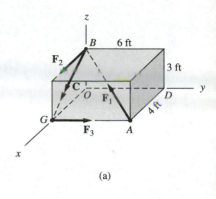

(a)

Solution

Because of the three-dimensional nature of this problem, we will use vector algebra to solve it. The first step is to express the three forces and the couple in vector form:

$$F_1 = 100\lambda_{AB} = 100\frac{\overrightarrow{AB}}{|\overrightarrow{AB}|} = 100\left(\frac{-4i - 6j + 3k}{\sqrt{(-4)^2 + (-6)^2 + 3^2}}\right)$$

$$= -51.22i - 76.82j + 38.41k \text{ lb}$$

$$F_2 = 90i \text{ lb}$$

$$F_3 = 120j \text{ lb}$$

$$C = 200\lambda_{BG} = 200\frac{\overrightarrow{BG}}{|\overrightarrow{BG}|} = 200\left(\frac{4i - 3k}{\sqrt{4^2 + (-3)^2}}\right)$$

$$= 160i - 120k \text{ lb} \cdot \text{ft}$$

Part 1 Force-Couple System with Force at Point O

To determine the resultant force **R**, we simply add the three original forces as if they were concurrent:

$$R = \Sigma F = F_1 + F_2 + F_3$$

$$= (-51.22i - 76.82j + 38.41k) + (90i) + (120j)$$

$$= 38.8i + 43.2j + 38.4k \text{ lb} \qquad\qquad Answer$$

The resultant couple-vector equals the sum of the moments of the original force system about point O. We denote this couple-vector by C_1^R in order to distinguish it from the resultant couple-vector to be determined in Part 2. Referring to Fig. (a), we have

$$C_1^R = \Sigma M_O = (r_{OA} \times F_1) + (r_{OB} \times F_2) + (r_{OG} \times F_3) + C$$

$$= \begin{vmatrix} i & j & k \\ 4 & 6 & 0 \\ -51.22 & -76.82 & 38.41 \end{vmatrix} + \begin{vmatrix} i & j & k \\ 0 & 0 & 3 \\ 90 & 0 & 0 \end{vmatrix}$$

$$+ \begin{vmatrix} i & j & k \\ 4 & 0 & 0 \\ 0 & 120 & 0 \end{vmatrix} + (160i - 120k)$$

Expanding the determinants and simplifying, we get

$$C_1^R = 390i + 116j + 360k \text{ lb} \cdot \text{ft} \qquad\qquad Answer$$

The force-couple system determined here is shown in Fig. (b).

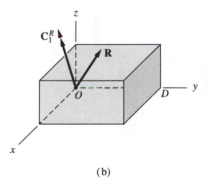

(b)

Part 2 Force-Couple System with Force at Point D

The resultant force that acts at point D is, of course, the same as the resultant force **R** found in Part 1.

$$R = 38.8i + 43.2j + 38.4k \text{ lb} \qquad\qquad Answer$$

The resultant couple-vector, which we denote as $\mathbf{C}_2^R$, is obtained by summing the moments of the original force system about point D. You may wish to verify that the result is

$$\mathbf{C}_2^R = \Sigma \mathbf{M}_D = 160\mathbf{i} + 116\mathbf{j} + 593\mathbf{k} \text{ lb} \cdot \text{ft} \qquad \textit{Answer}$$

An alternate method for determining the resultant couple-vector $\mathbf{C}_2^R$ is to equate it to the sum of the moments about point D for the force-couple system shown in Fig. (b), as follows:

$$\mathbf{C}_2^R = \mathbf{C}_1^R + (\mathbf{r}_{DO} \times \mathbf{R})$$

$$= (390\mathbf{i} + 116\mathbf{j} + 360\mathbf{k}) + \begin{vmatrix} \mathbf{i} & \mathbf{j} & \mathbf{k} \\ 0 & -6 & 0 \\ 38.8 & 43.2 & 38.4 \end{vmatrix}$$

which, on simplification, yields the same expression for $\mathbf{C}_2^R$ as given above.

Sample Problem 3.2

The coplanar force system in Fig. (a) consists of three forces and one couple. Determine the equivalent force-couple system with the force acting at (1) point O; and (2) point A.

Solution

Part 1 Force-Couple System with Force at Point O

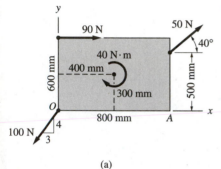

(a)

We will solve this problem with scalar algebra; however, the use of vector algebra would be almost as convenient.

The components of the resultant force $\mathbf{R}$ are given by

$$R_x = \Sigma F_x \quad \xrightarrow{+} \quad R_x = 50\cos 40° + 90 - \frac{3}{5}(100) = 68.3 \text{ N}$$

$$R_y = \Sigma F_y \quad +\uparrow \quad R_y = 50\sin 40° - \frac{4}{5}(100) = -47.9 \text{ N}$$

Thus the resultant force is

$$\mathbf{R} = 68.3\mathbf{i} - 47.9\mathbf{j} \text{ N} \qquad \textit{Answer}$$

The magnitude of $\mathbf{R}$ is

$$R = \sqrt{(68.3)^2 + (-47.9)^2} = 83.4 \text{ N}$$

and the angle that $\mathbf{R}$ makes with the x-axis is

$$\theta = \tan^{-1}\frac{47.9}{68.3} = 35.0°$$

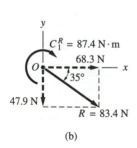

(b)

The force $\mathbf{R}$ acting at point O is shown in Fig. (b).

The magnitude of the resultant couple, which we label $\mathbf{C}_1^R$ to distinguish it from the resultant couple to be determined in Part 2, equals the magnitude of the total moment about point O of the original force system. Referring to Fig. (a), we write

$$C_1^R = \Sigma M_O \quad \overset{+}{\curvearrowleft} \quad C_1^R = 50\sin 40°(0.800) - 50\cos 40°(0.500)$$

$$- 90(0.600) - 40$$

$$= -87.4 \text{ N} \cdot \text{m}$$

Therefore,

$$C_1^R = 87.4 \text{ N} \cdot \text{m} \quad \text{clockwise} \qquad \textit{Answer}$$

as shown in Fig. (b)

Part 2 Force-Couple System with Force at Point A

The resultant force is equal to **R** determined in Part 1:

$$\mathbf{R} = 68.3\mathbf{i} - 47.9\mathbf{j} \text{ N} \qquad \textit{Answer}$$

The resultant couple, $\mathbf{C}_2^R$, is obtained by summing the moments of the original force system about point A. Referring to Fig. (a), we get

$$C_2^R = \Sigma M_A \qquad \overset{+}{\curvearrowleft} \qquad C_2^R = -50 \cos 40°(0.500) - 90(0.600)$$

$$+ \frac{4}{5}(100)(0.800) - 40$$

$$= -49.2 \text{ N} \cdot \text{m}$$

or

$$C_2^R = 49.2 \text{ N} \cdot \text{m} \quad \text{clockwise} \qquad \textit{Answer}$$

The force-couple system with the force at point A is shown in Fig. (c).

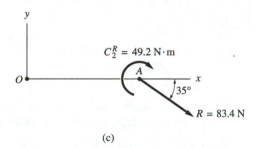

(c)

An alternative approach for determining C_2^R is to equate it to the total moment about point A for the force-couple system shown in Fig. (b).

Problems

3.1 Determine which of the force systems in Figs. (b) through (f) are equivalent to the force-couple system in (a).

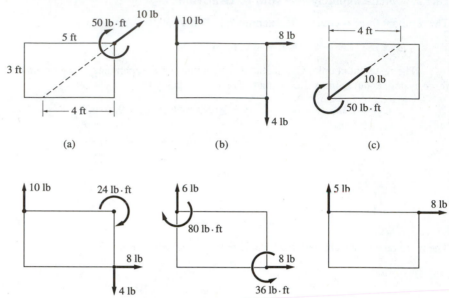

(a)   (b)   (c)

(d)   (e)   (f)

Fig. P3.1

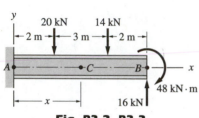

Fig. P3.2, P3.3

3.2 The cantilever beam is loaded by three forces and a couple. Determine the equivalent force-couple system with the force acting at (a) point A; and (b) point B.

3.3 The three forces and one couple acting on the cantilever beam can be replaced by an equivalent force-couple system with the force acting at point C. Determine the distance x for which the resultant couple will be zero.

3.4 The four forces shown act on the rollers of an in-line skate. Determine the equivalent force-couple system, with the force acting at O (the ankle joint of the skater).

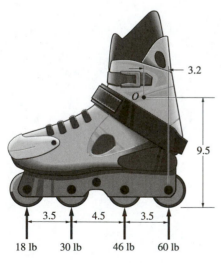

Dimensions in inches

Fig. P3.4

3.5 Replace the three forces with an equivalent force-couple system, with the force acting at O.

3.6 A force system consists of the three forces shown and a fourth force **P** (not shown). The equivalent force-couple system is known to be **R** = 90**i** + 300**j** lb acting at O and $\mathbf{C}^R = \mathbf{0}$. Determine **P** and the y-coordinate of the point where its line of action intersects the y-axis.

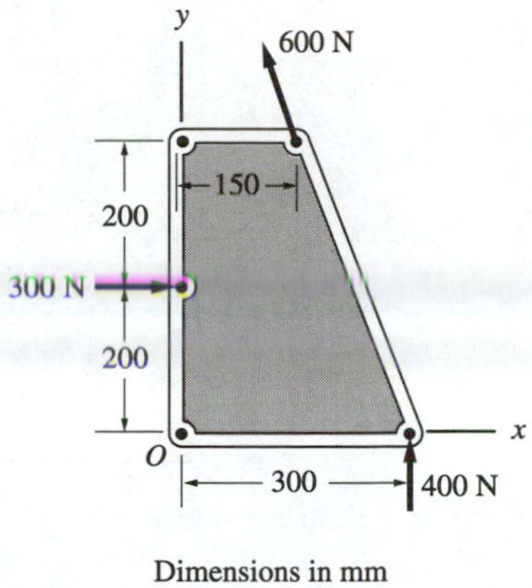

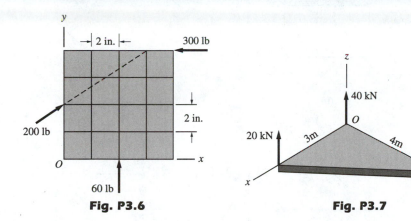

Dimensions in mm

Fig. P3.5

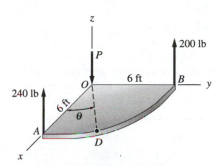

Fig. P3.6

Fig. P3.7

3.7 The three forces are perpendicular to the triangular plate. Find the equivalent force-couple system, with the force acting at O.

3.8 Replace the three forces acting on the quarter-circular plate with an equivalent force-couple system, having the force act at point D. Use P = 320 lb and θ = 30°.

Fig. P3.8, P3.9

3.9 When the three forces acting on the quarter-circular plate are replaced by an equivalent force-couple system with the force acting at point D, the resultant couple is zero. Determine P and the angle θ.

3.10 Represent each of the force systems with a force-couple system, having the force act at point *A*. Which systems are equivalent to each other?

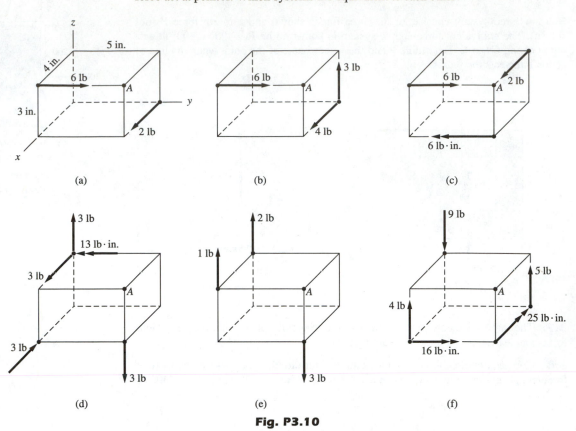

Fig. P3.10

3.11 A worker applies the forces **P** and **Q** = 10**i** lb to the handgrips of the electric drill. If the drill is to be held steady during operation, **P** and **Q** must be equivalent to the force-couple system acting at the drill bit. Determine **P** and the dimensions *x* and *z* if $R = 8$ lb and $C^R = 72$ lb · in.

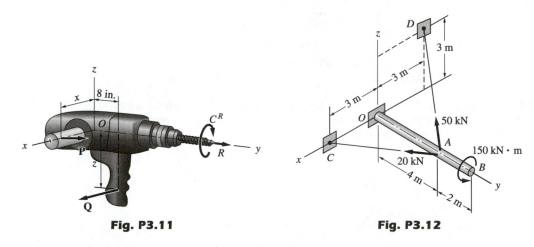

Fig. P3.11

Fig. P3.12

3.12 Two cable tensions and a couple act on the rod *OAB*. Determine the equivalent force-couple system, with the force acting at *O*.

3.13   The force system consists of the 120-N force, the force **Q** = 92**j** − 50**k** N, and the 180-N · m couple. Determine the equivalent force-couple system with the force acting at point *C*.

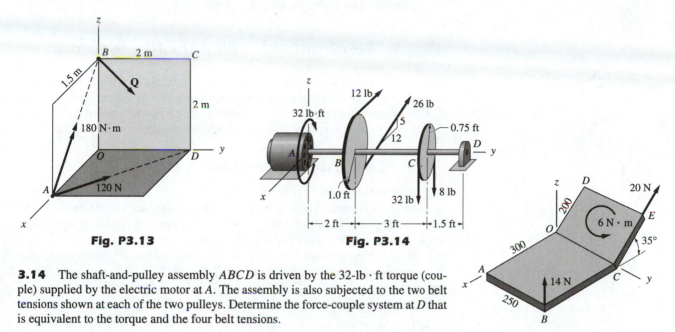

Fig. P3.13   **Fig. P3.14**   **Fig. P3.15**

Dimensions in mm

3.14   The shaft-and-pulley assembly *ABCD* is driven by the 32-lb · ft torque (couple) supplied by the electric motor at *A*. The assembly is also subjected to the two belt tensions shown at each of the two pulleys. Determine the force-couple system at *D* that is equivalent to the torque and the four belt tensions.

3.15   Replace the two forces and the couple with an equivalent force-couple system, with the force acting at *A*. Note that the 6-N · m couple lies in the plane *OCED*.

3.3   *Definition of Resultant*

> The *resultant of a force system* is defined to be the simplest system that can replace the original system without changing its external effect on a rigid body.

The word *simplest* is not precisely defined here; it is used in the sense that one force is simpler than two forces, one couple is simpler than two couples, a force is simpler than a force and a couple, and so on.

The external effect that a force system has on a body is determined by its resultant force-couple system **R** and **C**^{R}. As explained in Art. 2.8, if **R** and **C**^{R} are mutually perpendicular, they can be further reduced to a single force. It follows that the resultant of a force system must be one of the following:*

- A resultant force **R** (if **C**^{R} = **0** or if **R** and **C**^{R} are perpendicular)
- A resultant couple-vector **C**^{R} (if **R** = **0**)
- A resultant force-couple system (if **R** and **C**^{R} are not mutually perpendicular)

---

*It is important that you pay particular attention to the use of the terms *resultant, resultant force* **R**, and *resultant couple-vector* **C**^{R}.

Force systems that have the same resultant are called *statically equivalent*.

The remainder of this chapter discusses the procedures for determining the resultants of two- and three-dimensional force systems.

3.4   *Resultants of Coplanar Force Systems*

This article investigates the resultants of force systems in which all the forces lie in a single plane, chosen as the xy-coordinate plane. We begin with a discussion of the resultants of general coplanar force systems and then consider two special cases: concurrent force systems and parallel force systems.

a.   *General coplanar force system*

A general coplanar force system is shown in Fig. 3.2(a), with all the forces lying in the xy-plane. The origin O is located at any convenient point in the plane. The reduction of this force system to its resultant (simplest equivalent force system) is accomplished by the following procedure.

Replace the original force system with the equivalent system consisting of the resultant force $\mathbf{R} = \Sigma \mathbf{F}$ (or $R_x = \Sigma F_x$ and $R_y = \Sigma F_y$) acting at O and the resultant couple $C^R = \Sigma M_O$, as shown in Fig. 3.2(b). This procedure has three possible outcomes:

- $\mathbf{R} = \mathbf{0}$. The resultant is the couple C^R.
- $C^R = 0$. The resultant is the force $\mathbf{R}$ acting through O.
- $\mathbf{R} \neq \mathbf{0}$ and $C^R \neq 0$. Because $\mathbf{R}$ and C^R are perpendicular to each other, the system can be reduced to a single force $\mathbf{R}$ acting at a point different from O, as illustrated in Fig. 3.2(c). The perpendicular distance d between O and the line of action of $\mathbf{R}$ is determined by the requirement that moments about O of the force systems in Figs. 3.2(b) and (c) must be the same; that is, $\Sigma M_O = Rd$.

In summary, the resultant of the general coplanar force system shown in Fig. 3.2(a) is either a force or a couple. If $\Sigma \mathbf{F} \neq \mathbf{0}$, then the resultant is a force $\mathbf{R}$ determined by

$$R_x = \Sigma F_x \qquad R_y = \Sigma F_y \qquad \Sigma M_O = Rd \tag{3.5}$$

Note that the moment equation locates the line of action of $\mathbf{R}$.

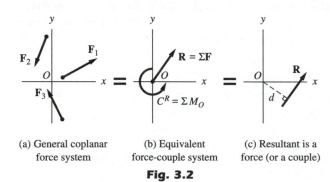

(a) General coplanar
force system

(b) Equivalent
force-couple system

(c) Resultant is a
force (or a couple)

Fig. 3.2

If $\Sigma \mathbf{F} = \mathbf{0}$ and $\Sigma M_O \neq 0$, then the resultant is the couple

$$C^R = \Sigma M_O \qquad (3.6)$$

b.   *Concurrent, coplanar force system*

The resultant of a concurrent, coplanar force system is the force $\mathbf{R} = \Sigma \mathbf{F}$ ($R_x = \Sigma F_x$, $R_y = \Sigma F_y$) acting through the point of concurrency O, as indicated in Fig. 3.3. This conclusion follows from Eq. (3.5): Because $\Sigma M_O = 0$ for a force system that is concurrent at O, the moment equation $\Sigma M_O = Rd$ yields $d = 0$.

c.   *Parallel, coplanar force system*

Figure 3.4(a) shows a coplanar force system, where the forces $F_1, F_2, F_3, \ldots$ are parallel to the y-axis. The equivalent force-couple system at point O is shown in Fig. 3.4(b), where

$$R = F_1 + F_2 + F_3 + \cdots = \Sigma F$$

$$C^R = F_1 x_1 + F_2 x_2 + F_3 x_3 + \cdots = \Sigma M_O$$

If $\Sigma F \neq 0$, the resultant is a force R located at the distance x from O, as indicated in Fig. 3.4(c). The value of x is obtained by equating the moments about O in Fig. 3.4(b) and (c):

$$\Sigma M_O = Rx \qquad (3.7)$$

If, on the other hand, $\Sigma F = 0$ and $\Sigma M_O \neq 0$, then the resultant is the couple $C^R = \Sigma M_O$.

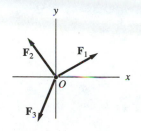

(a) Concurrent, coplanar force system

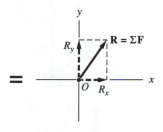

(b) Resultant is a force through point of concurrency

Fig. 3.3

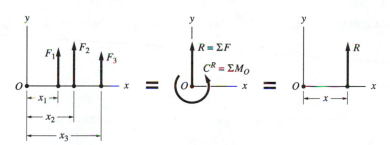

(a) Parallel, coplanar force system

(b) Equivalent force-couple system

(c) Resultant is a force (or a couple)

Fig. 3.4

Sample Problem **3.3**

The values of $R_x = \Sigma F_x$, $R_y = \Sigma F_y$, and ΣM_O for five force systems lying in the xy-plane are listed in the following table. Point O is the origin of the coordinate system, and positive moments are counterclockwise. Determine the resultant for each force system, and show it on a sketch of the coordinate system.

Part	R_x	R_y	ΣM_O
1	0	200 N	400 N · m
2	0	200 N	−400 N · m
3	300 lb	400 lb	600 lb · ft
4	400 N	−600 N	−900 N · m
5	0	0	−200 lb · ft

Solution

Part 1

$$R_x = 0 \qquad R_y = 200 \text{ N} \qquad \Sigma M_O = 400 \text{ N} \cdot \text{m}$$

The resultant is a 200-N force that is parallel to the y-axis, as shown in Fig. (a). Letting x be the distance from point O to the line of action of the resultant, as shown in Fig. (a), and using Eq. (3.7), we have

$$\Sigma M_O = Rx \qquad +\!\!\circlearrowleft \qquad 400 = 200x$$

which gives

$$x = 2 \text{ m}$$

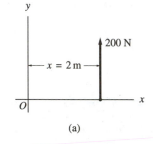

(a)

Part 2

$$R_x = 0 \qquad R_y = 200 \text{ N} \qquad \Sigma M_O = -400 \text{ N} \cdot \text{m}$$

The resultant is the same 200-N force as in Part 1, but here the moment equation gives

$$\Sigma M_O = Rx \qquad +\!\!\circlearrowleft \qquad -400 = 200x$$

or

$$x = -2 \text{ m}$$

The negative sign indicates that x lies to the left of point O, as shown in Fig. (b).

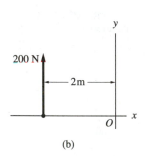

(b)

Part 3

$$R_x = 300 \text{ lb} \qquad R_y = 400 \text{ lb} \qquad \Sigma M_O = 600 \text{ lb} \cdot \text{ft}$$

The resultant is the force $\mathbf{R} = 300\mathbf{i} + 400\mathbf{j}$ lb. Its magnitude is $R = \sqrt{(300)^2 + (400)^2} = 500$ lb. The moment equation of Eq. (3.5) must be used to determine the line of action of $\mathbf{R}$. Letting d be the perpendicular distance from point O to the line of action of $\mathbf{R}$, as shown in Fig. (c), we have

$$\Sigma M_O = Rd \qquad +\!\!\circlearrowleft \qquad 600 = 500d$$

which yields

$$d = 1.2 \text{ ft}$$

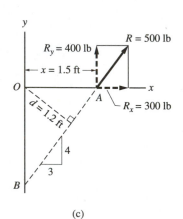

(c)

The points where **R** intersects the coordinate axes can then be determined by trigonometry or by using the principle of moments, as follows.

With R placed at A, as in Fig. (c):   With R placed at B, as in Fig. (d):

$$\Sigma M_O = R_y x \qquad\qquad \Sigma M_O = R_x y$$

$$\underset{+}{\circlearrowleft}\quad 600 = 400x \qquad\qquad \underset{+}{\circlearrowleft}\quad 600 = 300y$$

$$x = 1.5 \text{ ft} \qquad\qquad\qquad y = 2 \text{ ft}$$

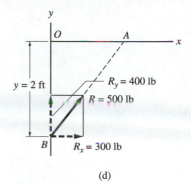

(d)

Part 4

$$R_x = 400 \text{ N} \qquad R_y = -600 \text{ N} \qquad \Sigma M_O = -900 \text{ N} \cdot \text{m}$$

The resultant is the force $\mathbf{R} = 400\mathbf{i} - 600\mathbf{j}$ N; its magnitude is $R = \sqrt{(400)^2 + (600)^2} = 721$ N. Letting d be the perpendicular distance from point O to the line of action of **R**, as shown in Fig. (e), we have

$$\Sigma M_O = Rd \quad \underset{+}{\circlearrowleft} \quad -900 = -721d$$

which gives

$$d = 1.248 \text{ m}$$

Note that the line of action of **R** must be placed to the right of the origin, so that its moment about point O has the same sense as ΣM_O—that is, clockwise.

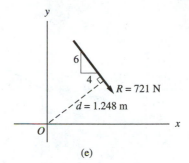

(e)

Part 5

$$R_x = 0 \qquad R_y = 0 \qquad \Sigma M_O = -200 \text{ lb} \cdot \text{ft}$$

Because the sum of the forces is zero, the resultant of this force system is a 200-lb · ft clockwise couple, as shown in Fig. (f).

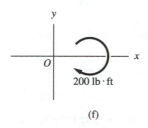

(f)

Sample Problem 3.4

The force **R** is the resultant of the other three concurrent forces shown. Determine **P** and **R**.

Solution

The three applied forces represent a concurrent, coplanar force system. Therefore, the components of the resultant force are determined by two scalar equations: $R_x = \Sigma F_x$ and $R_y = \Sigma F_y$. Because the directions of all the forces are known, there are two unknowns in this problem—the magnitudes P and R. The most direct method for determining these two unknowns is to solve the following two scalar equations (comparing the number of unknowns with the number of available equations is often a valuable aid in the solution of problems):

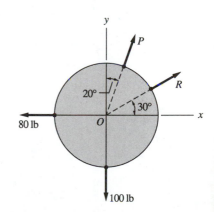

$$R_x = \Sigma F_x \quad \xrightarrow{+} \quad R\cos 30° = P\sin 20° - 80 \qquad (1)$$

$$R_y = \Sigma F_y \quad +\!\uparrow \quad R\sin 30° = P\cos 20° - 100 \qquad (2)$$

Solving Eqs. (1) and (2) simultaneously gives

$$P = 72.5 \text{ lb} \qquad \text{and} \qquad R = -63.7 \text{ lb}$$

The positive value of P indicates that $\mathbf{P}$ is directed as shown in the figure. The negative sign associated with R means that $\mathbf{R}$ acts in the direction opposite to that shown in the figure.

Therefore, the forces $\mathbf{P}$ and $\mathbf{R}$ are

$P = 72.5$ lb 20°   30° $R = 63.7$ lb   *Answer*

Of course, the lines of action of $\mathbf{P}$ and $\mathbf{R}$ pass through O, the point of concurrency.

Sample Problem 3.5

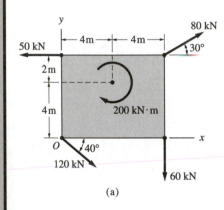

(a)

Determine the resultant of the coplanar force system shown in Fig. (a) that consists of four forces and one couple. Show your answer on a sketch of the coordinate system.

Solution

For a coplanar force system, the resultant is either a force or a couple. If the resultant is a force, then Eqs. (3.5) provide three scalar equations: $R_x = \Sigma F_x$, $R_y = \Sigma F_y$, and $\Sigma M_O = Rd$.

We see that there are no unknown quantities in the original force system. Therefore, if the resultant is a force, the three unknowns in this problem will be R_x, R_y, and d, which could be determined from the three scalar equations. Referring to Fig. (a), the three equations become

$$R_x = \Sigma F_x \quad \xrightarrow{+} \quad R_x = 80 \cos 30° + 120 \cos 40° - 50 = 111.2 \text{ kN}$$

This equation is sufficient to tell us that the resultant is a force, not a couple; if the resultant were a couple, ΣF_x would be zero.

$$R_y = \Sigma F_y \quad +\uparrow \quad R_y = 80 \sin 30° - 120 \sin 40° - 60 = -97.1 \text{ kN}$$

$$\Sigma M_O = Rd \quad \left(\stackrel{+}{\curvearrowleft}\right) \quad \Sigma M_O = 50(6) - 200 + 80 \sin 30° (8)$$
$$- 80 \cos 30°(6) - 60 (8) = -475.7 \text{ kN} \cdot \text{m}$$

Therefore, the resultant $\mathbf{R}$ is

111.2 kN θ 97.1kN $R = 147.6$ kN

$$\theta = \tan^{-1} \frac{97.1}{111.2} = 41.1°$$

Because ΣM_O is negative (i.e., clockwise), the resultant $\mathbf{R}$ must also provide a clockwise moment about O, as shown in Fig. (b). Therefore we obtain

$$\Sigma M_O = Rd \quad \left(\stackrel{+}{\curvearrowleft}\right) \quad 475.7 = 147.6d$$

which gives

$$d = 3.22 \text{ m}$$

The final result is shown in Fig. (b).

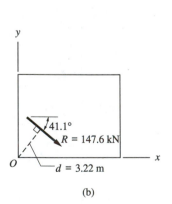

(b)

Sample Problem 3.6

The force system shown consists of the couple C and four forces. If the resultant of this system is a 500-lb · in. counterclockwise couple, determine P, Q, and C.

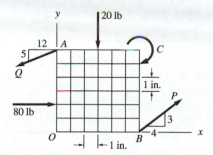

Solution

This problem contains three unknowns: P, Q, and C. Because the force system is the general coplanar case, three equations are available to determine the resultant. The resultant is a couple, so the first two of Eqs. (3.5) become

$$R_x = \Sigma F_x = 0 \qquad \xrightarrow{+} \qquad -\frac{12}{13}Q + \frac{4}{5}P + 80 = 0 \qquad \text{(a)}$$

$$R_y = \Sigma F_y = 0 \qquad +\uparrow \qquad -\frac{5}{13}Q + \frac{3}{5}P - 20 = 0 \qquad \text{(b)}$$

Solving Eqs. (a) and (b) simultaneously gives

$$P = 200 \text{ lb} \quad \text{and} \quad Q = 260 \text{ lb} \qquad \qquad \textit{Answer}$$

The third equation is Eq. (3.6), $C^R = \Sigma M_O$. Because a couple is a free vector, the moment center can be any suitable point. Given that $C^R = 500$ lb · in., counterclockwise, and choosing point A as the moment center, we have

$$C^R = \Sigma M_A \qquad \overset{+}{\curvearrowleft} \qquad 500 = -20(3) - C + 80(4) + \frac{3}{5}P(6) + \frac{4}{5}P(6)$$

Substituting $P = 200$ lb and solving yields

$$C = 1440 \text{ lb} \cdot \text{in.} \qquad \qquad \textit{Answer}$$

Because the values for P, Q, and C are positive, each acts in the direction in which it is shown in the figure.

Problems

3.16   The values of $R_x = \Sigma F_x$, $R_y = \Sigma F_y$, and ΣM_O for four force systems lying in the xy-plane are given in the following table. Point O is the origin of the coordinate system, and positive moments are counterclockwise. Determine the resultant of each force system, and show it on a sketch of the coordinate system.

Part	R_x	R_y	ΣM_O
1	300 lb	0	-900 lb · in.
2	200 N	-200 N	800 N · m
3	-600 kN	-400 kN	0
4	-600 lb	800 lb	$-24\,000$ lb · ft

3.17   Determine the resultant of the three concurrent forces acting on the circular arch.

3.18   The resultant of the three concurrent forces acting on the eyebolt is the force $\mathbf{R} = 800\mathbf{j}$ lb. Determine the magnitude of the force $\mathbf{P}$ and the angle θ that specifies the direction of the 900-lb force.

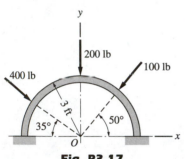

Fig. P3.17

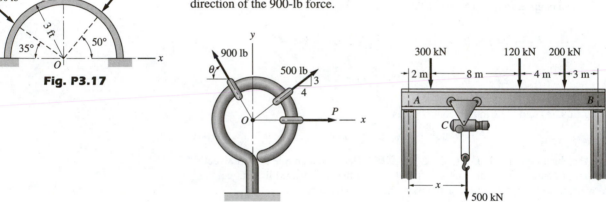

Fig. P3.18

Fig. P3.19

3.19   The overhead electric hoist C rides along a track on the horizontal beam AB. In addition to the 500-kN vertical force carried by the hoist, the beam also supports the three vertical forces shown. (a) If $x = 5$ m, determine the resultant of the four forces carried by the beam. (b) Determine the distance x for which the resultant of the four forces would act at the center of the span AB.

3.20   Determine the resultant of the force system if (a) $P = 120$ lb; and (b) $P = 80$ lb.

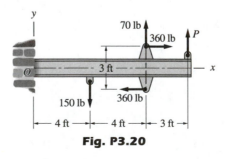

Fig. P3.20

3.21 Determine which of the force systems in Figs. (b) through (f) are equivalent to the 21-kN force in (a).

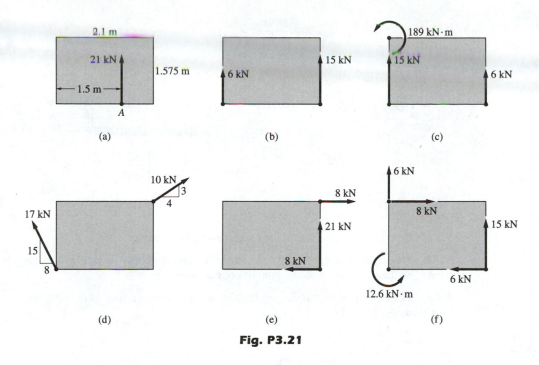

(a)   (b)   (c)

(d)   (e)   (f)

Fig. P3.21

3.22 Determine the resultant of the three forces if (a) $\theta = 30°$; and (b) $\theta = 45°$.

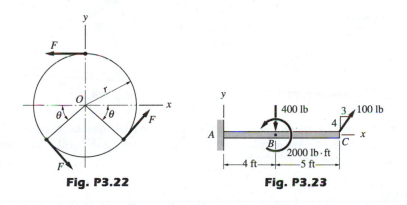

Fig. P3.22   **Fig. P3.23**

3.23 Determine the resultant of the two forces and the couple that act on the beam.

3.24 Determine the resultant of the three forces and the couple C, and show it on a sketch of the coordinate system if (a) $C = 0$; and (b) $C = 90 \text{ N} \cdot \text{m}$.

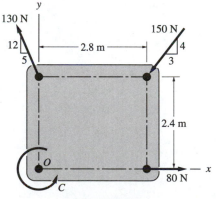

Fig. P3.24

3.25   The resultant of the three forces is a force **R** that passes through point *B*. Determine **R** and *F*.

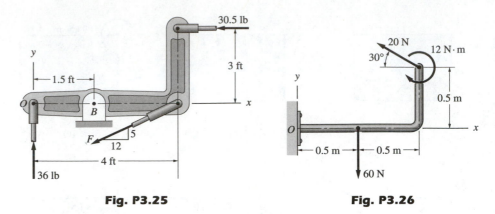

Fig. P3.25   Fig. P3.26

3.26   Determine the resultant of the two forces and the couple.

3.27   The resultant of the three forces shown is a counterclockwise couple of magnitude 150 lb · ft. Calculate the magnitudes of the forces.

3.28   The vertical force *R* is the resultant of the three forces acting on the bent bar. Determine P_1, P_2, and *R*.

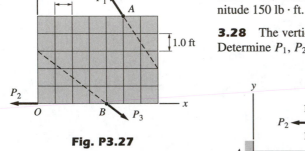

Fig. P3.27

Fig. P3.28   Fig. P3.29

*3.29   The bar *AB*, which is inclined at the angle θ to the horizontal, is subjected to the four forces shown. Knowing that these forces have no resultant (neither a force nor a couple), determine P_1, P_2, and θ.

3.5   *Resultants of Noncoplanar Force Systems*

In general, a three-dimensional force system cannot be simplified beyond a force-couple system. Exceptions are systems in which the forces are either concurrent or parallel. In this article, we discuss these two special cases, together with a special form of the force-couple system called the *wrench*.

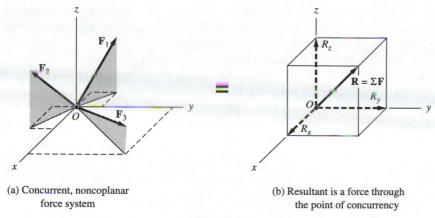

(a) Concurrent, noncoplanar
force system

(b) Resultant is a force through
the point of concurrency

Fig. 3.5

a.   *Concurrent, noncoplanar force system*

A concurrent, three-dimensional force system is shown in Fig. 3.5(a). As in the
case of concurrent, coplanar forces, this system can be reduced to the resultant
force $\mathbf{R} = \Sigma\mathbf{F}$ ($R_x = \Sigma F_x$, $R_y = \Sigma F_y$, $R_z = \Sigma F_z$) acting through the point of
concurrency O, as indicated in Fig. 3.5(b).

b.   *Parallel, noncoplanar force system*

Consider the force system in Fig. 3.6(a), where the forces F_1, F_2, F_3, . . . are
parallel to the z-axis. To find the resultant, we begin by replacing the forces
with an equivalent force-couple system, with the force acting at the origin O,
as shown in Fig. 3.6(b). The magnitude of the resultant force $\mathbf{R}$, which is also
parallel to the z-axis, and the resultant couple-vector $\mathbf{C}^R$ are given by

$$\boxed{R = \Sigma F} \tag{3.8}$$

and

$$\mathbf{C}^R = \Sigma\mathbf{M}_O \tag{3.9}$$

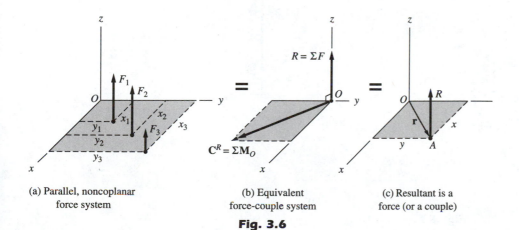

(a) Parallel, noncoplanar
force system

(b) Equivalent
force-couple system

(c) Resultant is a
force (or a couple)

Fig. 3.6

where $\Sigma\mathbf{M}_O$ is the sum of the moments of $F_1, F_2, F_3, \ldots$ about O. The resultant couple-vector $\mathbf{C}^R$ lies in the xy-plane ($\mathbf{C}^R$ has no z-component because forces parallel to an axis have no moment about that axis).

Because $\mathbf{R}$ and $\mathbf{C}^R$ are mutually perpendicular, the force system can be further simplified. If $\Sigma F = 0$, then the resultant is the couple $\mathbf{C}^R = \Sigma\mathbf{M}_O$. If $\Sigma F \neq 0$, the resultant is the force $\mathbf{R}$ acting through the unique point A in the xy-plane, as shown in Fig. 3.6(c). The vector $\mathbf{r} = x\mathbf{i} + y\mathbf{j}$ that locates this point is obtained by equating the moments about point O of the force-couple system in Fig. (a) and the force R in Fig. (c):

$$\Sigma\mathbf{M}_O = \mathbf{r} \times \mathbf{R} \tag{3.10}$$

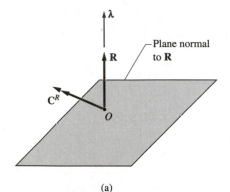

(a)

The scalar components of this moment equation are

$$\Sigma M_x = F_1 y_1 + F_2 y_2 + F_3 y_3 + \cdots = Ry$$

$$\Sigma M_y = -F_1 x_1 - F_2 x_2 - F_3 x_3 - \cdots = -Rx$$

where ΣM_x and ΣM_y are the moments of the original forces about the x- and y-axes, respectively. Therefore, the coordinates x and y become

$$x = -\frac{\Sigma M_y}{R} \qquad y = \frac{\Sigma M_x}{R} \tag{3.11}$$

c.   General noncoplanar force system: The wrench

As shown in Art. 3.2, a given force system can always be reduced to a force-couple system consisting of a resultant force $\mathbf{R} = \Sigma\mathbf{F}$, acting at an arbitrary point O, and a resultant couple-vector $\mathbf{C}^R = \Sigma\mathbf{M}_O$, as shown in Fig. 3.7(a). If $\mathbf{R}$ and $\mathbf{C}^R$ are mutually perpendicular, they can be reduced to a single force $\mathbf{R}$, acting through a unique point (this property was used in the special cases of coplanar and parallel force systems). In the general case, $\mathbf{R}$ and $\mathbf{C}^R$ will not be perpendicular to each other, and thus they will not be reducible to a single force. However, a general force system can always be represented by a force and a *parallel* couple-vector by the procedure described below.

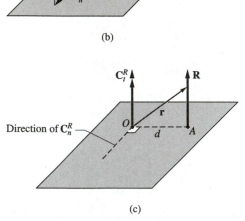

(b)

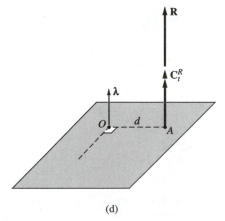

(c)

(d)

Fig. 3.7

- Resolve $\mathbf{C}^R$ into the orthogonal components $\mathbf{C}_t^R$ and $\mathbf{C}_n^R$, which are parallel and perpendicular to $\mathbf{R}$, respectively. The result is shown in Fig. 3.7(b). The parallel component can be computed from Eq. (2.11):

$$\mathbf{C}_t^R = (\mathbf{C}^R \cdot \boldsymbol{\lambda})\boldsymbol{\lambda} \qquad (3.12)$$

where $\boldsymbol{\lambda}$ is the unit vector in the direction of $\mathbf{R}$. The normal component is then found from

$$\mathbf{C}_n^R = \mathbf{C}^R - \mathbf{C}_t^R \qquad (3.13)$$

- Because $\mathbf{C}_n^R$ and $\mathbf{R}$ are mutually perpendicular, they can be replaced by the single force $\mathbf{R}$ acting at point A, as illustrated in Fig. 3.7(c). The line of action of this force is determined by the requirement that its moment about O must be equal to $\mathbf{C}_n^R$. In other words,

$$\mathbf{r} \times \mathbf{R} = \mathbf{C}_n^R \qquad (3.14)$$

where $\mathbf{r}$ is the vector drawn from O to any point on the new line of action of $\mathbf{R}$. The scalar form of Eq. (3.14) is $Rd = C_n^R$, where d is the distance between O and A, as indicated in Fig. 3.7(c). This equation yields

$$d = \frac{C_n^R}{R} \qquad (3.15)$$

Note that the line OA is perpendicular to $\mathbf{C}_n^R$.

- Move $\mathbf{C}_t^R$ to point A, as shown in Fig. 3.7(d) (we can do this because a couple is a free vector). The result is a collinear force-couple system, called the *wrench*. The direction of the wrench, also known as the *axis of the wrench,* is specified by the vector $\boldsymbol{\lambda}$.

An example of the wrench is the operation of a screwdriver. We exert a force along the axis of the screwdriver to hold its tip against the screw, while applying a couple about the same axis to turn the screw. Because the force and the couple-vector are parallel, they constitute a wrench.

Sample Problem 3.7

The values of ΣF_z, ΣM_x, and ΣM_y for three force systems that are parallel to the z-axis are as follows:

Part	ΣF_z	ΣM_x	ΣM_y
1	50 kN	60 kN · m	−125 kN · m
2	−600 lb	0	−1200 lb · ft
3	0	600 lb · in.	−800 lb · in.

Determine the resultant of each force system and show it on a sketch of the coordinate system.

Solution

Part 1

$$\Sigma F_z = 50 \text{ kN} \qquad \Sigma M_x = 60 \text{ kN} \cdot \text{m} \qquad \Sigma M_y = -125 \text{ kN} \cdot \text{m}$$

The resultant is the force $\mathbf{R} = 50\mathbf{k}$ kN. With $\Sigma \mathbf{M}_O = \Sigma M_x \mathbf{i} + \Sigma M_y \mathbf{j}$ and $\mathbf{r} = x\mathbf{i} + y\mathbf{j}$, Eq. (3.10) can be used to determine the line of action of $\mathbf{R}$:

$$\Sigma \mathbf{M}_O = \mathbf{r} \times \mathbf{R}$$

$$60\mathbf{i} - 125\mathbf{j} = \begin{vmatrix} \mathbf{i} & \mathbf{j} & \mathbf{k} \\ x & y & 0 \\ 0 & 0 & 50 \end{vmatrix} = 50y\mathbf{i} - 50x\mathbf{j}$$

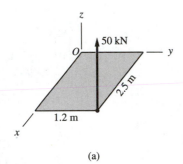

(a)

Equating like components gives the following scalar equations:

$$60 = 50y \quad \text{and} \quad -125 = -50x$$

which gives $x = 2.5$ m and $y = 1.2$ m. The sketch of the resultant is shown in Fig. (a). Identical results for x and y are obtained if one uses Eq. (3.11):

$$x = -\frac{\Sigma M_y}{R} = -\frac{-125}{50} = 2.5 \text{ m}$$

$$y = \frac{\Sigma M_x}{R} = \frac{60}{50} = 1.2 \text{ m}$$

Part 2

$$\Sigma F_z = -600 \text{ lb} \qquad \Sigma M_x = 0 \qquad \Sigma M_y = -1200 \text{ lb} \cdot \text{ft}$$

The resultant is the force $\mathbf{R} = -600\mathbf{k}$ lb. In this case, Eq. (3.10) yields

$$\Sigma \mathbf{M}_O = \mathbf{r} \times \mathbf{R}$$

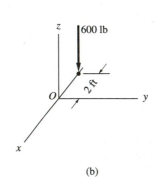

(b)

$$-1200\mathbf{j} = \begin{vmatrix} \mathbf{i} & \mathbf{j} & \mathbf{k} \\ x & y & 0 \\ 0 & 0 & -600 \end{vmatrix} = -600y\mathbf{i} + 600x\mathbf{j}$$

Equating like components gives $x = -2$ ft and $y = 0$. The resultant is shown in Fig. (b).

Part 3

$$\Sigma F_z = 0 \qquad \Sigma M_x = 600 \text{ lb} \cdot \text{in.} \qquad \Sigma M_y = -800 \text{ lb} \cdot \text{in.}$$

Because the sum of the forces is zero and the sum of the moments is not zero, the resultant is the couple-vector $\mathbf{C}^R = \Sigma M_x \mathbf{i} + \Sigma M_y \mathbf{j} = 600\mathbf{i} - 800\mathbf{j}$ lb $\cdot$ in., shown in Fig. (c). The magnitude of this couple-vector is 1000 lb $\cdot$ in.

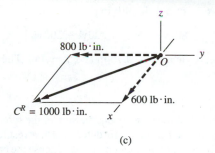

(c)

Sample Problem 3.8

The parallel force system in Fig. (a) consists of the three forces shown and the 1250-N $\cdot$ m couple-vector. (1) Determine the resultant, and show it on a sketch of the coordinate system. (2) Determine the resultant if the direction of the 100-N force is reversed.

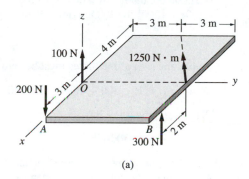

(a)

Solution

Part 1

The resultant of the force system shown in Fig. (a) will be either a force or a couple. We begin by summing the forces.

$$+\uparrow \quad R = \Sigma F_z = 100 - 200 + 300 = 200 \text{ N}$$

Therefore, the resultant is the force $\mathbf{R} = 200\mathbf{k}$ N.

We must use a moment equation to find the line of action of $\mathbf{R}$. Using the origin O as the moment center and assuming that $\mathbf{R}$ intersects the xy-plane at the point $(x, y, 0)$,

Eq. (3.10) becomes

$$\Sigma \mathbf{M}_O = \mathbf{r} \times \mathbf{R}$$

$$3\mathbf{i} \times (-200\mathbf{k}) + [(2\mathbf{i} + 6\mathbf{j}) \times 300\mathbf{k}]$$

$$-\left(\frac{4}{5}\right)1250\mathbf{i} - \left(\frac{3}{5}\right)1250\mathbf{j} = (x\mathbf{i} + y\mathbf{j}) \times 200\mathbf{k}$$

Expanding the cross products and simplifying, we obtain

$$800\mathbf{i} - 750\mathbf{j} = 200y\mathbf{i} - 200x\mathbf{j}$$

Equating like components yields $x = 3.75$ m and $y = 4$ m. The resultant is shown in Fig. (b).

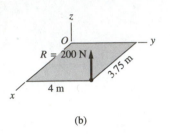

(b)

Part 2

If the direction of the 100-N force is reversed, the sum of the forces will be zero, which means that the resultant is not a force. To determine the resultant couple, we must compute the moment about any convenient point. We choose the origin O as the moment center. Because reversing the direction of the 100-N force has no effect on the moment about O, we conclude that the moment is the same as that found in the solution to Part 1. Therefore, the resultant is the couple-vector $\mathbf{C}^R = \Sigma \mathbf{M}_O = 800\mathbf{i} - 750\mathbf{j}$ N · m.

Sample Problem **3.9**

The plate is acted on by four parallel forces, three of which are shown in Fig. (a). The fourth force $\mathbf{P}$ and its line of action are unknown. The resultant of this force system is the couple-vector $\mathbf{C}^R = -1100\mathbf{i} + 1500\mathbf{j}$ lb · ft. Determine $\mathbf{P}$ and its line of action.

Solution

Because the resultant is a couple, the sum of the forces must be zero:

$$+\uparrow \quad R = \Sigma F_z = P + 300 + 400 - 200 = 0$$

from which $P = -500$ lb. Therefore, the force $\mathbf{P}$ is

$$\mathbf{P} = -500\mathbf{k} \text{ lb} \qquad\qquad\qquad \textit{Answer}$$

As shown in Fig. (b), we let A be the point where $\mathbf{P}$ intersects the xy-plane. To determine the location of A, we equate the sum of the moments of the original forces about any point to the moment of the resultant about that point (in this case, the moment of the resultant about every point is simply $\mathbf{C}^R$). Choosing point O as the moment center and noting that $x_A\mathbf{i} + y_A\mathbf{j}$ is the vector from point O to point A, the moment equation becomes

$$\mathbf{C}^R = \Sigma \mathbf{M}_O = \Sigma \mathbf{r} \times \mathbf{F}$$

$$-1100\mathbf{i} + 1500\mathbf{j} = \begin{vmatrix} \mathbf{i} & \mathbf{j} & \mathbf{k} \\ 3 & 0 & 0 \\ 0 & 0 & -200 \end{vmatrix} + \begin{vmatrix} \mathbf{i} & \mathbf{j} & \mathbf{k} \\ 2 & 3 & 0 \\ 0 & 0 & 300 \end{vmatrix}$$

$$+ \begin{vmatrix} \mathbf{i} & \mathbf{j} & \mathbf{k} \\ x_A & y_A & 0 \\ 0 & 0 & -500 \end{vmatrix}$$

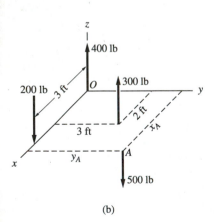

(a)

(b)

Expanding the above determinants and equating like components gives

$$-1100 = 900 - 500y_A$$

$$1500 = 600 - 600 + 500x_A$$

from which

$$x_A = 3 \text{ ft} \quad \text{and} \quad y_A = 4 \text{ ft} \qquad \textit{Answer}$$

Sample Problem 3.10

Determine the wrench that is equivalent to the force system described in Sample Problem 3.1. Find the coordinates of the point where the axis of the wrench crosses the xy-plane.

Solution

As explained in the solution to Sample Problem 3.1, the original force system can be reduced to the force-couple system shown in Fig. (a): the force **R**, acting at the origin O, and the couple $\mathbf{C}^R$, where

$$\mathbf{R} = 38.8\mathbf{i} + 43.2\mathbf{j} + 38.4\mathbf{k} \text{ lb} \qquad (a)$$

$$\mathbf{C}^R = 390\mathbf{i} + 116\mathbf{j} + 360\mathbf{k} \text{ lb} \cdot \text{ft}$$

The magnitude of **R** is

$$R = \sqrt{(38.8)^2 + (43.2)^2 + (38.4)^2} = 69.6 \text{ lb}$$

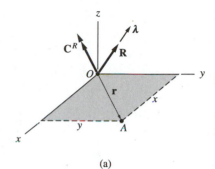

(a)

We begin by determining the axis of the wrench, defined by the unit vector $\boldsymbol{\lambda}$ in the direction of **R**:

$$\boldsymbol{\lambda} = \frac{\mathbf{R}}{R} = \frac{38.8\mathbf{i} + 43.2\mathbf{j} + 38.4\mathbf{k}}{69.6}$$

$$= 0.557\mathbf{i} + 0.621\mathbf{j} + 0.552\mathbf{k}$$

The component of $\mathbf{C}^R$ in the direction of $\boldsymbol{\lambda}$ can now be obtained from Eq. (3.12): $\mathbf{C}_t^R = (\mathbf{C}^R \cdot \boldsymbol{\lambda})\boldsymbol{\lambda}$. The magnitude of this vector is

$$C_t^R = \mathbf{C}^R \cdot \boldsymbol{\lambda}$$

$$= (390\mathbf{i} + 116\mathbf{j} + 360\mathbf{k}) \cdot (0.557\mathbf{i} + 0.621\mathbf{j} + 0.552\mathbf{k})$$

$$= 488 \text{ lb} \cdot \text{ft}$$

which gives

$$\mathbf{C}_t^R = C_t^R \boldsymbol{\lambda} = 488(0.557\mathbf{i} + 0.621\mathbf{j} + 0.552\mathbf{k})$$

$$= 272\mathbf{i} + 303\mathbf{j} + 269\mathbf{k} \text{ lb} \cdot \text{ft}$$

Therefore, the wrench consists of the force-couple system

$$\mathbf{R} = 38.8\mathbf{i} + 43.2\mathbf{j} + 38.4\mathbf{k} \text{ lb} \qquad \textit{Answer}$$

$$\mathbf{C}_t^R = 272\mathbf{i} + 303\mathbf{j} + 269\mathbf{k} \text{ lb} \cdot \text{ft} \qquad \textit{Answer}$$

To find the coordinates of the point where the axis of the wrench intersects the xy-plane, we must find $\mathbf{C}_n^R$, the component of $\mathbf{C}^R$ that is normal to $\boldsymbol{\lambda}$. From Eq. (3.13), we obtain

$$\mathbf{C}_n^R = \mathbf{C}^R - \mathbf{C}_t^R = (390\mathbf{i} + 116\mathbf{j} + 360\mathbf{k}) - (272\mathbf{i} + 303\mathbf{j} + 269\mathbf{k})$$

$$= 118\mathbf{i} - 187\mathbf{j} + 91\mathbf{k} \text{ lb} \cdot \text{ft}$$

Referring to Fig. (a), we let $\mathbf{r} = x\mathbf{i} + y\mathbf{j}$ be the vector from the origin O to A, the point where the wrench intersects the xy-plane. Using Eq. (3.14), we have

$$\mathbf{r} \times \mathbf{R} = \mathbf{C}_n^R$$

$$\begin{vmatrix} \mathbf{i} & \mathbf{j} & \mathbf{k} \\ x & y & 0 \\ 38.8 & 43.2 & 38.4 \end{vmatrix} = 118\mathbf{i} - 187\mathbf{j} + 91\mathbf{k}$$

After expanding the determinant, we get

$$38.4y\mathbf{i} - 38.4x\mathbf{j} + (43.2x - 38.8y)\mathbf{k} = 118\mathbf{i} - 187\mathbf{j} + 91\mathbf{k}$$

Equating the coefficients of $\mathbf{i}$ and $\mathbf{j}$ yields

$$38.4y = 118 \qquad y = 3.07 \text{ ft} \qquad\qquad \textit{Answer}$$

$$-38.4x = -187 \qquad x = 4.87 \text{ ft} \qquad\qquad \textit{Answer}$$

The third equation, obtained by equating the coefficients of $\mathbf{k}$, is not independent of the preceding two equations, as can be easily verified.

The resultant wrench is depicted in Fig. (b), which shows the magnitudes of the force and the couple-vector.

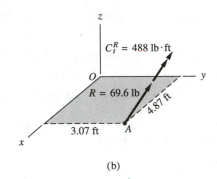

(b)

Problems

3.30 The values of ΣF_z, ΣM_x, and ΣM_y for four force systems that are parallel to the z-axis are as follows.

Part	ΣF_z	ΣM_x	ΣM_y
a	−80 lb	−400 lb · ft	320 lb · ft
b	50 kN	−300 kN · m	0
c	400 lb	−1200 lb · in.	1000 lb · in.
d	25 N	200 N · m	−250 N · m

Determine the resultant of each force system, and show it on a sketch of the coordinate system.

3.31 State whether the resultant of each force system shown is a force, a couple, or a wrench. *Do not determine the resultant.*

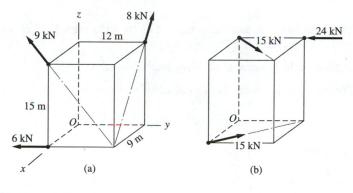

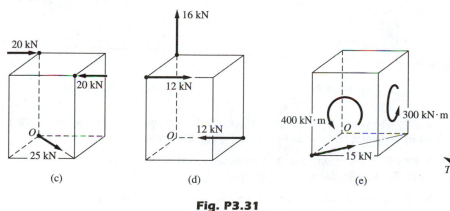

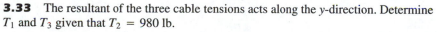

Fig. P3.31

3.32 Determine the resultant of the three cable tensions that act on the horizontal boom if $T_1 = 900$ lb, $T_2 = 500$ lb, and $T_3 = 300$ lb.

3.33 The resultant of the three cable tensions acts along the y-direction. Determine T_1 and T_3 given that $T_2 = 980$ lb.

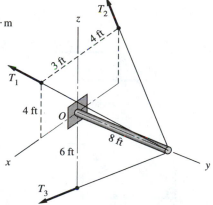

Fig. P3.32, P3.33

3.34   If the resultant of the three forces shown is $\mathbf{R} = 200\mathbf{k}$ kN, determine the magnitudes P_1, P_2, and P_3.

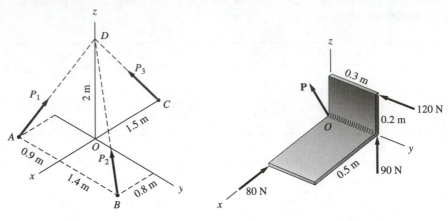

Fig. P3.34   **Fig. P3.35**

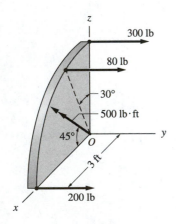

Fig. P3.36

3.35   The resultant of the four forces that act on the right-angle bracket is a couple $\mathbf{C}^R$. Determine $\mathbf{C}^R$ and the force $\mathbf{P}$.

3.36   Determine the resultant of the three forces shown.

3.37   Find the resultant of the four forces acting on the plate.

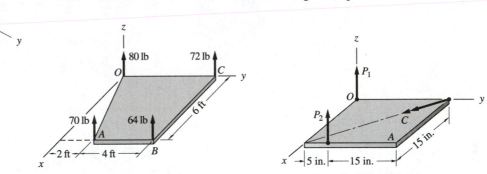

Fig. P3.37   **Fig. P3.38, P3.39**

3.38   The resultant of the forces P_1, P_2, and the couple C is the force $\mathbf{R} = 12\mathbf{k}$ lb acting at point A ($\mathbf{R}$ is not shown in the figure). Determine P_1, P_2, and C.

3.39   Find the resultant of the two forces and the couple shown, given that $P_1 = 20$ lb, $P_2 = 30$ lb, and $C = 100$ lb · in.

3.40   Determine the resultant of the force system that acts on the quarter-circular plate.

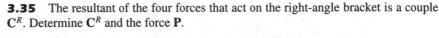

Fig. P3.40

3.41   The streetlight A is attached to the end of the horizontal boom ABO. The light, which weighs 100 N, is subjected to a wind load of 20 N acting in the negative y-direction. The forces **P** and **Q** represent the tensions in the two cables that are attached at point B. The resultant of the four forces shown is a force **R** acting at point O. Determine the tensions P and Q and the force **R**.

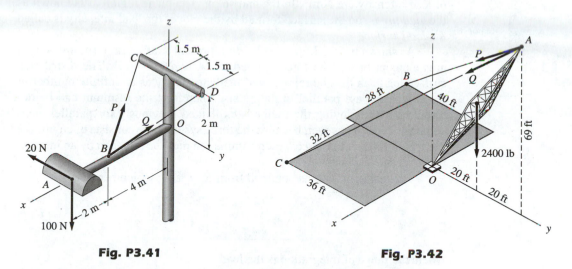

Fig. P3.41   **Fig. P3.42**

3.42   The transmission tower OA is being hoisted into position by the cables AB and AC. The resultant of the cable tensions P and Q, along with the 2400-lb weight of the tower, is a force **R** acting at point O. Determine P, Q, and **R**.

3.43   A force-couple system consists of the force **R** = 600**i** + 1400**j** + 700**k** lb, acting at the origin of a rectangular coordinate system, and the couple-vector $\mathbf{C}^R$ = −800**i** + 500**j** + 600**k** lb·ft. Determine the equivalent wrench, and find the coordinates of the point where the axis of the wrench crosses the yz-plane.

3.44   A force-couple system consists of the force **R** = 300**i** + 400**j** − 500**k** N, acting at the origin of a rectangular coordinate system, and the couple-vector $\mathbf{C}^R$ = 1400**i** + 800**j** + 600**k** N·m. Determine the equivalent wrench, and find the coordinates of the point where the axis of the wrench crosses the xy-plane.

3.45   (a) Replace the force system in Prob. 3.31(d) by an equivalent force-couple system with the force acting at point O. (b) Determine the equivalent wrench, and find the coordinates of the point where the axis of the wrench crosses the xy-plane.

3.6   *Introduction to Distributed Normal Loads*

All forces considered up to this point have been assumed to be concentrated. Here we consider distributed loads that are directed normal to the surface on which they act, such as *pressure*. Two examples of distributed normal loads are the wind pressure acting on the side of a building and the water pressure on a dam. The methods for determining the resultants of distributed normal loads are very similar to those used for concentrated loads. The only notable difference is that integration is used in place of summation.

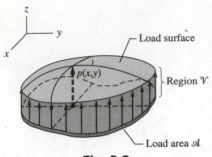

Load surface

Region $\mathcal{V}$

$p(x,y)$

Load area $\mathcal{A}$

Fig. 3.8

a. Surface loads

Consider the load shown in Fig. 3.8, which is parallel to the z-axis and distributed over the plane region $\mathcal{A}$ lying in the xy-plane.* The distribution of the load is specified by the function $p(x, y)$, called the *load intensity*. The units of load intensity are N/m^2, lb/ft^2, and so on. The plane region $\mathcal{A}$ is known as the *load area*, and the surface formed by the plot of the load intensity is called the *load surface*.

As shown in Fig. 3.9(a), we let dA represent a differential (infinitesimal) area element of $\mathcal{A}$. The force applied to dA is $dR = p\,dA$. The distributed surface load can thus be represented mathematically as an infinite number of forces dR that are parallel to the z-axis. Therefore, the resultant can be determined by employing the methods explained previously for parallel, noncoplanar forces. However, because the force system here consists of an infinite number of differential forces, each summation must be replaced by an integration over the load area $\mathcal{A}$.

The resultant force is obtained from $R = \Sigma F_z$, which becomes

$$R = \int_{\mathcal{A}} dR = \int_{\mathcal{A}} p\,dA \qquad (3.16)$$

where the range of integration is the load area $\mathcal{A}$.

The coordinates $\bar{x}$ and $\bar{y}$ that locate the line of action of R, shown in Fig. 3.9(b), are determined by Eqs. (3.11): $\bar{x} = -\Sigma M_y/R$ and $\bar{y} = \Sigma M_x/R$. After making the appropriate substitutions, these equations become

$$\bar{x} = \frac{\int_{\mathcal{A}} p x\,dA}{\int_{\mathcal{A}} p\,dA} \quad \text{and} \quad \bar{y} = \frac{\int_{\mathcal{A}} p y\,dA}{\int_{\mathcal{A}} p\,dA} \qquad (3.17)$$

Let us now consider Eqs. (3.16) and (3.17) from a geometrical viewpoint. By inspection of Fig. 3.9 we observe that $dR = p\,dA$ represents a differential volume between the load area $\mathcal{A}$ and the load surface. This volume has been denoted dV in the figure. Therefore, the resultant force R in Eq. (3.16) can also

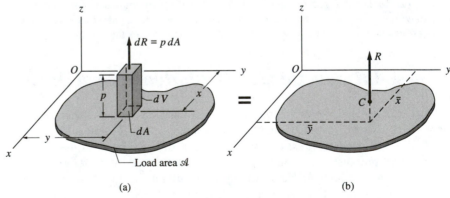

(a)   (b)

Fig. 3.9

---

*The more general case of a load distributed over a curved surface is discussed in Chapter 8.

be written as

$$R = \int_{\mathcal{V}} dV = V \qquad (3.18)$$

where V is the total volume of the region $\mathcal{V}$ that lies between the load area $\mathcal{A}$ and the load surface.

Replacing $p\,dA$ with dV in Eqs. (3.17), we get

$$\bar{x} = \frac{\int_{\mathcal{V}} x\,dV}{\int_{\mathcal{V}} dV} = \frac{\int_{\mathcal{V}} x\,dV}{V}$$

$$\bar{y} = \frac{\int_{\mathcal{V}} y\,dV}{\int_{\mathcal{V}} dV} = \frac{\int_{\mathcal{V}} y\,dV}{V} \qquad (3.19)$$

As will be explained in Chapter 8, Eqs. (3.19) define the coordinates of a point known as the *centroid* of the volume that occupies the region $\mathcal{V}$. This point is labeled C in Fig. 3.9(b). The z-coordinate of the centroid is of no concern here because $\bar{x}$ and $\bar{y}$ are sufficient to define the line of action of the resultant force.

The determination of the resultant force of a normal loading distributed over a plane area may thus be summarized as follows:

> - The magnitude of the resultant force is equal to the volume of the region between the load area and the load surface.
> - The line of action of the resultant force passes through the centroid of the volume bounded by the load area and the load surface.

b.   *Line loads*

Whenever the width of the loading area is negligible compared with its length, a distributed load can be represented as a line load. Loadings distributed along a plane curve and along a straight line are shown in Fig. 3.10(a) and (b), respectively. Line loads are characterized by the *load intensity w*, a function of the distance measured along the line of distribution. The plot of w is called the *load diagram*. The units of w are N/m, lb/ft, and so on. In this article, we consider only straight-line loads. Loads distributed along plane curves will be discussed in Chapter 8.

As shown in Fig. 3.11(a), a straight-line load is equivalent to an infinite number of differential forces, each of magnitude $dR = w\,dx$. Because these

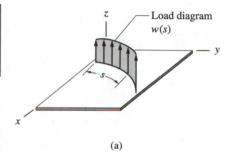

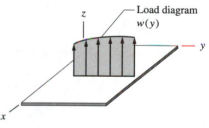

Fig. 3.10

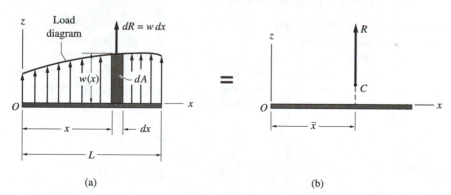

Fig. 3.11

forces are parallel, their resultant is $R = \Sigma F_z$, or

$$R = \int_{x=0}^{L} dR = \int_{0}^{L} w\,dx \tag{3.20}$$

directed parallel to the z-axis, as indicated in Fig. 3.11(b).

The line of action of R can be determined by equating the moments about point O for the two systems in Figs. 3.11(a) and (b):

$$\underset{+}{\curvearrowleft} \quad \Sigma M_O = \int_{x=0}^{L} x\,dR = \int_{0}^{L} wx\,dx = R\bar{x}$$

where we have used $dR = w\,dx$. Substituting the expression for R given in Eq. (3.20), and solving for $\bar{x}$, we obtain

$$\bar{x} = \frac{\int_{0}^{L} wx\,dx}{\int_{0}^{L} w\,dx} \tag{3.21}$$

Referring to Fig. 3.11(a), we observe that $dR = w\,dx$ equals the differential area dA under the load diagram. Therefore, Eq. (3.20) represents the total area A under that diagram. Substituting $w\,dx = dA$, Eq. (3.21) can be written as

$$\bar{x} = \frac{\int_{x=0}^{L} x\,dA}{\int_{x=0}^{L} dA} = \frac{\int_{x=0}^{L} x\,dA}{A} \tag{3.22}$$

It is shown in Chapter 8 that $\bar{x}$ locates the *centroid* of the area under the load diagram, labeled C in Fig. 3.11(b) (the z-coordinate of the centroid is not of interest in this case). Therefore, we may conclude the following for straight-line loads:

> • The magnitude of the resultant force is equal to the area under the load diagram.
> • The line of action of the resultant force passes through the centroid of the area under the load diagram.

c. Computation of resultants

Looking at Eqs. (3.16) through (3.22), we see that the computation of the resultant of distributed loading is essentially an integration problem. A discussion of the associated integration techniques is postponed until Chapter 8. However, if the load surface or the load diagram has a simple shape, then tables of centroids—such as Table 3.1—can be used to determine the resultant as illustrated in the following sample problems.

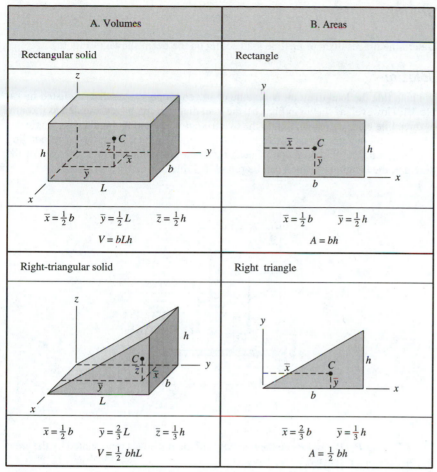

A. Volumes	B. Areas
Rectangular solid	Rectangle
$\bar{x} = \frac{1}{2}b$ $\bar{y} = \frac{1}{2}L$ $\bar{z} = \frac{1}{2}h$ $V = bLh$	$\bar{x} = \frac{1}{2}b$ $\bar{y} = \frac{1}{2}h$ $A = bh$
Right-triangular solid	Right triangle
$\bar{x} = \frac{1}{2}b$ $\bar{y} = \frac{2}{3}L$ $\bar{z} = \frac{1}{3}h$ $V = \frac{1}{2}bhL$	$\bar{x} = \frac{2}{3}b$ $\bar{y} = \frac{1}{3}h$ $A = \frac{1}{2}bh$

Table 3.1 *Centroids of Some Common Geometric Shapes (More tables are found in Chapter 8.)*

Sample Problem 3.11

Determine the resultant of the line load acting on the beam shown in Fig. (a).

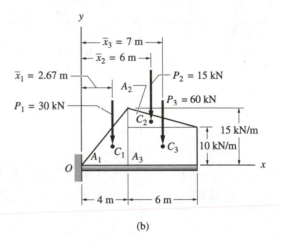

Solution

We note that the load diagram is not one of the common shapes that are listed in Table 3.1. However, as shown in Fig. (b), the load diagram can be represented as the sum of three line loads corresponding to the two triangles, A_1 and A_2, and the rectangle A_3. The resultant of each of these three line loads is equal to the area of the corresponding load diagram. The line of action of each resultant passes through the centroid of the diagram, the location of which can be found in Table 3.1.

(b)

Letting P_1, P_2, and P_3 be the resultants of the line loads represented by the areas A_1, A_2, and A_3, respectively, we have

$$P_1 = \frac{1}{2}(4 \times 15) = 30 \text{ kN}$$

$$P_2 = \frac{1}{2}(6 \times 5) = 15 \text{ kN}$$

$$P_3 = 6 \times 10 = 60 \text{ kN}$$

The line of action of each of these forces passes through the centroid of the corresponding load diagram, labeled C_1, C_2, and C_3 in Fig. (b). The x-coordinates of the centroids are obtained using Table 3.1:

$$\bar{x}_1 = \frac{2}{3}(4) = 2.67 \text{ m}$$

$$\bar{x}_2 = 4 + \frac{1}{3}(6) = 6 \text{ m}$$

$$\bar{x}_3 = 4 + \frac{1}{2}(6) = 7 \text{ m}$$

It follows that the magnitude of the resultant of the line load in Fig. (a) is given by

$$+\downarrow \quad R = P_1 + P_2 + P_3 = 30 + 15 + 60 = 105 \text{ kN} \qquad \textit{Answer}$$

To determine $\bar{x}$, the horizontal distance from point O to the line of action of R, we use the moment equation:

$$\Sigma M_O = R\bar{x} \quad \overset{+}{\curvearrowright} \quad 30(2.67) + 15(6) + 60(7) = 105\bar{x}$$

which gives

$$\bar{x} = 5.62 \text{ m} \qquad \textit{Answer}$$

The resultant is shown in Fig. (c).

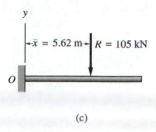

(c)

Sample Problem 3.12

After a severe rainstorm, the flat roof of the building shown in Fig. (a) is covered by 2.5 in. of rainwater. The specific weight of water is 62.4 lb/ft^3, so water at a depth of 2.5 in. causes a uniform pressure of $62.4 \times (2.5/12) = 13$ lb/ft^2. Determine the resultant force that the water exerts on the roof.

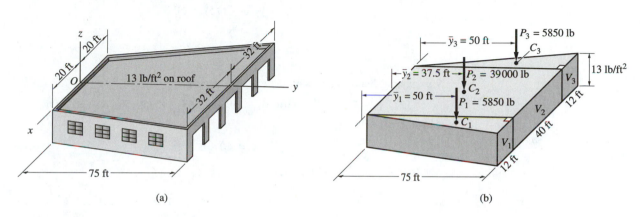

(a)   (b)

Solution

We begin by representing the volume under the load surface in Fig. (a) by the three solid shapes shown in Fig. (b): two right-triangular solids of volume V_1 and V_3 and a rectangular solid of volume V_2. The resultant force corresponding to each of these shapes is equal to the volume of the shape. Letting P_1, P_2, and P_3 be the resultants, we therefore have

$$P_1 = V_1 = 13\left[\frac{1}{2}(12)(75)\right] = 5850 \text{ lb}$$

$$P_2 = V_2 = 13[(40)(75)] = 39\,000 \text{ lb}$$

$$P_3 = V_3 = P_1 = 5850 \text{ lb}$$

The lines of action of these forces pass through the centroids of the corresponding volumes. The points where these forces intersect the roof of the building are labeled C_1, C_2, and C_3 in Fig. (b).

The magnitude of the resultant force is given by

$$+\downarrow \quad R = P_1 + P_2 + P_3$$

$$= 5850 + 39\,000 + 5850 = 50\,700 \text{ lb} \qquad \textit{Answer}$$

Because the load area (the roof of the building) is symmetrical about the y-axis and the pressure is uniform, the resultant will lie along the y-axis. Therefore, we need only calculate the distance $\bar{y}$ shown in Fig. (c).

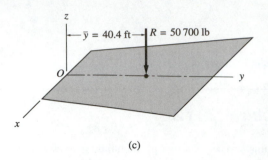

(c)

Using Table 3.1, the coordinates of C_1, C_2, and C_3 are

$$\bar{y}_1 = \bar{y}_3 = \frac{2}{3}(75) = 50 \text{ ft}$$

$$\bar{y}_2 = \frac{1}{2}(75) = 37.5 \text{ ft}$$

We can now determine $\bar{y}$ using the moment equation

$$\Sigma M_x = -R\bar{y} \qquad -5850(50) - 39\,000(37.5) - 5850(50) = -50\,700\,\bar{y}$$

which yields

$$\bar{y} = 40.4 \text{ ft} \qquad\qquad \textit{Answer}$$

The resultant is shown in Fig. (c).

Problems

3.46 During a storm, wind exerts a pressure of 2.3 lb/ft^2, normal to the surface of the stop sign. Determine the resultant force due to the wind.

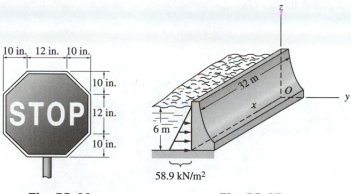

Fig. P3.46   **Fig. P3.47**

3.47 Water pressure acting on the vertical wall of the concrete dam varies linearly with the depth of the water as shown. Determine the resultant force caused by the water.

3.48 Determine the resultant of the line load acting on the bar ABC.

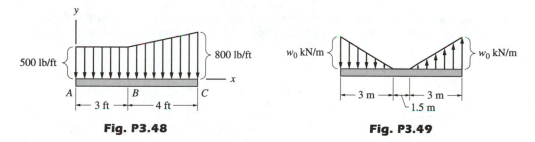

Fig. P3.48   **Fig. P3.49**

3.49 The resultant of the two triangular line loads is a 60-kN · m counterclockwise couple. Determine the load intensity w_0.

3.50 Determine the resultant of the line loads acting on the frame, and the x-coordinate of the point where the resultant intersects the x-axis.

3.51 Find the resultant of the distributed load acting on the flat plate.

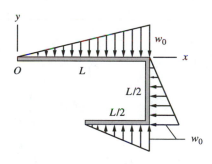

Fig. P3.50

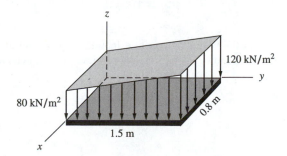

Fig. P3.51

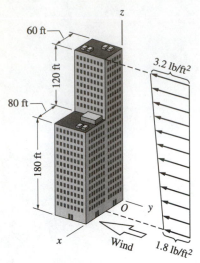

Fig. P3.52

3.52 At a certain time during a hurricane, the wind pressure acting on the wall of a high-rise building varies linearly as shown. Determine the resultant force caused by the wind.

3.53 A movable roof shield used in underground mining is subjected to the line loads shown, which are caused by the pressure of the overburden. (The *overburden* is the earth above the level being mined.) Determine the resultant of the overburden pressure.

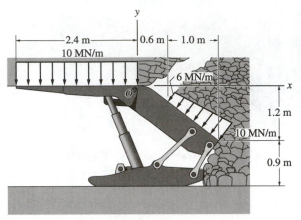

Fig. P3.53

3.54 The water pressure acting on a masonry dam varies as shown. If the dam is 20 ft wide, determine the resultant force of the water pressure acting on the dam.

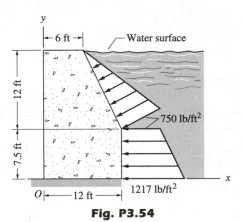

Fig. P3.54

3.55 The reaction at the base of a concrete footing may be approximated by the piecewise linear line load shown. Determine the resultant force and its line of action.

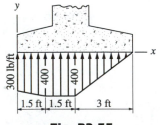

Fig. P3.55

Review Problems

3.56 The force $R = 400$ lb is the resultant of the forces T_1, T_2, and T_3. Knowing that $T_3 = 95$ lb, calculate T_1 and T_2.

3.57 The resultant of the force system shown is a 50-lb · ft counterclockwise couple. Find P, Q, and C.

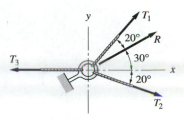

Fig. P3.56

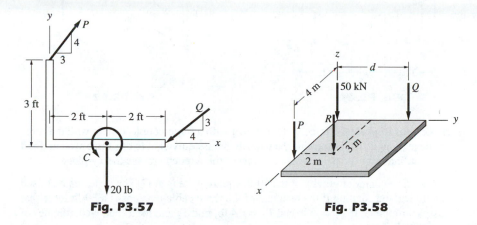

Fig. P3.57   **Fig. P3.58**

3.58 The force $R = 600$ kN is the resultant of the forces P, Q, and 50 kN. Calculate P, Q, and the distance d that specifies the line of action of Q.

3.59 The five forces act at end A of the boom. Determine T_1, T_2, and T_3 if the resultant of this force system is zero.

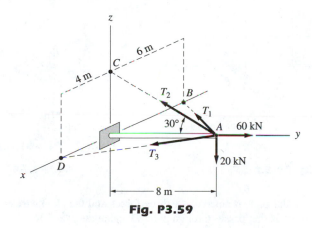

Fig. P3.59

3.60 A portion of the square plate is loaded by the uniformly distributed load $p = 20$ lb/ft^2. Find the coordinates of the point in the xy-plane through which the resultant passes.

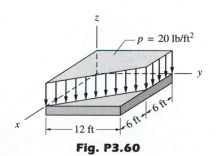

Fig. P3.60

3.61   Two distributed loads act on the beam *AB*. If the resultant of these loads is a 1200 lb · ft counterclockwise couple, determine the load intensity *w* at *B*, and the distance *a*.

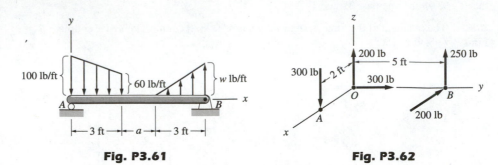

Fig. P3.61   **Fig. P3.62**

3.62   (a) Replace the force system shown with a force-couple system with the force acting at point *O*. (b) Determine the wrench that is equivalent to this force system. Find the coordinates of the point where the axis of the wrench crosses the *xy*-plane.

3.63   The center of gravity of the 30-lb square plate is at *G*. The plate can be raised slowly without rotating if the resultant of the three cable tensions is a 30-lb force that passes through *G*. If $T_1 = 6$ lb and $T_2 = 14$ lb, find T_3 and the *x*- and *y*-coordinates of its point of attachment.

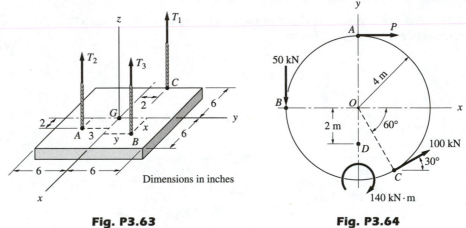

Dimensions in inches

Fig. P3.63   **Fig. P3.64**

3.64   The circular plate is acted on by three forces and the clockwise couple. If the resultant is a force **R** that passes through point *D*, calculate *P* and **R**.

3.65   Find the *x*- and *y*-coordinates of the point where the resultant of the three forces crosses the plate.

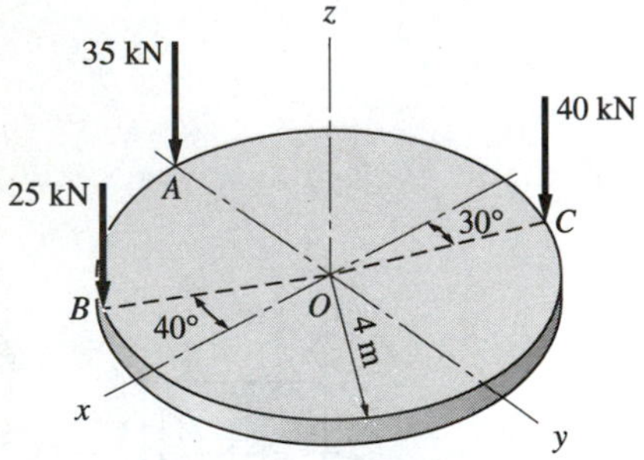

Fig. P3.65

3.66   The force system consists of the forces $\mathbf{F}_1$ and $\mathbf{F}_2$ and the couple-vector $\mathbf{C}$. (Note that $\mathbf{F}_1$ and $\mathbf{C}$ are oppositely directed along the line AB). Determine the equivalent force-couple system that consists of the force $\mathbf{R}$, acting at point E, and the couple $\mathbf{C}^R$. Use $F_1 = 260$ N, $F_2 = 300$ N, and $C = 780$ N $\cdot$ m.

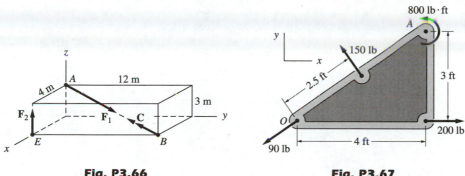

Fig. P3.66

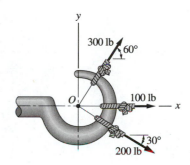

Fig. P3.67

3.67   Replace the coplanar force system that acts on the casting with an equivalent force-couple system, with the force acting at (a) point O; and (b) point A.

3.68   Determine the magnitude of the resultant of the three concurrent forces acting on the hook.

3.69   Determine the wrench that is equivalent to the force-couple system shown, and find the coordinates of the point where the axis of the wrench crosses the xz-plane.

Fig. P3.68

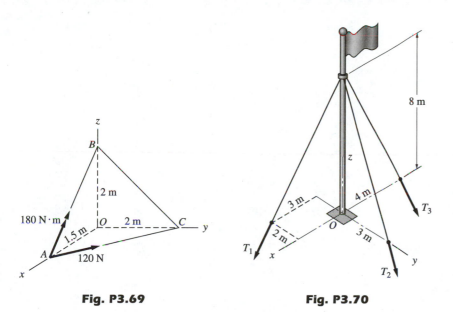

Fig. P3.69

Fig. P3.70

3.70   The resultant of the three cable tensions acting on the flagpole is the force $\mathbf{R} = R\mathbf{k}$. Find T_1, T_2, and R, given that $T_3 = 500$ N.

4

Coplanar Equilibrium Analysis

4.1 Introduction

The first three chapters of this text were devoted to mastering the elements of vector algebra, with emphasis on forces and couples. Proficiency in vector algebra is a prerequisite to the study of statics and most other areas of engineering mechanics.

With this chapter we begin the application of vector methods to the equilibrium analysis of engineering problems. We introduce the free-body diagram, which is perhaps the most important physical concept found in this text. We show how the free-body diagram is used to obtain the equations that relate the forces acting on a body in equilibrium.

For the present, we restrict our attention to the analysis of bodies that are held in equilibrium by coplanar force systems. The material of this chapter is divided into three parts: analysis of single bodies, analysis of composite bodies (called *frames and machines* in some texts), and analysis of plane trusses.

4.2 Definition of Equilibrium

A body is said to be in equilibrium if the resultant of the force system that acts on the body vanishes. Equilibrium means that both the resultant force and the resultant couple are zero.

When a force system acts on a body that is initially at rest, the absence of a resultant means that there is no tendency of the body to move. The analysis of problems of this type is the focus of statics; dynamics is concerned with the response of bodies to force systems that are not in equilibrium.

We showed in Chapter 3 that a coplanar force system can always be represented as a resultant force $\mathbf{R}$, passing through an arbitrarily chosen point O, and a couple C^R that lies in the plane of the forces. Assuming that the forces lie in the xy-plane, $\mathbf{R}$ and C^R can be determined from $R_x = \Sigma F_x$, $R_y = \Sigma F_y$, $C^R = \Sigma M_O$. Therefore, the equations of equilibrium are

$$\Sigma F_x = 0 \qquad \Sigma F_y = 0 \qquad \Sigma M_O = 0 \qquad (4.1)$$

The summations in Eq. (4.1) must, of course, include *all* the forces that act on the body—both the applied forces and the reactions (the forces provided by supports).

PART A: *Analysis of Single Bodies*

4.3   *Free-Body Diagram of a Body*

The first step in equilibrium analysis is to identify all the forces that act on the body. This is accomplished by means of a *free-body diagram.*

> The *free-body diagram* (FBD) of a body is a sketch of the body showing all forces that act on it. The term *free* implies that all supports have been removed and replaced by the forces (reactions) that they exert on the body.

The importance of mastering the FBD technique cannot be overemphasized. Free-body diagrams are fundamental to all engineering disciplines that are concerned with the effects that forces have on bodies. The construction of an FBD is the key step that translates a physical problem into a form that can be analyzed mathematically.

Forces that act on a body can be divided into two general categories—*reactive forces* (or, simply, *reactions*) and *applied forces.* Reactions are those forces that are exerted on a body by the supports to which it is attached. Forces acting on a body that are not provided by the supports are called *applied forces.* Of course, *all* forces, both reactive and applied, must be shown on free-body diagrams.

The following is the general procedure for constructing a free-body diagram.

1. A sketch of the body is drawn assuming that all supports (surfaces of contact, supporting cables, etc.) have been removed.
2. All applied forces are drawn and labeled on the sketch. The weight of the body is considered to be an applied force acting at the center of gravity. As shown in Ch. 8, the *center of gravity* of a homogeneous body coincides with the *centroid* of its volume.
3. The reactions due to each support are drawn and labeled on the sketch. (If the sense of a reaction is unknown, it should be assumed. The solution will determine the correct sense: A positive result indicates that the assumed sense is correct, whereas a negative result means that the correct sense is opposite to the assumed sense.)
4. All relevant angles and dimensions are shown on the sketch.

When you have completed this procedure, you will have a drawing (i.e., a free-body diagram) that contains all of the information necessary for writing the equilibrium equations of the body.

The most difficult step to master in the construction of FBDs is the determination of the support reactions. Table 4.1 shows the reactions exerted by various coplanar supports; it also lists the number of unknowns that are introduced on an FBD by the removal of each support. To be successful at drawing FBDs, you must be completely familiar with the contents of Table 4.1. It is also helpful to understand the physical reasoning that determines the reactions at each support, which are described below.

(a) *Flexible Cable (Negligible Weight).*   A flexible cable exerts a pull, or tensile force, in the direction of the cable. (With the weight of the cable

Support	Reaction(s)	Description of reaction(s)	Number of unknowns
(a) Flexible cable of negligible weight	T θ	Tension of unknown magnitude T in the direction of the cord or cable	One
(b) Frictionless surface (single point of contact)	N θ	Force of unknown magnitude N directed normal to the surface	One
(c) Roller support	N θ	Force of unknown magnitude N normal to the surface supporting the roller	One
(d) Surface with friction (single point of contact)	F N θ	Force of unknown magnitude N normal to the surface and a friction force of unknown magnitude F parallel to the surface	Two
(e) Pin support	R_x R_y	Unknown force $\mathbf{R}$	Two
(f) Built-in (cantilever) support	C R_x R_y	Unknown force $\mathbf{R}$ and a couple of unknown magnitude C	Three

Table 4.1   *Reactions of Coplanar Supports*

(a)

(b)

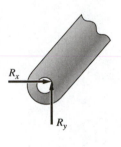

(c)

Fig. 4.1

neglected, the cable forms a straight line.) If its direction is known, removal of the cable introduces one unknown in a free-body diagram—the magnitude of the force exerted by the cable.

(b) *Frictionless Surface: Single Point of Contact.* When a body is in contact with a frictionless surface at only one point, the reaction is a force that is perpendicular to the surface, acting at the point of contact. This reaction is often referred to simply as the *normal force.* (Walking on an icy sidewalk is treacherous because it is difficult to generate a force in any direction except perpendicular to the sidewalk.) Therefore, removing such a surface introduces one unknown in a free-body diagram—the magnitude of the normal force. If contact between the body and the surface occurs across a finite area, rather than at one point, the line of action of the resultant normal force will also be unknown.

(c) *Roller Support.* A roller support is equivalent to a frictionless surface: It can only exert a force that is perpendicular to the supporting surface. The magnitude of the force is thus the only unknown introduced in a free-body diagram when the support is removed.

(d) *Surface with Friction: Single Point of Contact.* A friction surface can exert a force that acts at an angle to the surface. The unknowns may be taken to be the magnitude and direction of the force; however, it is usually advantageous to represent the unknowns as N and F, the components that are perpendicular and parallel to the surface, respectively. The component N is called the *normal force,* and F is known as the *friction force.* If there is an area of contact, the line of action of N will also be unknown.

(e) *Pin Support.* A pin is a cylinder that is slightly smaller than the hole into which it is inserted, as shown in Fig. 4.1. Neglecting friction, the pin can only exert a force that is normal to the contact surface, shown as **R** in Fig. 4.1(b). A pin support thus introduces two unknowns: the magnitude of **R** and the angle α that specifies the direction of **R** (α is unknown because the point where the pin contacts the surface of the hole is not known). More commonly, the two unknowns are chosen to be perpendicular components of **R**, such as R_x and R_y shown in Fig. 4.1(c).

(f) *Built-in (Cantilever) Support.* A built-in support, also known as a *cantilever support,* prevents all motion of the body at the support. Translation (horizontal or vertical movement) is prevented by a force, and a couple prohibits rotation. Therefore, a built-in support introduces three unknowns in a free-body diagram: the magnitude and direction of the reactive force **R** (these unknowns are commonly chosen to be two components of **R**, such as R_x and R_y) and the magnitude C of the reactive couple.

You should keep the following points in mind when you are drawing free-body diagrams.

1. Be neat. Because the equilibrium equations will be derived directly from the free-body diagram, it is essential that the diagram be readable.

2. Clearly label all forces, angles, and distances with values (if known) or symbols (if the values are not known).

3. The support reactions must be consistent with the information presented in Table 4.1.

4. Show only forces that are external to the body (this includes support reactions). Internal forces occur in equal and opposite pairs and thus will not appear on free-body diagrams.

Sample Problem 4.1

The homogeneous 6-m bar AB in Fig. (a) is supported in the vertical plane by rollers at A and B and by a cable at C. The mass of the bar is 50 kg. Draw the FBD of bar AB. Determine the number of unknowns on the FBD.

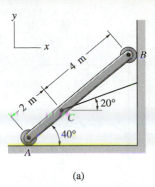

(a)

Solution

The FBD of bar AB is shown in Fig. (b). The first step in the construction of this diagram is to sketch the bar, assuming the supports have been removed. Then the following forces are added to the sketch.

W: The Weight of the Bar

The weight W is shown as a vertical force acting at G, the center of gravity of the bar. Because the bar is homogeneous, G is located at the center of the bar. The magnitude of the weight is $W = mg = (50)(9.81) = 491$ N.

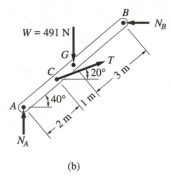

(b)

N_A: The Normal Reaction at A

Removal of the roller support at A dictates that we show the force that this support can exert on the bar. From Table 4.1, we note that a roller support can exert a single force that is normal to the supporting surface. Therefore, on the FBD we show the reaction at A as a vertical force and label its magnitude as N_A.

N_B: The Normal Reaction at B

Following an argument similar to that for N_A, we conclude that the removal of the roller support at B means that we must show a horizontal force at that point. On the FBD, we label this reaction as N_B.

T: The Tension in the Cable at C

From Table 4.1, the force exerted by a cable is a tensile force acting in the direction of the cable. Therefore, the force exerted on the bar by the cable is shown as a force of magnitude T, acting at 20° to the horizontal.

We note that there are three unknowns on the FBD: the magnitudes of the three reactions (N_A, N_B, and T).

Sample Problem 4.2

The homogeneous, 250-kg triangular plate in Fig. (a) is supported by a pin at A and a roller at C. Draw the FBD of the plate and determine the number of unknowns.

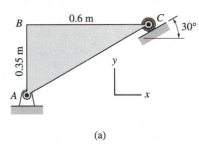

(a)

Solution

The FBD of the plate is shown in Fig. (b). The pin and roller supports have been removed and replaced by the reactive forces. The forces acting on the plate are described below.

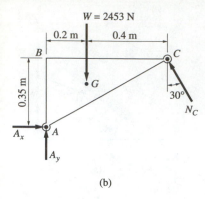

(b)

W: The Weight of the Plate

The weight of the plate is $W = mg = (250)(9.81) = 2453$ N. It acts at the centroid G of the triangle ABC, the location of which was determined from Table 3.1. Only the horizontal location of G is shown in the figure, because it is sufficient to determine the line of action of W.

A_x and A_y: The Components of the Pin Reaction at A

From Table 4.1, we see that a pin reaction can be shown as two components A_x and A_y, which are equivalent to an unknown force acting at an unknown angle. We have shown A_x acting to the right and A_y acting upward. These directions are chosen arbitrarily; the solution of the equilibrium equations will determine the correct sense for each force. Therefore, the free-body diagram would be correct even if A_x or A_y were chosen to act in directions opposite to those shown in Fig. (b).

N_C: The Normal Reaction at C

From Table 4.1, the force exerted by a roller support is normal to the inclined surface. Therefore, on the FBD we show the force N_C at C, inclined at 30° to the vertical.
The FBD contains three unknowns: A_x, A_y, and N_C.

Sample Problem 4.3

A rigid frame is fabricated by joining the three bars with pins at B, C, and D, as shown in Fig. (a). The frame is loaded by the 1000-lb force and the 1200-lb · ft couple; the supports consist of a pin at A and a roller support at E. Draw the FBD of the frame, neglecting the weights of the members. How many unknowns are there?

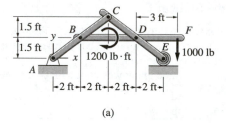

(a)

Solution

The FBD of the frame is shown in Fig. (b). In addition to the applied force and couple, the diagram shows the pin reaction at A (A_x and A_y) and the normal force at roller E (N_E).

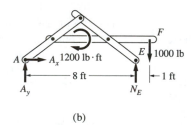

(b)

It is important to realize that the forces at pins B, C, and D do not appear on the FBD of the frame. These pin forces, as well as the forces inside the bars themselves, are internal to the frame (recall that only external forces are shown on FBDs).

We note that there are three unknowns on the FBD: A_x, A_y, and N_E.

Sample Problem 4.4

The beam ABC, built into the wall at A and supported by a cable at C, carries a distributed load over part of its length, as shown in Fig. (a). The weight of the beam is 70 lb/ft. Draw the FBD of the beam.

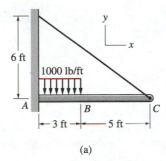

(a)

Solution

The FBD is shown in Fig. (b). Because a built-in, or cantilever, support can exert a force and a couple, the reactions at the wall are shown as the force components A_x and A_y and the couple C_A. The tension in the cable is labeled T. Also shown on the FBD are the weight of the beam (70 lb/ft $\times$ 8 ft = 560 lb) and the resultant of the distributed load (3000 lb, acting at the centroid of the loading diagram).

Observe that the FBD contains four unknowns whereas the number of equilibrium equations in Eq. (4.1) is three. Therefore, it would not be possible to calculate all of the unknowns using only equilibrium analysis. The reason for the indeterminacy is that the beam is oversupported; it would be in equilibrium even if the cable at C were removed or if the built-in support were replaced by a pin connection.

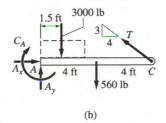

(b)

135

Problems

4.1–4.3 Each of the bodies shown is homogeneous and has a mass of 30 kg. Assume friction at all contact surfaces. Draw the fully dimensioned FBD for each body and determine the number of unknowns.

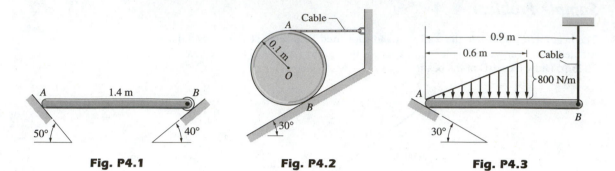

Fig. P4.1   Fig. P4.2   Fig. P4.3

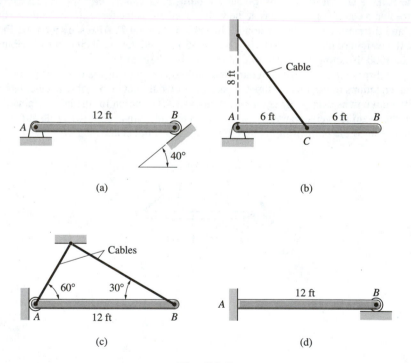

Fig. P4.4

4.4 The homogeneous triangular plate weighing 90 lb is supported by a friction surface at *A* and a roller at *B*. Draw the FBD of the plate, and determine the number of unknowns. (*Note:* The weight acts at the centroid of the plate.)

4.5 The homogeneous beam *AB* weighs 400 lb. For each support condition shown in Figs. (a) through (d), draw the FBD of the beam, and determine the number of unknowns.

Fig. P4.5

4.6 The weight W and radius R for each of the homogeneous cylinders in Figs. (a) through (d) are assumed known. Draw the FBD of each cylinder, and determine the number of unknowns.

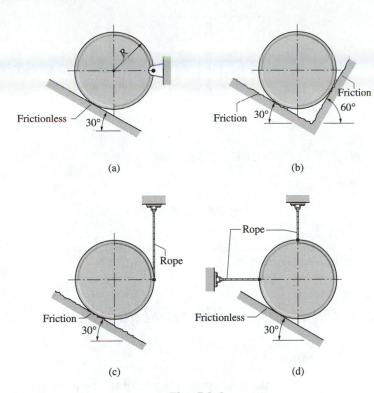

(a)

(b)

(c)

(d)

Fig. P4.6

4.7 The bracket of negligible weight is supported by a pin at A and by a frictionless peg at B, which can slide in the slot in the bracket. Draw the FBD of the bracket if (a) $\theta = 45°$; and (b) $\theta = 90°$. What are the unknowns?

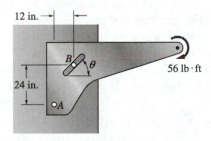

Fig. P4.7

4.8 To open the high-pressure water cock, a 12-lb horizontal force must be applied to the handle at A. Draw the FBD of the handle, neglecting its weight. Count the unknowns.

4.9 The high-pressure water cock is rigidly attached to the support at D. Neglecting the weights of the members, draw the FBD of the entire assembly, and count the unknowns.

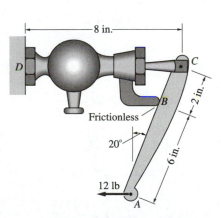

Fig. P4.8, P4.9

4.10 Draw the FBD of the entire frame, assuming that friction and the weights of the members are negligible. How many unknowns appear on this FBD?

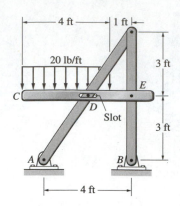

Fig. P4.10, P4.11

4.11 Draw an FBD of member *CE* of the frame described in the previous problem. How many unknowns appear on this FBD?

4.4 *Coplanar Equilibrium Equations*

a. *General case*

As explained in Art. 4.2, the following three equilibrium equations are available for a general coplanar force system.

$$\Sigma F_x = 0 \qquad \Sigma F_y = 0 \qquad \Sigma M_O = 0 \qquad \text{(4.1, repeated)}$$

It is worthwhile noting that point *O*, the moment center, and the orientation of the *x*- and *y*-axes can be chosen arbitrarily. One usually makes the choices that simplify the resulting algebra.

When analyzing equilibrium problems, it is often convenient to use a set of three independent equations different from the two force equations and one moment equation given in Eq. (4.1). However, before discussing other sets of independent equations, let us take a look at two rather self-evident properties of the resultant.

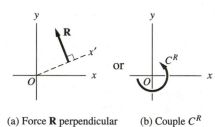

(a) Force **R** perpendicular to *x*′-direction   (b) Couple C^R

Fig. 4.2

Property 1: If $\Sigma F_{x'} = 0$, where *x*′ is a direction in the *xy*-plane, the resultant is either a force **R** perpendicular to the *x*′-direction or a couple C^R, as shown in Fig. 4.2(a) and (b).

Property 2: If $\Sigma M_A = 0$, where *A* is a point in the *xy*-plane, the resultant is a force **R** that passes through point *A*, as indicated in Fig. 4.3.

To explain Property 2, we note that the resultant cannot be a couple if $\Sigma M_A = 0$, because the moment of a couple is the same about every point. It follows that the resultant can only be a force that has zero moment about point *A*—that is, a force **R** that passes through *A*.

Force **R** passing through point *A*

Fig. 4.3

From Properties 1 and 2, we can derive the following three sets of independent equilibrium equations for a coplanar force system, each set containing three equations.

1. Two force equations and one moment equation.

$$\Sigma F_{x'} = 0 \qquad \Sigma F_{y'} = 0 \qquad \Sigma M_A = 0 \qquad (4.2)$$

where x' and y' are any two nonparallel directions in the xy-plane and A is any point in the xy-plane.

The moment equation $\Sigma M_A = 0$ ensures that the resultant cannot be a couple (Property 2); that is, $C^R = 0$. We use Property 1 to show that the resultant force **R** also vanishes. First, note that the equation $\Sigma F_{x'} = 0$ requires **R** to be perpendicular to the x'-direction. Similarly, the equation $\Sigma F_{y'} = 0$ requires **R** to be perpendicular to the y'-direction. Because **R** cannot be perpendicular to two different directions, we conclude that $\mathbf{R} = \mathbf{0}$.

2. Two moment equations and one force equation.

$$\Sigma M_A = 0 \qquad \Sigma M_B = 0 \qquad \Sigma F_{x'} = 0 \qquad (4.3)$$

where A and B are points in the xy-plane and x' is any direction in the xy-plane that is not perpendicular to line AB.

According to Property 2, the two moment equations can be satisfied only if the resultant is a force that passes through both A and B; that is, **R** and the line AB must be collinear. On the other hand, from Property 1 we know that the equation $\Sigma F_{x'} = 0$ requires **R** to be perpendicular to the x'-axis. Both of these conditions are shown in Fig. 4.4. If the x'-axis and the line AB are not perpendicular to each other, we find that **R** must act in two different directions, which gives $\mathbf{R} = \mathbf{0}$.

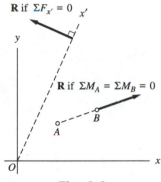

Fig. 4.4

3. Three moment equations.

$$\Sigma M_A = 0 \qquad \Sigma M_B = 0 \qquad \Sigma M_C = 0 \qquad (4.4)$$

where A, B, and C are any three noncollinear points in the xy-plane.

Using Property 2, we see that any one of the moment equations determines that the resultant is not a couple. Property 2 also states that if the resultant is a force **R**, it must pass through points A, B, and C. If A, B, and C do not lie on the same straight line, this is clearly impossible, which proves that $\mathbf{R} = \mathbf{0}$.

b. *Concurrent force system*

As shown before, the resultant of a concurrent force system is a force **R** that passes through the point of concurrency, which we take to be point O. The moment equation $\Sigma M_O = 0$ is trivially satisfied, so that the number of independent equilibrium equations is reduced from three to two. Using Properties 1 and 2, it is straightforward to verify that the following are the three sets of independent equilibrium equations, each set consisting of two equations.

1. Two force equations.

$$\Sigma F_{x'} = 0 \qquad \Sigma F_{y'} = 0 \qquad (4.5)$$

where x' and y' are any two nonparallel directions in the xy-plane.

2. Two moment equations.

$$\Sigma M_A = 0 \qquad \Sigma M_B = 0 \qquad (4.6)$$

where A and B are any two points in the xy-plane, except point O and those points for which A, B, and O lie on a straight line.

3. One force equation and one moment equation.

$$\Sigma F_{x'} = 0 \qquad \Sigma M_A = 0 \qquad (4.7)$$

where A is any point in the xy-plane except point O, and x' is any direction that is not perpendicular to the line OA.

c. Parallel force system

Assume that all the forces lying in the xy-plane are parallel to the y-axis. The equation $\Sigma F_x = 0$ is automatically satisfied, and the number of independent equilibrium equations is again reduced from three to two. Using Properties 1 and 2, it can be shown that the following are the two sets of independent equilibrium equations, each containing two equations.

1. One force equation and one moment equation.

$$\Sigma F_{y'} = 0 \qquad \Sigma M_A = 0 \qquad (4.8)$$

where y' is any direction in the xy-plane except the x-direction, and A is any point in the xy-plane.

2. Two moment equations.

$$\Sigma M_A = 0 \qquad \Sigma M_B = 0 \qquad (4.9)$$

where A and B are any two points in the xy-plane, provided that the line AB is not parallel to the y-axis.

4.5 Writing and Solving Equilibrium Equations

The three steps in the equilibrium analysis of a body are the following:

Step 1. Draw a free-body diagram (FBD) of the body that shows all of the forces and couples that act on the body.

Step 2. Write the equilibrium equations in terms of the forces and couples that appear on the free-body diagram.

Step 3. Solve the equilibrium equations for the unknowns.

In this article, we assume that the correct free-body diagram has already been drawn, so that we can concentrate on Steps 2 and 3—writing and solving the equilibrium equations.

The force system that holds a body in equilibrium is said to be *statically determinate* if the number of independent equilibrium equations equals the number of unknowns that appear on its free-body diagram. Statically determinate problems can therefore be solved by equilibrium analysis alone. If the number of unknowns exceeds the number of independent equilibrium equations, the problem is called *statically indeterminate*. The solution of statically indeterminate problems requires the use of additional principles that are beyond the scope of this text.

When analyzing a force system that holds a body in equilibrium, you should first determine the number of independent equilibrium equations and count the number of unknowns. If the force system is statically determinate, these two numbers will be equal. It is then best to outline a *method of analysis,* or plan of attack, which specifies the sequence in which the equations are to be written and lists the unknowns that will appear in the equations. Once you have determined a viable method of analysis, you can then proceed to the *mathematical details* of the solution.

One word of caution—the set of equilibrium equations used in the analysis *must* be independent. An attempt to solve a nonindependent set of equations will, at some stage, yield a useless identity, such as $0 = 0$.

By now you should realize that, although the solution of a statically determinate problem is unique, the set of equations used to determine that solution is not unique. For example, there is an infinite number of choices for point O in the equilibrium equation $\Sigma M_O = 0$.

With an infinite number of equilibrium equations from which to choose, how are you to decide which equations to use for a given problem? The answer is to base your choice on mathematical convenience. If you intend to solve the equations by hand, try to select equations that involve as few unknowns as possible, thus simplifying the algebraic manipulations required. However, if you have access to a computer or a programmable calculator with equation-solving software, the solution of simultaneous equations is not burdensome and the choice of equations is therefore not critical. However, it cannot be overemphasized that the set of chosen equations must be independent.

Sample Problem 4.5

The homogeneous, 120-kg wooden beam is suspended from ropes at A and B. A power wrench applies the 500-N $\cdot$ m clockwise couple to tighten a bolt at C. Use the given FBD to determine the tensions in the ropes.

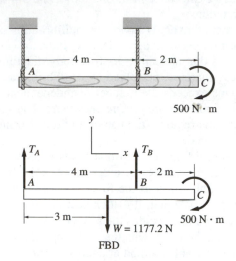

FBD

Solution

Method of Analysis

The FBD of the beam contains the weight $W = mg = (120)(9.81) = 1177.2$ N acting at the center of the beam, the couple applied by the wrench, and the unknown tensions T_A and T_B in the ropes. Because all the forces lie in the xy-plane and are parallel to the y-axis, there are two independent equilibrium equations. There are also two unknowns: T_A and T_B. Therefore, the problem is statically determinate.

It is convenient to start the analysis with the equilibrium equation $\Sigma M_A = 0$. Because this equation does not contain T_A, we can immediately solve it for T_B. We can then use the equation $\Sigma F_y = 0$ to calculate T_A.

Mathematical Details

$$\Sigma M_A = 0 \qquad \overset{+}{\curvearrowright} \qquad 4T_B - 1177.2(3) - 500 = 0$$

$$T_B = 1007.9 \text{ N} \qquad \qquad \textit{Answer}$$

$$\Sigma F_y = 0 \qquad +\uparrow \quad T_A + T_B - 1177.2 = 0$$

Substituting $T_B = 1007.9$ N and solving for T_A, we get

$$T_A = 169.3 \text{ N} \qquad \qquad \textit{Answer}$$

Other Methods of Analysis

Another, equally convenient option is to use the moment equations $\Sigma M_A = 0$, as above, and

$$\Sigma M_B = 0 \qquad \overset{+}{\curvearrowright} \qquad -4T_A + 1177.2(1.0) - 500 = 0$$

$$T_A = 169.3 \text{ N}$$

Sample Problem 4.6

The 420-lb homogeneous log is supported by a rope at A and loose-fitting rollers at B and C as it is being fed into a sawmill. Calculate the tension in the rope and the reactions at the rollers, using the given FBD. Which rollers are in contact with the log?

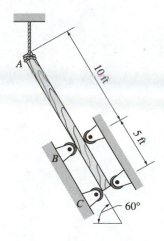

Solution

Method of Analysis

The FBD contains the weight W of the log and three unknown forces: the tension T in rope and the roller reactions N_B and N_C, which are perpendicular to the log. The sense of each roller reaction indicates that we have assumed the upper rollers to be in contact with the lumber.

The force system in the FBD is the general coplanar case, for which three independent equilibrium equations are available. Because there are also three unknowns, the problem is statically determinate.

The most inconvenient choice of equations would lead to the solution of three simultaneous equations, with all three unknowns appearing in each equation. With planning, it is often possible to reduce the number of unknowns that must be solved simultaneously. Referring to the FBD, we could start with $\Sigma F_x = 0$, which would contain only two unknowns: N_B and N_C. Then we would look for another equation that contains only these two unknowns. Inspection of the FBD reveals that the equation $\Sigma M_A = 0$ would not contain T, because this force passes through A. The equations ΣF_x and ΣM_A could thus be solved simultaneously for N_A and N_B. Finally, ΣF_y would be used to compute T.

Mathematical Details

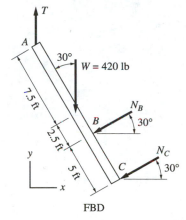

$$\Sigma F_x = 0 \qquad \overset{+}{\longrightarrow} \qquad -N_B \cos 30° - N_C \cos 30° = 0 \qquad \text{(a)}$$

$$\Sigma M_A = 0 \qquad \overset{+}{\curvearrowleft} \qquad 420\,(7.5 \sin 30°) + 10 N_B + 15 N_C = 0 \qquad \text{(b)}$$

The solution of Eqs. (a) and (b) is

$$N_B = 315.0 \text{ lb} \qquad \text{and} \qquad N_C = -315.0 \text{ lb} \qquad \textit{Answer}$$

The signs indicate that the sense of N_B is as shown on the FBD, whereas the sense of N_C is opposite to that shown. Therefore, the upper roller at B and the lower roller at C are in contact with the log.

$$\Sigma F_y = 0 \qquad +\!\uparrow \qquad T - 420 - N_B \sin 30° - N_C \sin 30° = 0$$

Because $N_B = -N_C$, this equation yields

$$T = 420 \text{ lb} \qquad \textit{Answer}$$

Other Methods of Analysis

The above solution used the equations $\Sigma F_x = 0$, $\Sigma F_y = 0$, and $\Sigma M_A = 0$. There are other sets of independent equilibrium equations that would serve equally well. For example, we could find T from just a single equation—summation of forces parallel to the log equals zero. Because N_B and N_C are perpendicular to the log, T would be the only unknown in this equation. The reaction N_C could also be computed independently from the other unknowns by setting the sum of the moments about the point where T and N_B intersect to zero. Similarly, we could find N_B from a single equation—summation of moments about the point where T and N_C intersect equals zero.

It is important to realize that the equilibrium equations must be independent. Referring to the FBD, you might be tempted to use the three moment equations $\Sigma M_A = 0$, $\Sigma M_B = 0$, and $\Sigma M_C = 0$. Although each is a valid equation, they are not independent of each other. Why not? What would happen if you tried to solve these equations for N_B, N_C, and T?

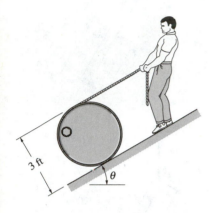

Sample Problem 4.7

A man keeps the homogeneous, 800-lb drum from rolling down the inclined surface by pulling on the rope wrapped around the drum. If the pull exerted by the man is 125 lb, determine the angle θ of the incline. (Assume that there is enough friction to prevent the drum and the man from slipping.)

Solution

Method of Analysis

The FBD of the drum contains the 125-lb pull on the rope, the 800-lb weight of the drum, and the unknown normal and friction forces at the contact point A (labeled N and F, respectively). The force system is the general, coplanar case, for which there are three independent equilibrium equations. Because the number of unknowns on the FBD is also three (N, F, and θ), the problem is statically determinate.

Because we are asked to find only the angle θ, we look for an equilibrium equation that does not contain N and F. Such an equation is $\Sigma M_A = 0$, which we can solve for θ. (If desired, the equations $\Sigma F_x = 0$ and $\Sigma F_y = 0$ could then be used to determine F and N.)

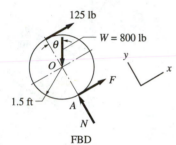

FBD

Mathematical Details

$$\Sigma M_A = 0 \qquad \overset{+}{\curvearrowleft} \qquad 800(1.5 \sin \theta) - 125(3) = 0$$

$$\sin \theta = \frac{125(3)}{800(1.5)} = 0.3125$$

$$\theta = 18.21° \qquad\qquad Answer$$

Other Methods of Analysis

Another approach is to start with the moment equation about O, the center of the drum:

$$\Sigma M_O = 0 \qquad \overset{+}{\curvearrowleft} \qquad F(1.5) - 125(1.5) = 0$$

$$F = 125 \text{ lb}$$

Then the following equation can be used to compute θ:

$$\Sigma F_x = 0 \qquad \overset{+}{\rightarrow} \qquad 125 + F - 800 \sin \theta = 0$$

After substituting for F, this equation yields $\theta = 18.21°$ as before.

Problems

In each of the following problems, the free-body diagram is given. Write the equilibrium equations, and compute the requested unknowns.

4.12   The cylinder of weight W rests in a frictionless right-angled corner. In the position shown, the relationship between the contact forces at A and B is $N_B = 1.5N_A$. Determine the angle θ and the contact forces N_A and N_B in terms of W.

4.13   Calculate the force P that is required to hold the 120-lb roller at rest on the rough incline.

4.14   Solve Prob. 4.13 if the force P pushes rather than pulls.

4.15   The homogeneous, 200-lb sign is suspended from three wires as shown. Determine the tension in each wire.

4.16   The table lamp consists of two uniform arms, each weighing 0.8 lb, and a 2-lb bulb fixture. If $\theta = 16°$, calculate the couple C_A that must be supplied by the friction in joint A.

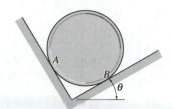

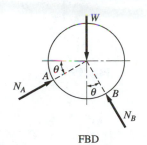

FBD

Fig. P4.12

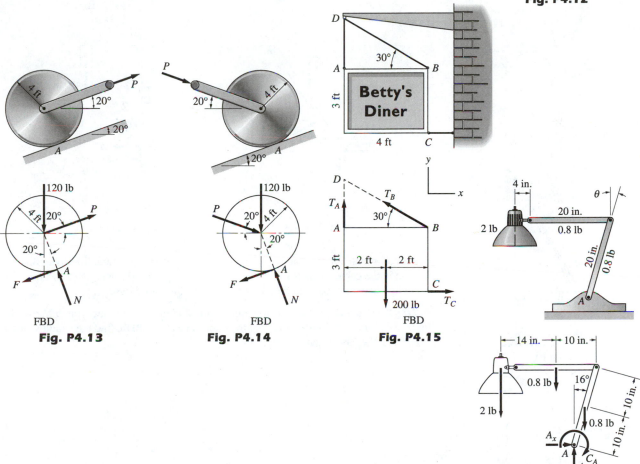

FBD

Fig. P4.13

FBD

Fig. P4.14

FBD

Fig. P4.15

FBD

Fig. P4.16

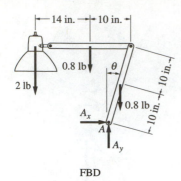

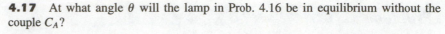

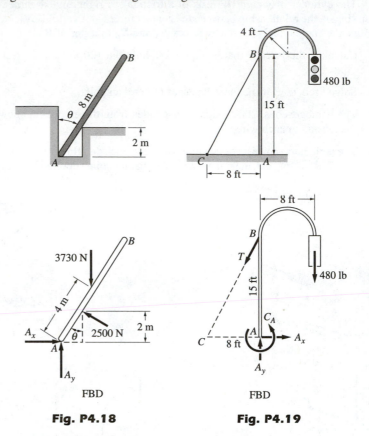

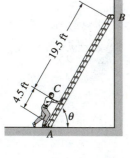

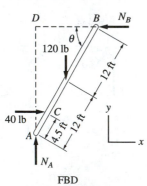

FBD

Fig. P4.17

4.17 At what angle θ will the lamp in Prob. 4.16 be in equilibrium without the couple C_A?

4.18 The telephone pole AB weighing 3730 N is resting in a hole that is 2 m deep. If the contact force between the pole and the edge of the hole is limited to 2500 N, what is the largest safe value of the angle θ? Neglect friction.

FBD

Fig. P4.18

FBD

Fig. P4.19

4.19 Compute all reactions at the base A of the traffic light standard, given that the tension in the cable BC is (a) $T = 544$ lb; and (b) $T = 0$. The weight of the standard is negligible compared with the 480-lb weight of the traffic light.

4.20 The man keeps the 120-lb ladder from sliding down by applying a horizontal force at C. If the maximum force that the man can exert is 40 lb, find the smallest angle θ at which the ladder can be held.

FBD

Fig. P4.20

4.21   The machine part of negligible weight is supported by a pin at *A* and a roller at *C*. Determine the magnitudes of the forces acting on the part at *A* and *C*.

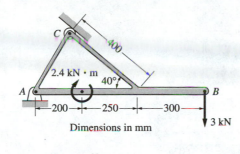

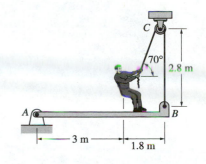

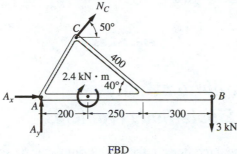

Fig. P4.21

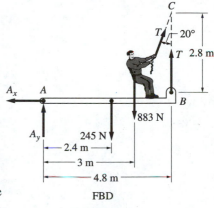

FBD

Fig. P4.22

4.22   A man supports himself and the uniform, horizontal beam by pulling on the rope with a force *T*. The weights of the man and the beam are 883 N and 245 N, respectively. Compute the tension *T* in the rope and the forces exerted by the pin at *A*.

4.23   The FBD of the man in Prob. 4.22 is shown in the figure. If the tension in the rope is 388 N, compute the distance *b* locating the man's feet. Also calculate N_D and F_D, the forces exerted on the man by the beam.

4.24   The uniform 60-lb bar *AB* is supported by a cable at *A* and a frictionless surface at *B*. Find the distance *x* where the 10-lb weight *C* must be positioned for the bar to be at rest in the horizontal position shown.

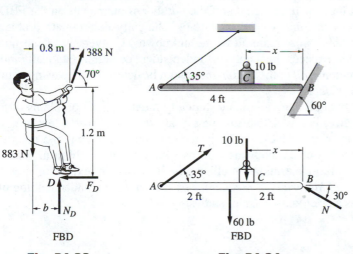

Fig. P4.23

Fig. P4.24

4.25 When the truck is empty, it weighs 6000 lb and its center of gravity is at *G*. Determine the total weight *W* of the logs, knowing that the load on the rear axle is twice the load on the front axle—that is, $N_A = 2N_B$.

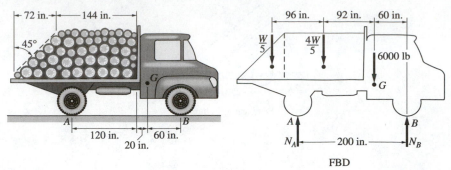

Fig. P4.25

4.6 *Equilibrium Analysis for Single-Body Problems*

We learned that the three steps in the equilibrium analysis of a body are:

1. Draw the free-body diagram (FBD).
2. Write the equilibrium equations.
3. Solve the equations for the unknowns.

The individual steps were introduced separately in the preceding articles. The purpose of this article is to give you experience in the entire process of equilibrium analysis.

Always begin by drawing the FBD; there are no exceptions. The FBD is the very key to equilibrium analysis, so it should be drawn with great care. We recommend that you use a straightedge and circle template. After the FBD has been drawn, the remainder of the solution, consisting of writing and solving equilibrium equations, should be relatively easy.

It must be reiterated that if the number of unknowns on the FBD equals the number of independent equations (statically determinate problem), you will be able to calculate all of the unknowns. Conversely, if the number of unknowns exceeds the number of independent equations (statically indeterminate problem), *all* of the unknowns cannot be determined by using equilibrium analysis alone.

Although there are many statically indeterminate problems of practical importance, you will find that nearly all problems in this text are statically determinate. To solve a statically indeterminate problem, one must consider deformations of the body, as well as equations of equilibrium. The solution of statically indeterminate problems is discussed in texts with such titles as *Strength of Materials* or *Mechanics of Materials,* the understanding of which requires a prior knowledge of statics.

Sample Problem 4.8

The telephone cable spool in Fig. (a) weighs 300 lb and is held at rest on a 40° incline by the horizontal cable. The cable is wound around the inner hub of the spool and attached to the support at B. Assume that G, the center of gravity of the spool, is located at the center of the spool. Find all forces acting on the spool.

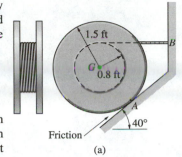

(a)

Solution

Method of Analysis

The first step is, of course, to draw the FBD of the spool, which is shown in Fig. (b). In addition to its weight, the spool is acted on by the normal contact force N and friction force F (both acting at the point of contact A) and by the cable tension T. Note that the magnitudes T, N, and F are the only unknowns and that there are three independent equilibrium equations (general coplanar force system). Therefore, the solution is statically determinate. We illustrate one method of solution in detail and then discuss several other methods that could be used.

We start with the equation

$$\Sigma M_A = 0$$

The tension T can be calculated using this equation because it will be the only unknown (N and F do not have moments about point A). The next equation is

$$\Sigma M_G = 0$$

The unknowns in this equation will be T and F, because N has no moment about G. Because T has already been found, this equation can be solved for F. Finally, we use the equation

$$\Sigma F_{y'} = 0$$

The unknowns in this equation will be T and N (F is perpendicular to the y'-direction). Again, with T already computed, N can be found.

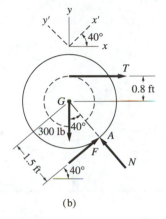

(b)

Mathematical Details

To help you follow the details of the preceding analysis, the FBD of the spool has been redrawn in Fig. (c). Note that the 300-lb weight of the spool has been replaced by its x'- and y'-components and that the vertical distance between A and G (1.5 cos 40° ft) has been added. The analysis now proceeds as follows:

$\Sigma M_A = 0$ $\quad\overset{+}{\curvearrowright}\quad$ $300 \sin 40°(1.5) - T(0.8 + 1.5 \cos 40°) = 0$

$\qquad\qquad\qquad\qquad T = 148.4 \text{ lb}$ *Answer*

$\Sigma M_G = 0$ $\quad\overset{+}{\curvearrowright}\quad$ $F(1.5) - T(0.8) = 0$

$$F = \frac{148.4(0.8)}{1.5} = 79.1 \text{ lb} \qquad \textit{Answer}$$

$\Sigma F_{y'} = 0$ $\quad\overset{+}{\nwarrow}\quad$ $N - 300 \cos 40° - T \sin 40° = 0$

$\qquad\qquad\qquad N = 300 \cos 40° + 148.4 \sin 40° = 325.2 \text{ lb}$ *Answer*

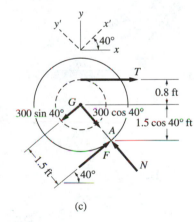

(c)

The positive signs determined for T, F, and N indicate that the correct sense for each force was assumed on the FBD.

As a check on this solution, we can verify that the above answers satisfy a fourth equilibrium equation. For example,

$\Sigma F_x = 0$ $\quad\overset{+}{\longrightarrow}\quad$ $F \cos 40° - N \sin 40° + T$

$\qquad\qquad\qquad\qquad = 79.1 \cos 40° - 325.2 \sin 40° + 148.4 \approx 0$ *Check*

Other Methods of Analysis

Two additional methods of analysis are outlined in the table below, with the mathematical details omitted.

Equation	Unknowns	Solution
$\Sigma M_G = 0$	T and F	Solve simultaneously for T and F
$\Sigma F_{x'} = 0$	T and F	
$\Sigma F_{y'} = 0$	T and N	Knowing T, solve for N
$\Sigma M_A = 0$	T	Solve for T
$\Sigma F_{x'} = 0$	T and F	Knowing T, solve for F
$\Sigma F_y = 0$	T, N, and F	Knowing T and F, solve for N

In this sample problem, we have illustrated only three of the many sets of equations that can be used to analyze this problem. You may find it beneficial to outline one or more additional analyses. Outlining the solution will permit you to concentrate on the method of analysis without becoming too involved with the mathematical details of the solution.

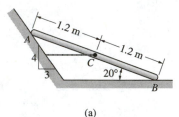

(a)

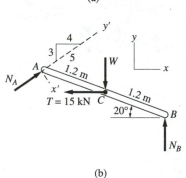

(b)

Sample Problem 4.9

Determine the mass of the heaviest uniform bar that can be supported in the position shown in Fig. (a) if the breaking strength of the horizontal cable attached at C is 15 kN. Neglect friction.

Solution

Method of Analysis

We begin by drawing the FBD of the bar as shown in Fig. (b). The weight W of the heaviest bar that can be supported will be obtained when the tension T is set equal to 15 kN. A heavier bar would result in a cable tension greater than 15 kN, and the cable would break.

There are three unknowns in the FBD: the normal contact forces N_A and N_B, and W. Note that the directions of all these forces are known. The unknowns can, therefore, be found using the three independent equilibrium equations that are available for a general coplanar force system. However, because we are seeking W only, it may not be necessary to use all three equations.

In our analysis, we will use the following two equations.

$$\Sigma F_x = 0$$

The force N_A can be found from this equation (W and N_B will not appear because they are perpendicular to the x-direction).

$$\Sigma M_B = 0$$

This equation will contain the two unknowns W and N_A. Because N_A has already been determined, W can now be found.

Mathematical Details

Referring to the FBD in Fig. (b), the mathematical details of the preceding analysis are as follows:

$$\Sigma F_x = 0 \qquad \xrightarrow{+} \qquad \frac{4}{5}N_A - 15 = 0$$

$$N_A = 18.75 \text{ kN} \qquad \qquad \text{(a)}$$

$$\Sigma M_B = 0 \qquad \overset{+}{\curvearrowright} \qquad W(1.2\cos 20°) + 15(1.2\sin 20°) - \frac{3}{5}N_A(2.4\cos 20°)$$

$$- \frac{4}{5}N_A(2.4\sin 20°) = 0 \qquad \qquad \text{(b)}$$

Substituting $N_A = 18.75$ kN from Eq. (a) into Eq. (b) gives $W = 28.0$ kN. Therefore, the mass of the heaviest bar that can be supported without breaking the cable is

$$m = \frac{W}{g} = \frac{28.0 \times 10^3}{9.81} = 2850 \text{ kg} \qquad \qquad \textit{Answer}$$

Other Methods of Analysis

Another method that could be used to calculate W is outlined in the following table.

Equation	Unknowns	Solution
$\Sigma M_A = 0$	N_B and W	Solve simultaneously
$\Sigma F_{x'} = 0$	N_B and W	for N_B and W

There are, of course, many other sets of equations that could be used to compute W. It is even possible to determine W using only one equilibrium equation—a moment equation taken about the point where N_A and N_B intersect.

Sample Problem 4.10

Figure (a) shows the distributed loading due to water pressure that is acting on the upstream side of the flood barrier. Determine the support reactions acting on the barrier at A and B. Neglect the weight of the barrier.

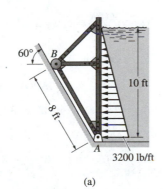

(a)

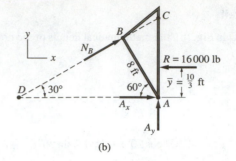

(b)

Solution

Method of Analysis

The FBD of the barrier is shown in Fig. (b), where N_B is the reaction at B, acting perpendicular to the inclined surface, and A_x and A_y are the components of the pin reaction at A. Any three independent equilibrium equations can be used to determine these three unknowns. As explained in Art. 3.6, the resultant of a distributed load is equal to the area under the loading diagram, acting at the centroid of that area. Therefore, we obtain

$$R = \frac{1}{2}(10)(3200) = 16\,000 \text{ lb}$$

and from Table 3.1, we find

$$\bar{y} = \frac{10}{3} \text{ ft}$$

Because the unknown forces A_x and A_y intersect at A, a convenient starting point is

$$\Sigma M_A = 0$$

This equation will determine N_B. We then use

$$\Sigma F_x = 0$$

Having previously determined N_B, this equation will give A_x. The final equation is

$$\Sigma F_y = 0$$

With N_B previously computed, A_y can be found from this equation.

Mathematical Details

$$\Sigma M_A = 0 \qquad \overset{\curvearrowleft}{+} \qquad 16\,000\left(\frac{10}{3}\right) - N_B(8) = 0$$

$$N_B = 6670 \text{ lb} \qquad\qquad\qquad Answer$$

$$\Sigma F_x = 0 \qquad \xrightarrow{+} \qquad N_B \cos 30° + A_x - 16\,000 = 0$$

$$A_x = 16\,000 - (6670)\cos 30° = 10\,220 \text{ lb} \qquad Answer$$

$$\Sigma F_y = 0 \qquad +\uparrow \qquad A_y + N_B \sin 30° = 0$$

$$A_y = -(6670)\sin 30° = -3340 \text{ lb} \qquad Answer$$

The signs indicate that N_B and A_x are directed as shown on the FBD, whereas the correct direction of A_y is opposite the direction shown on the FBD. Therefore, the force that acts on the barrier at A is

$$|\mathbf{A}| = \sqrt{(10\,220)^2 + (3340)^2} = 10\,750 \text{ lb}$$

Answer

$$\theta = \tan^{-1}\left(\frac{3340}{10\,220}\right) = 18.1°$$

and the force at B is

6670 lb

Answer

30°

Other Methods of Analysis

There are, of course, many other independent equations that could be used to solve this problem. Referring to the FBD in Fig. (b), the following set of equations has the advantage of determining each unknown independently of the other two.

Equation	Unknowns	Solution
$\Sigma M_A = 0$	N_B	Solve for N_B
$\Sigma M_C = 0$	A_x	Solve for A_x
$\Sigma M_D = 0$	A_y	Solve for A_y

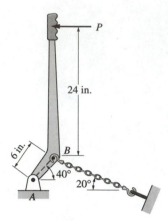

Fig. P4.30

Problems

In Probs. 4.26 through 4.29, determine the magnitudes and directions of the forces acting on the body at A and B.

4.26 See Prob. 4.1.

4.27 See Prob. 4.2.

4.28 See Prob. 4.3.

4.29 See Prob. 4.4.

4.30 The horizontal force P is applied to the handle of the puller. Determine the resulting tension T in the chain in terms of P.

4.31 The thin steel plate, weighing 82 lb/ft^2, is being lifted slowly by the cables AC and BC. Compute the distance x for which the plate will be in equilibrium, and find the corresponding tension in each of the cables.

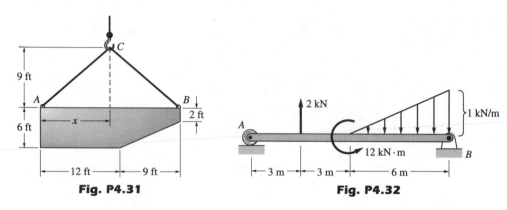

Fig. P4.31   **Fig. P4.32**

4.32 Neglecting the mass of the beam, compute the reactions.

4.33 The 1200-kg car is being lowered slowly onto the dock using the hoist A and winch C. Determine the forces in cables BA and BC for the position shown.

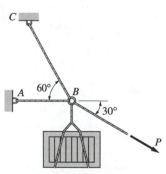

Fig. P4.34, P4.35

Fig. P4.33

4.34 The crate weighing 400 lb is supported by three ropes concurrent at B. Find the forces in ropes AB and BC if $P = 460$ lb.

4.35 Find the smallest value of P for which the crate in the Prob. 4.34 will be in equilibrium in the position shown. (*Hint:* A rope can only support a tensile force.)

4.36 (a) Show that the pulley, which is supported by a smooth pin, can be in equilibrium only if the rope tensions T_1 and T_2 are equal. (b) If T_2 is 200 N larger than T_1, determine the magnitude and direction of the couple that must be applied to the pulley to maintain equilibrium.

4.37 The 60-kg homogeneous disk is resting on an inclined friction surface. (a) Compute the magnitude of the horizontal force P. (b) Could the disk be in equilibrium if the inclined surface were frictionless?

4.38 The 60-kg homogeneous disk is placed on a frictionless inclined surface and held in equilibrium by the horizontal force P and a couple C (C is not shown on the figure). Find P and C.

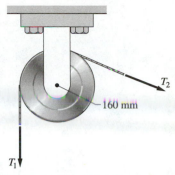

Fig. P4.36

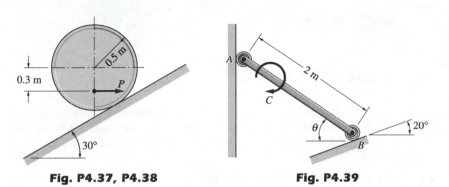

Fig. P4.37, P4.38   **Fig. P4.39**

4.39 The mass of the uniform bar AB is 40 kg. Calculate the couple C required for equilibrium if (a) $\theta = 0$; and (b) $\theta = 54°$.

4.40 The mechanism shown is a modified Geneva drive—a constant velocity input produces a varying velocity output with periods of dwell. The input torque is 120 N·m. For the position shown, compute the contact force at B and the magnitude of the bearing reaction at A. Neglect friction and the weight of the member.

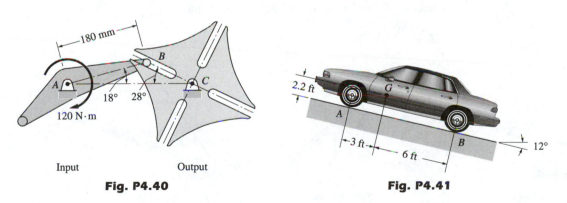

Input   Output

Fig. P4.40   **Fig. P4.41**

4.41 The center of gravity of the 3000-lb car is at G. The car is parked on an incline with the parking brake engaged, which locks the rear wheels. Find (a) the normal forces (perpendicular to the incline) acting under the front and rear pairs of wheels; and (b) the friction force (parallel to the incline) under the rear pair of wheels.

4.42 The 1800-kg boat is suspended from two parallel cables of equal length. The location of the center of gravity of the boat is not known. Calculate the force P required to hold the boat in the position shown.

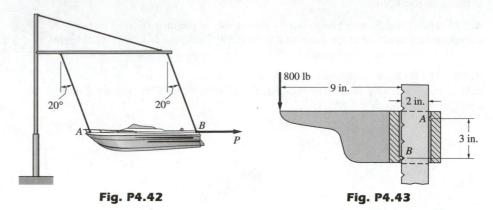

Fig. P4.42   **Fig. P4.43**

4.43 With no load applied to the bracket, its vertical position can be changed by sliding it along the notched post. Assuming that the contact surface at A is frictionless, calculate the magnitudes of the forces acting on the bracket at A and B when the 800-lb force is applied.

4.44 The uniform ladder of weight W is raised slowly by applying a vertical force P to the rope at A. Show that P is independent of the angle θ.

4.45 The uniform, 40-lb ladder is raised slowly by pulling on the rope attached at A. Determine the largest angle θ that the ladder can attain if the maximum allowable tension in rope BC is 330 lb.

4.46 The 90-kg person, whose center of gravity is at G, is climbing a uniform ladder. The length of the ladder is 5 m, and its mass is 20 kg. Friction may be neglected. (a) Compute the magnitudes of the reactions at A and B for $x = 1.5$ m. (b) Find the distance x for which the ladder will be ready to fall.

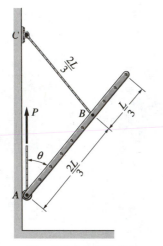

Fig. P4.44, P4.45

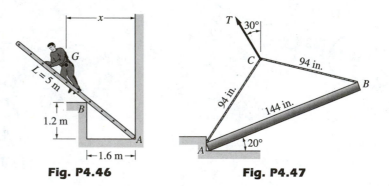

Fig. P4.46   **Fig. P4.47**

4.47 The 240-lb homogeneous pole AB is supported by cables AC and BC. Determine the force T and the magnitude of the reaction at A if the pole is in equilibrium in the position shown.

4.48 The tensioning mechanism of a magnetic tape drive has a mass of 0.4 kg, and its center of gravity is at G. The tension T in the tape is maintained by presetting the tensile force in the spring at B to 14 N. Calculate T and the magnitude of the pin reaction at A.

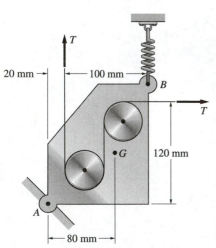

Fig. P4.48

4.49 The homogeneous 300-kg cylinder is pulled over the 100-mm step by the horizontal force P. Find the smallest P that would raise the cylinder off the surface at A. Assume sufficient friction at corner B to prevent slipping.

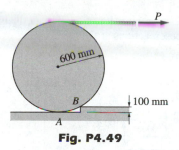

Fig. P4.49

4.50 The circular flange of a pipe is clamped between two rough surfaces by the 1.5-kN force in the screw jack at A. An automatic drilling machine exerts the 40-N · m torques shown as it simultaneously drills six holes in the flange. If the mass of the flange is 12 kg, compute all unknown forces that act on the flange during the drilling operation.

4.51 Each of the sandbags piled on the 250-lb uniform beam weighs 12 lb. Determine the support reactions at A and C.

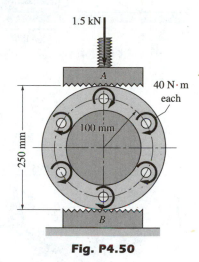

Fig. P4.50

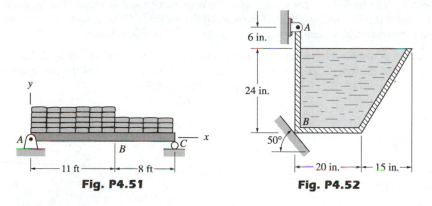

Fig. P4.51   **Fig. P4.52**

4.52 The water trough, shown in cross section, is pinned at A and rests against a frictionless surface at B. The ends of the trough are flat plates, parallel to the cross section. If the water in the trough weighs 1500 lb, compute the magnitudes of the reactions at A and B. The weight of the trough may be neglected.

4.53 The supporting structure of a billboard is attached to the ground by a pin at B, and its rear leg rests on the ground at A. Friction may be neglected. Point G is the center of gravity of the billboard and structure, which together weigh 2800 lb. To prevent tipping over in high winds, a 2370-lb weight is placed on the structure near A, as shown. (a) Compute the magnitudes of the reactions at A and B if the wind load on the billboard is $q = 120$ lb/ft. (b) Find the smallest wind load q that would cause the structure to tip over.

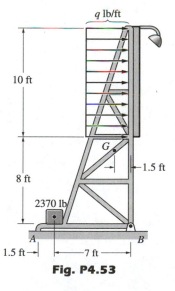

Fig. P4.53

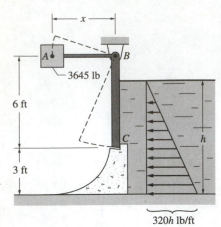

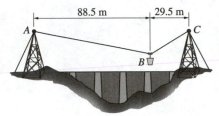

Fig. P4.54

Fig. P4.57

4.54 The self-regulating floodgate ABC, pinned at B, is pressed against the lip of the spillway at C by the action of the 3645-lb weight A. If the gate is to open when the water level reaches a height $h = 6$ ft, determine the distance x locating the weight A. Neglect the weight of the gate.

4.55 The cantilever beam is built into a wall at O. Neglecting the weight of the beam, determine the support reactions at O.

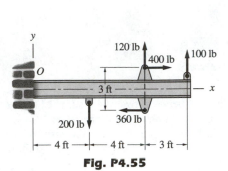

Fig. P4.55

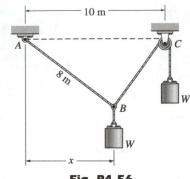

Fig. P4.56

*4.56 Two identical weights are supported by a cable that runs over a frictionless pulley at C. Compute the distance x for which the system will remain at rest.

*4.57 The bucket B carries concrete at the construction site of a dam. The length of the support cable ABC is 120 m. If the mass of the bucket is 5100 kg, compute the force in each portion of the cable when the bucket is in the position shown.

4.58 A machine operator produces the tension T in the control rod by applying the force P to the foot pedal. Determine the largest P if the magnitude of the pin reaction at B is limited to 1.8 kN. Neglect the mass of the mechanism.

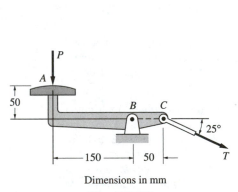

Dimensions in mm
Fig. P4.58

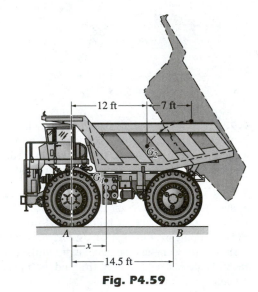

Fig. P4.59

*4.59 The dump truck consists of a chassis and a tray, with centers of gravity at G_1 and G_2, respectively. With the tray down, the axle loads (normal forces at A and B) are 41 900 lb each. When the tray is in the raised position, the rear axle load increases to 48 700 lb. Compute the weight of the chassis, the weight of the tray, and the distance x.

*4.60   The centers of gravity of the 50-kg lift truck and the 120-kg box are at G_1 and G_2, respectively. The truck must be able to negotiate the 5-mm step when the pushing force P is 600 N. Find the smallest allowable radius of the wheel at A. Be sure to check whether the truck will tip.

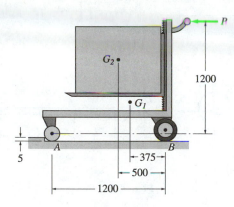

Dimensions in mm

Fig. P4.60

PART B: *Analysis of Composite Bodies*

4.7   *Free-Body Diagrams Involving Internal Reactions*

Up to now, we have been considering "one-body" problems. Because we have been concerned primarily with calculating external reactions, each problem has required the use of only one free-body diagram (FBD) and the solution of one set of equilibrium equations. We now begin a study of the forces that act at connections that are internal to the body, called *internal reactions*. The calculation of internal reactions often requires the use of more than one FBD.

In this article, attention is focused on the drawing of FBDs of the various parts that together form a composite body. Frames and machines are examples of connected bodies that are commonly used in engineering applications. *Frames* are rigid structures that are designed to carry load in a fixed position. *Machines* contain moving parts and are usually designed to convert an input force to an output force.

The construction of FBDs that involve internal forces relies on Newton's third law: For every action there is an equal and opposite reaction. Strict adherence to this principle is the key to the construction of FBDs.

a.   *Internal forces in members*

Consider the beam in Fig. 4.5(a), which carries the load P acting at its center (P and θ are assumed known). In the FBD of the entire beam, Fig. 4.5(b), there are three unknown external reactions (A_x, A_y, and N_B) and three independent equilibrium equations. Therefore, the beam is statically determinate, and the three unknowns could be easily calculated, although we will not do so here.

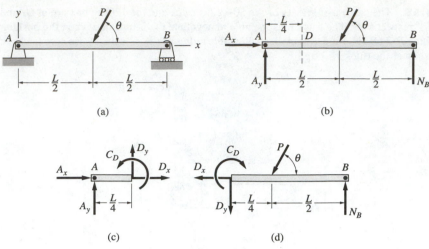

Fig. 4.5

Now suppose that we are asked to determine the force system that acts on the internal cross section at D, located a distance $L/4$ to the right of end A. We begin by isolating the parts of the beam that lie to the left and right of D. In effect, we imagine that the beam is cut open at the section of interest. Thus the cross section that was initially an internal section now becomes an external section. We then draw the FBDs of both parts of the beam, as shown in Fig. 4.5(c) and (d).

Consider the FBD for the left portion of the beam, Fig. 4.5(c). The right portion of the beam has been removed, and its effect is shown as an unknown force (represented by the independent components D_x and D_y) and an unknown couple (C_D), the senses of which are assumed.

On the FBD for the right portion of the beam, Fig. 4.5(d), the effect of the removed left portion is likewise an unknown force and an unknown couple. However, Newton's third law prescribes that the effect that the right part of the beam has on the left part is equal and opposite to the effect that the left part has on the right. Therefore, on the FBD in Fig. 4.5(d), the force system at D consists of the forces D_x and D_y and the couple C_D, each equal in magnitude, but opposite in direction, to its counterpart in Fig. 4.5(c). That is the key to understanding FBDs! When isolating two parts of a body in order to expose the internal reactions, these reactions must be shown as equal and opposite force systems on the FBDs of the respective parts.

Note that we are using scalar representation for the forces and couples in the FBDs in Fig. 4.5. For example, the magnitude of the x-component of the force at D is labeled D_x, and its direction is indicated by an arrow. If a vector representation is used, one force of the pair would be labeled $\mathbf{D}_x$ and the other force $-\mathbf{D}_x$. Because there is no advantage to using vector notation here, we continue to use the scalar representation.

Finally, note that if A_x, A_y, and N_B had been previously computed from the FBD for the entire beam, the FBD in either Fig. 4.5(c) or (d) could be used to calculate the three unknowns D_x, D_y, and C_D.

Observe that internal forces do not appear on the FBD of the entire beam, Fig. 4.5(b). The reason is that there are two internal force systems acting on every section of the beam, each system being equal and opposite to the other. Therefore, internal reactions have no effect on the force or moment equations for the entire beam.

b. *Internal forces at connections*

Consider the frame shown in Fig. 4.6(a), which consists of two identical, homogeneous bars AB and BC, each of weight W and length L. The bars are pinned together at B and are attached to the supports with pins at A and C. Two forces P and Q are applied directly to the pin at B. We assume that L, W, P, Q, and θ are known quantities. Furthermore, throughout this text *we neglect the weights of pins and other connectors,* unless stated otherwise.

The FBD of the structure shown in Fig. 4.6(b) contains four unknown pin reactions: A_x and A_y (the forces exerted on bar AB by the pin A) and C_x and C_y (the forces exerted on bar BC by the pin C). The senses of these forces have been chosen arbitrarily. Because only three independent equilibrium equations are available from this FBD, you might presume that the problem is statically indeterminate. Indeed, this would be the correct conclusion if ABC were a single rigid unit, rather than two rigid bars joined by a pin. If the system is "taken apart" and an FBD is drawn for each component, it will be seen that the problem is statically determinate. As explained in the following, drawing the FBD of each component increases the number of unknowns, but the number of independent equations also increases.

The FBD of each component is shown in Fig. 4.7.

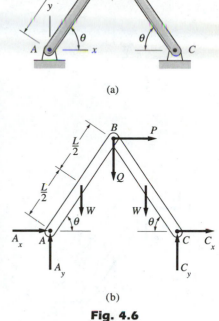

(a)

(b)

Fig. 4.6

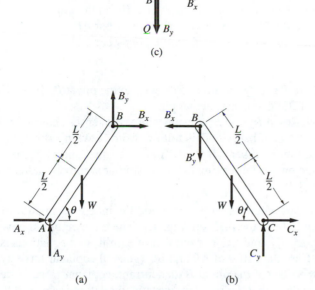

(c)

(a)   (b)

Fig. 4.7

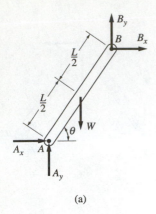

(a)

Figure 4.7(a)   FBD of bar *AB* with pins at *A* and *B* removed.

W is the weight of bar *AB* acting at the center of the bar.
A_x and A_y are the forces exerted *on* bar *AB* *by* the pin at *A*.
B_x and B_y are the forces exerted *on* bar *AB* *by* the pin at *B*.

Notes

1. The senses of A_x and A_y cannot be chosen arbitrarily here. These senses were already assumed when the FBD of the system, Fig. 4.6(b), was drawn. Initially, the sense of an unknown force may be chosen arbitrarily, but if that force appears on more than one FBD, its sense must be consistent with the original assumption.
2. The senses of B_x and B_y were chosen arbitrarily, because this is the first FBD on which these forces appear.
3. *P* and *Q* are applied directly to the pin at *B*, so they do not appear on this FBD (recall that the pin at *B* has been removed).

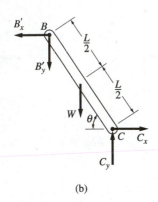

(b)

Figure 4.7(b)   FBD of *BC* with pins at *B* and *C* removed.

W is the weight of *BC* acting at the center of the bar.
C_x and C_y are the forces exerted *on* bar *BC* *by* the pin at *C*.
B'_x and B'_y are the forces exerted *on* bar *BC* *by* the pin at *B*.

Notes

4. The directions of C_x and C_y must be the same as shown in Fig. 4.6(b). (See Note 1.)
5. The forces exerted *by* the pin *B on* member *BC* are labeled B'_x and B'_y. Because this is the first FBD on which B'_x and B'_y have appeared, their senses have been chosen arbitrarily.

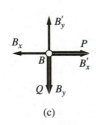

(c)

Fig. 4.7, repeated

Figure 4.7(c)   FBD of pin *B* with bars *AB* and *BC* removed.

P and *Q* are the external forces acting directly on the pin.
B_x and B_y are the forces exerted *on* the pin *by* bar *AB*.
B'_x and B'_y are the forces exerted *on* the pin *by* bar *BC*.

Notes

6. Because *P* and *Q* are applied directly to the pin at *B*, they will appear on every FBD that contains that pin.
7. The senses of B_x and B_y are opposite to the senses chosen for these forces on the FBD of bar *AB*. This follows from Newton's third law: The force exerted on member *AB* by the pin *B* is equal and opposite to the force exerted on pin *B* by member *AB*. A similar argument holds for the directions of B'_x and B'_y.

Let us now count the unknowns and the independent equilibrium equations available from the FBDs in Fig. 4.7. There are eight unknowns: A_x, A_y, B_x, B_y, B'_x, B'_y, C_x, and C_y. The number of equilibrium equations is also eight: three each from the FBDs of *AB* and *BC* (general coplanar force systems) and two from the FBD of the pin at *B* (concurrent, coplanar force system). Therefore, we conclude that the problem is statically determinate, and the eight unknowns are solvable from the eight independent equilibrium equations.

As mentioned previously, there are also three independent equations for the FBD of the entire body, shown in Fig. 4.6. Does this mean that we have a total of $8 + 3 = 11$ independent equations? The answer is no! The FBD for the entire system is not independent of the FBDs for all of its parts—the FBDs in Fig. 4.7 could be put back together again to form the FBD in Fig. 4.6. In other words, if each part of the body is in equilibrium, then equilibrium of the entire body is guaranteed. This means that of the eleven equations just cited, only eight will be independent.

Let us now change the problem by assuming *ABC* to be a single rigid unit rather than two bars pinned together at *B*. In this case, the body would be able to transmit a force *and* a couple at *B*. Consequently, the number of unknowns would be increased by one (the magnitude of the couple), but the number of independent equations would remain at eight. Hence, this problem would be statically indeterminate, as noted previously.

So far, we have drawn the FBD for the entire system and the FBDs for each of its parts. There are two other FBDs that could be constructed—the FBDs with the pin *B* left inside bar *AB* and those with pin *B* left inside bar *BC*. These FBDs are shown in Figs. 4.8 and Fig. 4.9, respectively. As you study each of the FBDs, note that B'_x and B'_y do not appear in Fig. 4.8 because they are now internal forces. For the same reason, B_x and B_y do not appear in Fig. 4.9. It should be noted again that, although we have drawn additional FBDs, the total number of independent equilibrium equations remains eight.

The following special case is extensively used in the construction of FBDs of bodies that are joined by pins.

Special Case: Equal and Opposite Pin Reactions
If two members are joined by a pin and *if* there are no external forces applied to the pin, then the forces that the pin exerts on each member are equal in magnitude and oppositely directed.

It is relatively easy to verify this statement by referring again to the FBD of the pin *B*, Fig. 4.7(c). Note that if there are no forces applied to the pin (i.e., if $P = Q = 0$), the equilibrium equations for the pin dictate that $B_x = B'_x$ and $B_y = B'_y$—the pin reactions are equal in magnitude and oppositely directed.

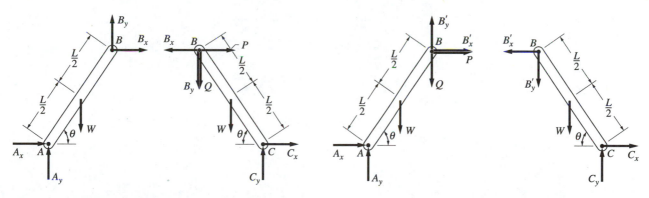

FBDs of *AB* and *BC* with pin *B* left in *BC*

Fig. 4.8

FBDs of *AB* and *BC* with pin *B* left in *AB*

Fig. 4.9

Using this special case, the FBDs for members *AB* and *BC*, and pin *B* would be as shown in Fig. 4.10. Here, the total number of unknowns is six, and the total number of independent equations is six—three for each bar. Obviously, two of the original eight equations have been used up in proving that the pin reactions at *B* are equal and opposite.

In this text we utilize *equal and opposite pin reactions* wherever applicable. That is, our FBDs of members display equal and opposite reactions at the pins, as shown at *B* in Fig. 4.10. The FBDs of the pins are not drawn. The two most common situations where the pin reactions are *not equal and opposite* are:

- External force is applied to the pin.
- More than two members are connected by the pin.

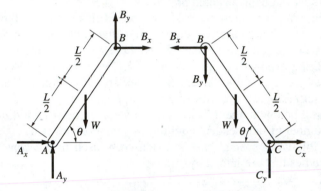

FBDs with no forces applied to pin *B*

Fig. 4.10

Sample Problem 4.11

(1) Referring to Fig. (a), draw the FBD for the entire frame and for each of its parts, neglecting the weights of the members. (2) Determine the total number of unknowns and the total number of independent equilibrium equations, assuming that the force P and couple C_0 are known.

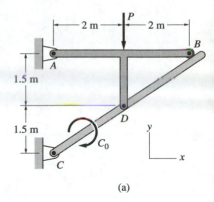

(a)

Solution

Part 1

The force system on each of the FBDs is described below.

FBD of Entire Frame—Fig. (b)

P and C_0: applied force and applied couple
A_x and A_y: components of the force exerted on the frame by pin A (directions are assumed)
C_x and C_y: components of the force exerted on the frame by pin C (directions are assumed)

FBD of Member ABD—Fig. (c)

P: applied force
A_x and A_y: components of the force exerted on member ABD by pin A [must be shown acting in the same directions as in Fig. (b)]
D_x and D_y: components of the force exerted on member ABD by pin D (directions are assumed)
N_B: force exerted on member ABD by the roller at B (must be perpendicular to member CDB)

FBD of Member CDB—Fig. (d)

C_0: applied couple
C_x and C_y: components of the force exerted on member CDB by pin C [must be shown acting in the same directions as in Fig. (a)]
D_x and D_y: components of the force exerted on member CDB by pin D [must be equal and opposite to the corresponding components in Fig. (c)]
N_B: force exerted on the member by the roller at B [must be equal and opposite to the corresponding force in Fig. (c)]

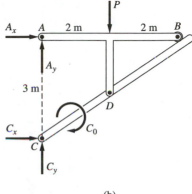

(b)

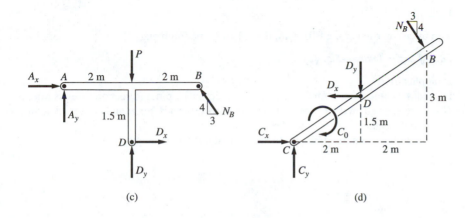

(c)   (d)

165

Part 2

Three independent equilibrium equations are available from the FBD of member *ABD*, and three from the FBD of member *CDB*, which gives a total of six independent equilibrium equations (recall that the FBD for the entire frame is not independent of the FBDs for its two composite members). The total number of unknowns is seven: two unknowns each at *A*, *C*, and *D* and one unknown at *B*.

Because the number of unknowns exceeds the number of independent equilibrium equations, we conclude that this problem is statically indeterminate; that is, all unknowns cannot be determined from equilibrium analysis alone.

Sample Problem 4.12

(1) Draw the FBDs for the entire frame in Fig. (a) and for each of its parts. The weights of the members are negligible. The cable at *C* is attached directly to the pin. (2) Determine the total number of unknowns and the total number of independent equilibrium equations, assuming that *P* is known.

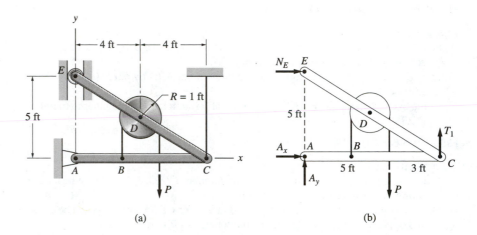

(a)   (b)

Solution

Part 1

The forces on each of the FBDs are described in the following.

FBD of Entire Frame—Fig. (b)

P: applied force
A_x and A_y: components of force exerted on the frame by pin *A* (directions are assumed)
N_E: force exerted on the frame by the roller *E* (direction is horizontal, assumed acting
   to the right)
T_1: force exerted on the frame by the cable that is attached to pin *C*

FBD of Member EDC—Fig. (c)

N_E: force exerted on member EDC by the roller at E [must be shown acting in the same direction as in Fig. (b)]

D_x and D_y: components of the force exerted on member EDC by pin D (directions are assumed)

C_x and C_y: components of the force exerted on member EDC by pin C (directions are assumed)

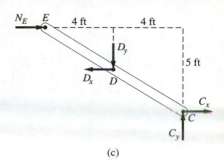

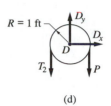

(c)

FBD of the Pulley—Fig. (d)

P: applied force

D_x and D_y: components of the force exerted on the pulley by pin D [must be shown equal and opposite to the corresponding components in Fig. (c)]

T_2: tension in the cable on the left side of the pulley

FBD of Member ABC—Fig. (e)

A_x and A_y: components of the force exerted on member ABC by pin A [must be shown acting in the same directions as in Fig. (b)]

T_2: force exerted on member ABC by the cable that is attached at B [must be equal and opposite to the corresponding force in Fig. (d)]

C'_x and C'_y: components of the force exerted on member ABC by pin C (directions are assumed)

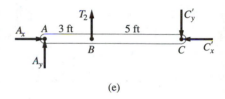

(d)

FBD of Pin C—Fig. (f)   This FBD is necessary because a cable is attached directly to pin C.

T_1: force exerted on pin C by the cable [must be shown acting in the same direction as in Fig. (b)]

C_x and C_y: components of the force exerted on pin C by member EDC [must be shown equal and opposite to the corresponding components in Fig. (c)]

C'_x and C'_y: components of the force exerted on pin C by member ABC [must be shown equal and opposite to the corresponding components in Fig. (e)]

(e)

Part 2

There are a total of eleven independent equilibrium equations: three for each of the two bars, three for the pulley, and two for pin C (the force system acting on the pin is concurrent, coplanar). Recall that the FBD for the entire frame is not independent of the FBDs of its members.

The problem is statically determinate because the total number of unknowns is also eleven: A_x and A_y, C_x and C_y, C'_x and C'_y, D_x and D_y, N_E, T_1, and T_2.

(f)

Problems

For Probs. 4.61–4.68, (a) draw the free-body diagrams for the entire assembly (or structure) and each of its parts. Neglect friction and the weights of the members unless specified otherwise. Be sure to indicate all relevant dimensions. For each problem, (b) determine the total number of unknown forces and the total number of independent equilibrium equations.

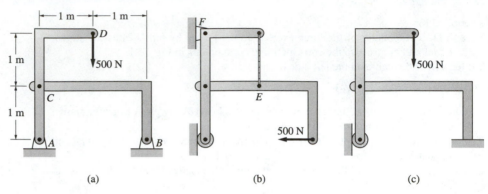

(a)   (b)   (c)

Fig. P4.61

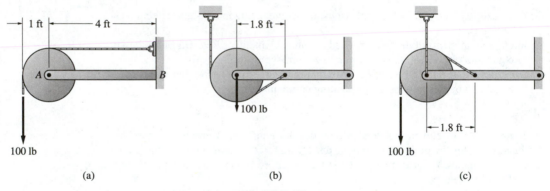

(a)   (b)   (c)

Fig. P4.62

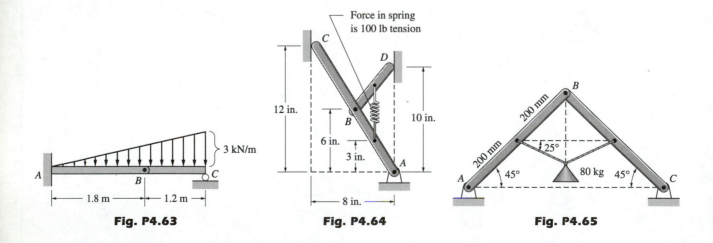

Fig. P4.63   **Fig. P4.64**   **Fig. P4.65**

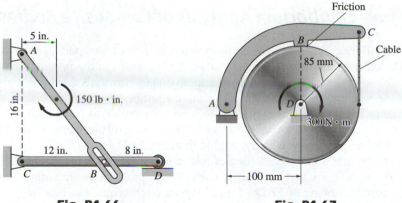

Fig. P4.66

Fig. P4.67

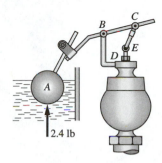

Fig. P4.68

4.69 The fluid control valve *D* is controlled by the float *A*. Draw FBDs for the float-arm assembly *ABC*, the link *CE,* the support arm *BD,* and the assembly composed of all three of these components. The upward thrust on the float is 2.4 lb. Neglect the weights of the components. Assume that all dimensions are known.

4.70 Draw the FBDs for the following: (a) bar *ABC* with pin *A* inside the bar; (b) bar *ABC* with pin *A* removed; and (c) pin *A*. Neglect the weights of the members.

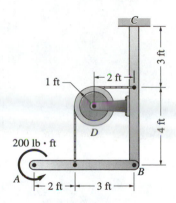

Fig. P4.69

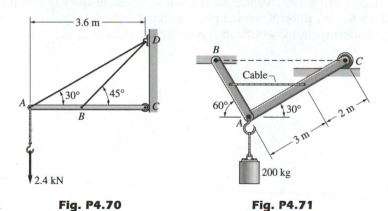

Fig. P4.70   **Fig. P4.71**

4.71 Bars *AB* and *AC* are joined by a smooth pin *A*. The hook supporting the 200-kg mass is also attached to pin *A*. Draw the FBDs for (a) the entire structure; (b) pin *A*; and (c) each of the bars (with pin *A* removed). Neglect the masses of the bars.

4.72 The 200-lb steel ball *B* is supported by a cantilever beam and a cable that passes over the pulley at *D*. Draw FBDs for the following: (a) the entire assembly; (b) the pulley; (c) the ball; and (d) the beam. The weights of the pulley and beam are negligible.

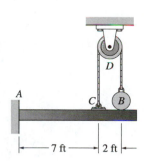

Fig. P4.72

4.8 *Equilibrium Analysis of Composite Bodies*

In the equilibrium analysis of a composite body and its various parts, you must be able to construct the appropriate FBDs. As explained in the previous article, this ability depends on the correct application of Newton's third law. Furthermore, you must be able to write and solve equilibrium equations based on FBDs, a technique that was explained for one-body problems in Art. 4.6. The primary difference between one-body and composite-body problems is that the latter often require that you analyze more than one FBD.

The problems in the preceding article required the construction of FBDs for a composite body and each of its parts. These problems were simply exercises in the drawing of FBDs. Beginning an equilibrium analysis by constructing all possible FBDs is inefficient; in fact, it can lead to confusion. You should begin by drawing the FBD of the entire body and, if possible, calculate the external reactions. Then, and only then, should you consider the analysis of one or more parts of the body. The advantages of this technique are the following: First, because only external reactions appear on the FBD of the entire body, some or all of them can often be calculated without referring to internal forces. A second advantage is that FBDs are drawn only as needed, thereby reducing the amount of labor. Most of the time it will not be necessary to draw all possible FBDs and compute all internal reactions in order to find the desired unknowns. Knowing which FBDs to draw and what equations to write are undoubtedly the most difficult parts of equilibrium analysis.

Sample Problem 4.13

The structure in Fig. (a) is loaded by the 240-lb · in. counterclockwise couple applied to member AB. Neglecting the weights of the members, determine all forces acting on member BCD.

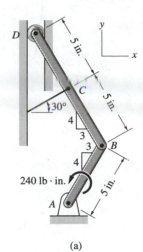

(a)

Solution

The solution of a problem involving a composite body such as this must be approached with caution. Unless an efficient method of analysis is planned from the outset, it is easy to be overwhelmed by the number of FBDs that can be drawn and the number of equilibrium equations that can be written.

Method of Analysis

Although not absolutely necessary, considering the FBD of the entire structure is often a good starting point. The FBD shown in Fig. (b) contains the four unknowns N_D, T_C, A_x, and A_y. With four unknowns and three independent equilibrium equations (general coplanar force system), we cannot determine all unknowns on this FBD (one more independent equation is required). Therefore, without writing a single equation from the FBD in Fig. (b), we turn our attention to another FBD.

Because we are seeking the forces acting on member BCD, let us next consider its FBD, shown in Fig. (c). This FBD contains four unknowns: N_D, T_C, B_x, and B_y. Again, however, there are only three independent equations (general coplanar force system), but a study of Fig. (c) reveals that the equation $\Sigma M_B = 0$ will relate the unknowns N_D and T_C. Additionally, from Fig. (b) we see that N_D and T_C are also related by the equation $\Sigma M_A = 0$. Therefore, these two moment equations can be solved simultaneously for N_D and T_C. Once those two unknowns have been found, the calculation of B_x and B_y, which are the remaining unknown forces acting on BCD, is straightforward.

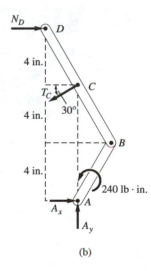

(b)

Mathematical Details

Referring to the FBD of the entire structure, in Fig. (b),

$$\Sigma M_A = 0 \quad \overset{+}{\curvearrowleft} \quad T_C \cos 30°(8) - N_D(12) + 240 = 0$$

$$N_D = 0.5774\, T_C + 20 \qquad \text{(a)}$$

From the FBD of member BCD in Fig. (c),

$$\Sigma M_B = 0 \quad \overset{+}{\curvearrowleft} \quad T_C \cos 30°(4) + T_C \sin 30°(3) - 8\,N_D = 0$$

$$N_D = 0.6205\, T_C \qquad \text{(b)}$$

Solving Eqs. (a) and (b) simultaneously yields

$$T_C = 464 \text{ lb} \quad \text{and} \quad N_D = 288 \text{ lb} \qquad \textit{Answer}$$

Also from the FBD of member BCD in Fig. (c),

$$\Sigma F_x = 0 \quad \overset{+}{\longrightarrow} \quad N_D - T_C \cos 30° + B_x = 0$$

$$B_x = 464 \cos 30° - 288 = 114 \text{ lb} \qquad \textit{Answer}$$

and

$$\Sigma F_y = 0 \quad +\!\uparrow \quad B_y - T_C \sin 30° = 0$$

$$B_y = 464 \sin 30° = 232 \text{ lb} \qquad \textit{Answer}$$

Because the solution yields positive numbers for the unknowns, each force is directed as shown on the FBDs.

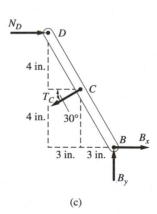

(c)

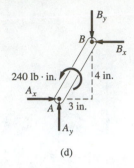

(d)

The FBD of member AB, although not required in the foregoing analysis, is shown in Fig. (d) for future reference.

Other Methods of Analysis

Note that the FBDs for the two members that make up the structure, Figs. (c) and (d), contain a total of six unknowns: A_x, A_y, B_x, B_y, N_D, and T_C. There are also six independent equilibrium equations—three for each member. (Thus you see that it is not absolutely necessary to use the FBD of the entire assembly.) There are many combinations of equations that could be used to determine the forces acting on member BCD. It is recommended that you practice your skills by outlining one or more additional methods of analysis.

Sample Problem 4.14

An 80-N box is placed on a folding table as shown in Fig. (a). Neglecting friction and the weights of the members, determine all forces acting on member EFG and the tension in the cable connecting points B and D.

Solution

Method of Analysis

We begin by considering the FBD of the entire table, Fig. (b). Because this FBD contains three unknowns (N_G, H_x, and H_y), it will be possible to compute all of them from this FBD. In particular, N_G can be found using the equation $\Sigma M_H = 0$.

Next, we turn our attention to the FBD of member EFG by skipping ahead to Fig. (d). We note that there are five unknowns on this FBD (N_G, F_x, F_y, E_x, and E_y). Although we have already found a way to find N_G, four unknowns remain—with only three independent equations.

Therefore, without writing any equations for the time being, we consider another FBD—the FBD of the tabletop, shown in Fig. (c). Although this FBD also contains four unknown forces, we see that three of them (A_x, A_y, and E_x) pass through point A. Therefore, the fourth force, E_y, which is one of the forces we are seeking, can be determined from the equation $\Sigma M_A = 0$.

Having computed E_y, only three unknowns remain on the FBD in Fig. (d). These unknowns can now be readily found by using the three available equilibrium equations.

Thus far, our analysis has explained how to determine the five forces acting on member EFG. All that remains is to find the tension in the cable connected between B and D. This force has not yet appeared on any of the FBDs, so we must draw another FBD.

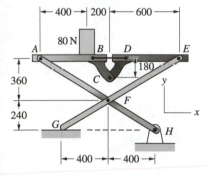

Dimensions in mm

(a)

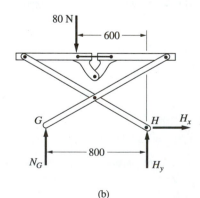

(b)

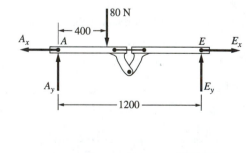

(c)

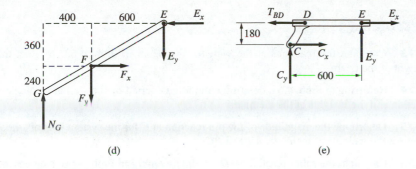

(d)   (e)

We choose to draw the FBD of the right half of the tabletop, shown in Fig. (e). The right half is chosen instead of the left because the pin reactions at E have already been determined. With E_x and E_y previously found, the remaining three unknowns (C_x, C_y, and T_{BD}) can be computed. In particular, the tension T_{BD} can be found from the equation $\Sigma M_C = 0$.

Note that we did not find it necessary to draw the FBD for member AFH or for the left half of the tabletop.

Mathematical Details

From the FBD of the entire table, Fig. (b),

$$\Sigma M_H = 0 \quad \curvearrowleft+ \qquad 80(600) - N_G(800) = 0 \qquad \text{(a)}$$

$$N_G = 60 \text{ N} \qquad \qquad \textit{Answer}$$

From the FBD of the tabletop, Fig. (c),

$$\Sigma M_A = 0 \quad \curvearrowleft+ \qquad -80(400) + E_y(1200) = 0 \qquad \text{(b)}$$

$$E_y = 26.67 \text{ N} \qquad \qquad \textit{Answer}$$

From the FBD of member EFG, Fig. (d),

$$\Sigma M_F = 0 \quad \curvearrowleft+ \qquad E_x(360) - E_y(600) - N_G(400) = 0 \qquad \text{(c)}$$

$$E_x(360) = 26.67(600) + 60(400)$$

$$E_x = 111.12 \text{ N} \qquad \qquad \textit{Answer}$$

$$\Sigma F_x = 0 \quad \xrightarrow{+} \qquad F_x - E_x = 0 \qquad \text{(d)}$$

$$F_x = E_x = 111.12 \text{ N} \qquad \qquad \textit{Answer}$$

$$\Sigma F_y = 0 \quad +\uparrow \qquad N_G - F_y - E_y = 0 \qquad \text{(e)}$$

$$F_y = 60 - 26.67 = 33.33 \text{ N} \qquad \qquad \textit{Answer}$$

From the FBD of the right half of the tabletop, Fig. (e),

$$\Sigma M_C = 0 \quad \curvearrowleft+ \qquad E_y(600) - E_x(180) + T_{BD}(180) = 0 \qquad \text{(f)}$$

$$T_{BD}(180) = 111.12(180) - 26.67(600)$$

$$T_{BD} = 22.22 \text{ N} \qquad \qquad \textit{Answer}$$

Other Methods of Analysis

Our analysis was based on the six independent equilibrium equations, Eqs. (a)–(f). For a structure as complex as the one shown in Fig. (a), there are many other methods of analysis that could be used. For example, a different set of equations would result if we chose to consider the left side of the tabletop instead of the right, as was done in Fig. (e).

Problems

4.73 Compute the reactions at the wall for the beam in Prob. 4.63. Neglect the weight of the beam.

4.74 Referring to Prob. 4.64, determine the force exerted by the wall at *C*. Neglect friction and the weights of the members.

4.75 Determine the magnitude of the pin reaction at *C* for the system in Prob. 4.65. Neglect the weights of members *AB* and *BC*.

4.76 Calculate the roller reaction at *D* for the assembly in Prob. 4.66. Neglect the weights of the members.

4.77 Neglecting the weights of the members, determine the magnitude of the pin reaction at *D* when the frame is loaded by the 200-N · m couple.

4.78 The bars *AB* and *BC* of the structure are each of length *L* and weigh *W* and 2*W*, respectively. Find the tension in cable *DE* in terms of *W*, *L*, and the angle *θ*.

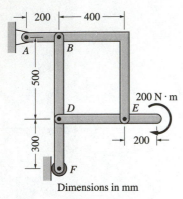

Dimensions in mm

Fig. P4.77

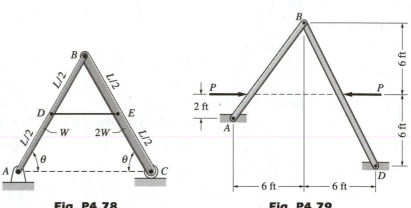

Fig. P4.78   **Fig. P4.79**

4.79 Determine the magnitude of the pin reaction at *A* as a function of *P*. The weights of the members are negligible.

4.80 Neglecting friction and the weights of the members, compute the magnitudes of the pin reactions at *A* and *C* for the folding table shown.

4.81 When activated by the force *P*, the gripper on a robotic arm is able to pick up objects by applying the gripping force *F*. Given that *P* = 120 N, calculate the gripping force for the position shown.

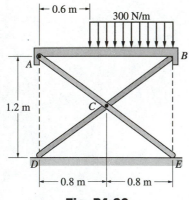

Fig. P4.80

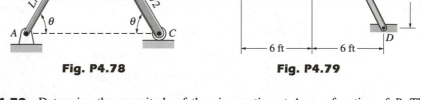

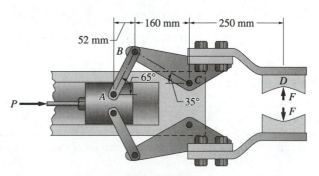

Fig. P4.81

4.82 Determine the axle loads (normal forces at A, B, and C) for the ore hauler when it is parked on a horizontal roadway with its brakes off. The masses of the cab and trailer are 4000 kg and 6000 kg, respectively, with centers of gravity at D and E. Assume that the connection at F is equivalent to a smooth pin.

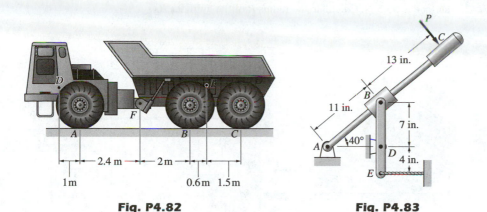

Fig. P4.82

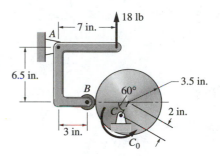

Fig. P4.83

4.83 Determine the force P that would produce a tensile force of 25 lb in the cable at E. Neglect the weights of the members.

4.84 Calculate the couple C_0 required for equilibrium if the weights of the members are negligible.

4.85 Determine the contact force between the smooth 200-lb ball B and the horizontal bar, and the magnitude of the pin reaction at A. Neglect the weights of the bar and the pulley.

Fig. P4.84

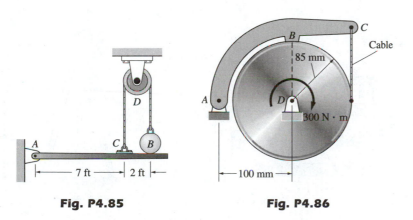

Fig. P4.85   **Fig. P4.86**

4.86 Compute the tension in the cable and the contact force at the smooth surface B when the 300-N · m couple is applied to the cylinder. Neglect the weights of the members.

4.87 The centers of gravity of the 40-kg arms of the lifting device are at G_1 and G_2. Determine the magnitude of the force that acts at D on the 60-kg drum B.

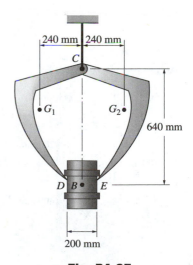

Fig. P4.87

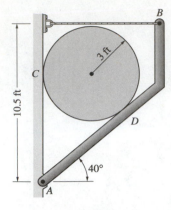

Fig. P4.88

4.88 Determine the tension in the cable at B, given that the uniform cylinder weighs 350 lb. Neglect friction and the weight of bar AB.

4.89 The masses of the frictionless cylinders A and B are 2.0 kg and 1.0 kg, respectively. The smallest value of the force P that will lift cylinder A off the horizontal surface is 55.5 N. Calculate the radius R of the cylinder B.

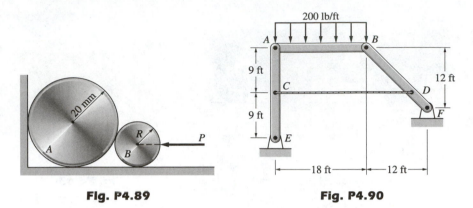

Fig. P4.89

Fig. P4.90

4.90 Neglecting the weight of the frame, find the tension in cable CD.

4.91 When the plunger A is pushed between the rollers B by the force P, the gripper can produce a large gripping force Q. The springs C return the fingers to the open position when the plunger is withdrawn. Assuming that the forces in the springs are relatively small, determine the relation between the activating force P and the gripping force Q.

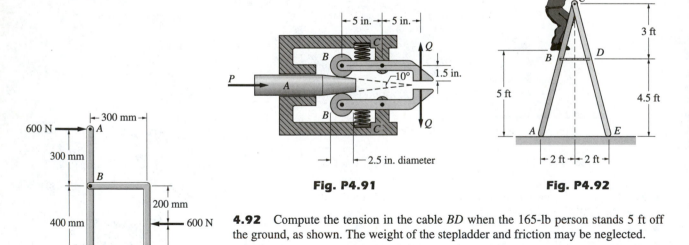

Fig. P4.91

Fig. P4.92

4.92 Compute the tension in the cable BD when the 165-lb person stands 5 ft off the ground, as shown. The weight of the stepladder and friction may be neglected.

4.93 Calculate the reactions at the built-in support at C, neglecting the weights of the members.

Fig. P4.93

4.94 In the angular motion amplifier shown, the oscillatory motion of *AC* is amplified by the oscillatory motion of *BC*. Neglecting friction and the weights of the members, determine the output torque C_0, given that the input torque is 36 N · m.

4.95 The linkage of the braking system consists of the pedal arm *DAB*, the connecting rod *BC*, and the hydraulic cylinder *C*. At what angle θ will the force *Q* be four times greater than the force *P* applied to the pedal? Neglect friction and the weight of the linkage.

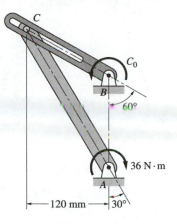

Fig. P4.94

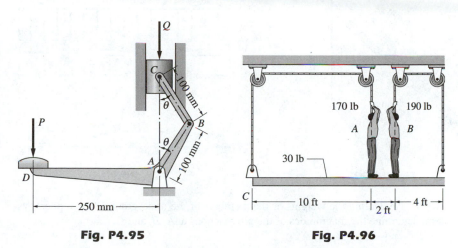

Fig. P4.95   **Fig. P4.96**

4.96 The window washers *A* and *B* support themselves and the uniform plank *CD* by pulling down on the two ropes. Determine (a) the tension in each rope and (b) the vertical force that each man exerts on the plank.

4.97 The scoop of the mechanical shovel, rigidly attached to the pulley *D*, is raised and lowered by the control cable that is wrapped around the pulley and attached to the cab at *C*. Compute the tensions in the control cable and cable *AB* that are due only to the 3200-lb load in the scoop.

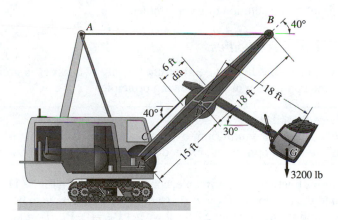

Fig. P4.97

4.98 Find the tension in the cable at *B* when the 180-N force is applied to the pedal at *E*. Neglect friction and the weights of the parts.

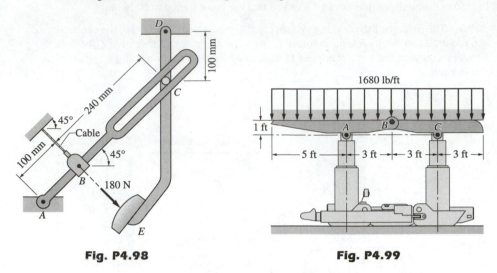

Fig. P4.98　　　　　　　　　**Fig. P4.99**

4.99 Neglecting the weights of the members of the underground roof support system, determine the magnitudes of the pin reactions at *A*, *B*, and *C*.

4.9 *Special Cases: Two-Force and Three-Force Bodies*

Up to now, we have been emphasizing a general approach to the solution of equilibrium problems. Special cases, with the exception of equal and opposite pin reactions, have been avoided so as not to interfere with our discussion of the general principles of equilibrium analysis. Here we study two special cases that can simplify the solution of some problems.

a. *Two-force bodies*

The analysis of bodies held in equilibrium by only two forces is greatly simplified by the application of the following principle.

> **Two-Force Principle**
> If a body is held in equilibrium by two forces, the forces must be equal in magnitude and oppositely directed along the same line of action.

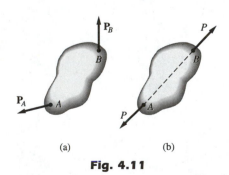

(a)　　　　　　　(b)

Fig. 4.11

To prove the two-force principle, consider the body in Fig. 4.11 (a) that is subjected to the two forces $\mathbf{P}_A$ and $\mathbf{P}_B$ (the forces do not have to be coplanar). From the equilibrium equation $\Sigma\mathbf{F} = \mathbf{0}$ we get $\mathbf{P}_A = -\mathbf{P}_B$. That is, the forces must be equal in magnitude and of opposite sense; they must form a couple. Because the second equilibrium equation, $\Sigma\mathbf{M}_O = \mathbf{0}$ (*O* is an arbitrary point), requires the magnitude of the couple to be zero, $\mathbf{P}_A$ and $\mathbf{P}_B$ must be collinear. We conclude that a two-force body can be in equilibrium only if the two forces are as shown in Fig. 4.11(b).

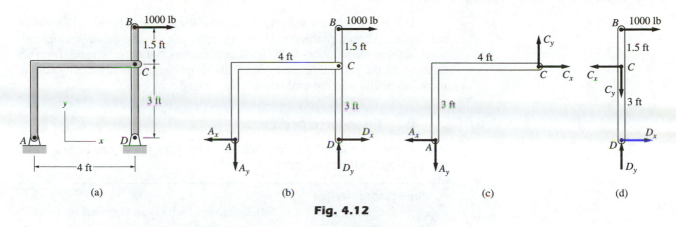

Fig. 4.12

To illustrate the use of the two-force principle, consider the frame shown in Fig. 4.12(a). Neglecting the weights of the members, the FBDs for the entire frame and each of its parts are as shown in Figs. 4.12(b) through (d). There are six unknowns (A_x, A_y, C_x, C_y, D_x, and D_y) and six independent equilibrium equations (three each for the two members). Therefore, the problem is statically determinate.

An efficient analysis is obtained if we recognize that member AC is a two-force body; that is, it is held in equilibrium by two forces—one acting at A (A_x and A_y are its components) and the other acting at C (C_x and C_y are its components). Using the two-force principle, we know—without writing any equilibrium equations—that the resultant forces at A and C are equal in magnitude and oppositely directed along the line joining A and C. The magnitude of these forces is labeled P_{AC} in Fig. 4.13.

Therefore, if we recognize that AC is a two-force body, either of the FBDs in Fig. 4.14 can be used to replace the FBDs in Fig. 4.12. Because each of the FBDs in Fig. 4.14 contains three unknowns (P_{AC}, D_x, and D_y) and provides us with three independent equilibrium equations, either could be solved completely.

Fig. 4.13

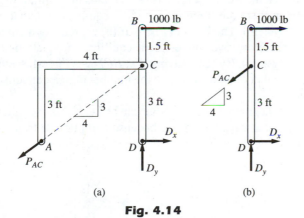

(a)

(b)

Fig. 4.14

It is not absolutely necessary that two-force bodies be identified when solving equilibrium problems. However, applying the two-force principle always reduces the number of equilibrium equations that must be used (from six to three, in the preceding example). This simplification is invariably convenient, particularly in the analysis of complicated problems.

b. Three-force bodies

The analysis of a body held in equilibrium by three forces can be facilitated by applying the following principle.

> **Three-Force Principle**
> Three nonparallel, coplanar forces that hold a body in equilibrium must be concurrent.

The proof of this principle can be obtained by referring to Fig. 4.15, which shows a body subjected to the three nonparallel, coplanar forces P_A, P_B, and P_C. Because the forces are not parallel, two of them—say, P_A and P_B—must intersect at some point, such as O. For the body to be in equilibrium, we must have $\Sigma M_O = 0$. Therefore, the third force, P_C, must also pass through O, as shown in Fig. 4.15. This completes the proof of the principle.

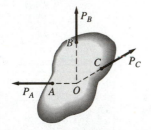

Fig. 4.15

As an example of the use of the three-force principle, consider once again the frame shown in Fig. 4.12. We have already shown how the analysis is simplified by recognizing that member AC is a two-force body. A further simplification can be made if we utilize the fact that member BCD is a three-force body.

The FBD of member BCD, repeated in Fig. 4.16(a), shows that the member is held in equilibrium by three nonparallel, coplanar forces. Knowing that the three forces must be concurrent, we could draw the FBD of BCD as shown in Fig. 4.16(b). Because the 1000-lb force and P_{AC} intersect at point E, the pin reaction at D must also pass through that point. Therefore, the two components D_x and D_y can be replaced by a force R_D with the slope 11/4.

Observe that the FBD in Fig. 4.16(a) contains three unknowns (P_{AC}, D_x, and D_y) and that there are three independent equilibrium equations (general coplanar force system). The FBD in Fig. 4.16(b) contains two unknowns (P_{AC} and R_D), and there are two independent equilibrium equations (concurrent, coplanar force system). By recognizing that BCD is a three-force body, we reduce both the number of unknowns and the number of independent equilibrium equations by one.

The use of the three-force principle can be helpful in the solution of some problems. It is not always beneficial, because complicated trigonometry may be required to locate the point where the three forces intersect.

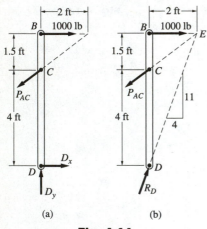

(a)   (b)

Fig. 4.16

Sample Problem 4.15

Determine the pin reactions at A and all forces acting on member DEF of the frame shown in Fig. (a). Neglect the weights of the members and use the two-force principle wherever applicable.

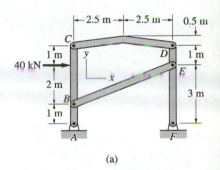

(a)

Solution

Method of Analysis

We begin by considering the FBD of the entire frame, Fig. (b). Because there are only three independent equilibrium equations, it will not be possible to find all four unknowns (A_x, A_y, F_x, and F_y) from this FBD alone, but it is possible to compute F_y from $\Sigma M_A = 0$, because it is the only unknown force that has a moment about point A. Similarly, $\Sigma M_F = 0$ will give A_y. However, in order to calculate A_x and F_x, we must consider the FBD of at least one member of the frame.

Note that members CD and BE are two-force bodies, because the only forces acting on them are the pin reactions at each end (the weights of the members are neglected). Therefore, the FBD of member DEF is as shown in Fig. (c). The forces P_{CD} and P_{BE} act along the lines CD and BE, respectively, as determined by the two-force principle. With F_y having been previously computed, the remaining three unknowns in the FBD (P_{CD}, P_{BE}, and F_x) can then be calculated. Returning to the FBD of the entire frame, Fig. (b), we can then find A_x from $\Sigma F_x = 0$.

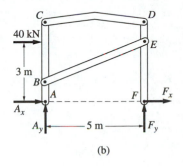

(b)

Mathematical Details

From the FBD of the entire frame, Fig. (b),

$$\Sigma M_A = 0 \quad \overset{+}{\curvearrowleft} \quad -40(3) + F_y(5) = 0$$

$$F_y = 24.0 \text{ kN} \qquad \textit{Answer}$$

$$\Sigma M_F = 0 \quad \overset{+}{\curvearrowleft} \quad -40(3) + A_y(5) = 0$$

$$A_y = -24.0 \text{ kN} \qquad \textit{Answer}$$

From the FBD of member DEF, Fig. (c),

$$\Sigma F_y = 0 \quad +\uparrow \quad F_y - \frac{2}{\sqrt{29}} P_{BE} = 0$$

$$P_{BE} = \frac{\sqrt{29}}{2}(24.0) = 64.6 \text{ kN} \qquad \textit{Answer}$$

$$\Sigma M_F = 0 \quad \overset{+}{\curvearrowleft} \quad P_{CD}(4) + \frac{5}{\sqrt{29}} P_{BE}(3) = 0$$

$$P_{CD} = -\frac{15}{4\sqrt{29}}(64.6) = -45.0 \text{ kN} \qquad \textit{Answer}$$

$$\Sigma F_x = 0 \quad \overset{+}{\rightarrow} \quad -P_{CD} - \frac{5}{\sqrt{29}} P_{BE} + F_x = 0$$

$$F_x = \frac{5}{\sqrt{29}}(64.6) + (-45.0) = 15.0 \text{ kN} \qquad \textit{Answer}$$

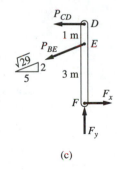

(c)

From the FBD of the entire frame, Fig. (b),

$$\Sigma F_x = 0 \quad \overset{+}{\rightarrow} \quad A_x + F_x + 40.0 = 0$$

$$A_x = -15.0 - 40.0 = -55.0 \text{ kN} \qquad \textit{Answer}$$

Other Methods of Analysis

There are, of course, many other methods of analysis that could be used. For example, we could analyze the FBDs of the members *ABC* and *DEF*, without considering the FBD of the entire frame.

Sample Problem 4.16

Neglecting the weights of the members in Fig. (a), determine the forces acting on the cylinder at *A* and *B*. Apply the two-force and three-force principles where appropriate. Use two methods of solution: utilizing (1) conventional equilibrium equations; and (2) the force triangle.

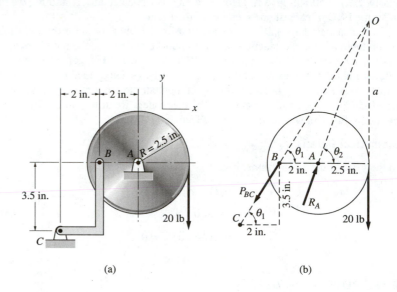

(a)   (b)

Solution

We begin by drawing the FBD of the cylinder, Fig. (b). Because bar *BC* is acted on only by the pin reactions at its ends, it is a two-force body. Therefore, the force at *B*, labeled P_{BC}, is directed along the line *BC*. The point where the line of action of P_{BC} intersects the 20-lb force is labeled *O*.

Next, we note that the cylinder is acted on by three forces: P_{BC}, the 20-lb applied force, and the pin reaction R_A. From the three-force principle, the line of action of R_A must also pass through point *O*, as shown in Fig. (b).

The angles θ_1 and θ_2, locating the lines of action of P_{BC} and R_A, respectively, can be found from trigonometry. Referring to Fig. (b), we obtain

$$\theta_1 = \tan^{-1}\left(\frac{3.5}{2}\right) = 60.3°$$

$$a = (2 + 2.5)\tan\theta_1 = 4.5\tan 60.3° = 7.89 \text{ in.}$$

$$\theta_2 = \tan^{-1}\left(\frac{a}{2.5}\right) = \tan^{-1}\left(\frac{7.89}{2.5}\right) = 72.4°$$

Part 1

The force system acting on the cylinder is concurrent and coplanar, yielding two independent equilibrium equations. Therefore, referring to the FBD in Fig. (b), the unknowns P_{BC} and R_A can be determined as follows:

$$\Sigma F_x = 0 \quad \xrightarrow{+} \quad -P_{BC}\cos\theta_1 + R_A\cos\theta_2 = 0$$

$$R_A = \frac{\cos\theta_1}{\cos\theta_2}P_{BC} = \frac{\cos 60.3°}{\cos 72.4°}P_{BC}$$

$$R_A = 1.639P_{BC} \qquad \text{(a)}$$

$$\Sigma F_y = 0 \quad +\uparrow \quad -P_{BC}\sin\theta_1 + R_A\sin\theta_2 - 20 = 0$$

$$-P_{BC}\sin 60.3° + R_A\sin 72.4° - 20 = 0 \qquad \text{(b)}$$

Solving Eqs. (a) and (b) simultaneously yields

$$R_A = 47.2 \text{ lb} \quad \text{and} \quad P_{BC} = 28.8 \text{ lb} \qquad \textit{Answer}$$

Part 2

Because the three forces acting on the cylinder are concurrent, the unknowns P_{BC} and R_A can be found by applying the law of sines to the force triangle in Fig. (c).

The angles in Fig. (c) may be computed as follows:

$$\alpha = 72.4° - 60.3° = 12.1°$$

$$\theta_3 = 90° - 60.3° = 29.7°$$

$$\beta = 180° - \theta_3 = 180° - 29.7° = 150.3°$$

$$\gamma = 180° - (\alpha + \beta) = 180° - (12.1° + 150.3°) = 17.6°$$

Applying the law of sines, we obtain

$$\frac{20}{\sin\alpha} = \frac{R_A}{\sin\beta} = \frac{P_{BC}}{\sin\gamma}$$

Substituting the values for α, β, and γ into this equation yields the same values for P_{BC} and R_A as given in Part 1.

The force triangle that results from the application of the three-force principle, Fig. (c), can also be solved graphically. If the triangle is drawn to a suitable scale, the unknown forces and angles can be measured directly.

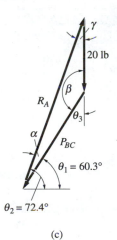

(c)

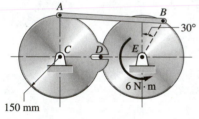

Fig. P4.101

Problems

Problems 4.100–4.120 each contain at least one two-force member. Solve by utilizing the two-force principle, where appropriate. If the weight of a body is not specifically stated, it can be neglected.

4.100 Solve Sample Problem 4.15 if bar *BE* is homogeneous and weighs 20 kN.

4.101 Calculate all forces acting on member *CDB*.

4.102 The automatic drilling robot must sustain a thrust of 38 lb at the tip of the drill bit. Determine the couple C_A that must be developed by the electric motor to resist this thrust.

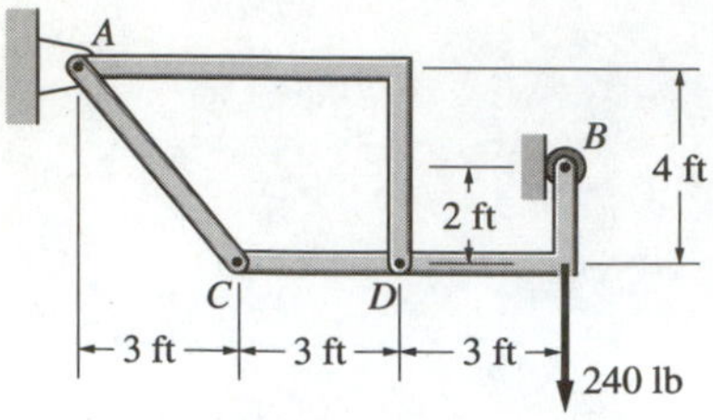

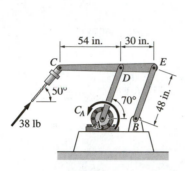

Fig. P4.102

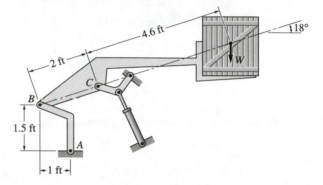

Fig. P4.103

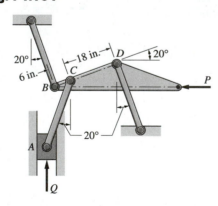

Fig. P4.104

4.103 Determine the magnitudes of the pin reactions at *B* and *C* as functions of *W* for the lifting mechanism shown.

4.104 The two disks are connected by the bar *AB* and the smooth peg in the slot at *D*. Compute the magnitude of the pin reaction at *A*.

4.105 Neglecting friction, determine the relationship between *P* and *Q*, assuming that the mechanism is in equilibrium in the position shown.

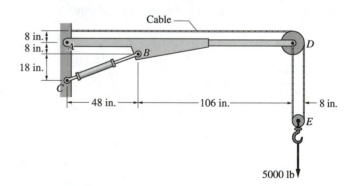

Fig. P4.105

Fig. P4.106

4.106 Calculate the magnitudes of the pin reactions acting on the crane at *A* and *C* due to the 5000-lb load.

4.107 The arm of the excavator is controlled by three hydraulic cylinders. If the soil in the scoop weighs 780 lb, with center of gravity at G, find the magnitude of the force applied by the cylinder AB to the arm BC.

4.108 In Prob. 4.107, calculate the pin reactions at E and F acting on the scoop.

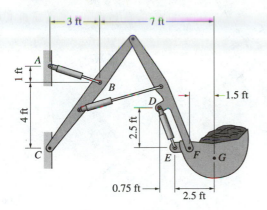

Fig. P4.107, P4.108

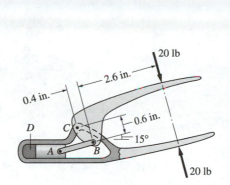

Fig. P4.109

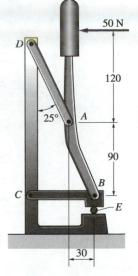

Dimensions in mm

Fig. P4.110

4.109 The tool shown is used to crimp terminals onto electric wires. The wire and terminal are inserted into the space D and are squeezed together by the motion of slider A. Compute the magnitude of the crimping force.

4.110 The 50-N force is applied to the handle of the toggle cutter. Determine the force exerted by the cutting blade CB on the workpiece E.

4.111 The blade of the bulldozer is rigidly attached to a linkage consisting of the arm AB, which is controlled by the hydraulic cylinder BC. There is an identical linkage on the other side of the bulldozer. Determine the magnitudes of the pin reactions at A, B, and C.

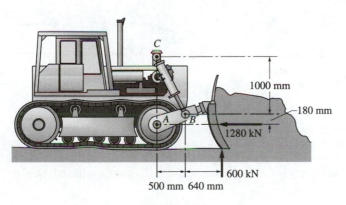

Fig. P4.111

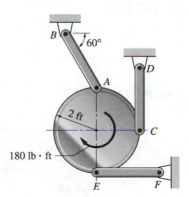

Fig. P4.112

4.112 Find the magnitudes of the pin reactions at A, C, and E caused by the 180-lb · ft couple.

4.113 Determine the forces exerted by the nutcracker at C and D.

4.114 When the C-shaped member is suspended from the edge of a frictionless table, it assumes the position shown. Use a graphical construction to find the distance x locating the center of gravity G.

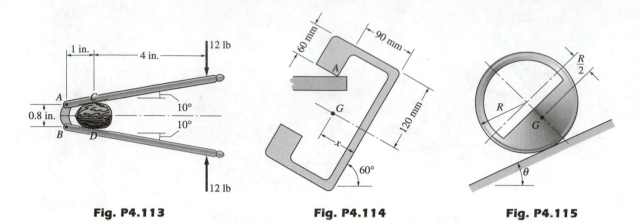

Fig. P4.113   **Fig. P4.114**   **Fig. P4.115**

4.115 The center of gravity of the eccentric wheel is at point G. Determine the largest angle θ for which the wheel will be at rest on a rough inclined surface.

4.116 For the pliers shown, determine the relationship between the magnitudes of the applied forces P and the gripping forces at E.

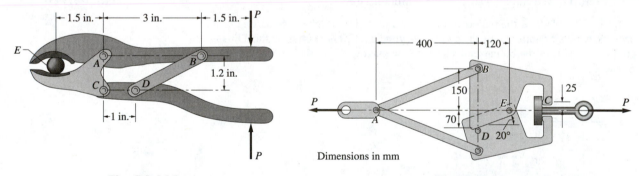

Fig. P4.116   **Fig. P4.117**

4.117 The device shown is an overload prevention mechanism. When the force acting on the smooth peg at D reaches 1.0 kN, the peg will be sheared, allowing the jaws at C to open and thereby releasing the eyebolt. Determine the maximum value of the tension P that can be applied without causing the eyebolt to be released. Neglect friction.

4.118 The figure represents the head of a pole-mounted tree pruner. Determine the force applied by the cutting blade *ED* on the tree branch when the vertical rope attached at *A* is pulled with the force *P*.

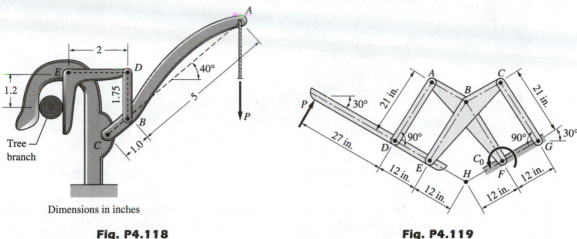

Fig. P4.118   **Fig. P4.119**

*4.119 The hinge shown is the type used on the doors of some automobiles. If a spring at *F* applies the constant couple $C_0 = 20$ lb · ft to member *ABF*, calculate the force *P* required to hold the door open in the position shown.

*4.120 The mechanism shown controls the movement of the fingers of a robotic gripper. The activating force *P* causes parallel movement of the fingers. Determine the relation between *P* and the gripping force *Q* for the position shown.

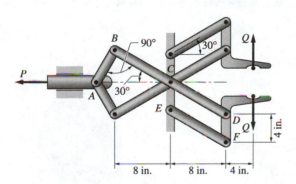

Fig. P4.120

Solve Probs. 4.121–4.130, using the two-force and three-force principles where appropriate.

4.121 Determine the horizontal force *P* that would keep the 15-kg rectangular plate in the position shown.

4.122 See Prob. 4.43.

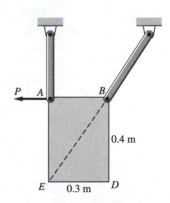

Fig. P4.121

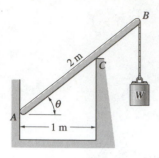

Fig. P4.123

4.123   Determine the angle θ at which the bar AB is in equilibrium. Neglect friction and the weight of the bar.

4.124   The automobile, with center of gravity at G, is parked on an 18° slope with its brakes off. Determine the height h of the smallest curb that will prevent the automobile from rolling down the plane.

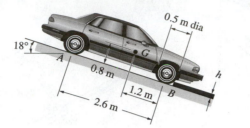

Fig. P4.124

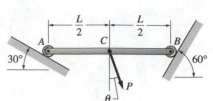

Fig. P4.125

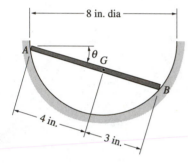

Fig. P4.126

4.125   Compute the angle θ for which the bar will be in equilibrium. Assume that the weight of the bar is negligible compared with the applied load P.

4.126   The center of gravity of the nonhomogeneous bar AB is located at G. Find the angle θ at which the bar will be in equilibrium if it is free to slide on the frictionless cylindrical surface.

4.127   When suspended from two cables, the rocket assumes the equilibrium position shown. Determine the distance x that locates G, the center of gravity of the rocket.

4.128   The pump oiler is operated by pressing on the handle at D, causing the plunger to raise and force out the oil. Determine the distance d of link BC so that the horizontal pin reaction at A is zero. Neglect the weights of the members.

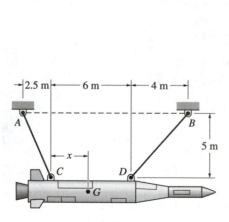

Fig. P4.127

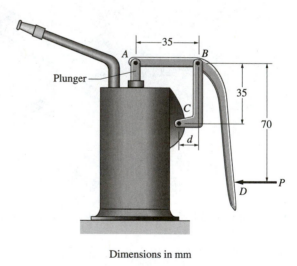

Dimensions in mm

Fig. P4.128

4.129   The uniform 320-lb bar is held in the position shown by the cable *AC*. Compute the tension in the cable.

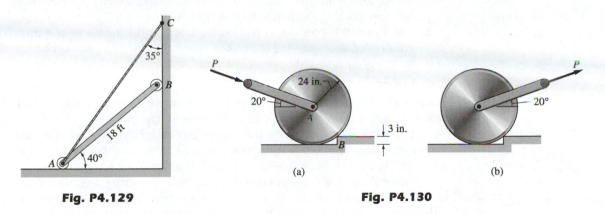

Fig. P4.129   **Fig. P4.130**

4.130   Find the force *P* required to (a) push; and (b) pull the 80-lb roller over the 3-in. curb.

PART C: *Analysis of Plane Trusses*

4.10   *Description of a Truss*

A *truss* is a structure that is made of straight, slender bars that are joined together to form a pattern of triangles. Trusses are usually designed to transmit forces over relatively long spans; common examples are bridge trusses and roof trusses. A typical bridge truss is shown in Fig. 4.17(a).

The analysis of trusses is based on the following three assumptions:

1. **The weights of the members are negligible.** A truss can be classified as a lightweight structure, meaning that the weights of its members are generally much smaller than the loads that it is designed to carry.
2. **All joints are pins.** In practice, the members at each joint are usually riveted or welded to a plate, called a *gusset plate,* as shown in Fig. 4.17(b).

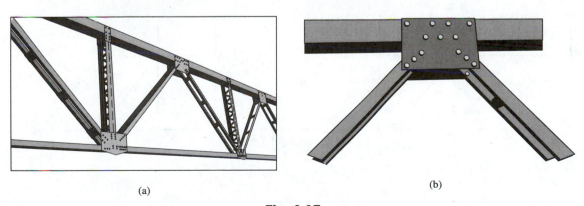

(a)   (b)

Fig. 4.17

However, if the members at a joint are aligned so that their centroidal axes (axes that pass through the centroids of the cross-sectional areas of the members) intersect at a common point, advanced methods of analysis indicate that the assumption of pins is justified.

3. **The applied forces act at the joints.** Because the members of a truss are slender, they may fail in bending when subjected to loads applied at locations other than the joints. Therefore, trusses are designed so that the major applied loads must act at the joints.

Although these assumptions may appear to oversimplify the real situation, they lead to results that are more than adequate in most applications.

Using the assumptions, the free-body diagram for any member of a truss will contain only two forces—the forces exerted on the member by the pin at each end. Therefore, *each member of a truss is a two-force body.*

When dealing with the internal force in a two-force body, engineers commonly distinguish between *tension* and *compression*. Figure 4.18 shows the external and internal forces in tension and compression. Tensile forces elongate (stretch) the member, whereas compressive forces compress (shorten) it. Because the forces act along the longitudinal axis of the member, they are often called *axial forces*. Note that internal forces always occur as equal and opposite pairs on the two faces of an internal cross section.

The two common techniques for computing the internal forces in a truss are the method of joints and the method of sections, each of which is discussed in the following articles.

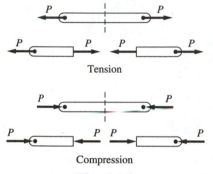

Fig. 4.18

4.11 *Method of Joints*

When using the *method of joints* to calculate the forces in the members of a truss, the equilibrium equations are applied to individual joints (or pins) of the truss. Because the members are two-force bodies, the forces in the FBD of a joint are concurrent. Consequently, two independent equilibrium equations are available for each joint.

To illustrate this method of analysis, consider the truss shown in Fig. 4.19(a). The supports consist of a pin at A and a roller at E (one of the supports is usually designed to be equivalent to a roller, in order to permit the elongation and contraction of the truss with temperature changes).

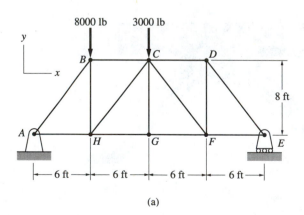

(a)

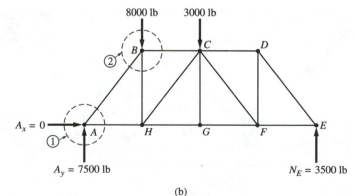

(b)

Fig. 4.19

a. Support reactions

It is usually a good idea to start the analysis by determining the reactions at the supports using the FBD of the entire truss. The FBD of the truss in Fig. 14.9(b) contains three unknown reactions: A_x, A_y, and N_E, which can be found from the three available equilibrium equations. The results of the computation are shown in Fig. 4.19(b).

Note that A_x, the horizontal reaction at A, is zero. This result indicates that the truss would be in equilibrium under the given loading even if the pin at A were replaced by a roller. However, we would then have an *improper constraint*, because an incidental horizontal force would cause the truss to move horizontally. Therefore, a pin support at A (or B) is necessary to properly constrain the truss.

Sometimes the number of unknown reactions on the FBD of the entire truss is greater than three. In this case, all the reactions cannot be found at the outset, but we may still be able to calculate some of them.

b. Equilibrium analysis of joints

Let us now determine the forces in the individual members of the truss. Because the force in a member is internal to the truss, it will appear on a FBD only if the FBD "cuts" the member, thereby separating it from the rest of the truss. For example, in order to determine the force in members AB and AH, we can draw the FBD of joint A—that is, the portion of the truss encircled by the dashed line ① in Fig. 4.19(b). This FBD, shown in Fig. 4.20(a), contains the external reactions A_x and A_y and the member forces P_{AB} and P_{AH} (the subscripts identify the member). Note that we have assumed the forces in the members to be tensile. If the solution yields a negative value for a force, the force is compressive. By assuming the members to be in tension, we are using an established convention for which positive results indicate tension and negative results indicate compression.

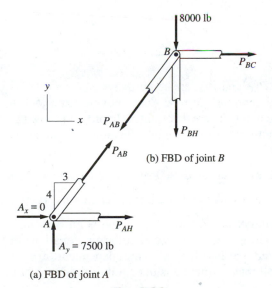

(a) FBD of joint A

(b) FBD of joint B

Fig. 4.20

Having previously computed A_x and A_y, the forces P_{AB} and P_{AH} are the only unknowns in the FBD for joint A. Therefore, they can be determined from the two independent equilibrium equations for the joint, as follows.

$$\Sigma F_y = 0 \quad +\uparrow \quad 7500 + \frac{4}{5}P_{AB} = 0$$

$$P_{AB} = -\frac{5}{4}(7500) = -9375 \text{ lb}$$

$$\Sigma F_x = 0 \quad \xrightarrow{+} \quad \frac{3}{5}P_{AB} + P_{AH} + A_x = 0$$

$$\frac{3}{5}(-9375) + P_{AH} + 0 = 0$$

$$P_{AH} = -\frac{3}{5}(-9375) = 5625 \text{ lb}$$

The negative value for P_{AB} indicates that the force in member AB is compressive; the positive value for P_{AH} means that the force in member AH is tensile.

To compute the forces in members BC and BH, we draw the FBD of joint B—the portion of the truss encircled by the dashed line ② in Fig. 4.19(b). This FBD is shown in Fig. 4.20(b). Note that the force P_{AB} is equal and opposite to the corresponding force in Fig. 4.20(a), and that we again assumed P_{BC} and P_{BH} to be tensile. Knowing that $P_{AB} = -9375$ lb, P_{BC} and P_{BH} are the only unknowns in this FBD. The equilibrium equations of the joint yield

$$\Sigma F_x = 0 \quad \xrightarrow{+} \quad P_{BC} - \frac{3}{5}P_{AB} = 0$$

$$P_{BC} = \frac{3}{5}P_{AB} = \frac{3}{5}(-9375) = -5625 \text{ lb}$$

$$\Sigma F_y = 0 \quad +\uparrow \quad -\frac{4}{5}P_{AB} - P_{BH} - 8000 = 0$$

$$P_{BH} = -8000 - \frac{4}{5}P_{AB}$$

$$= -8000 - \frac{4}{5}(-9375) = -500 \text{ lb}$$

Note that both P_{BC} and P_{BH} are compressive.

We could continue the procedure, moving from joint to joint, until the forces in all the members are determined. In order to show that this is feasible, we count the number of unknowns and the number of independent equilibrium equations:

$$13 \text{ member forces} + 3 \text{ support reactions} = 16 \text{ unknowns}$$

$$8 \text{ joints, each yielding 2 equilibrium equations} = 16 \text{ equations}$$

Because the number of equations equals the number of unknowns, the truss is statically determinate. The three equilibrium equations of the entire truss were not counted, because they are not independent of the joint equilibrium equations (recall that a structure is in equilibrium if each of its components is in equilibrium).

c.   Equilibrium analysis of pins

In the above example, the FBD of a joint contained a finite portion of the truss surrounding the joint. This required "cutting" the members attached to the joint, so that the internal forces in the members would appear on the FBD. An alternative approach, preferred by many engineers, is to draw the FBDs of the "pins," as illustrated in Fig. 4.21. In this case, the internal forces in the members appear as forces acting on the pin. For all practical purposes, the FBDs in Figs. 4.20 and 4.21 are identical. The FBD of a pin is easier to draw, but the FBD of a joint is somewhat more meaningful, particularly when it comes to determining whether the member forces are tensile or compressive.

d.   Zero-force members

There is a special case that occurs frequently enough to warrant special attention. Figure 4.22(a) shows the FBD for joint G of the truss in Fig. 4.19. Because no external loads are applied at G, the joint equilibrium equations $\Sigma F_x = 0$ and $\Sigma F_y = 0$ yield $P_{GH} = P_{GF}$ and $P_{GC} = 0$. Because member GC does not carry a force, it is called a *zero-force member*. It is easily verified that the results remain unchanged if member GC is inclined to GH and GF, as shown in Fig. 4.22(b). When analyzing a truss, it is often advantageous to begin by identifying zero-force members, thereby simplifying the solution.

You may wonder why a member, such as GC, is included in the truss if it carries no force. The explanation is the same as the one given for providing a pin support—rather than a roller—at A for the truss in Fig. 4.19(a): It is necessary to ensure the proper constraint of joint G. If member GC were removed, the truss would theoretically remain in equilibrium for the loading shown.* However, the slightest vertical load applied to the joint at G would cause the truss to deform excessively, or even collapse. Moreover, it is unlikely that the loads shown in Fig. 4.19(a) will be the only forces acting on the truss during its lifetime. Should a vertical load be suspended from joint G at some future time, member GC would be essential for equilibrium.

---

*The word *theoretically* is to be interpreted as "in accordance with the assumptions." Our mathematical model for a truss assumes that the weights of the members are negligible. In practice, the force in a so-called zero-force member is not exactly zero but is determined by the weights of the members.

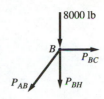

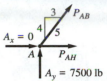

(b) FBD of pin B

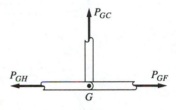

(a) FBD of pin A

Fig. 4.21

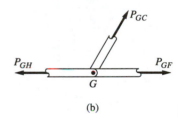

(a) FBD of joint G

(b)

Fig. 4.22

Sample Problem 4.17

Using the method of joints, determine the force in each member of the truss shown in Fig. (a). Indicate whether the members are in tension or compression.

Solution

The FBD of the entire truss is shown in Fig. (b). The three unknowns (N_A, C_x, and C_y) can be computed from the three equilibrium equations

$$\Sigma M_C = 0 \qquad \overset{+}{\curvearrowleft} \qquad -N_A(6) + 60(3) - 10(6) = 0$$

$$N_A = 20 \text{ kN}$$

$$\Sigma F_y = 0 \qquad +\uparrow \qquad N_A - 60 + C_y = 0$$

$$C_y = 60 - N_A = 60 - 20 = 40 \text{ kN}$$

$$\Sigma F_x = 0 \qquad \xrightarrow{+} \qquad 10 - C_x = 0$$

$$C_x = 10 \text{ kN}$$

We now proceed to the computation of the internal forces by analyzing the FBDs of various pins.

Method of Analysis

In the following discussion, the external reactions are treated as knowns, because they have already been calculated. It is convenient to assume the force in each member to be tensile. Therefore, positive values of the forces indicate tension, and negative values denote compression.

The FBD of pin A, shown in Fig. (c), contains two unknowns: P_{AB} and P_{AD}. We can compute these two forces immediately, because two independent equilibrium equations are available from this FBD.

The FBD of pin D, in Fig. (d), contains the forces P_{AD}, P_{BD}, and P_{CD}. Because P_{AD} has already been found, we have once again two equations that can be solved for the two unknowns.

Figure (e) shows the FBD of pin C. With P_{CD} previously found, the only remaining unknown is P_{BC}, which can be easily computed.

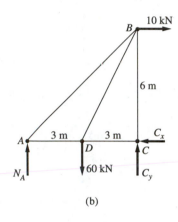

(a)

(b)

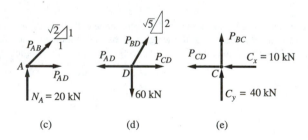

(c)   (d)   (e)

Mathematical Details

From the FBD of pin A, Fig. (c),

$$\Sigma F_y = 0 \quad +\uparrow \qquad \frac{1}{\sqrt{2}}P_{AB} + N_A = 0$$

$$P_{AB} = -\sqrt{2}(20) = -28.3 \text{ kN}$$

$$P_{AB} = 28.3 \text{ kN} \quad \text{(compression)} \qquad \textit{Answer}$$

$$\Sigma F_x = 0 \quad \xrightarrow{+} \qquad \frac{1}{\sqrt{2}}P_{AB} + P_{AD} = 0$$

$$P_{AD} = -\frac{1}{\sqrt{2}}P_{AB} = -\frac{1}{\sqrt{2}}(-28.3)$$

$$= 20.0 \text{ kN} \quad \text{(tension)} \qquad \textit{Answer}$$

From the FBD of pin D, Fig. (d),

$$\Sigma F_y = 0 \quad +\uparrow \qquad \frac{2}{\sqrt{5}}P_{BD} - 60 = 0$$

$$P_{BD} = \frac{\sqrt{5}}{2}(60) = 67.1 \text{ kN} \quad \text{(tension)} \qquad \textit{Answer}$$

$$\Sigma F_x = 0 \quad \xrightarrow{+} \qquad -P_{AD} + \frac{1}{\sqrt{5}}P_{BD} + P_{CD} = 0$$

$$P_{CD} = P_{AD} - \frac{1}{\sqrt{5}}P_{BD}$$

$$P_{CD} = 20.0 - \frac{1}{\sqrt{5}}(67.1) = -10.0 \text{ kN}$$

$$P_{CD} = 10.0 \text{ kN} \quad \text{(compression)} \qquad \textit{Answer}$$

From the FBD of pin C, Fig. (e),

$$\Sigma F_y = 0 \quad +\uparrow \quad C_y + P_{BC} = 0$$

$$P_{BC} = -C_y = -40 \text{ kN}$$

$$P_{BC} = 40 \text{ kN} \quad \text{(compression)} \qquad \textit{Answer}$$

Note that the equation $\Sigma F_x = 0$ yields $P_{CD} = -10.0$ kN, a value that has been found before. Therefore, this equation is not independent of the equations used previously. The reason for the dependence is that the external reactions were determined by analyzing the FBD of the entire truss. As discussed previously, however, the equations for the pins and those for the entire truss are not independent of each other.

Other Methods of Analysis

In the preceding analysis, the pins were considered in the following order: A, D, and C (the FBD of pin B was not used). Another sequence that could be used is

1. FBD of pin A: Calculate P_{AB} and P_{BD} (as before).
2. FBD of pin B: With P_{AB} already found, calculate P_{BD} and P_{BC}.
3. FBD of pin C: Calculate P_{CD}.

In this analysis, the FBD of pin D would not be used.

Yet another approach would be to compute the three external reactions and the forces in the five members (a total of eight unknowns) by using the equilibrium equations for all the pins (a total of eight equations, two for each pin).

Problems

4.131–4.140 Using the method of joints, calculate the force in each member of the trusses shown. State whether each member is in tension or compression.

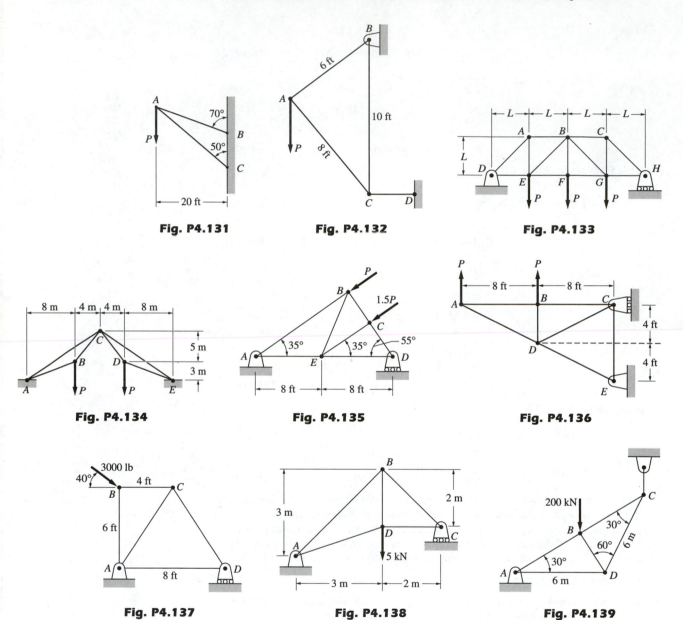

Fig. P4.131

Fig. P4.132

Fig. P4.133

Fig. P4.134

Fig. P4.135

Fig. P4.136

Fig. P4.137

Fig. P4.138

Fig. P4.139

4.141 If the load *P* acting at *B* is removed from each of the trusses in Probs. 4.134–4.136, identify the zero-force members.

4.142 The walkway *ABC* of the footbridge is stiffened by adding the cable *ADC* and the short post of length *L*. If the tension in the cable is not to exceed 500 lb, what is the smallest value of *L* for which the 185-lb person can be supported at *B*?

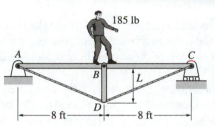

185 lb

Fig. P4.142

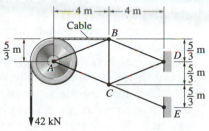

Fig. P4.143

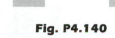

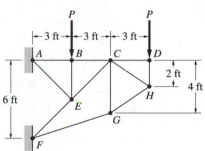

Fig. P4.140

4.143 Calculate the forces in members *AB*, *AC*, and *BD*.

4.144 Find the forces in members *HC* and *HG* in terms of *P*.

4.145 Determine the reaction at *E* and the force in each member of the right half of the truss.

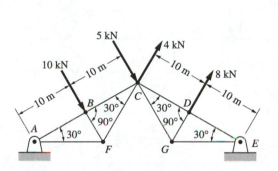

Fig. P4.144

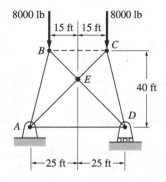

Fig. P4.145

Fig. P4.146

4.146 Determine the force in member *AD* of the truss.

4.147 Determine the force in member *BE* of the truss.

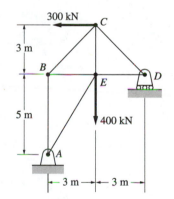

Fig. P4.147

4.12 *Method of Sections*

Truss analysis by the method of joints is based on the FBDs of individual joints. Analyzing the free-body diagram of a part of a truss that contains two or more joints is called the *method of sections*. The FBD for a single joint results in a concurrent, coplanar force system (two independent equilibrium equations). When applying the method of sections, the force system will generally be non-concurrent, coplanar (three independent equilibrium equations).

In the method of sections, a part of the truss is isolated on an FBD so that it exposes the forces to be computed. If the FBD for the isolated portion contains three unknowns, all of them can usually be computed from the three available equilibrium equations. If the number of unknowns exceeds three, one or more of the unknowns must be found by analyzing a different part of the truss. If you are skillful in writing and solving equilibrium equations, the only challenge in using the method of sections is selecting a convenient part of the truss for the FBD.

Consider once again the truss discussed in the preceding article [its FBD is repeated in Fig. 4.23(a)]. We now use the method of sections to determine the forces in members BC, HC, HG, and DF—each of these members is identified by two short parallel lines in Fig. 4.23(a).

Assuming that the external reactions have been previously computed, the first and most important step is the selection of the part of the truss to be analyzed. We note that the section labeled ① in Fig. 4.23(a) passes through members BC, HC, and HG. The forces in these three members are the only unknowns if the FBD is drawn for a part of the truss that is isolated by this section. Note that after the section has been chosen, the portion of the truss on either side of the cut may be used for the FBD. The forces inside the members occur in equal and opposite pairs, so the same results will be obtained regardless of which part is analyzed. Of course, given a choice, one would naturally select the less complicated part.

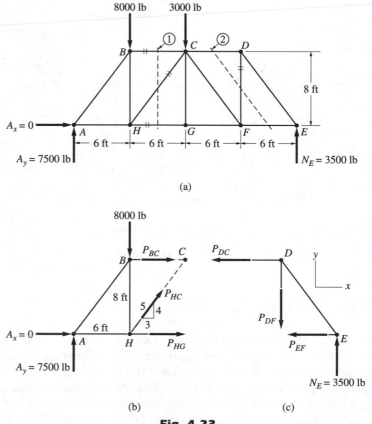

(a)

(b)   (c)

Fig. 4.23

For our truss, using either the portion to the left or to the right of section ① is equally convenient. For no particular reason, we choose to analyze the left portion, with its FBD shown in Fig. 4.23(b) (we have again assumed the members to be in tension). Note that the force system is nonconcurrent and coplanar, so that any set of three independent equations can be used to compute the three unknown forces.

The following is an efficient method of solution.

$\Sigma F_y = 0$—determines P_{HC}, because P_{BC} and P_{HG} have no y-components
$\Sigma M_H = 0$—determines P_{BC}, because P_{HC} and P_{HG} have no moment about point H
$\Sigma M_C = 0$—determines P_{HG}, because P_{BC} and P_{HC} have no moment about point C

The details of this analysis are

$$\Sigma F_y = 0 \qquad +\uparrow \qquad \frac{4}{5}P_{HC} - 8000 + 7500 = 0$$

$$P_{HC} = \frac{5}{4}(8000 - 7500)$$

$$= 625 \text{ lb} \quad \text{(tension)} \qquad \textit{Answer}$$

$$\Sigma M_H = 0 \qquad \overset{+}{\curvearrowright} \qquad -7500(6) - P_{BC}(8) = 0$$

$$P_{BC} = -5625 \text{ lb}$$

$$= 5625 \text{ lb} \quad \text{(compression)} \qquad \textit{Answer}$$

$$\Sigma M_C = 0 \qquad \overset{+}{\curvearrowright} \qquad 8000(6) - 7500(12) + P_{HG}(8) = 0$$

$$P_{HG} = \frac{1}{8}[7500(12) - 8000(6)]$$

$$= 5250 \text{ lb} \quad \text{(tension)} \qquad \textit{Answer}$$

To determine the force in member *DF* by the method of sections, consider once again the FBD of the entire truss in Fig. 4.23(a). Our intention is to isolate a part of the truss by cutting only three members, one of which is the member *DF*. It can be seen that section ② accomplishes this task. We choose to analyze the part to the right of this section because it contains fewer forces. The FBD is shown in Fig. 4.23(c), with the unknown forces again assumed to be tensile. Note that P_{DF} can be computed from the equation

$$\Sigma F_y = 0 \qquad +\uparrow \qquad 3500 - P_{DF} = 0$$

$$P_{DF} = 3500 \text{ lb} \quad \text{(tension)} \qquad \textit{Answer}$$

If desired, P_{EF} could now be calculated using $\Sigma M_D = 0$, and $\Sigma M_E = 0$ would give P_{DC}.

As you see, the forces in the members of a truss can be found by either the method of joints or the method of sections. Selecting the method that results in the most straightforward analysis is usually not difficult. For example, for the truss shown in Fig. 4.23(a), the FBD of joint *A* is convenient for computing P_{AB}, whereas the method of sections is more advantageous for calculating P_{BC}.

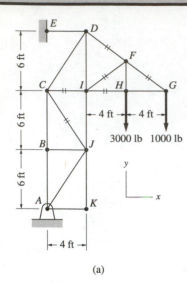

(a)

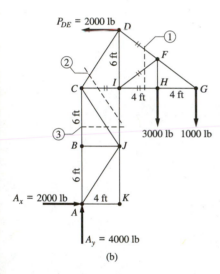

(b)

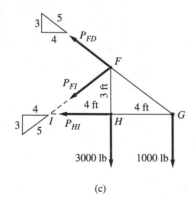

(c)

Sample Problem 4.18

Using the method of sections, determine the forces in the following members of the truss in Fig. (a): *FD*, *FI*, *HI*, *CI*, and *JC* (each member of interest is marked by two short, parallel lines). Indicate tension or compression.

Solution

The FBD of the entire truss is shown in Fig. (b). The three external reactions (A_x, A_y, and P_{DE}) can be calculated using the following equations (mathematical details have been omitted):

$$\Sigma M_A = 0 \quad \text{gives} \quad P_{DE} = 2000 \text{ lb}$$
$$\Sigma F_x = 0 \quad \text{gives} \quad A_x = 2000 \text{ lb}$$
$$\Sigma F_y = 0 \quad \text{gives} \quad A_y = 4000 \text{ lb}$$

These forces are shown in Fig. (b), and from now on we treat them as known quantities.

Members *FD*, *FI*, and *HI*

Method of Analysis

Referring to the FBD in Fig. (b), it can be seen that section ① cuts through members *FD*, *FI*, and *HI*. Because there are three equilibrium equations available for a portion of the truss separated by this section, we can find the forces in all three members.

Mathematical Details

Having chosen section ①, we must now decide which portion of the truss to analyze. We select the portion lying to the right of the section, because it is somewhat less complicated than the portion on the left. Of course, identical results will be obtained regardless of which part of the truss is analyzed.

The FBD for the part of the truss lying to the right of section ① is shown in Fig. (c), with the three unknowns being P_{FD}, P_{FI}, and P_{HI}. Any set of three independent equations can be used to solve for these unknowns. A convenient solution is the following:

$$\Sigma M_G = 0 \quad \text{gives } P_{FI}, \text{ because } P_{FD} \text{ and } P_{HI} \text{ pass through point } G.$$

$$\overset{+}{\curvearrowright} \quad 3000(4) + \frac{3}{5}P_{FI}(8) = 0$$

$$P_{FI} = -2500 \text{ lb}$$

$$= 2500 \text{ lb} \quad \text{(compression)} \qquad \qquad \textit{Answer}$$

Note: The moment of P_{FI} about point *G* was computed by replacing P_{FI} with its *x*- and *y*-components acting at point *I*.

$$\Sigma M_I = 0 \quad \text{gives } P_{FD}, \text{ because } P_{FI} \text{ and } P_{HI} \text{ pass through point } I.$$

$$\overset{+}{\curvearrowright} \quad \frac{3}{5}P_{FD}(8) - 1000(8) - 3000(4) = 0$$

$$P_{FD} = 4170 \text{ lb} \quad \text{(tension)} \qquad \qquad \textit{Answer}$$

Note: When computing the moment of P_{FD} about point *I*, we considered the components of P_{FD} to be acting at point *G*.

$$\Sigma M_F = 0 \quad \text{gives } P_{HI}, \text{ because } P_{FD} \text{ and } P_{FI} \text{ pass through point } F.$$

$$\overset{+}{\curvearrowright} \quad P_{HI}(3) + 1000(4) = 0$$

$$P_{HI} = -1333 \text{ lb} = 1333 \text{ lb} \quad \text{(compression)} \qquad \textit{Answer}$$

The three equations used to compute the forces in members FD, FI, and HI were convenient, because each equation involved only one unknown. Of course, any three independent equations can be used, but they may require the simultaneous solution of two or more equations.

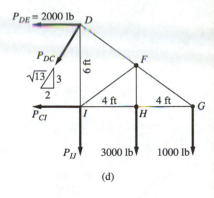

(d)

Member CI

Method of Analysis

Referring to Fig. (b), we see that section ② also passes through three members, which means that the forces in all three members can be obtained from a single FBD.

Mathematical Details

We choose to analyze the portion of the truss lying to the right of section ②. The FBD is shown in Fig. (d), and the three unknowns are P_{DC}, P_{CI}, and P_{IJ}. The force of interest, P_{CI}, can be found by using the following moment equation:

$$\Sigma M_D = 0 \quad \text{gives } P_{CI}, \text{ because } P_{DC} \text{ and } P_{IJ} \text{ pass through point } D.$$

$$\overset{+}{\curvearrowleft} \quad 3000(4) + 1000(8) + P_{CI}(6) = 0$$

$$P_{CI} = -3330 \text{ lb} = 3330 \text{ lb} \quad \text{(compression)} \qquad \textit{Answer}$$

Member JC

Method of Analysis

In Fig. (b), section ③ passes through only three members, one of which is member JC. Therefore, we can find the force in that member (and the forces in the other two members if desired) from one FBD.

Mathematical Details

The FBD for the portion of the truss lying below section ③ is shown in Fig. (e). The three unknowns are P_{BC}, P_{JC}, and P_{IJ}. The force P_{JC} can be obtained from the following equation:

$$\Sigma F_x = 0 \quad \overset{+}{\longrightarrow} \quad A_x - \frac{2}{\sqrt{13}}P_{JC} = 0$$

$$P_{JC} = \frac{\sqrt{13}}{2}A_x = \frac{\sqrt{13}}{2}(2000)$$

$$= 3610 \text{ lb} \quad \text{(tension)} \qquad \textit{Answer}$$

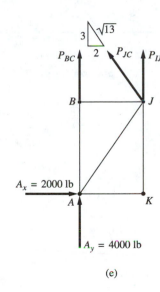

(e)

Problems

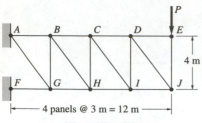

Fig. P4.148

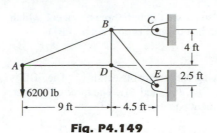

Fig. P4.149

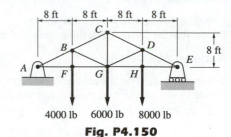

Fig. P4.150

4.148 Show that all diagonal members of the truss carry the same force, and find the magnitude of this force.

4.149 Find the forces in members *AB*, *BD*, and *DE*.

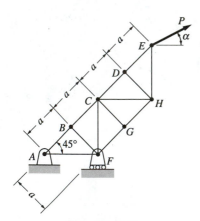

Fig. P4.151

4.150 Determine the forces in members *BC*, *BG*, and *FG*.

4.151 Calculate the force in member *CE*.

4.152 Compute the forces in members *EF*, *NF*, and *NO*.

4.153 Repeat Prob. 4.152 assuming that the 300-kN force is applied at *O* instead of *L*.

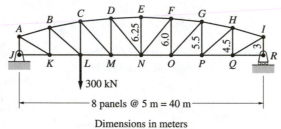

Dimensions in meters

Fig. P4.152, P4.153

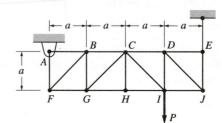

Fig. P4.154, P4.155

4.154 Determine the forces in members *BG*, *CI*, and *CD*.

4.155 Assuming that $P = 48\,000$ lb and that it may be applied at any joint on the line *FJ*, determine the location of *P* that would cause (a) maximum tension in member *HI*; (b) maximum compression in member *CI*; and (c) maximum tension in member *CI*. Also determine the magnitude of the indicated force in each case.

4.156 Given that $P_{GH} = 0.6$ kN and $P_{BC} = 4.8$ kN, both tensile, determine *P* and α.

Fig. P4.156

4.157 The platform carrying the uniformly distributed load is supported by the truss shown. Calculate the forces in members *DE*, *EJ*, and *JK*.

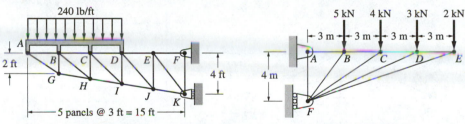

Fig. P4.157   **Fig. P4.158**

4.158 Determine the forces in members *CD* and *DF*.

4.159 Compute the forces in members *CD* and *JK*, given that $P = 3000$ lb and $Q = 1000$ lb. (*Hint:* Use the section indicated by the dashed line.)

4.160 If $P_{CD} = 6000$ lb and $P_{GD} = 1000$ lb (both compression), find *P* and *Q*.

4.161 Determine the forces in members *BC*, *EC*, and *FG* for the truss in Prob. 4.144 in terms of *P*.

4.162 The capacity of the crane hoist is 200 kN, and the mass of the counterweight at *D* is 38 Mg. Determine the range of the force in member *GK* as the hoist travels between *G* and *I*. (*Hint:* Be sure to consider the case $P = 0$.)

4.163 Compute the force in member *CG* of the crane in Prob. 4.162 if $P = 180$ kN and the hoist is at (a) joint *I*; and (b) joint *G*.

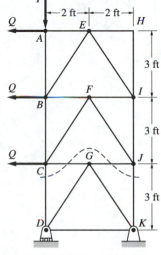

Fig. P4.159, P4.160

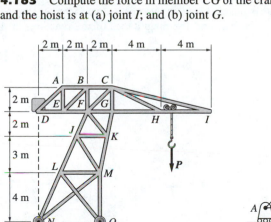

Fig. P4.162, P4.163

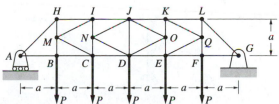

Fig. P4.164

4.164 Determine the forces in members *CD*, *IJ*, and *NJ* of the K-truss in terms of *P*.

4.165 Determine the largest allowable value for the angle θ if the magnitude of the force in member *BC* is not to exceed 5*P*.

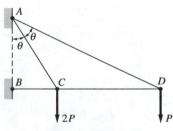

Fig. P4.165

4.166 Find the forces in members *BC* and *BG*.

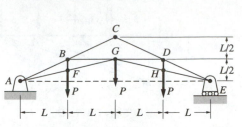

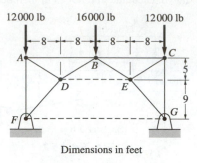

12000 lb   16000 lb   12000 lb

Dimensions in feet

Fig. P4.166   **Fig. P4.167**

4.167 Determine the forces in members *BC* and *BE* and the horizontal pin reaction at *G*.

4.168 A couple acting on the winch at *G* slowly raises the load *W* by means of a rope that runs around the pulleys attached to the derrick at *A* and *B*. Determine the forces in members *EF* and *KL* of the derrick, assuming the diameters of the pulleys and the winch to be negligible.

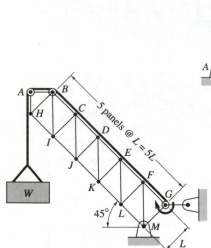

Fig. P4.168

Review Problems

4.169 The uniform, 20-kg bar is placed between two vertical surfaces. Assuming sufficient friction at *A* to support the bar, find the magnitudes of the reactions at *A* and *B*.

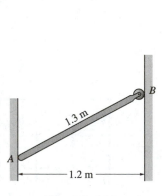

Fig. P4.169

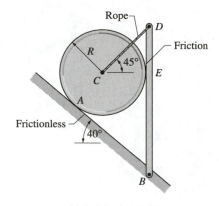

Fig. P4.170

4.170 The homogeneous cylinder of weight *W* is placed on an inclined surface of negligible friction. The cylinder is prevented from moving by the vertical rod *DB*, which is connected to the center of the cylinder by a rope. Assuming sufficient friction between the bar and the cylinder, determine the tension in the rope and the magnitudes of the contact forces at *A* and *E*.

4.171 Determine the magnitude of the pin reaction at *A*, assuming the weight of bar *ABC* to be negligible.

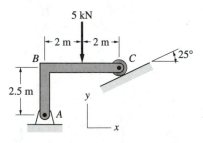

Fig. P4.171

4.172 Neglecting the weight of the boom *ABC*, determine the tension in the supporting cable *BD* as a function of the angle θ. For what value of θ will the tension be zero?

4.173 Calculate the magnitudes of the pin reactions at *A*, *B*, and *C* for the frame shown. Neglect the weights of the members.

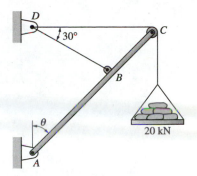

Fig. P4.172

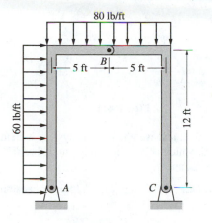

Fig. P4.173

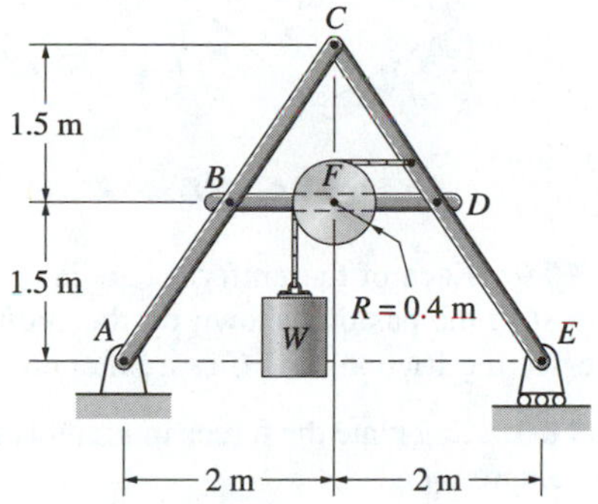

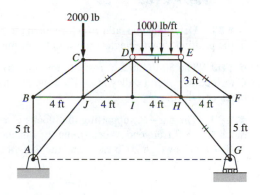

Fig. P4.174

4.174 The weight $W = 6$ kN hangs from the cable, which passes over the pulley at *F*. Neglecting the weights of the bars and the pulley, determine the magnitude of the pin reaction at *D*.

4.175 The 10-kN and 40-kN forces are applied to the pins at *B* and *C*, respectively. Calculate the magnitudes of the pin reactions at *A* and *F*. Neglect the weights of the members.

4.176 The two couples act at the midpoints of bars *AB* and *BD*. Determine the magnitudes of the pin reactions at *A* and *D*. Neglect the weights of the members.

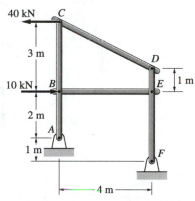

Fig. P4.175

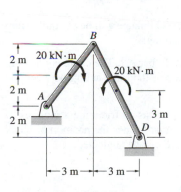

Fig. P4.176

Fig. P4.177

4.177 Determine the forces in members (a) *DJ*; (b) *DE*; (c) *EF*; and (d) *GH*. Indicate tension or compression.

4.178   Determine the angle θ for which the uniform bar of length L and weight W will be in equilibrium. Neglect friction.

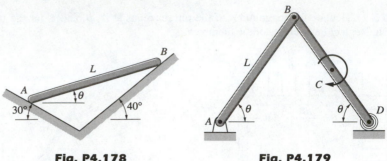

Fig. P4.178   **Fig. P4.179**

4.179   Each of the uniform bars is of length L and weight W. The system is held at rest in the position shown by the couple C, which acts at the midpoint of bar BD. Neglecting friction, find C as a function of W, L, and θ.

4.180   Calculate the forces in members (a) DE; (b) BE; and (c) BC. Indicate tension or compression.

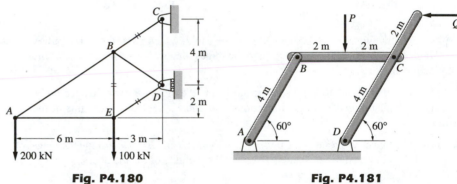

Fig. P4.180   **Fig. P4.181**

4.181   Determine the ratio P/Q for which the parallel linkage will be in equilibrium in the position shown. Neglect the weights of the members.

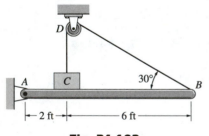

Fig. P4.182

4.182   The 50-lb block C rests on the uniform 20-lb bar AB. The cable connecting C to B passes over a pulley at D. Find the magnitude of the force acting between the block and the bar.

4.183   The bar AB, supporting the blocks C and D, rests on frictionless inclines. If block C is 5 ft from end A, determine the distance x that locates the position of block D. The blocks weigh 20 lb each, and the weight of AB can be neglected.

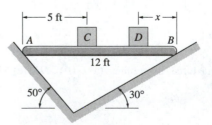

Fig. P4.183

4.184 Determine the forces in members (a) *BF*; and (b) *EF*. Indicate tension or compression.

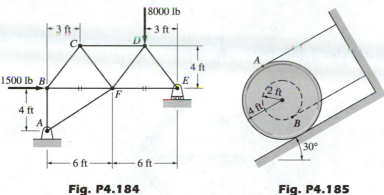

Fig. P4.184

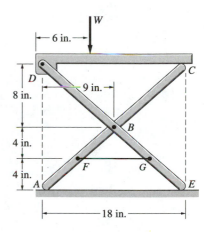

Fig. P4.185

4.185 The homogeneous spool weighing 200 lb is at rest on the frictionless incline. Calculate the tensions in the cables attached at *A* and *B*, respectively.

4.186 The breaking strength of the cable *FG* that supports the portable camping stool is 400 lb. Determine the maximum weight *W* that can be supported. Neglect friction and the weights of the members.

4.187 For the truss shown, determine the forces in members (a) *BD*; and (b) *BF*.

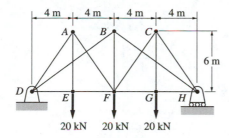

Fig. P4.187

Fig. P4.186

4.188 The 80-N force is applied to the handle of the embosser at *E*. Determine the resulting normal force exerted on the workpiece at *D*. Neglect the weights of the members.

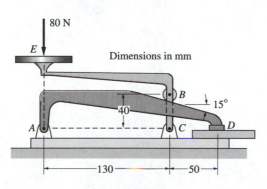

Fig. P4.188

4.189   The tongs shown are designed for lifting blocks of ice. If the weight of the iceblock is W, find the horizontal force between the tongs and the iceblock at C and D.

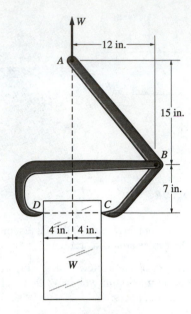

Fig. P4.189

5

Noncoplanar Equilibrium

5.1 Introduction

In this chapter, we discuss the analysis of bodies that are held in equilibrium by noncoplanar (three-dimensional) force systems. The emphasis on free-body diagrams and the number of independent equations, begun in Chapter 4, is continued here.

In the analysis of coplanar force systems, there was little advantage in using vector notation. This is not true for noncoplanar equilibrium analysis, where vector notation frequently has a decided advantage over scalar notation.

5.2 Definition of Equilibrium

By definition, a body is in equilibrium if the resultant of the force system acting on the body vanishes. A general force system can always be reduced to an equivalent force-couple system $\mathbf{R} = \Sigma \mathbf{F}$ and $\mathbf{C}^R = \Sigma \mathbf{M}_O$ (where O is any point). Therefore, for a body to be in equilibrium, the following two vector equations must be satisfied:

$$\Sigma \mathbf{F} = \mathbf{0} \qquad \Sigma \mathbf{M}_O = \mathbf{0} \qquad (5.1)$$

The equivalent six scalar equations are

$$\begin{aligned} \Sigma F_x &= 0 & \Sigma F_y &= 0 & \Sigma F_z &= 0 \\ \Sigma M_x &= 0 & \Sigma M_y &= 0 & \Sigma M_z &= 0 \end{aligned} \qquad (5.2)$$

where the x-, y-, and z-axes are assumed to intersect at point O. It is important to recall that the summations must be taken over *all* forces acting on the body, including the support reactions. This, of course, leads us to a discussion of free-body diagrams.

5.3 Free-Body Diagrams

Our study of coplanar equilibrium in Chapter 4 has demonstrated the importance of correctly drawn free-body diagrams (FBDs) in the solution of equilibrium problems. To extend the free-body diagram concept to problems in which the loads are noncoplanar, we must again investigate the reactions that are applied by the various connections and supports.

Support	Reaction(s)	Description of reaction(s)	Number of unknowns
(a) Flexible cord or cable of negligible weight	T	Tension of unknown magnitude T in the direction of cord or cable	One
(b) Spherical roller on any surface or a single point of contact on frictionless surface	N	Force of unknown magnitude N directed normal to the surface	One
(c) Cylindrical roller on friction surface or guide rail	F N	Force of unknown magnitude N normal to the surface and a force of unknown magnitude F in the direction of the axis of the roller	Two
(d) Ball-and-socket joint	R_x R_y R_z	Unknown force **R**	Three
(e) Single point of contact on friction surface	F_x F_y N	Force of unknown magnitude N normal to the surface and an unknown friction force **F** in the plane of the surface	Three

Table 5.1   *Common Supports for Noncoplanar Loads (Table continues on p. 211.)*

As mentioned in Chapter 4, the reactions that a connection is capable of exerting on the body can be derived from the following rule: A support that prevents translation in a given direction must apply a force in that direction and a support that prevents rotation about an axis must apply a couple about that axis. Some common supports for noncoplanar loads are illustrated in Table 5.1 and are described in the following paragraphs.

Support	Reaction(s)	Description of reaction(s)	Number of unknowns
(f) Slider (radial) bearing or hinge	R_z, R_y / R_z, R_y	Unknown force **R** directed normal to the axis of the bearing or hinge	Two
(g) Thrust bearing or hinge	R_z, R_x, R_y / R_z, R_x, R_y	Unknown force **R**	Three
(h) Universal joint	R_z, R_x, R_y, C	Unknown force **R** and a couple-vector of unknown magnitude C directed along the axis of the joint	Four
(i) Built-in (cantilever) support	R_z, C_z, R_x, R_y, C_x, C_y	Unknown force **R** and an unknown couple-vector **C**	Six

Table 5.1 (*continued*)

(a) Flexible Cable (Negligible Weight).   A flexible cable can exert a tensile force only in the direction of the cable. (With the weight of the cable neglected, the cable can be shown as a straight line.) Assuming that the direction of the cable is known, the removal of a cable introduces one unknown in the free-body diagram—the magnitude of the tension.

(b) Spherical Roller or Single Point of Contact on Frictionless Surface.   A spherical roller, or a frictionless surface with a single point of contact, can exert only a force that acts normal to the surface. Consequently, the magnitude of the force is the only unknown.

(c) Cylindrical Roller on Friction Surface, or on Guide Rail.   A cylindrical roller placed on a friction surface can exert a force that is normal to the surface and a friction force that is perpendicular to the plane of the roller. If a cylindrical roller is placed on a guide rail, the force perpendicular to

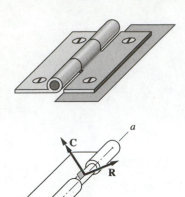

(a)

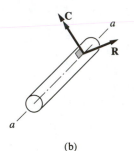

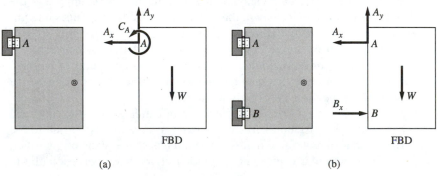

(b)

Fig. 5.1

the plane of the roller is provided by the side of the guide rail. In either case, two unknowns are introduced in the FBD.

(d) Ball-and-Socket Joint. A ball-and-socket joint prevents translational movement but not rotation. Therefore, a connection of this type exerts an unknown force, usually shown as three independent components; it does not apply a couple.

(e) Friction Surface: Single Point of Contact. A friction surface in contact with a body at one point can exert an unknown force, acting through the point of contact. This force is usually shown as three components—a normal force and two components of the friction force acting parallel to the surface.

Before describing the hinge and bearing supports, shown in Table 5.1 parts (f) and (g), it is necessary to discuss how these supports are designed and used in practice.

A hinge and a thrust bearing are illustrated in Fig 5.1(a) and 5.1(b), respectively. Each of these can exert both a force and a couple on the body it supports. Because hinges and bearings prevent translation completely, the force **R** can act in any direction. Because these connections are designed to allow rotation about the *a-a* axis, the reactive couple-vector **C** is always perpendicular to that axis. However, hinges and bearings are seldom strong enough to exert the couples without failure. Consequently, they must be arranged so that the reactive couples are not needed to support the body. For this reason, we consider all hinges and bearings capable of providing *only reactive forces*.

To further illustrate the physical reasoning behind omitting couples at hinges and bearings, consider the door shown in Fig. 5.2. If the door is supported by a single hinge, as shown in Fig. 5.2(a), the reactive couple C_A is essential for equilibrium; otherwise the moment equation $\Sigma M_A = 0$ could not be satisfied. However, we know from experience that this is not the way to suspend a door if we expect it to last for any length of time. The screws used to attach the hinge to the door and doorframe would soon pull out as a result of the large forces that are necessary to provide the couple C_A. Figure 5.2(b) shows the conventional method for supporting a door. Two hinges are aligned along a common axis with the hinge at *B* assumed to be free to slide, so that it does not provide an axial thrust—that is, a force along the axis of the hinge. The FBD shows that equilibrium can be satisfied without developing reactive couples at the hinges. The reactive couple, identified as C_A in Fig. 5.2(a), is now provided by the reactive forces A_x and B_x in Fig. 5.2(b). Any small

Fig. 5.2

misalignment between the axes of the hinges, which could also give rise to reactive couples, is usually accommodated by the slack that is present in most hinges and bearings.

(f) Slider Bearing or Hinge.   A slider bearing, or slider hinge, can exert only a force normal to the axis of the shaft passing through it. Therefore, two unknown force components are introduced into the FBD by this support. A slider bearing is also called a *radial bearing,* because it is designed to carry loads acting in the radial direction only.

(g) Thrust Bearing or Hinge.   In thrust bearings and thrust hinges, the sliding motion of the shaft along its axis is prevented by an end cap or equivalent support. Consequently, this type of support results in three unknown force components on an FBD. The force component R_x, acting parallel to the axis of the shaft, is called the *axial thrust.*

(h) Universal Joint.   A universal joint prevents all translation, and rotation about the axis of the joint. A universal joint can, therefore, apply an unknown force, usually shown as three independent components, and a couple-vector acting along the axis of the joint. Consequently, four unknowns are introduced.

(i) Built-in (Cantilever) Support.   A built-in support, also called a cantilever support, prevents all translational and rotational movement of the body. Therefore, the support can exert an unknown force and an unknown couple-vector, introducing six unknowns in the FBD.

The procedure for constructing a free-body diagram involving noncoplanar forces is identical to that used for a coplanar force system:

1. A sketch of the body (or part of a body) is drawn with all supports removed.

2. All applied forces are shown on the sketch.

3. The reactions are shown for each support that was removed.

When analyzing connected bodies, it is again important that you adhere to Newton's third law: For every action, there is an equal and opposite reaction.

If a problem contains two-force members, the FBD can be simplified considerably by recalling the two-force principle: Two forces in equilibrium must be equal, opposite, and collinear. This principle is illustrated in Fig. 5.3.

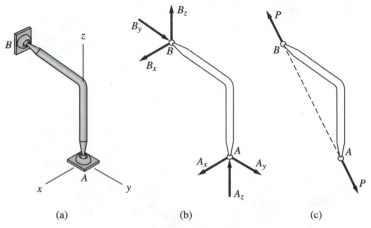

(a)   (b)   (c)

Fig. 5.3

The bar in Fig. 5.3(a) is supported by a ball-and-socket joint at each end and is not subjected to any forces other than the joint reactions (the weight is assumed negligible). The FBD of the bar, Fig. 5.3(b), shows the joint reactions—one force at A with components A_x, A_y, and A_z and the other at B with components B_x, B_y, and B_z. Therefore, we see that the bar is a two-force member. After invoking the two-force principle, the FBD of the bar simplifies to that shown in Fig. 5.3(c)—the forces at A and B are equal, opposite, and collinear.

Sample Problem 5.1

The 2-Mg uniform pole in Fig. (a) is supported by a ball-and-socket joint at O and two cables. Draw the FBD for the pole, and determine the number of unknowns.

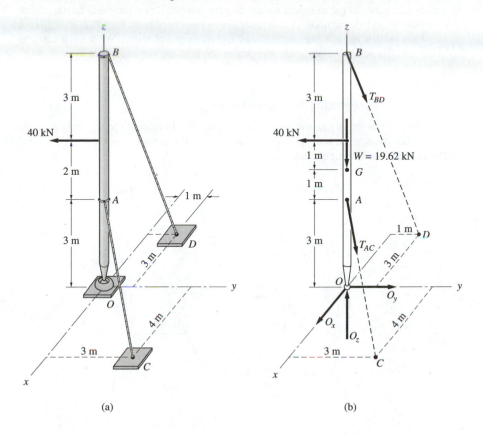

(a) (b)

Solution

The FBD for the pole is shown in Fig. (b). Observe that, in addition to the 40-kN applied load, the pole is subjected to the following forces:

- The tensions in the two cables: The magnitudes of the tensions are labeled T_{AC} and T_{BD}. Because the direction of each cable is known, the force in each cable introduces only one unknown on the FBD—its magnitude.
- The reaction at O: Because the support at O is a ball-and-socket joint, the reaction at O is an unknown force, which we show as the three independent components: O_x, O_y, and O_z.
- The weight of the pole: The center of gravity is at G, the midpoint of the pole. The weight is

$$W = mg = (2 \times 10^3 \text{ kg})(9.81 \text{ m/s}^2) = 19\,620 \text{ N}$$

Inspection of Fig. (b) reveals that there are five unknowns on the FBD: the magnitude of the tension in each of the two cables and the three force components at O.

Sample Problem 5.2

The 20-lb surveyor's transit in Fig. (a) is supported by a tripod of negligible weight that is resting on a horizontal friction surface. The legs of the tripod are connected by ball-and-socket joints to the platform supporting the transit. Draw the FBD for the entire assembly using two methods: (1) not recognizing two-force bodies; and (2) recognizing two-force bodies. In each case, determine the number of unknowns. What modifications to these FBDs are necessary if the weights of the legs are not negligible?

Solution

Part 1   Not Recognizing Two-Force Bodies

The FBD of the entire assembly is shown in Fig. (b). In addition to the weight of the transit, we show three independent components of the ground reactions at B, C, and D—giving a total of nine unknowns.

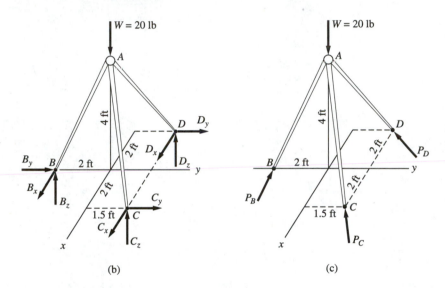

(b)   (c)

Part 2   Recognizing Two-Force Bodies

When each leg of the tripod is recognized to be a two-force body, the FBD of the assembly can be drawn as shown in Fig. (c). The forces at B, C, and D act in the direction of the corresponding leg of the tripod. Therefore, the three leg reactions are the unknowns.

If the weights of the legs are not negligible, the FBD in Fig. (b) can be modified by simply including the weight of each leg. However, the FBD in Fig. (c) cannot be corrected in the same manner. Because the legs are no longer two-force bodies, the forces at B, C, and D cannot be assumed to act in the directions of the respective legs.

Sample Problem 5.3

The space (noncoplanar) structure shown in Fig. (a) is supported by ball-and-socket joints at O and D, and by a slider bearing at C. The two members $OABC$ and AD, connected by a ball-and-socket joint at A, each weigh 20 lb/ft. (1) Draw the FBD for the entire structure, and count the unknowns. (2) Draw the FBD for each of the members, and count the total number of unknowns.

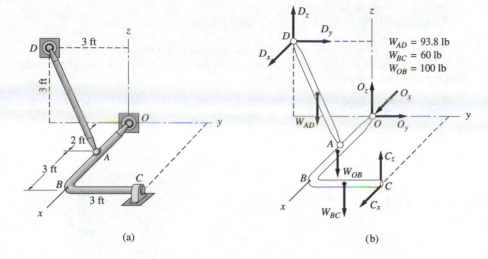

$W_{AD} = 93.8$ lb
$W_{BC} = 60$ lb
$W_{OB} = 100$ lb

(a)   (b)

Solution

Part 1   FBD for the Entire Structure

The FBD for the entire structure is shown in Fig. (b); the dimensions have been omitted for the sake of clarity.

    The weight of the structure can be conveniently represented using the weights of the segments OB, BC, and AD at their respective midpoints. The weights have been computed by multiplying the weight per unit length (20 lb/ft) by the lengths of the respective segments. (Note that the length of AD is $\sqrt{2^2 + 3^2 + 3^2} = 4.69$ ft.)

    In Fig. (b) we also show the three independent components of the forces at the ball-and-socket joints at both O and D. Note that member AD is not a two-force body (because its weight is not negligible), and thus we cannot assume that the force at D acts along the line AD. The FBD includes the two force components exerted on the structure by the slider bearing at C. The FBD of the entire structure in Fig. (b) contains eight unknowns: three forces at O, three forces at D, and two forces at C.

Part 2   FBD for Each of the Members

The FBDs for the members $OABC$ and AD are shown in Fig. (c); the dimensions are again omitted for clarity.

    The directions of the ball-and-socket reactions at O and D are shown in the same directions as assumed in Fig. (b).

    We must also include the force exerted on each member by the ball-and-socket joint at A [because this force is internal to the FBD of the entire structure, it does not appear in Fig. (b)]. Note that A_x, A_y, and A_z must be shown to be equal and opposite on the two members.

    Finally, the reactions at C—shown in the same directions as in Fig. (b)—as well as the weights of the segments are included on the FBDs in Fig. (c).

    When the composite structure is subdivided into its two constituent bodies, the total number of unknowns is eleven—three at O, three at D, three at A, and two at C.

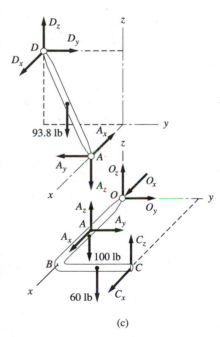

(c)

Problems

5.1 Bar *AB* of negligible weight is supported by a ball-and-socket joint at *B* and two cables attached at *A*. Draw the FBD for the bar, recognizing that it is a two-force body. Determine the number of unknowns.

5.2 Draw the FBD for the bar described in Prob. 5.1 if the bar is homogeneous and weighs 180 lb. Count the number of unknowns.

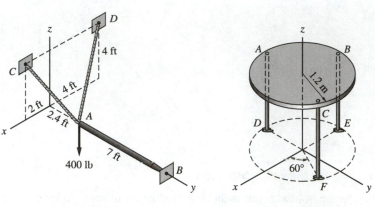

Fig. P5.1, P5.2   **Fig. P5.3**

5.3 The homogeneous top of the three-legged table has a mass of 40 kg. Draw the FBD of the table, neglecting friction and the weights of the legs. Determine the number of unknowns.

5.4 The 600-lb uniform log *OGA*—*G* being its center of gravity—is held in the position shown by the two cables and the light bar *BG*. Draw the FBD for the log, assuming friction at all contact surfaces and noting that *BG* is a two-force body. Count the unknowns.

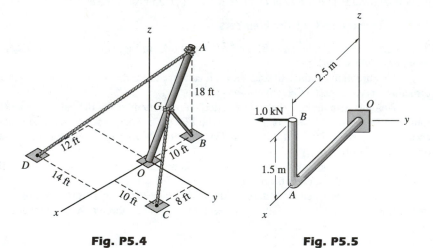

Fig. P5.4   **Fig. P5.5**

5.5 The bent bar has a mass of 40 kg and is subjected to the 1.0-kN force parallel to the *y*-axis. There is a built-in support at *O*. Draw the FBD of the bar, and determine the number of unknowns.

5.6 The pipe—supported by slider bearings at *A*, *B*, and *C*—is loaded by the 120-lb · ft couple acting about the axis *OA*. Neglecting the weight of the pipe, draw its FBD, and count the unknowns.

5.7 The shaft-pulley assembly is supported by the universal joint at *O* and by the slider bearing at *A*. The pulley, which has a mass of 7 kg, is subjected to the belt tensions shown. The mass of the shaft may be neglected. Draw an FBD that consists of the pulley and shaft *AO*. Determine the number of unknowns on this FBD.

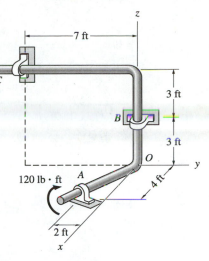

Fig. P5.6

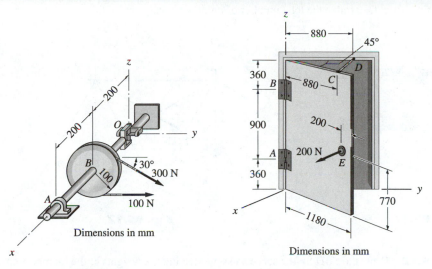

Fig. P5.7   Fig. P5.8

5.8 The 30-kg homogeneous door is supported by hinges at *A* and *B*, with only the hinge at *B* being capable of providing axial thrust. The cable *CD* prevents the door from fully opening when it is pulled by the 200-N force acting perpendicular to the door. Draw the FBD for the door, and count the unknowns.

5.9 Draw the FBD for bar *BCD*. The connections at *A* and *B* are ball-and-socket joints, *C* is a slider bearing, and *D* is a thrust bearing. Assume that the weights of members are negligible and recognize that *AB* is a two-force member. How many unknowns appear on the FBD?

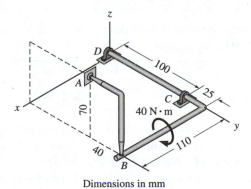

Dimensions in mm

Fig. P5.9

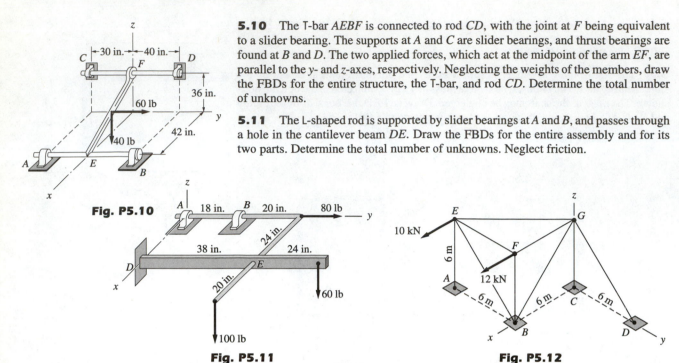

Fig. P5.10

Fig. P5.11

Fig. P5.12

5.10   The T-bar *AEBF* is connected to rod *CD*, with the joint at *F* being equivalent to a slider bearing. The supports at *A* and *C* are slider bearings, and thrust bearings are found at *B* and *D*. The two applied forces, which act at the midpoint of the arm *EF*, are parallel to the y- and z-axes, respectively. Neglecting the weights of the members, draw the FBDs for the entire structure, the T-bar, and rod *CD*. Determine the total number of unknowns.

5.11   The L-shaped rod is supported by slider bearings at *A* and *B*, and passes through a hole in the cantilever beam *DE*. Draw the FBDs for the entire assembly and for its two parts. Determine the total number of unknowns. Neglect friction.

5.12   Draw the FBD for the space truss, neglecting the weights of the members and assuming all connections to be ball-and-socket joints. (Use the fact that all members are two-force bodies.) How many unknowns are there?

5.4   *Independent Equilibrium Equations*

a.   General case

The equilibrium equations for a body subjected to a general force system have been given in Art 5.2:

$$\Sigma \mathbf{F} = 0 \qquad \Sigma \mathbf{M}_O = 0 \qquad \text{(5.1, repeated)}$$

where *O* is an arbitrary point. If *O* is the origin of the *x*, *y*, and *z* coordinate axes, the equivalent scalar equations are

$$\Sigma F_x = 0 \qquad \Sigma F_y = 0 \qquad \Sigma F_z = 0$$
$$\Sigma M_x = 0 \qquad \Sigma M_y = 0 \qquad \Sigma M_z = 0 \qquad \text{(5.2, repeated)}$$

As was the case for the coplanar force systems, alternate sets of independent equilibrium equations can be used in place of the above equations. Regrettably, the restrictions ensuring the independence of the equations for noncoplanar force systems are so numerous (and often fairly complicated) that a complete listing of the restrictions is of little practical value. It is much better to rely on logic rather than a long list of complex rules.

As an example of an alternate set of independent equilibrium equations, consider the six scalar moment equilibrium equations that result from summing the moments about two arbitrary points, say A and B. If these six equations are satisfied, there can be no resultant couple. However, there could still be a resultant force $\mathbf{R} = \Sigma\mathbf{F}$ with the line of action passing through points A and B. Therefore, only five of the moment equations are independent. An additional scalar equation (a carefully chosen force or moment equation) must be used to guarantee that $\mathbf{R}$ vanishes.

When considering a general force system, remember that the number of independent scalar equations is six. Although various combinations of force and moment equations may be used, at least three must be moment equations. The reason is that couples do not appear in force equations, so that the only way to guarantee that the resultant couple vanishes is to satisfy three independent moment equations of equilibrium. However, if properly chosen, the six independent equations could be three force and three moment equations, two force and four moment equations, one force and five moment equations, or even six moment equations.

Three special cases, occurring frequently enough to warrant special attention, are discussed in the next three sections and summarized in Fig. 5.4.

b. Concurrent, noncoplanar force system

In Chapter 3 the resultant of a concurrent, noncoplanar force system was found to be a force $\mathbf{R}$ passing through the point of concurrency. The components of $\mathbf{R}$ were given by $R_x = \Sigma F_x$, $R_y = \Sigma F_y$, and $R_z = \Sigma F_z$. It follows that there are only three independent equilibrium equations:

$$\Sigma F_x = 0 \qquad \Sigma F_y = 0 \qquad \Sigma F_z = 0 \tag{5.3}$$

The x-, y-, and z-axes do not have to be the coordinate axes; they can represent any three arbitrary directions, not necessarily perpendicular.

Note that the six independent equations for the general case are reduced to three for this special case. Alternate sets of equations are one moment and two force equations, one force and two moment equations, or three moment equations, each with its own restrictions to ensure independence.

c. Parallel, noncoplanar force system

It has been shown in Chapter 3 that, if all the forces are parallel to the z-axis, the resultant is either a force parallel to the z-axis or a couple-vector perpendicular to the z-axis. Therefore, the number of independent equilibrium equations is again reduced to three.

$$\Sigma F_z = 0 \qquad \Sigma M_x = 0 \qquad \Sigma M_y = 0 \tag{5.4}$$

The force equation eliminates the possibility of a resultant force, and the two moment equations ensure that there is no resultant couple.

In Eqs. (5.4), the moments can be summed about any two axes that lie in the xy plane. The three equations in Eqs. (5.4) can be replaced by three moment equations, with various restrictions required to guarantee their independence.

Type of force system		No. of independent equil. eqs.	A set of independent equations
General noncoplanar	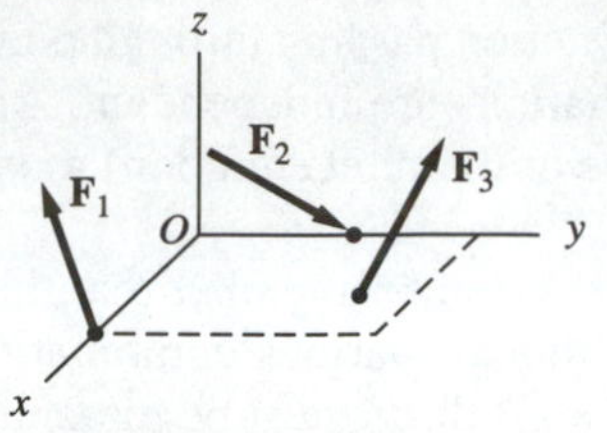	Six	$\Sigma F_x = 0 \qquad \Sigma F_y = 0$ $\Sigma F_z = 0$ $\Sigma M_x = 0 \qquad \Sigma M_y = 0$ $\Sigma M_z = 0$
Concurrent noncoplanar	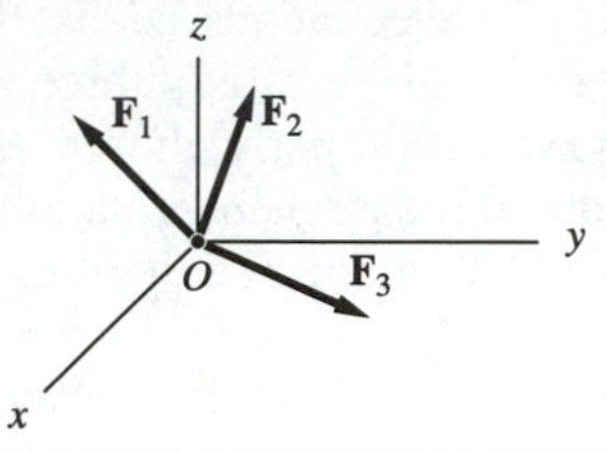	Three	$\Sigma F_x = 0 \qquad \Sigma F_y = 0$ $\Sigma F_z = 0$
Parallel noncoplanar	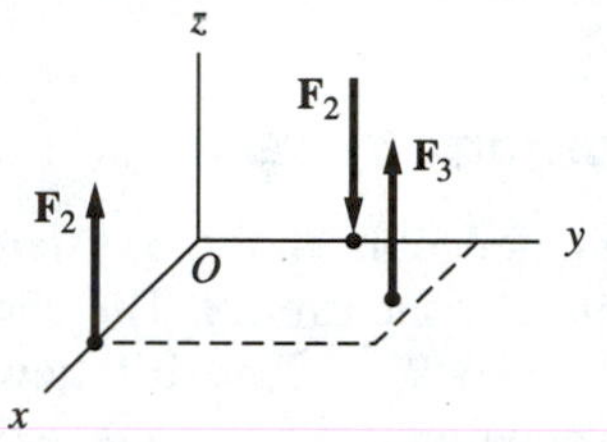	Three	$\Sigma F_z = 0$ $\Sigma M_x = 0 \qquad \Sigma M_y = 0$
All forces intersect an axis	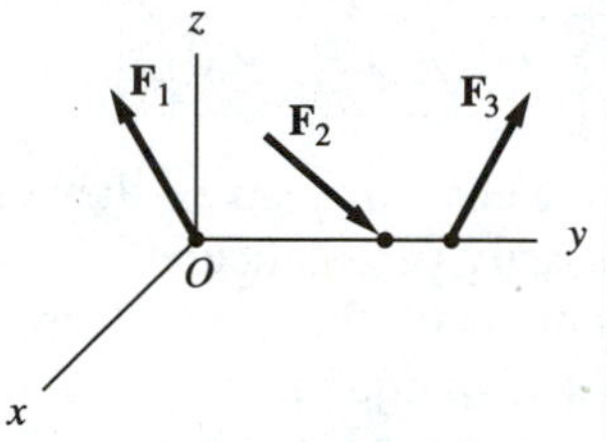	Five	$\Sigma F_x = 0 \qquad \Sigma F_y = 0$ $\Sigma F_z = 0$ $\Sigma M_x = 0 \qquad \Sigma M_z = 0$

Fig. 5.4

d. All forces intersect an axis

If all the forces intersect an axis—say, the y-axis, as shown in Fig. 5.4—the moment equation $\Sigma M_y = 0$ is trivially satisfied, and we are left with the following five independent equilibrium equations.

$$\Sigma F_x = 0 \qquad \Sigma F_y = 0 \qquad \Sigma F_z = 0$$
$$\Sigma M_x = 0 \qquad \qquad \Sigma M_z = 0$$

(5.5)

Of course, alternate sets of independent equations can be used—two force and three moment equations, one force and four moment equations, or five moment equations.

5.5 *Improper Constraints*

Even if the number of equilibrium equations equals the number of unknowns, we cannot always conclude that a solution exists. As we have mentioned several times, this is the predicament when the equilibrium equations are not independent. In such a case, the fault lies with the analyst who chooses the equations, not with the physical problem. But another situation exists, in which the problem itself precludes a solution of the equilibrium equations; it is known as the case of *improper constraints*.

As an example of improper constraints, consider the plate of weight W suspended from six parallel wires and pushed by the horizontal force P, as shown in Fig. 5.5(a). The free-body diagram of the plate shows that there are six unknowns (the forces in the wires). Because the two equations $\Sigma F_y = 0$ and $\Sigma M_z = 0$ are trivially satisfied, the number of independent equilibrium equations is reduced to four in this case. Moreover, the equation $\Sigma F_x = 0$ yields $P = 0$. From all this we conclude that the plate can be in equilibrium *in the position shown* only if $P = 0$, and then the problem is statically indeterminate (there are three equilibrium equations left with six unknowns).

The trouble with this problem is that the supports are not capable of resisting the applied load P in the given position; that is, they cannot provide the proper constraints that prevent motion. We encounter this situation whenever the *support reactions* form one of the special cases described in the preceding article: concurrent, parallel (as in the present example), or intersecting a common axis. An example of the latter is shown in Fig. 5.5(b), in which the plate is supported by three sliding hinges. Again we have six unknown reactions and ostensibly six independent equilibrium equations, but equilibrium is clearly impossible in the position shown, unless $P = 0$.

In summary, the support constraints are said to be improper if they are not capable of supporting an arbitrary load system (this does not preclude equilibrium under certain loads, e.g, when $P = 0$ in the examples shown in Fig. 5.5).

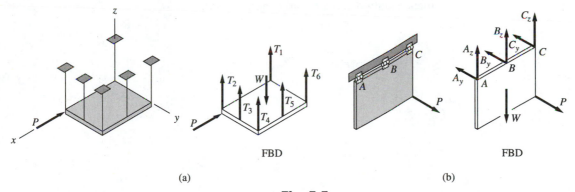

(a)   (b)

Fig. 5.5

5.6 *Writing and Solving Equilibrium Equations*

The steps in the analysis of noncoplanar equilibrium problems are identical to those we used in coplanar analysis.

1. Draw the free-body diagrams (FBDs).
2. Write the equilibrium equations.
3. Solve the equations for the unknowns.

The first step, the construction of FBDs for noncoplanar problems, was discussed in Art 5.3. In this article we assume that the FBDs are given, permitting us to concentrate on the second and third steps—writing and solving the equilibrium equations.

The solution of three-dimensional problems requires careful planning before any equilibrium equations are written. As recommended in Chapter 4, you should prepare a method of analysis or plan of attack that specifies the equations to be written and identifies the unknowns that appear in the equations. Comparing the number of unknowns with the number of independent equilibrium equations lets you determine if the problem is statistically determinate or indeterminate. With a stated plan, you are able to maintain control of the solution; without it, you can easily be overwhelmed with the complexity of the problem. After you adopt a workable method of analysis, you can then proceed to the mathematical details of the solution.

In the solution of coplanar equilibrium problems, the method of analysis frequently centers on a moment equation. The idea is to find a moment center A so that the equation $\Sigma M_A = 0$ involves the fewest possible number of unknowns (ideally only one unknown). This strategy is also convenient for analyzing three-dimensional problems. In most problems, you should look for moment equations that simplify the solution. A moment equation about an axis is frequently useful because it eliminates forces that pass through the axis. In many problems, it is possible to find an axis for which the corresponding moment equation contains only one unknown.

Sample Problem 5.4

Calculate the tension in each of the three cables that support the 1500-kN weight, using the given FBD.

Solution

Method of Analysis

As shown in the FBD, the forces acting on the weight are concurrent (all the forces intersect at A). Therefore, there are three independent equilibrium equations. Because there are also three unknowns (the tensions T_{AB}, T_{AC}, and T_{AD}), we conclude that the problem is statically determinate.

The most straightforward solution is obtained from the three scalar force equations, $\Sigma F_x = 0$, $\Sigma F_y = 0$, and $\Sigma F_z = 0$ (or the equivalent vector equation, $\Sigma \mathbf{F} = \mathbf{0}$).

Mathematical Details

The first step is to write the forces in vector form, as follows:

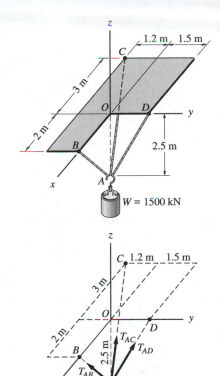

W = 1500 kN

FBD

$$\mathbf{T}_{AB} = T_{AB}\boldsymbol{\lambda}_{AB} = T_{AB}\frac{\overrightarrow{AB}}{|\overrightarrow{AB}|} = T_{AB}\left(\frac{2\mathbf{i} + 2.5\mathbf{k}}{3.202}\right)$$

$$= T_{AB}(0.6246\mathbf{i} + 0.7808\mathbf{k})$$

$$\mathbf{T}_{AC} = T_{AC}\boldsymbol{\lambda}_{AC} = T_{AC}\frac{\overrightarrow{AC}}{|\overrightarrow{AC}|}$$

$$= T_{AC}\left(\frac{-3\mathbf{i} - 1.2\mathbf{j} + 2.5\mathbf{k}}{4.085}\right)$$

$$= T_{AC}(-0.7344\mathbf{i} - 0.2938\mathbf{j} + 0.6120\mathbf{k})$$

$$\mathbf{T}_{AD} = T_{AD}\boldsymbol{\lambda}_{AD} = T_{AD}\frac{\overrightarrow{AD}}{|\overrightarrow{AD}|} = T_{AD}\left(\frac{1.5\mathbf{j} + 2.5\mathbf{k}}{2.916}\right)$$

$$= T_{AD}(0.5144\mathbf{j} + 0.8573\mathbf{k})$$

$$\mathbf{W} = -1500\mathbf{k} \text{ kN}$$

Summing the x-, y-, and z-components and setting the results equal to zero, we have

$$\Sigma F_x = 0 \qquad 0.6246T_{AB} - 0.7344T_{AC} = 0$$

$$\Sigma F_y = 0 \qquad -0.2938T_{AC} + 0.5144T_{AD} = 0$$

$$\Sigma F_z = 0 \qquad 0.7808T_{AB} + 0.6120T_{AC} + 0.8573T_{AD} - 1500 = 0$$

Solving these equations simultaneously gives

$$T_{AB} = 873 \text{ kN} \qquad T_{AC} = 743 \text{ kN} \qquad T_{AD} = 424 \text{ kN} \qquad \textit{Answer}$$

As you can see, the use of three force equations results in a straightforward method of analysis for a concurrent, noncoplanar force system. However, there are other sets of equilibrium equations that could have been used just as effectively.

Another Method of Analysis

Note that the tensions T_{AC} and T_{AD} intersect the line CD and thus have no moment about that line. Therefore, T_{AB} can be calculated from only one equation: $\Sigma M_{CD} = 0$. Similar arguments can be used to show that $\Sigma M_{DB} = 0$ yields T_{AC}, and $\Sigma M_{BC} = 0$ gives T_{AD}.

Mathematical Details

$$\Sigma M_{CD} = 0 \qquad (\mathbf{r}_{CB} \times \mathbf{T}_{AB} \cdot \boldsymbol{\lambda}_{CD}) + (\mathbf{r}_{CO} \times \mathbf{W} \cdot \boldsymbol{\lambda}_{CD}) = 0$$

From the figure, we note that $\mathbf{r}_{CB} = 5\mathbf{i} + 1.2\mathbf{j}$ m, $\mathbf{r}_{CO} = 3\mathbf{i} + 1.2\mathbf{j}$ m, and the unit vector $\boldsymbol{\lambda}_{CD}$ is given by

$$\boldsymbol{\lambda}_{CD} = \frac{\overrightarrow{CD}}{|\overrightarrow{CD}|} = \frac{3\mathbf{i} + 2.7\mathbf{j}}{4.036}$$

Using the vector expressions for $\mathbf{T}_{AB}$ and $\mathbf{W}$ determined in the foregoing, and using the determinant form of the scalar triple product, the moment equation $\Sigma M_{CD} = 0$ becomes

$$\frac{T_{AB}}{4.036}\begin{vmatrix} 5 & 1.2 & 0 \\ 0.6246 & 0 & 0.7808 \\ 3 & 2.7 & 0 \end{vmatrix} + \frac{1}{4.036}\begin{vmatrix} 3 & 1.2 & 0 \\ 0 & 0 & -1500 \\ 3 & 2.7 & 0 \end{vmatrix} = 0$$

Expanding the determinants and solving the resulting equation yields $T_{AB} = 873$ kN, the same answer determined in the preceding analysis.

As mentioned, the tensions in the other two cables could be obtained from $\Sigma M_{DB} = 0$ and $\Sigma M_{BC} = 0$.

Sample Problem 5.5

The horizontal boom OC, which is supported by a ball-and-socket joint and two cables, carries the vertical force $P = 8000$ lb. Calculate T_{AD} and T_{CE}, the tensions in the cables, and the components of the force exerted on the boom by the joint at O. Use the given FBD (the weight of the boom is negligible).

Solution

Method of Analysis

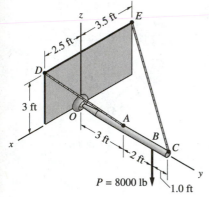

As shown in the FBD, the force system acting on the boom is the special case in which all forces intersect an axis, namely, the y-axis. Therefore, there are five independent equilibrium equations. Because there are also five unknowns in the FBD (T_{AD}, T_{CE}, O_x, O_y, and O_z), the problem is statically determinate.

Consider the moment equations about the x- and z-axes ($\Sigma M_y = 0$ is trivially satisfied):

$$\Sigma M_x = 0\text{—contains the unknowns } T_{AD} \text{ and } T_{CE}$$

$$\Sigma M_z = 0\text{—contains the unknowns } T_{AD} \text{ and } T_{CE}$$

These two equations can, therefore, be solved simultaneously for T_{AD} and T_{CE}. Once these tensions have been found, the reactions at O can be determined using the force equations of equilibrium.

Mathematical Details

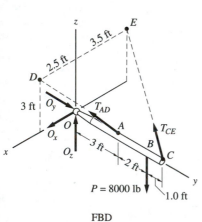

FBD

A convenient method of writing $\Sigma M_x = 0$ and $\Sigma M_z = 0$ is to sum the moments about point O using vector representation (recall that $\mathbf{M}_O = M_x\mathbf{i} + M_y\mathbf{j} + M_z\mathbf{k}$). Referring to the FBD, we have

$$\Sigma \mathbf{M}_O = (\mathbf{r}_{OA} \times \mathbf{T}_{AD}) + (\mathbf{r}_{OC} \times \mathbf{T}_{CE}) + (\mathbf{r}_{OB} \times \mathbf{P}) = \mathbf{0}$$

where

$$\mathbf{r}_{OA} = 3\mathbf{j} \text{ ft} \qquad \mathbf{r}_{OC} = 6\mathbf{j} \text{ ft} \qquad \mathbf{r}_{OB} = 5\mathbf{j} \text{ ft}$$

$$\mathbf{T}_{AD} = T_{AD}\boldsymbol{\lambda}_{AD} = T_{AD}\frac{\overrightarrow{AD}}{|\overrightarrow{AD}|} = T_{AD}\left(\frac{2.5\mathbf{i} - 3\mathbf{j} + 3\mathbf{k}}{4.924}\right)$$

$$= T_{AD}(0.5077\mathbf{i} - 0.6093\mathbf{j} + 0.6093\mathbf{k})$$

$$\mathbf{T}_{CE} = T_{CE}\boldsymbol{\lambda}_{CE} = T_{CE}\frac{\overrightarrow{CE}}{|\overrightarrow{CE}|} = T_{CE}\left(\frac{-3.5\mathbf{i} - 6\mathbf{j} + 3\mathbf{k}}{7.566}\right)$$

$$= T_{CE}(-0.4626\mathbf{i} - 0.7930\mathbf{j} + 0.3965\mathbf{k})$$

$$\mathbf{P} = -8000\mathbf{k} \text{ lb}$$

Using the determinant form for the cross products, we have

$$\Sigma\mathbf{M}_O = T_{AD}\begin{vmatrix} \mathbf{i} & \mathbf{j} & \mathbf{k} \\ 0 & 3 & 0 \\ 0.5077 & -0.6093 & 0.6093 \end{vmatrix}$$

$$+ T_{CE}\begin{vmatrix} \mathbf{i} & \mathbf{j} & \mathbf{k} \\ 0 & 6 & 0 \\ -0.4626 & -0.7930 & 0.3965 \end{vmatrix}$$

$$+ \begin{vmatrix} \mathbf{i} & \mathbf{j} & \mathbf{k} \\ 0 & 5 & 0 \\ 0 & 0 & -8000 \end{vmatrix} = \mathbf{0}$$

Expanding the determinants and equating the x- and z-components (the y-components are identically zero, as expected), we get

$$\Sigma M_x = 0 \qquad 1.828T_{AD} + 2.379T_{CE} - 40\,000 = 0$$
$$\Sigma M_z = 0 \qquad -1.523T_{AD} + 2.776T_{CE} = 0$$

from which we find

$$T_{AD} = 12\,770 \text{ lb} \qquad T_{CE} = 7010 \text{ lb} \qquad\qquad \textit{Answer}$$

After the tensions have been computed, the reactions at O can be calculated by using the force equation $\Sigma\mathbf{F} = \mathbf{0}$:

$$\Sigma F_x = 0 \qquad O_x + 0.5077T_{AD} - 0.4626T_{CE} = 0$$

$$\Sigma F_y = 0 \qquad O_y - 0.6093T_{AD} - 0.7930T_{CE} = 0$$

$$\Sigma F_z = 0 \qquad O_z + 0.6093T_{AD} + 0.3965T_{CE} - 8000 = 0$$

Substituting the previously found values for T_{AD} and T_{CE}, we obtain

$$O_x = -3240 \text{ lb} \qquad O_y = 13\,340 \text{ lb} \qquad O_z = -2560 \text{ lb} \qquad \textit{Answer}$$

The negative values for O_x and O_z indicate that the directions of these components are opposite to the directions shown in the FBD.

Sample Problem 5.6

The nonhomogeneous plate weighing 60 kN has center of gravity at G. It is supported in the horizontal plane by three vertical cables. Compute the tension in each cable using the given FBD.

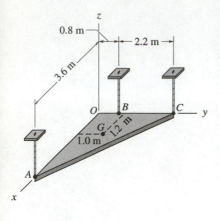

Solution

Method of Analysis

As shown in the FBD, the forces holding the plate in equilibrium form a parallel system, which has three independent equilibrium equations. Because there are also three unknowns (T_A, T_B, and T_C), the problem is statically determinate.

One method of analysis considers the moment equations about the x- and y-axes ($\Sigma M_z = 0$ is trivially satisfied because the forces are parallel to the z-axis) and the force equation in the z-direction.

$$\Sigma M_x = 0 \text{—contains the unknowns } T_B \text{ and } T_C$$

$$\Sigma M_y = 0 \text{—contains the unknown } T_A$$

$$\Sigma F_z = 0 \text{—contains the unknowns } T_A, T_B, \text{ and } T_C$$

First, the equation $\Sigma M_y = 0$ can be used to find T_A. Then, the other two equations can be solved simultaneously for T_B and T_C. The details of this analysis, using scalar representation, are shown in the following.

Mathematical Details

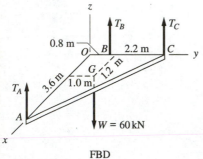

FBD

$$\Sigma M_y = 0 \qquad 60(1.2) - 3.6T_A = 0 \tag{1}$$

which gives

$$T_A = 20.0 \text{ kN} \qquad\qquad Answer$$

$$\Sigma M_x = 0 \qquad 0.8T_B + 3.0T_C - 60(1.0) = 0 \tag{2}$$

$$\Sigma F_z = 0 \qquad +\uparrow \qquad T_A + T_B + T_C - 60 = 0 \tag{3}$$

Substituting $T_A = 20.0$ kN, and solving Eqs. (2) and (3) yields

$$T_B = 27.3 \text{ kN} \qquad T_C = 12.7 \text{ kN} \qquad\qquad Answer$$

Other Methods of Analysis

In the above solution, we were able to find T_A using the equation $\Sigma M_y = 0$ because T_B and T_C have no moment about the y-axis. By studying the FBD, you will see that it is also possible to calculate T_B using one equation, and T_C using one equation.

Sample Problem 5.7

The bent bar of negligible weight is supported by a ball-and-socket joint at O, a cable connected between A and E, and a slider bearing at D. The bar is acted on by a wrench consisting of the force **P** and couple **C**, both parallel to the z-axis. Determine the components of bearing reaction at D and the force in the cable using the given FBD.

Solution

Method of Analysis

The force system in the FBD is the general case. Therefore, there are six independent equilibrium equations available for computing the six unknowns (O_x, O_y, O_z, T_{AE}, D_x, and D_z).

Referring to the FBD, we consider the moment equation about each of the coordinate axes:

$$\Sigma M_x = 0\text{—contains the unknown } D_z$$

$$\Sigma M_y = 0\text{—contains the unknowns } T_{AE}, D_x, \text{ and } D_z$$

$$\Sigma M_z = 0\text{—contains the unknowns } T_{AE} \text{ and } D_x$$

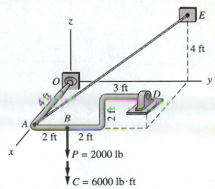

Therefore, the three unknowns T_{AE}, D_x, and D_z can be computed from these equations according to the following scheme: First find D_z from $\Sigma M_x = 0$, and then solve the other two equations simultaneously for D_x and T_{AE}.

Mathematical Details

When utilizing the above analysis, it is convenient to use the vector approach to derive the moment equations $\Sigma \mathbf{M}_O = \Sigma M_x\mathbf{i} + \Sigma M_y\mathbf{j} + \Sigma M_z\mathbf{k} = \mathbf{0}$. The details are as follows:

$$\Sigma \mathbf{M}_O = \mathbf{0} \qquad (\mathbf{r}_{OA} \times \mathbf{T}_{AE}) + (\mathbf{r}_{OD} \times \mathbf{D}) + (\mathbf{r}_{OB} \times \mathbf{P}) + \mathbf{C} = \mathbf{0}$$

Referring to the FBD, the vectors in the above equation are

$$\mathbf{r}_{OA} = 4\mathbf{i} \text{ ft} \qquad \mathbf{r}_{OD} = 4\mathbf{i} + 7\mathbf{j} + 2\mathbf{k} \text{ ft}$$

$$\mathbf{r}_{OB} = 4\mathbf{i} + 2\mathbf{j} \text{ ft}$$

$$\mathbf{P} = -2000\mathbf{k} \text{ lb} \qquad \mathbf{C} = -6000\mathbf{k} \text{ lb} \cdot \text{ft}$$

$$\mathbf{T}_{AE} = T_{AE}\boldsymbol{\lambda}_{AE} = T_{AE}\frac{\overrightarrow{AE}}{|\overrightarrow{AE}|} = T_{AE}\left(\frac{-4\mathbf{i} + 7\mathbf{j} + 4\mathbf{k}}{9}\right)$$

$$\mathbf{D} = D_x\mathbf{i} + D_z\mathbf{k}$$

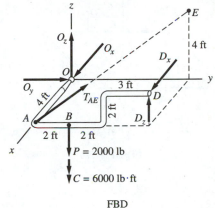

FBD

Therefore, $\Sigma \mathbf{M}_O = \mathbf{0}$ can be written in determinant form as

$$\frac{T_{AE}}{9}\begin{vmatrix} \mathbf{i} & \mathbf{j} & \mathbf{k} \\ 4 & 0 & 0 \\ -4 & 7 & 4 \end{vmatrix} + \begin{vmatrix} \mathbf{i} & \mathbf{j} & \mathbf{k} \\ 4 & 7 & 2 \\ D_x & 0 & D_z \end{vmatrix} + \begin{vmatrix} \mathbf{i} & \mathbf{j} & \mathbf{k} \\ 4 & 2 & 0 \\ 0 & 0 & -2000 \end{vmatrix} - 6000\mathbf{k} = \mathbf{0}$$

Expanding the determinants and equating the x-, y-, and z-components yields the equations

(x-component)	$7D_z - 4000 = 0$
(y-component)	$-1.778T_{AE} + 2D_x - 4D_z + 8000 = 0$
(z-component)	$3.111T_{AE} - 7D_x - 6000 = 0$

The solution of these equations yields

$$D_z = 571 \text{ lb} \qquad T_{AE} = 4500 \text{ lb} \qquad D_x = 1140 \text{ lb} \qquad \textit{Answer}$$

If desired, the reactions O_x, O_y, and O_z could now be found from the force equation $\Sigma \mathbf{F} = \mathbf{0}$.

It should be noted that T_{AE} could also be obtained from a single scalar equilibrium equation $\Sigma M_{OD} = 0$.

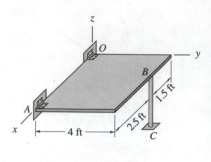

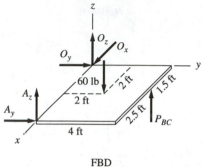

FBD

Fig. P5.21

Problems

5.13 In Sample Problem 5.4, determine the tension T_{AC} using the equation $\Sigma M_{DB} = 0$.

5.14 Use the equation $\Sigma M_{BC} = 0$ to calculate the tension T_{AD} in Sample Problem 5.4.

5.15 In Sample Problem 5.5, compute the tension T_{AD} using one scalar equilibrium equation.

5.16 Calculate the tension T_{CE} in Sample Problem 5.5 using one scalar equilibrium equation.

5.17 In Sample Problem 5.5, determine O_y with one scalar equilibrium equation.

5.18 Determine the tension T_B in Sample Problem 5.6 using one scalar equilibrium equation.

5.19 In Sample Problem 5.6, determine the tension T_C with one scalar equilibrium equation.

5.20 Compute the tension T_{AE} in Sample Problem 5.7 using just one scalar equilibrium equation.

5.21 The homogeneous door of weight $W = 60$ lb is held in the horizontal plane by a thrust hinge at O, a hinge at A, and the vertical prop BC. Determine all forces acting on the door.

5.22 The light boom AB is attached to the vertical wall by a ball-and-socket joint at A and supported by two cables at B. A force $\mathbf{P} = 12\mathbf{i} - 16\mathbf{k}$ kN is applied at B. Note that R_A, the reaction at A, acts along the boom, because it is a two-force body. Compute the cable tensions and R_A.

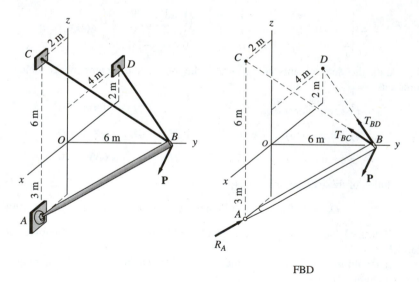

FBD

Fig. P5.22

5.23 The homogeneous 120-lb sign is suspended from a ball-and-socket joint at O, and cables AD and BC. Determine the forces in the cables.

5.24 The 600-N forces are applied to the ends of a cable that passes through a bent pipe. Determine the magnitudes of the reactions exerted on the pipe by the slider bearings at A, B, and C. Neglect the weight of the pipe.

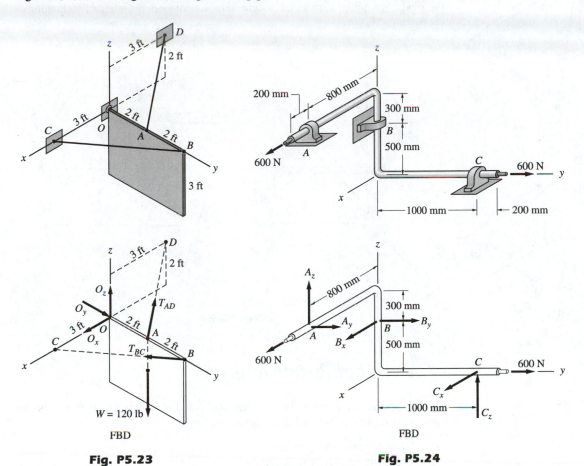

Fig. P5.23   Fig. P5.24

5.25 Determine the forces in the five wires that support the uniform 80-kg plate (weight = 784.8 N). Note that due to symmetry about the *yz* plane, $T_A = T_E$ and $T_B = T_D$.

5.26 The support for the T-shaped bar consists of a thrust bearing at *O* and a slider bearing at *B*. When a weight *W* is suspended from *D*, the force $P = W/2$, parallel to the *x*-axis, is required to maintain equilibrium. Calculate θ, the angle of inclination of the bearing axis *OB*.

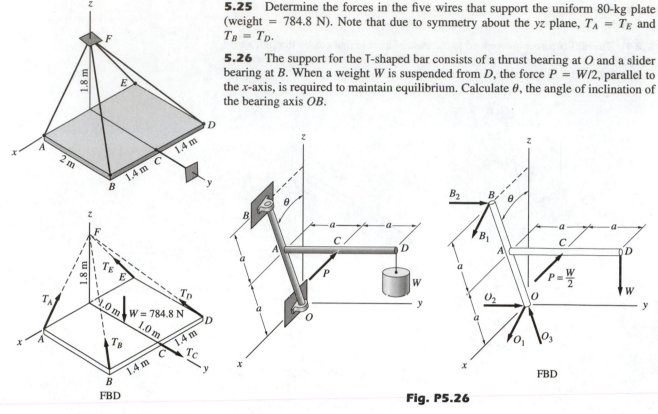

Fig. P5.25

Fig. P5.26

5.7 *Equilibrium Analysis*

The method for analyzing rigid bodies subjected to noncoplanar forces is the same as used in Chapter 4 for coplanar loadings.

1. Draw the free-body diagrams.
2. Write the equilibrium equations.
3. Solve the equations for the unknowns.

Article 5.3 concentrated on the construction of FBDs. Article 5.6 was devoted to writing and solving the equilibrium equations from given FBDs. The sample problems that follow this article illustrate the complete analysis of noncoplanar equilibrium problems, beginning with the construction of the FBDs and ending with the solution. Analyses of both single and connected bodies are considered.

We reiterate that you must be careful when drawing free-body diagrams. Sloppy sketches of three-dimensional problems are notoriously difficult to read; consequently, they are a major source of errors in the derivation of equilibrium equations.

Sample Problem 5.8

Determine the forces acting on the bent bar *OBD* in Fig. (a). The bar is loaded by the wrench consisting of the force *P* and couple *C*. Neglect the weights of the members, and assume that all connections are ball-and-socket joints.

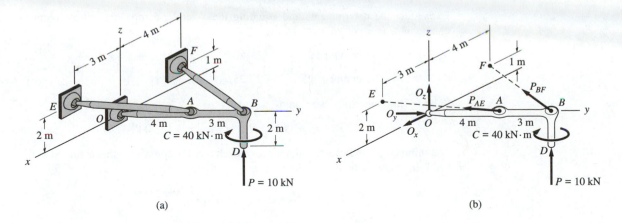

(a)   (b)

Solution

Method of Analysis

The first step is to draw the FBD of bar *OBD*, which is shown in Fig. (b). The reactions at the ball-and-socket at *O* are labeled O_x, O_y, and O_z. Note that we have used the fact that the struts *AE* and *BF* are two-force bodies, each assumed to act in tension.

We see that the FBD contains five unknowns (P_{AE}, P_{BF}, O_x, O_y, and O_z). Because there are also five independent equilibrium equations (all forces intersect the *y*-axis), the problem is statically determinate.

Referring to the FBD in Fig. (b), we consider the moment equations about the *x*- and *z*-axes ($\Sigma M_y = 0$ is trivially satisfied):

$$\Sigma M_x = 0 \text{—contains the unknowns } P_{AE} \text{ and } P_{BF}$$

$$\Sigma M_z = 0 \text{—contains the unknowns } P_{AE} \text{ and } P_{BF}$$

Therefore, these two equations can be solved simultaneously for P_{AE} and P_{BF}. Once these two unknowns have been found, the force equation $\Sigma \mathbf{F} = \mathbf{0}$ can be used to find the remaining three unknowns: O_x, O_y, and O_z.

Mathematical Details

We choose to write the moments about the *x*- and *z*-axes using the vector expression $\Sigma \mathbf{M}_O = \Sigma M_x \mathbf{i} + \Sigma M_y \mathbf{j} + \Sigma M_z \mathbf{k}$.

$$\Sigma \mathbf{M}_O = \mathbf{0} \qquad (\mathbf{r}_{OA} \times \mathbf{P}_{AE}) + (\mathbf{r}_{OB} \times \mathbf{P}_{BF}) + (\mathbf{r}_{OB} \times \mathbf{P}) + \mathbf{C} = \mathbf{0}$$

The vectors that appear in this equation are

$$\mathbf{r}_{OA} = 4\mathbf{j} \text{ m} \qquad \mathbf{r}_{OB} = 7\mathbf{j} \text{ m} \qquad \mathbf{P} = 10\mathbf{k} \text{ kN} \qquad \mathbf{C} = -40\mathbf{k} \text{ kN} \cdot \text{m}$$

$$\mathbf{P}_{AE} = P_{AE} \boldsymbol{\lambda}_{AE} = P_{AE} \frac{\overrightarrow{AE}}{|\overrightarrow{AE}|} = P_{AE} \left(\frac{3\mathbf{i} - 4\mathbf{j} + 2\mathbf{k}}{5.385} \right)$$

$$\mathbf{P}_{BF} = P_{BF} \boldsymbol{\lambda}_{BF} = P_{BF} \frac{\overrightarrow{BF}}{|\overrightarrow{BF}|} = P_{BF} \left(\frac{-4\mathbf{i} - 7\mathbf{j} + \mathbf{k}}{8.124} \right)$$

233

Expressing the cross products in determinant form, the equilibrium equation $\Sigma M_O = 0$ then becomes

$$\frac{P_{AE}}{5.385} \begin{vmatrix} \mathbf{i} & \mathbf{j} & \mathbf{k} \\ 0 & 4 & 0 \\ 3 & -4 & 2 \end{vmatrix} + \frac{P_{BF}}{8.124} \begin{vmatrix} \mathbf{i} & \mathbf{j} & \mathbf{k} \\ 0 & 7 & 0 \\ -4 & -7 & 1 \end{vmatrix} + \begin{vmatrix} \mathbf{i} & \mathbf{j} & \mathbf{k} \\ 0 & 7 & 0 \\ 0 & 0 & 10 \end{vmatrix} - 40\mathbf{k} = \mathbf{0}$$

Expanding the determinants and equating the x- and z-components (there is no y-component, as expected), we obtain

(x-component) $\qquad 1.486 P_{AE} + 0.862 P_{BF} + 70 = 0$

(z-component) $\qquad -2.228 P_{AE} + 3.447 P_{BF} - 40 = 0$

Solving simultaneously, we obtain

$$P_{AE} = -39.16 \text{ kN} \qquad P_{BF} = -13.70 \text{ kN} \qquad\qquad \textit{Answer}$$

The minus signs indicate that the sense of each force is opposite to that assumed in the FBD. In vector form, the two forces are

$$\mathbf{P}_{AE} = P_{AE}\boldsymbol{\lambda}_{AE} = -39.16 \left(\frac{3\mathbf{i} - 4\mathbf{j} + 2\mathbf{k}}{5.385} \right)$$

$$= -21.82\mathbf{i} + 29.09\mathbf{j} - 14.54\mathbf{k} \text{ kN}$$

$$\mathbf{P}_{BF} = P_{BF}\boldsymbol{\lambda}_{BF} = -13.70 \left(\frac{-4\mathbf{i} - 7\mathbf{j} + \mathbf{k}}{8.124} \right)$$

$$= 6.75\mathbf{i} + 11.80\mathbf{j} - 1.69\mathbf{k} \text{ kN}$$

Summing forces, we have

$$\Sigma\mathbf{F} = \mathbf{0} \qquad \mathbf{P}_{AE} + \mathbf{P}_{BF} + \mathbf{P} + (O_x\mathbf{i} + O_y\mathbf{j} + O_z\mathbf{k}) = \mathbf{0}$$

Substituting the expressions for $\mathbf{P}_{AE}$, $\mathbf{P}_{BF}$, and $\mathbf{P}$, and solving, yields

$$O_x = 15.1 \text{ kN} \qquad O_y = -40.9 \text{ kN} \qquad O_z = 6.2 \text{ kN} \qquad \textit{Answer}$$

Sample Problem 5.9

The window in Fig. (a) weighs 40 lb; its center of gravity G is located at the geometric center. Find all forces acting on the window when it is held open in the position shown by the rope attached to C. Assume that the hinge at A can provide an axial thrust whereas the hinge at B cannot.

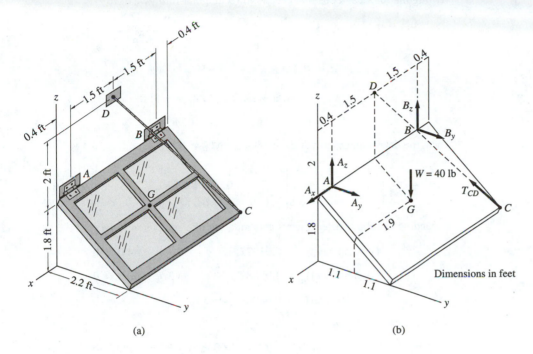

(a)   (b)

Solution

Method of Analysis

We begin by drawing the FBD of the window—see Fig. (b). In addition to its 40-lb weight and the tension T_{CD} in the rope, the window is acted on by the hinge reactions at A and B. Note that an axial thrust (force component in the x-direction) is shown only for the hinge at A.

The FBD contains six unknowns: A_x, A_y, A_z, B_y, B_z, and T_{CD}. Because the force system is the general case, there are also six independent equilibrium equations, which means that the problem is statically determinate.

Point A plays an important role in the analysis of this problem, because three of the unknowns (A_x, A_y, and A_z) pass through A. Referring to the FBD, we examine the moment equations for each of the axes passing through A that is parallel to a coordinate axis:

$\Sigma(M_A)_x = 0$—contains the unknown T_{CD}

$\Sigma(M_A)_y = 0$—contains the unknowns T_{CD} and B_z (B_y is parallel to this axis)

$\Sigma(M_A)_z = 0$—contains the unknowns T_{CD} and B_y (B_z is parallel to this axis)

These three scalar equations can be solved for the unknowns T_{CD}, B_y, and B_z. Once these values are known, the three force components at A (A_x, A_y, and A_z) can be found from the force equation $\Sigma \mathbf{F} = \mathbf{0}$.

Mathematical Details

We will use the vector representation to find the moment equations about the axes passing through A; that is, $\Sigma\mathbf{M}_A = (\Sigma M_A)_x\mathbf{i} + (\Sigma M_A)_y\mathbf{j} + (\Sigma M_A)_z\mathbf{k} = \mathbf{0}$.

$$\Sigma\mathbf{M}_A = \mathbf{0} \qquad (\mathbf{r}_{AD} \times \mathbf{T}_{CD}) + (\mathbf{r}_{AB} \times \mathbf{B}) + (\mathbf{r}_{AG} \times \mathbf{W}) = \mathbf{0}$$

Writing the forces and position vectors in rectangular form, we have

$$\mathbf{T}_{CD} = T_{CD}\boldsymbol{\lambda}_{CD} = T_{CD}\frac{\overrightarrow{CD}}{|\overrightarrow{CD}|} = T_{CD}\left(\frac{1.9\mathbf{i} - 2.2\mathbf{j} + 3.8\mathbf{k}}{4.784}\right)$$

$$\mathbf{B} = B_y\mathbf{j} + B_z\mathbf{k} \qquad\qquad \mathbf{W} = -40\mathbf{k}\ \text{lb}$$

$$\mathbf{r}_{AD} = -1.5\mathbf{i} + 2\mathbf{k}\ \text{ft} \qquad \mathbf{r}_{AB} = -3\mathbf{i}\ \text{ft}$$

$$\mathbf{r}_{AG} = -1.5\mathbf{i} + 1.1\mathbf{j} - 0.9\mathbf{k}\ \text{ft}$$

Then, the determinant form of the equation $\Sigma\mathbf{M}_A = \mathbf{0}$ is

$$\frac{T_{CD}}{4.784}\begin{vmatrix} \mathbf{i} & \mathbf{j} & \mathbf{k} \\ -1.5 & 0 & 2 \\ 1.9 & -2.2 & 3.8 \end{vmatrix} + \begin{vmatrix} \mathbf{i} & \mathbf{j} & \mathbf{k} \\ -3 & 0 & 0 \\ 0 & B_y & B_z \end{vmatrix} + \begin{vmatrix} \mathbf{i} & \mathbf{j} & \mathbf{k} \\ -1.5 & 1.1 & -0.9 \\ 0 & 0 & -40 \end{vmatrix} = \mathbf{0}$$

Expanding the above determinants and equating like components, we get

$$(x\text{-component}) \qquad 0.9197T_{CD} \qquad\qquad - 44.0 = 0$$

$$(y\text{-component}) \qquad 1.9858T_{CD} \qquad\qquad + 3B_z - 60.0 = 0$$

$$(z\text{-component}) \qquad 0.6898T_{CD} - 3B_y \qquad\qquad = 0$$

Solving these equations gives

$$T_{CD} = 47.84\ \text{lb} \qquad B_y = 11.00\ \text{lb} \qquad B_z = -11.67\ \text{lb} \qquad\qquad \textit{Answer}$$

Omitting the details, the remaining three unknowns are found from the force equation $\Sigma\mathbf{F} = \mathbf{0}$ to be

$$A_x = -19.00\ \text{lb} \qquad A_y = 11.00\ \text{lb} \qquad A_z = 13.67\ \text{lb} \qquad\qquad \textit{Answer}$$

Sample Problem 5.10

Determine all forces that act on bar AC. The two bars AC and CD are homogeneous and weigh 200 N/m. Joints A, C, and D are ball-and-sockets, and a cable is connected between B and E.

Solution

Method of Analysis

As you know, there are many ways in which one can calculate the unknown forces acting on bodies that are connected together. However, considering the FBD of the entire assembly is usually a good place to begin.

The FBD of the entire assembly is shown in Fig. (b). The weights of the bars, W_{AC} and W_{CD}, were calculated by multiplying the weight per unit length (200 N/m) by the length of each bar. The components of the reaction at A are A_x, A_y, and A_z; the components of the reaction at D are D_x, D_y, and D_z; T_{BE} is the tension in the cable.

We see that the FBD in Fig. (b) contains seven unknowns (three force components each at A and D, and the tension T_{BE}). Because there are only six independent equilibrium equations (the force system represents the general case), we cannot calculate all of the unknowns without taking the assembly apart. However, we see that T_{BE} is the only unknown that does not intersect the axis AD. Therefore, we can find T_{BE} from the moment equation $\Sigma M_{AD} = 0$.

We next draw the FBDs of bars AC and CD separately—see Figs. (c) and (d). The force components at A in Fig. (c) and the force components at D in Fig. (d) must act in the same directions as in Fig. (b). Furthermore, the components of the reaction at C (C_x, C_y, and C_z) in Fig. (c) must be equal in magnitude, but oppositely directed, to the corresponding components in Fig. (d). We note that there are ten unknowns in this problem: three each at A, C, and D, and the tension T_{BE}. The total number of independent equilibrium equations is also ten: five each for the two bars (the force system acting on each bar represents the special case in which the forces intersect an axis). This problem is therefore statically determinate.

Referring to the FBD in Fig. (d), C_y can be computed using the moment equation $\Sigma(M_D)_z = 0$. Next, consider the FBD in Fig. (c). Because we have already determined T_{BE} and C_y, only five unknowns remain: A_x, A_y, A_z, C_x, and C_z. Therefore, any of the five independent equations for this FBD can be used to find these unknowns.

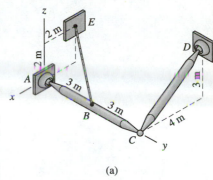

(a)

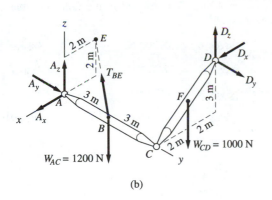

(b)

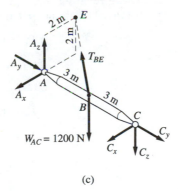

(c)

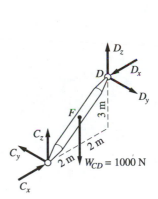

(d)

Mathematical Details

Referring to the FBD of the entire assembly in Fig. (b), and using scalar triple products to evaluate the moments about the axis AD, we obtain

$$\Sigma M_{AD} = 0$$

$$(\mathbf{r}_{AB} \times \mathbf{T}_{BE} \cdot \boldsymbol{\lambda}_{AD}) + (\mathbf{r}_{AB} \times \mathbf{W}_{AC} \cdot \boldsymbol{\lambda}_{AD}) + (\mathbf{r}_{AF} \times \mathbf{W}_{CD} \cdot \boldsymbol{\lambda}_{AD}) = 0$$

The vectors appearing in this equation are

$$\mathbf{r}_{AB} = 3\mathbf{j} \text{ m} \qquad\qquad \mathbf{r}_{AF} = -2\mathbf{i} + 6\mathbf{j} + 1.5\mathbf{k} \text{ m}$$

$$\mathbf{W}_{AC} = -1200\mathbf{k} \text{ N} \qquad \mathbf{W}_{CD} = -1000\mathbf{k} \text{ N}$$

$$\mathbf{T}_{BE} = T_{BE}\boldsymbol{\lambda}_{BE} = T_{BE}\frac{\overrightarrow{BE}}{|\overrightarrow{BE}|} = T_{BE}\left(\frac{-2\mathbf{i} - 3\mathbf{j} + 2\mathbf{k}}{4.123}\right)$$

$$\boldsymbol{\lambda}_{AD} = \frac{\overrightarrow{AD}}{|\overrightarrow{AD}|} = \frac{-4\mathbf{i} + 6\mathbf{j} + 3\mathbf{k}}{7.810}$$

The equation $\Sigma M_{AD} = 0$ thus becomes

$$\frac{1}{7.810}\frac{T_{BE}}{4.123}\begin{vmatrix} 0 & 3 & 0 \\ -2 & -3 & 2 \\ -4 & 6 & 3 \end{vmatrix} + \frac{1}{7.810}\begin{vmatrix} 0 & 3 & 0 \\ 0 & 0 & -1200 \\ -4 & 6 & 3 \end{vmatrix}$$

$$+ \frac{1}{7.810}\begin{vmatrix} -2 & 6 & 1.5 \\ 0 & 0 & -1000 \\ -4 & 6 & 3 \end{vmatrix} = 0$$

Expanding the determinants and solving yields

$$T_{BE} = 18\,140 \text{ N} \qquad\qquad\qquad \textit{Answer}$$

Using the FBD of bar CD in Fig. (d),

$$\Sigma(M_D)_z = 0 \qquad \text{gives } C_y = 0 \qquad\qquad \textit{Answer}$$

As mentioned, with T_{BE} and C_y already computed, we can use any five available equations to find the five remaining unknown forces on the FBD of bar AC in Fig. (c). One method for finding the forces at A and C is outlined in the following; the mathematical details are left as an exercise.

$$\Sigma F_y = 0 \qquad\qquad \text{gives } A_y = 13\,200 \text{ N} \qquad\qquad \textit{Answer}$$

$$\Sigma(M_C)_x = 0 \qquad\qquad \text{gives } A_z = 3800 \text{ N} \qquad\qquad \textit{Answer}$$

$$\Sigma(M_C)_z = 0 \qquad\qquad \text{gives } A_x = 4400 \text{ N} \qquad\qquad \textit{Answer}$$

$$\Sigma F_x = 0 \qquad\qquad \text{gives } C_x = 4400 \text{ N} \qquad\qquad \textit{Answer}$$

$$\Sigma F_z = 0 \qquad\qquad \text{gives } C_z = 3800 \text{ N} \qquad\qquad \textit{Answer}$$

Problems

5.27 Calculate all forces acting on the bar *AB* described in Prob. 5.1.

5.28 Determine the reactions at *D*, *E*, and *F* for the table in Prob. 5.3.

5.29 Find the reaction at *O* on the bar described in Prob. 5.5.

5.30 For the structure in Prob. 5.9, determine the reactions at *C* and *D*.

5.31 Calculate the reaction at *D* for the structure described in Prob. 5.11.

5.32 The frame is supported by a ball-and-socket joint at *A* and a pin at *C*. The strut *EF* has a ball-and-socket joint at each end. The cable *EBD* runs over a small pulley at *B* and carries a 600-lb weight at *D*. Neglecting the weights of the members, determine the force in *EF* and the magnitude of the reaction at *C*.

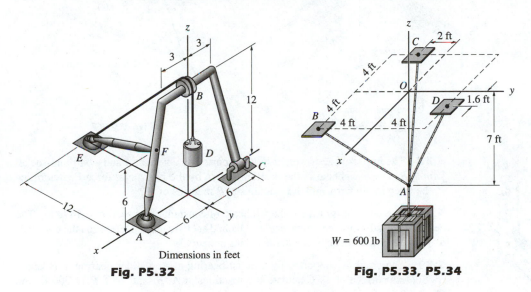

Fig. P5.32

Fig. P5.33, P5.34

5.33 Determine the tension in each of the three ropes supporting the 600-lb crate.

5.34 Using only one equilibrium equation, compute the force in rope *AD* of Prob. 5.33.

5.35 The 40-lb homogeneous pole *AB* rests on a horizontal surface at *A* and leans against a frictionless vertical wall at *B*. Determine the force *P* that must be applied at *B* to prevent the pole from falling and the corresponding reactions at *A* and *B*, if (a) $\theta = 0$; and (b) $\theta = 90°$. Assume that friction at *A* is sufficient to keep the lower end of the bar from sliding.

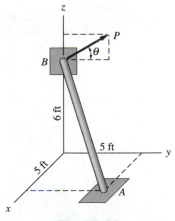

Fig. P5.35

5.36   The shaft AB is supported by a thrust bearing at A and a slider bearing at B. Determine the force in cable CD, and the bearing reactions at A and B caused by the 90-N vertical force applied at E. Neglect weights.

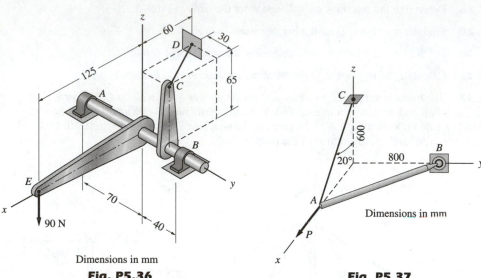

Dimensions in mm

Fig. P5.36

Fig. P5.37

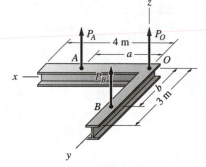

Fig. P5.38

5.37   The uniform bar of weight W is supported by a wire at A and a ball-and-socket joint at B. The bar is held in the position shown by the horizontal force P. Determine P, the force in the wire, and the reactions at B in terms of W.

5.38   The total mass of the L-shaped beam of constant cross section is 1470 kg. The beam is hoisted by three vertical cables attached at O, A, and B. Determine the distances a and b for which the tensions in the cables are equal.

5.39   The crank is supported by a thrust bearing at A, a slider bearing at B, and a frictionless surface at D. Calculate the reactions at A, B, and D if $P = 200$ lb and $C = 800$ lb · ft. The weight of the crank may be neglected.

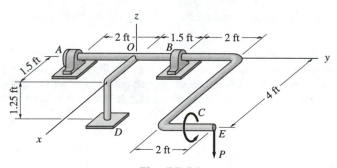

Fig. P5.39

5.40   A 120-lb weight is attached to the cable that is wrapped around the 50-lb homogeneous drum. The shaft attached to the drum is supported by a thrust bearing at A and a slider bearing at B. The drum is kept in equilibrium by the vertical force P acting on the handle of the crank. Determine P and the reactions at A and B. Neglect the weights of the crank and the shaft.

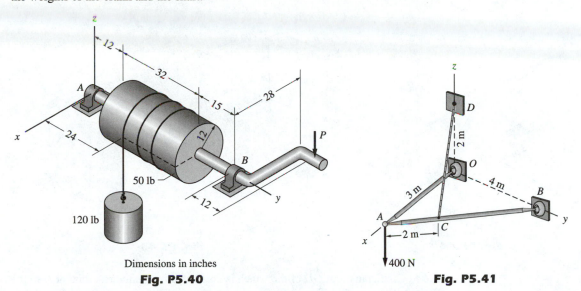

Dimensions in inches

Fig. P5.40

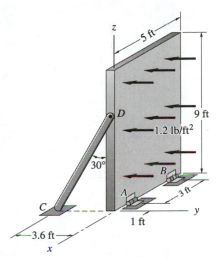

Fig. P5.41

5.41   Calculate the force in cable CD and the reaction at O. Assume that O, A, and B are ball-and-socket joints, and neglect the weights of the members.

5.42   The homogeneous rectangular panel weighs 64 lb and is attached to the floor by hinges at A and B. The panel is held in the vertical position by the light bar CD, which is pinned to the panel at D and is resting on a friction surface at C. Determine all forces acting on the panel if it is subjected to the uniform wind pressure of 1.2 lb/ft².

5.43   The arm ADE of the boring machine is attached to a rigid support by pins at A and A′. The arm is also supported by the hydraulic cylinder BD, which has a ball-and-socket joint at each end. The applied loading consists of a wrench acting at the boring tool E. Neglecting the weights of the members, find the force in the cylinder BD if $P = 1400$ lb and $C = 3600$ lb · ft.

Fig. P5.42

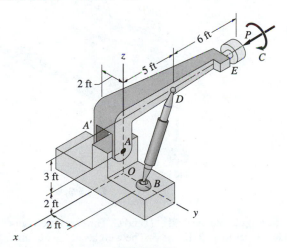

Fig. P5.43

5.44 A hoist is formed by connecting bars *BD* and *BE* to member *ABC*. Neglecting the weights of the members and assuming that all connections are ball-and-socket joints, determine the magnitudes of the forces in bars *BD* and *BE* in terms of the applied load *P*.

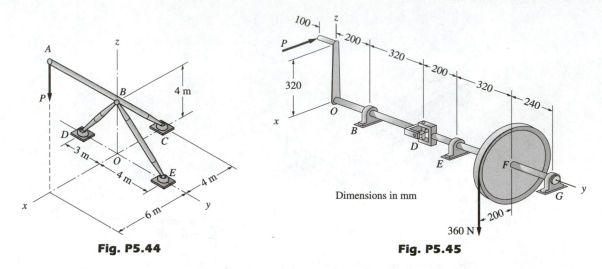

Fig. P5.44   **Fig. P5.45**

5.45 The crank arm *OD* of the winch is connected by a universal joint at *D* to the shaft-pulley assembly. The winch is supported by slider bearings at *B* and *E*, and by a thrust bearing at *G*. Determine the force *P* that will hold the winch at rest, and calculate the magnitudes of the corresponding bearing reactions. Neglect the weights of the members.

5.46 The cart weighs 80 lb; its center of gravity is at *G*. The roller at *A* is supported by a vertical surface, and the grooved rollers at *B* and *C* run along the horizontal rail. Neglecting friction, find the magnitudes of the reactions at the rollers.

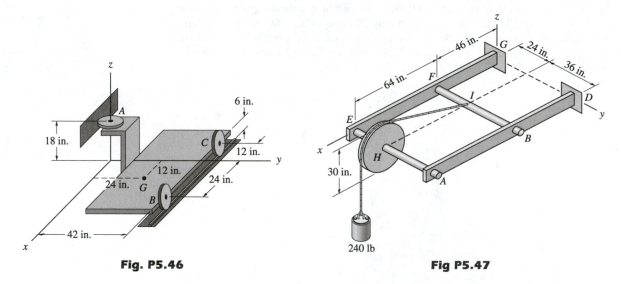

Fig. P5.46   **Fig P5.47**

5.47 The frame is built into the wall at *D* and *G*. The cross-members *AE* and *BF* pass through frictionless holes at *A*, *B*, *E*, and *F*. The weights of the members are negligible. Determine the reactions at *D*.

5.48 All connections of the structure are ball-and-socket joints, except for the slider bearings at *A* and *O*. The weights of the members may be neglected. Calculate the forces in members *BE* and *CF*.

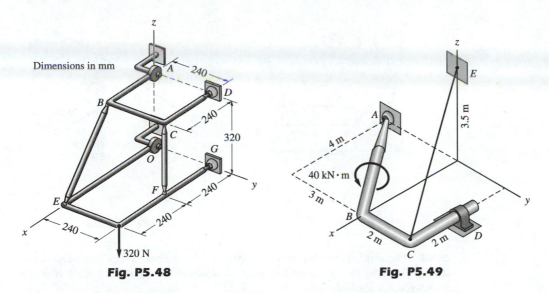

Fig. P5.48   Fig. P5.49

5.49 The bent bar is supported by a ball-and-socket joint at *A*, a cable at *C*, and a slider bearing at *D*. Determine the force in the cable due to the 40-kN · m couple acting on the bar (the couple-vector is directed along *AB*). Neglect the weight of the bar.

Review Problems

5.50 Three vertical cables are attached to the uniform, quarter circular plate. Determine the weight of the heaviest plate that can be supported if the tensile strength of each cable is 500 lb.

5.51 The bent rod is supported by a ball-and-socket joint at *O*, a cable at *B*, and a slider bearing at *D*. Neglecting the weight of the rod, calculate the tension in the cable and the magnitude of the bearing reaction at *D*.

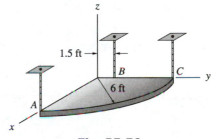

Fig. P5.50

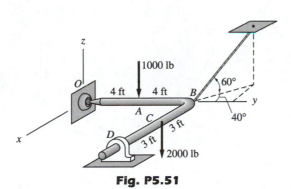

Fig. P5.51

5.52 Find the maximum load P that can be supported by the tripod if the force in any leg is limited to 2000 lb. Assume that the legs are two-force bodies.

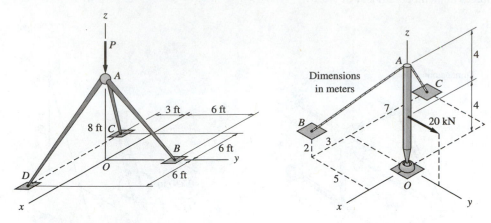

Fig. P5.52 **Fig. P5.53**

5.53 The vertical mast OA, which weighs 1.5 kN, is supported by a ball-and-socket joint at O and by the cables AB and AC. Calculate the tension in each cable when the 20-kN force is applied.

5.54 The upper end of the homogeneous slender bar AB leans against the smooth vertical wall. If the bar weighs 80 lb, find the tension in the cable BC, and the magnitude of the force exerted by the ball-and-socket joint at A.

5.55 The 500-kg crate is supported by the three cables. Find the tension in cable AD.

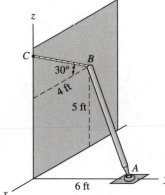

Fig. P5.54

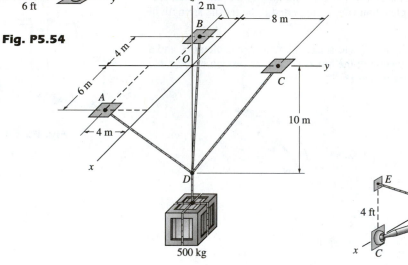

Fig. P5.55 **Fig. P5.56**

5.56 The uniform bars AB and BC each weigh 4 lb/ft. Calculate the tension in cable DE, and the magnitudes of the ball-and-socket reactions at A, B, and C.

5.57   The turnbuckle is tightened until the tension in cable *AD* is 1200 lb. Find the forces in cables *BF* and *CE*. The support at *O* is a ball-and-socket joint. Neglect the weight of bar *OAC*.

5.58   The turnbuckle is tightened until the *y*-component of the ball-and-socket reaction at *O* is $\mathbf{O}_y = 600\mathbf{j}$ lb. Determine the tension in cable *AD*. Neglect the weight of bar *OAC*.

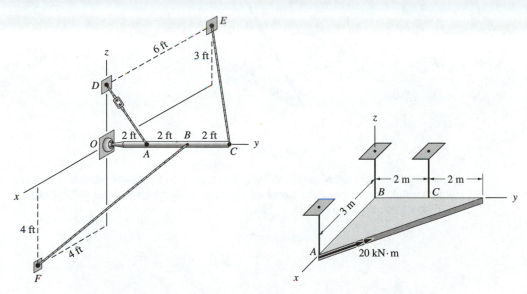

Fig. P5.57, P5.58   **Fig. P5.59**

5.59   The triangular plate is supported by three vertical rods, each of which is able to carry a tensile or compressive force. Calculate the force in each rod when the 20-kN · m couple is applied. Neglect the weight of the plate.

5.60   The connections at the ends of bars *AB* and *BC* are ball-and-socket joints. Neglecting the weights of the bars, determine the force in cable *DE* and the reaction at *A*.

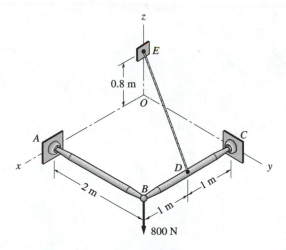

Fig. P5.60

5.61   The 150-kg bar *ABO* is supported by two cables at *A* and a slider bearing at *B*. The end of the bar presses against a frictionless surface at *O*. Find the tensions in the cables and the contact force at *O*.

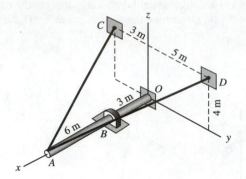

Fig. P5.61

5.62   The bent rod of negligible weight is supported by a ball-and-socket joint at *A* and a slider bearing at *B*. End *D* of the rod rests against a frictionless vertical surface. Find the reactions at *B* and *D*.

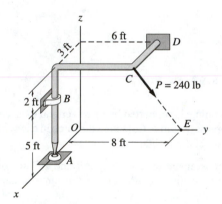

Fig. P5.62

6

Beams and Cables

*6.1 Introduction

In this chapter we introduce the analyses of beams and flexible cables, two important topics of structural mechanics. The analysis of beams that carry transverse loads deals with the computation of internal forces and couples. Because the internal forces and couples may vary in a complicated manner with the distance along the beam, we place considerable emphasis on methods of computation and graphical displays of the results.

The analysis of flexible cables can also become quite involved; the source of the difficulty lies in the geometry of the cable. Because a cable can carry only a tensile force, it must adjust its shape so that the internal tension is in equilibrium with the applied loads. Therefore, the geometry of the cable is not always known at the beginning of the analysis. When the shape of the cable is unknown, the solution invariably leads to nonlinear equations, which can only be solved numerically.

PART A: Beams

*6.2 Internal Force Systems

The determination of internal forces is a fundamental step in the design of members that carry loads. Only after this computation has been made can the design engineer select the proper dimensions for a member or choose the material from which the member should be fabricated.

If the external forces that hold a member in equilibrium are known, we can compute the internal forces by straightforward equilibrium analysis. For example, consider the bar in Fig. 6.1(a) that is loaded by the external forces $\mathbf{F}_1$, $\mathbf{F}_2, \ldots, \mathbf{F}_5$. To determine the internal force system acting on the cross section labeled ① (perpendicular to the axis of the bar), we must first isolate the portions of the bar lying on either side of section ①. The free-body diagram (FBD) of the portion to the left of section ① is shown in Fig. 6.1(b). In addition to the external forces $\mathbf{F}_1$, $\mathbf{F}_2$, and $\mathbf{F}_3$, this FBD shows the resultant force-couple system of the internal forces that are distributed over the cross section: the

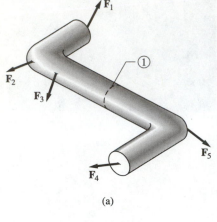

(a)

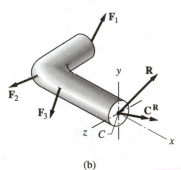

(b)

resultant force **R** acting at the centroid C of the cross section and the resultant couple $\mathbf{C}^R$. As explained in Chapter 3, we can place the resultant force **R** at any point, provided that we introduce the proper resultant couple. However, locating **R** at the centroid of the cross section is the standard engineering practice. If $\mathbf{F}_1$, $\mathbf{F}_2$, and $\mathbf{F}_3$ are known, the equilibrium equations $\Sigma\mathbf{F} = \mathbf{0}$ and $\Sigma\mathbf{M}_C = \mathbf{0}$ can be used to compute **R** and $\mathbf{C}^R$.

It is conventional to introduce the centroidal coordinate system shown in Fig. 6.1(b). The axis that is perpendicular to the cross section and passes through the centroid (*x*-axis) is called the *centroidal axis*. The components of **R** and $\mathbf{C}^R$ relative to this coordinate system are identified by the labels shown in Fig. 6.1(c) and are given the following physically meaningful names.

P: The force component that is perpendicular to the cross section, tending to elongate or shorten the bar, is called the *normal force*.

V_y and V_z: The force components lying in the plane of the cross section, tending to slide (shear) the parts of the bar lying on either side of the cross section relative to one another, are called *shear forces*.

T: The component of the resultant couple that tends to twist the bar is called *twisting moment,* or *torque.*

M_y and M_z: The components of the resultant couple that tend to bend the bar are called *bending moments*.

The deformations produced by these internal forces and couples are illustrated in Fig. 6.2.

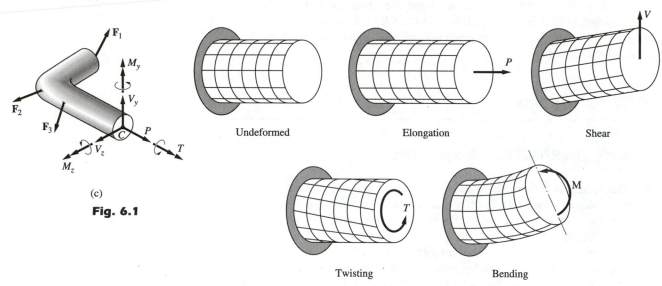

(c)

Fig. 6.1

Undeformed   Elongation   Shear

Twisting   Bending

Fig. 6.2

In many applications the external forces are coplanar and lie in a plane that contains the centroidal axis. Figure 6.3(a) illustrates the case in which all the external forces lie in the xy-plane, where the x-axis coincides with the centroidal axis of the bar. In this special case, the only nonzero components of the internal force system acting on any cross section—for example, section ①—are the normal force P, the shear force V, and the bending moment M, as shown in Fig. 6.3(b).

Thus far, we have concentrated on the internal force system acting on the portion of the bar lying to the left of section ①. Using Newton's third law, these internal forces occur in equal and opposite pairs on the two sides of the cross section, as shown in Fig. 6.3(c). In the following articles, we confine our attention to calculating the internal forces and couples in members subjected to *coplanar forces*.

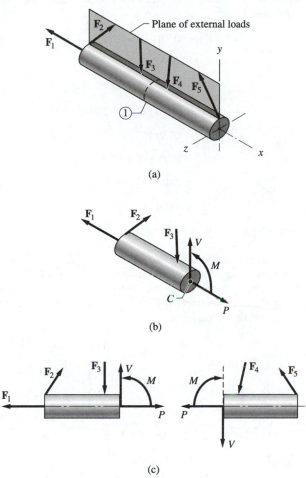

(a)

(b)

(c)

Fig. 6.3

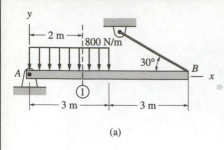

(a)

(b)

Sample Problem 6.1

The bar in Fig. (a), supported by a pin at A and a cable at B, carries a uniformly distributed load over its left half. Neglecting the weight of the bar, determine the normal force, shear force, and bending moment acting on the cross section at ① by analyzing (1) the bar segment on the left of section ①; and (2) the bar segment on the right of section ①.

Solution

Preliminary Calculations

We must calculate the external reactions before we can find the internal force system. As shown in the FBD in Fig. (b), the bar is subjected to the following forces: the components A_x and A_y of the pin reaction at A, the tension T in the cable at B, and the 2400-N resultant of the uniformly distributed load. Equilibrium analysis determines the reactions as follows:

$$\Sigma M_A = 0 \qquad \overset{+}{\curvearrowleft} \qquad T \sin 30°(6) - 2400(1.5) = 0$$
$$T = 1200 \text{ N}$$

$$\Sigma F_x = 0 \qquad \xrightarrow{+} \qquad A_x - T \cos 30° = 0$$
$$A_x = T \cos 30° = 1200 \cos 30°$$
$$A_x = 1039 \text{ N}$$

$$\Sigma F_y = 0 \qquad +\uparrow \quad A_y - 2400 + T \sin 30° = 0$$
$$A_y = 2400 - T \sin 30° = 2400 - 1200 \sin 30°$$
$$A_y = 1800 \text{ N}$$

Because these answers are positive, each of the reactions is directed as assumed in Fig. (b).

To find the internal force system acting on the cross section at ①, we must isolate the segments of the bar lying on either side of section ①. The FBDs of the segments on the left and on the right of section ① are shown in Figs. (c) and (d), respectively. Note that in determining the resultants of distributed loading, we considered only that part of the load that acts on the segment.

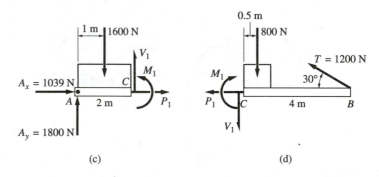

(c)   (d)

The force system acting on the cross section at ① consists of the normal force P_1, the shear force V_1, and the bending moment M_1. To be consistent with Newton's third law (equal and opposite reactions), P_1, V_1, and M_1 in Fig. (c) are shown equal in magnitude but oppositely directed to their counterparts in Fig. (d). We can use either FBD to compute P_1, V_1, and M_1.

Part 1

Applying the equilibrium equations to the FBD of the bar segment lying to the left of section ①, Fig. (c), we obtain

$$\Sigma F_x = 0 \quad \xrightarrow{+} \quad P_1 + 1039 = 0$$
$$P_1 = -1039 \text{ N} \qquad Answer$$

$$\Sigma F_y = 0 \quad +\uparrow \quad 1800 - 1600 + V_1 = 0$$
$$V_1 = -1800 + 1600 = -200 \text{ N} \qquad Answer$$

$$\Sigma M_C = 0 \quad \overset{+}{\frown} \quad -1800(2) + 1600(1) + M_1 = 0$$
$$M_1 = 3600 - 1600 = 2000 \text{ N} \cdot \text{m} \qquad Answer$$

The negative signs in P_1 and V_1 indicate that their senses are opposite to what is shown in the FBD.

Part 2

Applying the equilibrium equations to the FBD of the bar segment on the right of section ①, Fig. (d), yields

$$\Sigma F_x = 0 \quad \xrightarrow{+} \quad -P_1 - 1200 \cos 30° = 0$$
$$P_1 = -1200 \cos 30° = -1039 \text{ N} \qquad Answer$$

$$\Sigma F_y = 0 \quad +\uparrow \quad -V_1 + 1200 \sin 30° - 800 = 0$$
$$V_1 = 1200 \sin 30° - 800 = -200 \text{ N} \qquad Answer$$

$$\Sigma M_C = 0 \quad \overset{+}{\frown} \quad -M_1 - 800(0.5) + 1200 \sin 30°(4) = 0$$
$$M_1 = -800(0.5) + 1200 \sin 30°(4) = 2000 \text{ N} \cdot \text{m} \qquad Answer$$

These answers agree, of course, with those obtained in Part 1.

Sample Problem 6.2

A pin-connected circular arch supports a 5000-lb vertical load as shown in Fig. (a). Neglecting the weights of the members, determine the normal force, shear force, and bending moment that act on the cross section at ①.

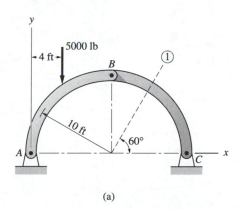

(a)

251

Solution

The FBD of the entire arch is shown in Fig. (b). The forces A_x and A_y are the components of the pin reactions at A, and R_C is the pin reaction at C. Recognizing that member BC is a two-force member, we know that R_C is directed along the line BC. In general, all the external reactions should be computed before the internal force systems are found. However, in this problem, we need only calculate R_C. From the FBD in Fig. (b) we obtain

$$\Sigma M_A = 0 \quad \overset{+}{\curvearrowleft} \quad R_C \sin 45°(20) - 5000(4) = 0$$
$$R_C = 1414 \text{ lb}$$

We next consider the FBD of the portion CD shown in Fig. (c). The forces D_x and D_y are the horizontal and vertical components of the resultant force acting on the cross section, and M_1 is the bending moment. We could compute D_x, D_y, and M_1 by recognizing that their resultant is a single force that is equal and opposite to R_C. However, it is simpler to compute these unknowns using the following equilibrium equations:

$$\Sigma F_x = 0 \quad \overset{+}{\rightarrow} \quad D_x - 1414 \cos 45° = 0 \qquad D_x = 1000 \text{ lb}$$

$$\Sigma F_y = 0 \quad +\uparrow \quad -D_y + 1414 \sin 45° = 0 \qquad D_y = 1000 \text{ lb}$$

$$\Sigma M_D = 0 \quad \overset{+}{\curvearrowleft} \quad M_1 - 1414 \cos 45°(8.66) - 1414 \sin 45°(5.00) = 0$$
$$M_1 = 3660 \text{ lb} \cdot \text{ft} \qquad \qquad \textit{Answer}$$

The FBD in Fig. (d) shows the resultant force acting on the cross section in terms of its normal component P_1 and shear component V_1. Comparing Figs. (c) and (d), we obtain

$$P_1 = D_y \cos 60° + D_x \sin 60°$$
$$= 1000 \cos 60° + 1000 \sin 60° = 1366 \text{ lb} \qquad \textit{Answer}$$

and

$$V_1 = D_y \sin 60° - D_x \cos 60°$$
$$= 1000 \sin 60° - 1000 \cos 60° = 366 \text{ lb} \qquad \textit{Answer}$$

Because P_1, V_1, and M_1 turned out to be positive, each of them is directed as shown in Fig. (d).

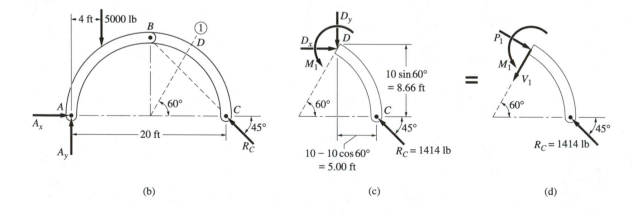

(b)　　　　　　　　　(c)　　　　　　　　　(d)

Problems

In the following problems the internal force system is to be represented as a normal force P, a shear force V, and a bending moment M. Neglect the weights of the members.

6.1–6.3 Determine the internal force system acting on section ① by analyzing the FBD of (a) segment *AD*; and (b) segment *DB*.

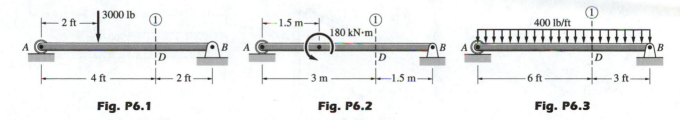

| Fig. P6.1 | Fig. P6.2 | Fig. P6.3 |

6.4–6.6 Find the internal force systems acting on sections ① and ②.

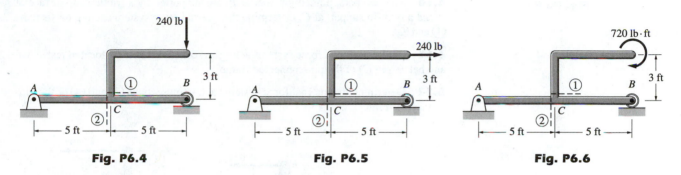

| Fig. P6.4 | Fig. P6.5 | Fig. P6.6 |

6.7 The three identical cantilever beams carry vertical loads that are distributed in a different manner. It is known that beam (a) fails because the maximum internal bending moment reaches its critical value when $P_1 = 360$ lb. Compute the values of P_2 and P_3 that would cause the failure of the other two beams.

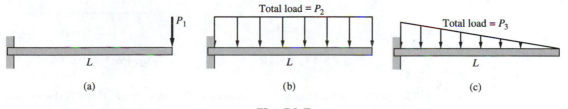

Fig. P6.7

6.8 Find the internal force systems acting on sections ① and ② for the eyebolt shown.

6.9 The homogeneous 24.5-kg bar is supported by a roller at *B* and a rough wall at *A*. Calculate the internal force system acting on a cross section at the middle of the bar.

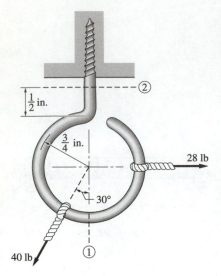

Fig. P6.8

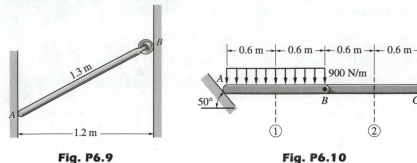

Fig. P6.9   **Fig. P6.10**

6.10 The two bars, pinned together at *B*, are supported by a frictionless surface at *A* and a built-in support at *C*. Determine the internal force systems acting on sections ① and ②.

6.11 Find the internal force systems acting on sections ① and ② (located just above and below pin *C*) of the pin-connected frame.

6.12 Determine the internal force systems acting on sections ③ and ④ of the pin-connected frame.

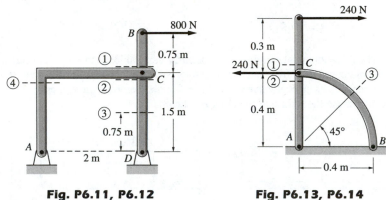

Fig. P6.11, P6.12   **Fig. P6.13, P6.14**

6.13 Determine the internal force systems acting on sections ① and ② for the pin-connected frame. The sections are located just above and just below pin *C*.

6.14 Find the internal force system acting on section ③ for the pin-connected frame.

6.15 Calculate the internal force systems acting on sections ① and ②, which are adjacent to point *C*.

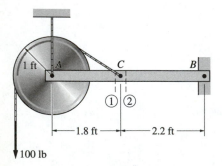

Fig. P6.15

6.16   The 600 lb · in couple is applied to member *DEF* of the pin-connected frame. Find the internal force systems acting on sections ① and ②.

6.17   A person of weight *W* climbs a ladder that has been placed on a frictionless horizontal surface. Find the internal force system acting on section ① as a function of *x* (the position coordinate of the person).

6.18   For the ladder in Prob. 6.17, find the internal force system acting on section ②, assuming that *x* < *a*/2.

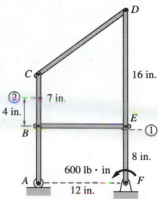

Fig. P6.16

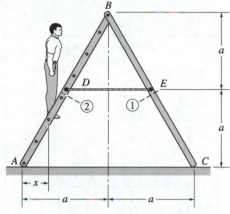

Fig. P6.17, P6.18

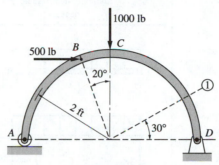

Fig. P6.19

6.19   Determine the internal force system acting on section ① of the circular arch.

*6.20   The equation of the parabolic arch is $y = (36 - x^2)/6$, where *x* and *y* are measured in feet. Compute the internal force system acting on section ①.

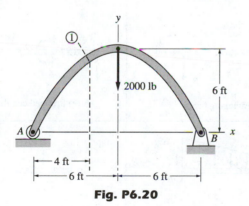

Fig. P6.20

*6.3   *Analysis of Internal Forces*

a.   *Loading and supports*

The term *beam* is reserved for a slender bar that is subjected to transverse loading (the applied forces are perpendicular to the bar). In this text, we consider only loadings that are also coplanar. As explained in Art. 6.2, the internal force system caused by coplanar loads can be represented as a normal force, a shear force, and a bending moment acting on the cross section.

Several examples of coplanar beam supports and loadings encountered in structural design are depicted in Fig. 6.4. Also shown are the free-body diagrams of the beams, which display both the applied loads and the support reactions. The reactions for statically determinate beams, Fig. 6.4(a)–(c), can be found from equilibrium analysis. The computation of the reactions for statically indeterminate beams, Fig. 6.4(d)–(f), requires analyses that are beyond the scope of this text.

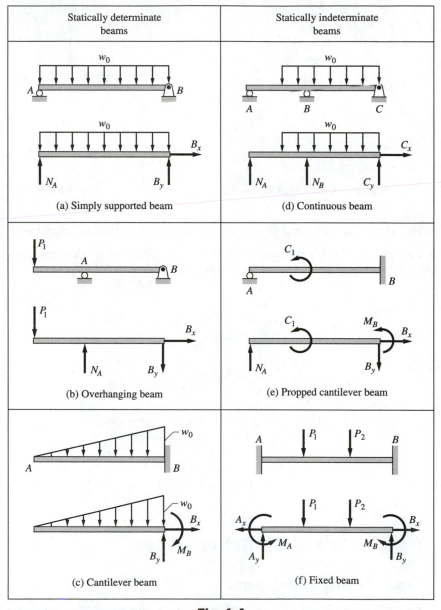

Statically determinate beams	Statically indeterminate beams
(a) Simply supported beam	(d) Continuous beam
(b) Overhanging beam	(e) Propped cantilever beam
(c) Cantilever beam	(f) Fixed beam

Fig. 6.4

b.   Sign convention

Before proceeding to the analysis of beams, we must establish a sign convention for the external as well as the internal forces and couples. (Unfortunately, there is no universally accepted sign convention for beams.) The sign convention that we adopt must be simple, clear, and unambiguous. It must work equally well for beams drawn horizontally, vertically, or even upside down; in other words, the sign convention should be independent of the observer's viewpoint.

Our sign convention is defined in terms of a coordinate system that is attached to the beam, called a *local coordinate system*. The local coordinate system that we use is shown in Fig. 6.5(a). The local x-axis is the *centroidal axis of the beam* (recall that the centroidal axis is the line that passes through the centroids of all the cross-sectional areas). The local y-axis is chosen so that

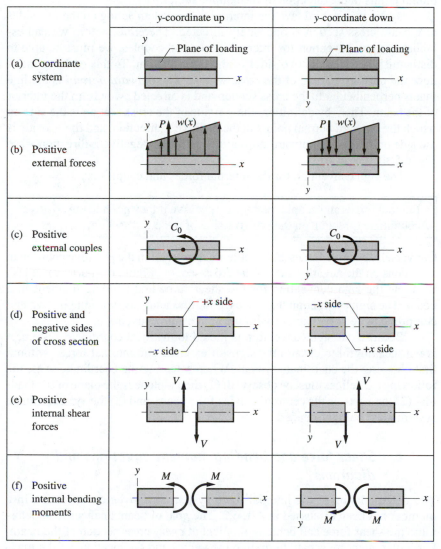

Fig. 6.5

the xy-plane coincides with the plane of the loading. Note that there are two choices for the positive direction of the y-axis (e.g., up or down)—it does not matter which direction you select, as long as it is shown in your drawing. The local z-axis (not shown in Fig. 6.5) is then determined by the right-hand rule.

The sign convention for the external loads is as follows:

> External forces and couple-vectors are positive if they point in the positive coordinate directions.

External forces that are positive (both concentrated and distributed)—shown in Fig. 6.5(b)—act in the positive y-direction. Positive external couple-vectors point in the positive z-direction, as illustrated in Fig. 6.5(c). (Recall that the direction of the couple-vector is determined by the right-hand rule: If you curl the fingers of your right hand in the direction of the couple, your outstretched thumb points in the direction of the couple-vector.)

By Newton's third law, the internal force system acting on the two sides of a beam cross section is oppositely directed. Therefore, before we can establish a sign convention for internal forces and couples, we must be able to distinguish between the two sides of the cross section. To this end, we introduce the outward normal of the cross section: The *outward normal* is the line that is perpendicular to the cross section and is directed away from the interior of the beam. Then, by definition, the $+x$ side of the cross section is the side on which the outward normal points in the positive x-direction, and the $-x$ side is the side on which the outward normal points in the negative x-direction—see Fig. 6.5(d).

The sign convention for the internal forces and couples is:

> Internal forces and couple-vectors are positive if they point in the positive coordinate direction on the positive side of the cross section.

Conversely, positive forces and couple-vectors point in the negative coordinate directions on the negative side of the cross section. Figure 6.5 parts (e) and (f) illustrate the sign convention for the shear force and bending moment in a beam. The sign convention that we used in truss analysis (tension positive and compression negative) agrees with the above sign convention.

Because our sign convention is linked to the local coordinate system, it is not meaningful to discuss the signs of external and internal forces without first choosing the coordinate system. When selecting the coordinate axes, the following guidelines must be observed: (1) the xy-plane is the plane of the loading, (2) the x-axis is the centroidal axis of the beam, and (3) the xyz-coordinate system must be right-handed.

c. Shear force and bending moment equations and diagrams

The determination of the internal force system at a *given* cross section in a member has been discussed in Art. 6.2. The goal of beam analysis is to determine the shear force and bending moment at *every* cross section of the beam. Particular attention is paid to finding the values and the locations of the maximum shear force and the maximum bending moment. The results enable the structural engineer to select a suitable beam that is capable of supporting the applied loads.

The equations that describe the variation of the shear force (V) and the bending moment (M) with the location of the cross section are called the *shear force and bending moment equations,* or simply, the V- and M-equations. These equations are always dependent on sign conventions, such as shown in Fig. 6.5.

When the V- and M-equations are drawn to scale, the results are called the *shear force and bending moment diagrams,* or simply, the V- and M-diagrams. After these diagrams have been drawn, the maximum shear force and the maximum bending moment can usually be found by inspection or with minimal computation.

In the following sample problems, we explain the procedures for deriving the V- and M-equations and for plotting the V- and M-diagrams.

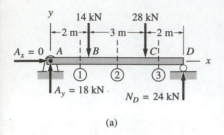

(a)

Sample Problem 6.3

For the simply supported beam shown in Fig. (a), (1) derive the shear force and bending moment equations for each segment of the beam; and (2) sketch the shear force and bending moment diagrams. Use the coordinate system shown, and neglect the weight of the beam. Note that the support reactions A_x, A_y, and N_D have already been computed, and are shown in Fig. (a).

Solution

Part 1

The determination of the V- and M-equations for each of the three beam segments (AB, BC, and CD) is explained in the following.

Segment AB ($0 < x < 2$ m)

Figure (b) shows the FBDs of the two segments of the beam that are separated by section ①, located within segment AB. Note that we have assumed V and M to be acting in their positive directions according to the sign convention illustrated in Fig. 6.5. Because V and M are equal in magnitude and oppositely directed on the two FBDs, they can be computed using either FBD.

The analysis of the FBD of the beam segment to the left of section ① yields

$$\Sigma F_y = 0 \quad +\uparrow \quad 18 + V = 0$$
$$V = -18 \text{ kN} \qquad \textit{Answer}$$

$$\Sigma M_E = 0 \quad \overset{+}{\curvearrowright} \quad -18x + M = 0$$
$$M = +18x \text{ kN} \cdot \text{m} \qquad \textit{Answer}$$

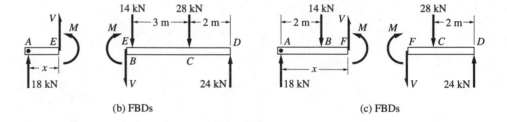

(b) FBDs (c) FBDs

Segment BC (2 m $< x < 5$ m)

Figure (c) shows the FBDs of the two segments of beam separated by section ②, an arbitrary section within segment BC. Once again, V and M have been assumed to be positive. The analysis of the segment to the left of section ② gives us

$$\Sigma F_y = 0 \quad +\uparrow \quad 18 - 14 + V = 0$$
$$V = -18 + 14 = -4 \text{ kN} \qquad \textit{Answer}$$

$$\Sigma M_F = 0 \quad \overset{+}{\curvearrowright} \quad -18x + 14(x - 2) + M = 0$$
$$M = 18x - 14(x - 2) = 4x + 28 \text{ kN} \cdot \text{m} \qquad \textit{Answer}$$

260

Segment CD (5 m $< x <$ 7 m)

Section ③ is used to find the shear force and bending moment in segment CD. The FBDs in Fig. (d) again show V and M acting in their positive directions. Analyzing the beam segment to the left of section ③, we obtain

$$\Sigma F_y = 0 \quad +\uparrow \qquad 18 - 14 - 28 + V = 0$$

$$V = -18 + 14 + 28 = +24 \text{ kN} \qquad \textit{Answer}$$

$$\Sigma M_G = 0 \quad \underset{+}{\curvearrowright} \qquad -18x + 14(x-2) + 28(x-5) + M = 0$$

$$M = 18x - 14(x-2) - 28(x-5)$$

$$= -24x + 168 \text{ kN} \cdot \text{m} \qquad \textit{Answer}$$

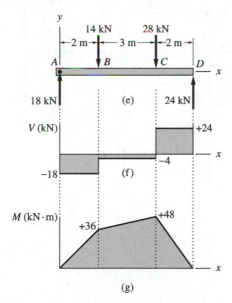

(d) FBDs

Part 2

The V- and M-diagrams—Figs. (f) and (g), respectively—are the plots of the V- and M-equations that have been derived in Part 1. By placing these plots directly below the sketch of the beam, Fig. (e), we establish a clear visual relationship between the diagrams and locations on the beam.

An inspection of the V-diagram reveals that the maximum shear force in the beam is +24 kN and that it occurs at every cross section of the beam in segment CD. From the M-diagram we see that the maximum bending moment is +48 kN · m, occurring under the 28-kN load at C. Note that at each concentrated force the V-diagram "jumps" by an amount equal to the force. Furthermore, there is a discontinuity in the slope of the M-diagram at each concentrated force.

Shear force and bending moment diagrams

261

Sample Problem 6.4

The simply supported beam shown in Fig. (a) carries the clockwise couple C_0 at B. Using the coordinate system shown, (1) derive the shear and moment equations; and (2) draw the shear force and bending moment diagrams for the beam. Neglect the weight of the beam. The support reactions A_x, A_y, and N_C have already been computed, and the results are shown in Fig. (a).

Solution

Part 1

Due to the presence of the couple, we must analyze beam segments AB and BC separately.

Segment AB ($0 < x < 3L/4$)

Figure (b) shows the FBD of the portion of the beam to the left of section ①. Note that V and M are assumed to be positive according to the sign convention in Fig. 6.5. Equilibrium equations of the segment yield

$$\Sigma F_y = 0 \quad +\!\downarrow \quad V + \frac{C_0}{L} = 0$$

$$V = -\frac{C_0}{L} \qquad\qquad \textit{Answer}$$

$$\Sigma M_D = 0 \quad \overset{+}{\curvearrowright} \quad -M + \frac{C_0}{L}x = 0$$

$$M = \frac{C_0 x}{L} \qquad\qquad \textit{Answer}$$

Segment BC ($3L/4 < x < L$)

Figure (c) shows the FBD of the portion of the beam to the left of section ②. Once again, V and M have been assumed to be positive. Applying the equilibrium equations

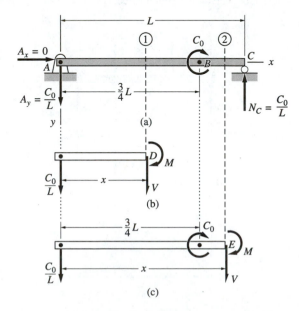

(a)

(b)

(c)

Free-body diagrams

to the segment, we obtain

$$\Sigma F_y = 0 \quad +\downarrow \quad V + \frac{C_0}{L} = 0$$

$$V = -\frac{C_0}{L} \qquad\qquad \textit{Answer}$$

$$\Sigma M_E = 0 \quad \overset{+}{\frown} \quad -M + \frac{C_0}{L}x - C_0 = 0$$

$$M = \frac{C_0 x}{L} - C_0 \qquad\qquad \textit{Answer}$$

Part 2

The V- and M-diagrams, shown in Figs. (d) and (e), respectively, were obtained by plotting the V- and M-equations that were found in Part 1. From the V-diagram we see that the shear force is the same for all cross sections of the beam. The M-diagram shows a jump of magnitude C_0 at the point of application of the couple.

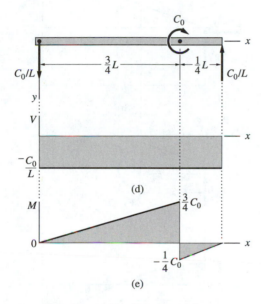

Shear force and bending moment diagrams

Sample Problem 6.5

The simply supported beam in Fig. (a) carries a distributed load, the intensity of which varies from zero at the left end to w_0 at the right end. Using the coordinate system shown, (1) derive the shear force and bending moment equations; and (2) draw the shear force and bending moment diagrams. Neglect the weight of the beam.

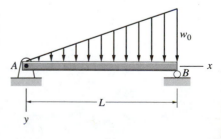

(a)

Solution

The FBD of the beam is shown in Fig. (b). Note that the triangular load diagram has been replaced by its resultant force $w_0 L/2$ (area under the load diagram), acting at the centroid of the load diagram. The support reactions at A and B can now be computed from the equilibrium equations; the results are shown in Fig. (b).

Part 1

Figure (c) shows the FBD of the portion of the beam that lies to the left of section ①, located at an arbitrary distance x from end A. Letting w be the intensity of the loading at section ①, as shown in Fig. (b), we have from similar triangles $w/x = w_0/L$, or $w = w_0 x/L$. Now the triangular load in Fig. (c) can be replaced by the resultant force $w_0 x^2/2L$ acting at the distance $x/3$ from section ①. The shear force V and the bending moment M acting at section ① were drawn in the positive directions in accordance with the sign convention shown in Fig. 6.5. Equilibrium analysis of the FBD in

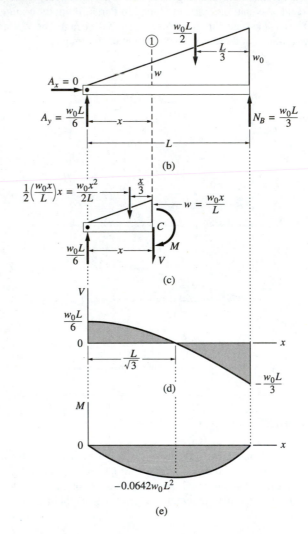

(b)

(c)

(d)

(e)

264

Fig. (c) then yields

$$\Sigma F_y = 0 \quad +\downarrow \qquad V + \frac{w_0 x^2}{2L} - \frac{w_0 L}{6} = 0$$

$$V = \frac{w_0 L}{6}\left(1 - \frac{3x^2}{L^2}\right) \qquad\qquad Answer \quad (a)$$

$$\Sigma M_C = 0 \quad \widehat{+} \qquad -M - \frac{w_0 L}{6}x + \frac{w_0 x^2}{2L}\left(\frac{x}{3}\right) = 0$$

$$M = \frac{w_0 L x}{6}\left(\frac{x^2}{L^2} - 1\right) \qquad\qquad Answer \quad (b)$$

Part 2

Plotting the V- and M-equations in Part 1 results in the V- and M-diagrams shown in Figs. (d) and (e), respectively. Observe that the shear force diagram is a parabola and that the bending moment diagram is a cubic polynomial in x.

The location of the section where the shear force is zero is found from Eq. (a):

$$V = \frac{w_0 L}{6}\left(1 - \frac{3x^2}{L^2}\right) = 0$$

which gives

$$x = \frac{L}{\sqrt{3}} \qquad\qquad (c)$$

The maximum bending moment occurs where the slope of the M-diagram is zero—that is, where $dM/dx = 0$. From Eq. (b), we obtain

$$\frac{dM}{dx} = \frac{w_0 L}{6}\left(\frac{3x^2}{L^2} - 1\right) = 0$$

which again yields $x = L/\sqrt{3}$. (In the next article, we show that the slope of the bending moment is always zero at a section where the shear force vanishes.) Substituting $x = L/\sqrt{3}$ into Eq. (b), we find the maximum bending moment to be

$$M_{max} = \frac{w_0 L}{6}\frac{L}{\sqrt{3}}\left(\frac{L^2}{3L^2} - 1\right) = -0.0642 w_0 L^2$$

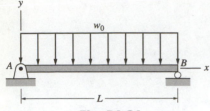

Fig. P6.21

Problems

In the following problems, assume the weight of each beam to be negligible, unless specified otherwise.

6.21–6.29   Derive the shear force and bending moment equations for the beam *AB*, and draw the *V*- and *M*-diagrams.

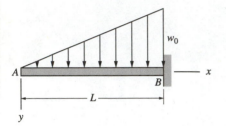

Fig. P6.22

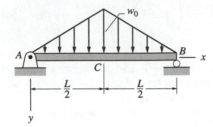

Fig. P6.23

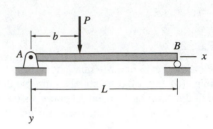

Fig. P6.24

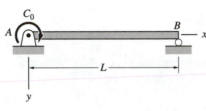

(a)

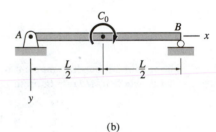

(b)

Fig. P6.25

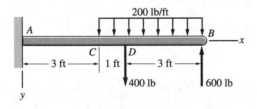

Fig. P6.26

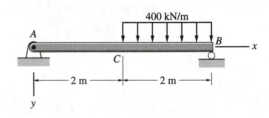

Fig. P6.27

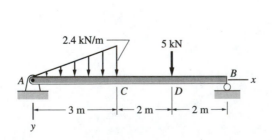

Fig. P6.28

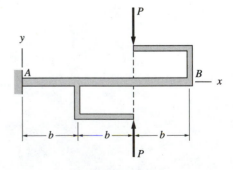

Fig. P6.29

6.30–6.32 Derive the shear force and bending moment equations for the beam *ABC*, and draw the *V*- and *M*-diagrams.

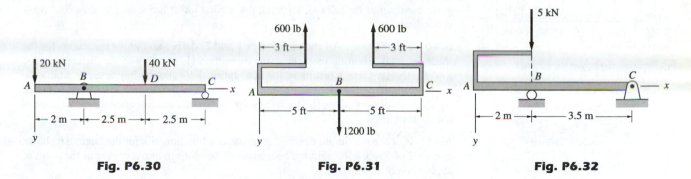

Fig. P6.30   **Fig. P6.31**   **Fig. P6.32**

6.33 Draw the shear force and bending moment diagrams for the beam *AB* if it weighs 14 lb/ft and (a) $C_0 = 0$; and (b) $C_0 = 220$ lb · ft.

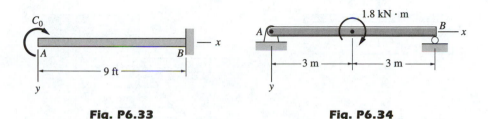

Fig. P6.33   **Fig. P6.34**

6.34 Draw the shear force and bending moment diagrams for the beam *AB* if the beam weighs 0.4 kN/m.

6.35 Draw the shear force and bending moment diagrams for the beam if $b = 4$ ft.

6.36 (a) Plot the bending moments acting at the cross sections *A* and *C* as functions of the overhang length *b*. (b) Find the value of *b* that will give the smallest magnitude for the maximum bending moment in the beam. What is the value of this bending moment? (Hint: The maximum bending moment occurs at either *A* or *C*, depending on the value of *b*.)

6.37 The trolley carrying the 6-kN load rides along the beam that is supported by cables at *A* and *B*. Draw the shear force and bending moment diagrams for the beam if $b = 2$ m.

6.38 (a) For the beam described in Problem 6.37, find the value of *b* for which the bending moment at the cross section at *C* reaches its largest value. (b) Plot the shear force and bending moment diagrams for the beam when the trolley is in the position determined in part (a).

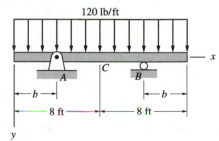

Fig. P6.35, P6.36

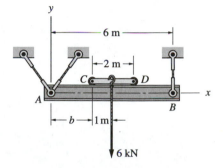

Fig. P6.37, P6.38

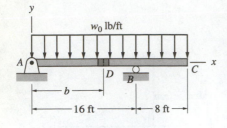

Detail of joint at D

Fig. P6.39

6.39 The 24-ft-long timber floor joist is designed to carry a uniformly distributed load. Because only 16-ft timbers are available, the joist is to be fabricated from two pieces connected by a nailed joint D. Determine the distance b for the most advantageous position of the joint D, knowing that nailed joints are strong in shear but weak in bending.

6.40 For the beam AB shown in cases 1 and 2, derive and plot expressions for the shear force and bending moment acting on section ① in terms of the distance x ($0 < x < L$). [Note: Case 1 results in the conventional V- and M-diagrams, in which the loads are fixed and the location of the section varies; the diagrams for case 2 (called *influence diagrams*) show the variation of V and M at a fixed section as the location of the load is varied.]

6.41 Plot the maximum bending moment as a function of x for the beam AB shown in case 2 of Prob. 6.40. (Hint: The maximum bending moment occurs at the point of application of the force P.)

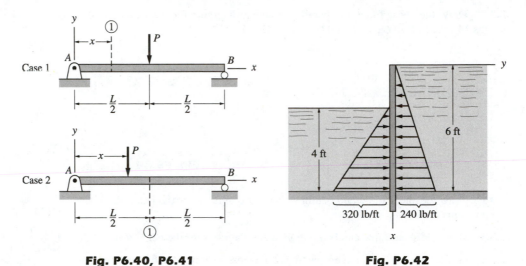

Fig. P6.40, P6.41   **Fig. P6.42**

6.42 Draw the shear force and bending moment diagrams for the piling that separates liquids of different densities.

*6.4 *Area Method for Drawing V- and M-Diagrams*

There are useful relationships between the load diagram, the shear force diagram, and the bending moment diagram, which are derivable from the equilibrium equations. Utilizing these relationships, we can plot the shear force diagram directly from the load diagram, and then sketch the bending moment diagram directly from the shear force diagram. This technique, called the *area method,* enables us to draw the V- and M-diagrams without having to go through the tedium of writing the V- and M-equations. We first consider beams subjected to distributed loading and then discuss concentrated forces and couples.

a.   *Distributed loading*

Consider the beam in Fig. 6.6(a) that is subjected to a line load of intensity $w(x)$, where $w(x)$ is assumed to be a continuous function. The free-body diagram of an infinitesimal segment of the beam, located at the distance x from the left end, is shown in Fig. 6.6(b). In addition to the distributed load $w(x)$, the segment is acted on by shear forces and bending moments at both ends. Let V and M be the shear force and bending moment at the left end, and let $(V + dV)$ and $(M + dM)$ be the shear force and bending moment at the right end. The infinitesimal differences dV and dM represent the changes in V and M, respectively, that occur over the differential length dx. Observe that all forces and bending moments are assumed to act in the positive directions defined in Fig. 6.5.

The force equation of equilibrium for the segment is

$$\Sigma F_y = 0 \quad +\!\downarrow \quad -V + w\,dx + (V + dV) = 0$$

from which

$$\boxed{w = -\frac{dV}{dx}} \tag{6.1}$$

The moment equilibrium yields

$$\Sigma M_O = 0 \quad \overset{+}{\frown} \quad M - V\,dx - (M + dM) + w\,dx\frac{dx}{2} = 0$$

After canceling M and dividing by dx, we get

$$-V - \frac{dM}{dx} + \frac{w\,dx}{2} = 0$$

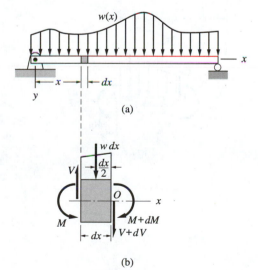

(a)

(b)

Fig. 6.6

Because dx is infinitesimally small, the last term is infinitesimal and can be dropped (this is not an approximation), yielding

$$V = -\frac{dM}{dx}$$

(6.2)

Equations (6.1) and (6.2) are called the *differential equilibrium equations* for beams. The following five theorems relating the load, shear force, and bending moment diagrams follow from these equations.

1. The value of the load intensity at any section of a beam is equal to the negative of the slope of the shear force diagram at that section.
   Proof—follows directly from Eq. (6.1).
2. The value of the shear force at any section of a beam is equal to the negative of the slope of the bending moment diagram at that section.
   Proof—follows directly from Eq. (6.2).
3. The difference between the shear force at two sections of a beam is equal to the negative of the area under the load diagram between those two sections.
   Proof—integrating Eq. (6.1) between sections at A and B (see Fig. 6.7), we obtain

$$\int_{x_A}^{x_B} dV = -\int_{x_A}^{x_B} w\,dx$$

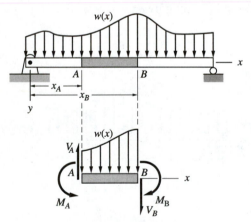

Fig. 6.7

Recognizing that the right-hand side of this equation represents the area under the load diagram between A and B, we get

$$V_B - V_A = -\text{area of } w\text{-diagram} \Big]_A^B \qquad \text{Q.E.D.}$$

For computational purposes, a more convenient form of this equation is

$$V_B = V_A - \text{area of } w\text{-diagram} \Big]_A^B \tag{6.3}$$

It should be noted that the signs in Eq. (6.3) are correct only if $x_B > x_A$.

4. The difference between the bending moments at two sections of a beam is equal to the negative of the area of the shear force diagram between those two sections.

Proof—integrating Eq. (6.2) between sections at A and B (see Fig. 6.7), we have

$$\int_{x_A}^{x_B} dM = -\int_{x_A}^{x_B} V\,dx$$

Because the right-hand side of this equation is the area of the shear diagram between A and B, we obtain

$$M_B - M_A = -\text{area of } V\text{-diagram} \Big]_A^B \qquad \text{Q.E.D.}$$

We find it convenient to use this last equation in the form

$$M_B = M_A - \text{area of } V\text{-diagram} \Big]_A^B \tag{6.4}$$

The signs in Eq. (6.4) are correct only if $x_B > x_A$.

5. If the load diagram is a polynomial of degree n, then the shear force diagram is a polynomial of degree $(n+1)$, and the bending moment diagram is a polynomial of degree $(n+2)$.

Proof—follows directly from the integration of Eqs. (6.1) and (6.2).

The area method for drawing V- and M-diagrams is simply an application of the foregoing theorems. For example, consider the beam segment shown in Fig. 6.8, which is 2 m long and is subjected to a uniformly distributed load $w(x)$. Following the sign convention of Fig. 6.5, the load intensity

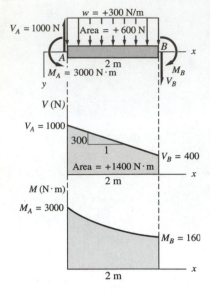

Fig. 6.8

Load Intensity	$w(x) = +300$ N/m
Degree of Load Diagram	0 (horizontal straight line)
Area under $w]_A^B$	$+600$ N
$V_B = V_A -$ area under $w]_A^B$ [Eq. (6.3)]	$V_B = 1000 - 600 = +400$ N
Slope of V-diagram: $dV/dx = -w(x)$ [Eq. (6.1)]	-300 N/m (constant)
Degree of V-diagram	1 (inclined straight line)
Area under $V]_A^B$	$+1400$ N $\cdot$ m
$M_B = M_A -$ area under $V]_A^B$ [Eq. (6.4)]	$M_B = 3000 - 1400 = +1600$ N $\cdot$ m
Slope of M-diagram: $dM/dx = -V(x)$ [Eq. (6.2)]	-1000 N/m at A -400 N/m at B
Degree of M-diagram	2 (parabola)
Shape of M-diagram	By inspection of the slopes at A and B, the parabolic M-diagram is concave upward.

Table 6.1   *Computation of Bending Moment and Shear Force Diagrams by the Area Method*

is $w(x) = +300$ N/m. The shear force and bending moment at the left end are known to be $V_A = +1000$ N and $M_A = +3000$ N $\cdot$ m. Table 6.1 gives the information required to construct the V- and M-diagrams for the segment by the area method. To master this technique, you must understand how each entry in the table is used in sketching the V- and M-diagrams in Fig. 6.8.

b.   *Concentrated forces and couples*

The area method for constructing V- and M-diagrams can be extended to beams that are loaded by concentrated forces and/or couples.

Figure 6.9 shows the free-body diagram of a beam segment of infinitesimal length dx containing point A where a concentrated force P_A and a concentrated couple C_A are applied. The shear force and bending moment acting on the negative cross section are denoted by V_A^- and M_A^-, whereas the notation V_A^+ and M_A^+ is used for the positive cross section. Observe that the directions of all forces and moments in Fig. 6.9 agree with the positive directions adopted in Fig. 6.5.

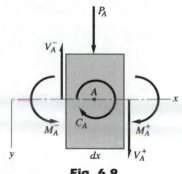

Fig. 6.9

The force equation of equilibrium gives

$$\Sigma F_y = 0 \quad +\!\downarrow \quad V_A^+ - V_A^- + P_A = 0$$

$$\boxed{V_A^+ = V_A^- - P_A} \tag{6.5}$$

Note that the concentrated force P_A causes a jump discontinuity in the V-diagram at A.

The moment equilibrium yields

$$\Sigma M_A = 0 \quad \underset{+}{\curvearrowleft} \quad M_A^+ - M_A^- + C_A + V_A^+ \frac{dx}{2} + V_A^- \frac{dx}{2} = 0$$

Dropping the last two terms because they are infinitesimals (this is not an approximation), we have

$$\boxed{M_A^+ = M_A^- - C_A} \tag{6.6}$$

Observe that the concentrated couple C_A gives rise to a jump discontinuity in the M-diagram at A.

c. Summary

Equations (6.1)–(6.6), which are repeated below, form the basis of the area method for constructing V- and M-diagrams without deriving the V- and M-equations. The area method is useful if the area under the loading and shear force diagrams can be easily computed.

$$w = -\frac{dV}{dx} \tag{6.1}$$

$$V = -\frac{dM}{dx} \tag{6.2}$$

$$V_B = V_A - \text{area of } w\text{-diagram} \Big|_A^B \tag{6.3}$$

$$M_B = M_A - \text{area of } V\text{-diagram} \Big|_A^B \tag{6.4}$$

$$V_A^+ = V_A^- - P_A \tag{6.5}$$

$$M_A^+ = M_A^- - C_A \tag{6.6}$$

Sample Problem 6.6

The simply supported beam in Fig. (a) supports a 30-kN concentrated force at B and a 40-kN · m couple at D. Using the xy-coordinate system shown, draw the shear force and bending moment diagrams by the area method.

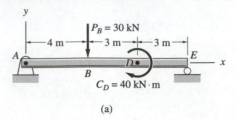

(a)

Solution

Load Diagram

The load diagram for the beam is shown in Fig. (b). The reactions at A and E were found from equilibrium analysis. The numerical value of each load is followed by a plus or minus in parentheses, indicating its sign as established by the sign conventions in Fig. 6.5.

Shear Force Diagram

We now explain the steps that have been used to construct the V-diagram, which is shown in Fig. (c). From the load diagram, we see that there are concentrated forces at A, B, and E that cause jumps in the shear force diagram at those points. Therefore, our discussion of shear force must distinguish between cross sections immediately to the left and to the right of each of these points.

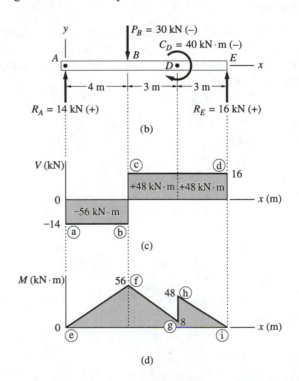

(b)

(c)

(d)

We begin by noting that $V_A^- = 0$, because there is no loading applied to the left of A. We then proceed across the beam from left to right, constructing the V-diagram as we go.

$$V_A^+ = V_A^- - R_A = 0 - (+14) = -14 \text{ kN}$$

Plot point ⓐ.

$$V_B^- = V_A^+ - \text{area of } w\text{-diagram}\Big]_A^B = -14 - 0 = -14 \text{ kN}$$

Plot point ⓑ.

Because $w = -dV/dx = 0$ between A and B, the slope of the V-diagram is zero between A and B.

Connect ⓐ *and* ⓑ *with a horizontal straight line.*

$$V_B^+ = V_B^- - P_B = -14 - (-30) = 16 \text{ kN}$$

Plot point ⓒ.

$$V_E^- = V_B^+ - \text{area of } w\text{-diagram}\Big]_B^E = 16 - 0 = 16 \text{ kN}$$

Plot point ⓓ.

Because $w = -dV/dx = 0$ between B and E, the slope of the V-diagram is zero for this segment.

Connect ⓒ *and* ⓓ *with a horizontal straight line.*

Note that because there is no loading to the right of E, we should find that $V_E^+ = 0$.

$$V_E^+ = V_E^- - R_E = 16 - (+16) = 0 \qquad\qquad \textit{Checks!}$$

Bending Moment Diagram

We now explain the steps required to construct the bending moment diagram, which is shown in Fig. (d). Because the applied couple is known to cause a jump in the M-diagram at D, we must distinguish between the bending moments just to the left and to the right of D. Before proceeding, we compute the areas under the shear force diagram for the different beam segments. The results of these computations are shown in Fig. (c). Observe that the areas may be positive or negative, depending on the sign of the shear force V.

We begin our construction of the M-diagram by noting that $M_A = 0$ (because there is no moment to the left of A and no couple applied at A).

Plot point ⓔ.

Proceeding across the beam from left to right, we generate the bending moment diagram in Fig. (d) as follows:

$$M_B = M_A - \text{area of } V\text{-diagram}\Big]_A^B = 0 - (-56) = 56 \text{ kN} \cdot \text{m}$$

Plot point ⓕ.

The V-diagram shows that the shear force between A and B is constant and negative. Therefore, the slope of the M-diagram between these two cross sections is constant and positive. (Recall that $dM/dx = -V$.)

Connect ⓔ and ⓕ with a straight line.

$$M_D^- = M_B - \text{area of } V\text{-diagram}\Big]_B^D = 56 - (+48) = 8 \text{ kN} \cdot \text{m}$$

Plot point ⓖ.

Because the slope of the V-diagram between B and D is positive and constant, the M-diagram has a negative, constant slope.

Connect ⓕ and ⓖ with a straight line.

$$M_D^+ = M_D^- - C_D = 8 - (-40) = 48 \text{ kN} \cdot \text{m}$$

Plot point ⓗ.

Next we note that $M_E = 0$ (because there is no moment to the right of E and no couple applied at E). Our computations based on the area of the V-diagram should verify this result.

$$M_E = M_D^+ - \text{area of } V\text{-diagram}\Big]_D^E = 48 - (+48) = 0 \qquad \textit{Checks!}$$

Plot point ⓘ.

The shear force between D and E is positive and constant, which means that the slope of the M-diagram for this segment is constant and negative.

Connect ⓗ and ⓘ with a straight line.

At first glance, the area method for constructing the V- and M-diagrams may appear to be more cumbersome than plotting the shear force and bending moment equations. However, with a little practice you will find that the area method is not only much faster but also less susceptible to numerical errors owing to the self-checking nature of the computations.

Sample Problem **6.7**

The simply supported beam in Fig. (a) carries two uniformly distributed loads and a concentrated load. Using the xy-coordinate system shown, draw the shear force and bending moment diagrams for the beam by the area method.

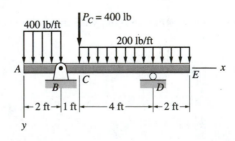

(a)

Solution

Load Diagram

The load diagram for the beam is shown in Fig. (b); the reactions at B and D have been determined by equilibrium analysis. Each of the numerical values is followed by a plus or a minus in parentheses, determined by the sign conventions established in Fig. 6.5. The significance of the section labeled F is explained in the following discussion.

Shear Force Diagram

The steps required to construct the V-diagram shown in Fig. (c) are now detailed. From the loading diagram, we see that there are concentrated forces at B, C, and D, which means that there will be jumps in the shear diagram at these points. Therefore, we have to differentiate between the shear force immediately to the left and to the right of each of these points.

We start our construction of the V-diagram by observing that $V_A = 0$ (because there is no load to the left of A and no force applied at A).

Plot point (a).

$$V_B^- = V_A - \text{area of } w\text{-diagram}\Big|_A^B = 0 - (+400)2 = -800 \text{ lb}$$

Plot point (b).

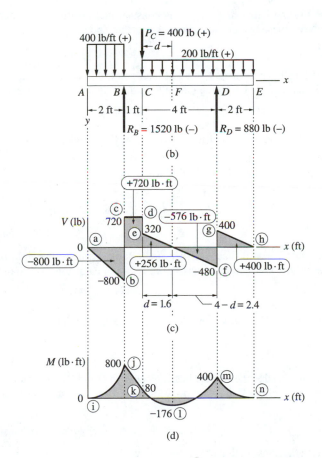

(b)

(c)

(d)

Observing from Fig. (b) that the applied loading between A and B is constant and positive, the slope of the shear diagram between these cross sections is constant and negative. (Recall that $dV/dx = -w$.)

Connect (a) *and* (b) *with a straight line.*

$$V_B^+ = V_B^- - R_B = -800 - (-1520) = 720 \text{ lb}$$

Plot point (c).

$$V_C^- = V_B^+ - \text{area of } w\text{-diagram} \Big|_B^C = 720 - 0 = 720 \text{ lb}$$

Plot point (d).

Because $w = -dV/dx = 0$ between B and C, the slope of the V-diagram is zero for this segment.

Connect (c) *and* (d) *with a horizontal straight line.*

$$V_C^+ = V_C^- - P_C = 720 - (+400) = 320 \text{ lb}$$

Plot point (e).

$$V_D^- = V_C^+ - \text{area of } w\text{-diagram} \Big|_C^D = 320 - (+200)4 = -480 \text{ lb}$$

Plot point (f).

Because the loading between C and D is constant and positive, the slope of the V-diagram between these two cross sections is constant and negative.

Connect (e) *and* (f) *with a straight line.*

Our computations have identified an additional point of interest—the point where the shear force is zero, labeled F on the load diagram in Fig. (b). Its location can be found from

$$V_F = V_C^+ - \text{area of } w\text{-diagram} \Big|_C^F = 320 - (+200)d = 0$$

which gives $d = 1.60$ ft, as shown in Fig. (c).

Continuing across the beam,

$$V_D^+ = V_D^- - R_D = -480 - (-880) = 400 \text{ lb}$$

Plot point (g).

Next we note that $V_E = 0$ (because there is no force acting to the right of E and no force applied at E). Our computations based on the area of the load diagram should verify this result:

$$V_E = V_D^+ - \text{area of } w\text{-diagram} \Big|_D^E = 400 - (+200)2 = 0 \qquad \textit{Checks!}$$

Plot point (h).

From Fig. (b), we see that the applied loading between D and E is constant and positive. Therefore the slope of the V-diagram between these two cross sections is constant and negative.

Connect (g) *and* (h) *with a straight line.*

This completes the construction of the V-diagram.

Bending Moment Diagram

We will now detail the steps required to construct the bending moment diagram shown in Fig. (d). Because there are no applied couples, there will be no jumps in the M-diagram. The areas of the V-diagram for the different segments of the beam are shown in Fig. (c).

We begin by noting that $M_A = 0$ (because there is no moment to the left of A and no couple applied at A).

Plot point (i).

Proceeding from left to right across the beam, the M-diagram is constructed as follows:

$$M_B = M_A - \text{area of } V\text{-diagram} \Big]_A^B = 0 - (-800) = 800 \text{ lb} \cdot \text{ft}$$

Plot point (j).

We note from Fig. (c) that the V-diagram between A and B is a first-degree polynomial (inclined straight line). Therefore, the M-diagram between these two cross sections will be a second-degree polynomial—that is, a parabola. From $dM/dx = -V$, we see that the slope of the M-diagram is zero at A and $+800$ lb/ft at B.

Connect (i) *and* (j) *with a parabola that has zero slope at* (i)
and positive slope at (j). *(The parabola will be concave upward.)*

$$M_C = M_B - \text{area of } V\text{-diagram} \Big]_B^C = 800 - (+720) = 80 \text{ lb} \cdot \text{ft}$$

Plot point (k).

Because the V-diagram is constant and positive between B and C, the slope of the M-diagram is constant and negative between those two sections.

Connect (j) *and* (k) *with a straight line.*

$$M_F = M_C - \text{area of } V\text{-diagram} \Big]_C^F = 80 - (+256) = -176 \text{ lb} \cdot \text{ft}$$

Plot point (l).

Using $dM/dx = -V$, we know that the slope of the M-diagram is -320 lb/ft at C and zero at F and that the curve is a parabola.

Connect (k) *and* (l) *with a parabola that has negative slope at* (k)
and zero slope at (l). *(The parabola will be concave upward.)*

$$M_D = M_F - \text{area of } V\text{-diagram} \Big]_F^D = -176 - (-576) = 400 \text{ lb} \cdot \text{ft}$$

Plot point (m).

The M-diagram between F and D is again a parabola, the slope of which is zero at F and $+480$ lb/ft at D.

Connect (l) *and* (m) *with a parabola that has zero slope at* (l)
and positive slope at (m). *(The parabola will be concave upward.)*

Next, we note that $M_E = 0$ (because there is no moment to the right of E and no couple applied at E). Our computations based on the area method should verify this result.

$$M_E = M_D - \text{area of } V\text{-diagram} \Big|_D^E = 400 - (+400) = 0 \qquad \textit{Checks!}$$

Plot point (n).

Using the familiar arguments, the M-diagram between D and E is a parabola with a slope equal to -400 lb/ft at D and zero slope at E.

Connect (m) *and* (n) *with a parabola that has negative slope at* (m) *and zero slope at* (n). *(The parabola will be concave upward.)*

This completes the construction of the M-diagram. It is obvious in Fig. (d) that the slope of the M-diagram is discontinuous at (j) and (m). Not as obvious is the slope discontinuity at (k): From $dM/dx = -V$ we see that the slope of the M-diagram to the left of (j) equals -720 lb/ft, whereas to the right of (j) the slope equals -320 lb/ft. But observe that the slope of the M-diagram is continuous at (l) because the shear force has the same value (zero) to the left and to the right of (l).

Problems

6.43 The uniformly distributed loading in (a) is approximated by the four forces in (b). Using the area method, draw the shear force and bending moment diagrams for each beam and compare the results. Neglect the weights of the beams.

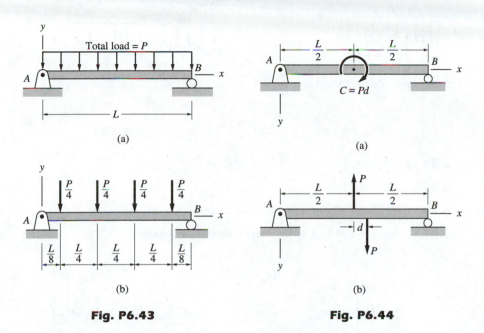

Fig. P6.43   Fig. P6.44

6.44 The couple in (a) is statically equivalent to the two forces in (b). Using the area method, draw the shear force and bending moment diagrams for each beam and compare the results when $d/L \rightarrow 0$ in (b). Neglect the weights of the beams.

Using the area method, draw the shear force and bending moment diagrams for the beams in Probs. 6.45–6.61. Neglect the weights of the beams unless stated otherwise.

6.45 See Problem 6.26.

6.46 See Problem 6.27.

6.47 See Problem 6.28.

6.48 See Problem 6.30.

6.49 See Problem 6.34.

6.50 See Problem 6.35.

6.51 See Fig. P6.51.

6.52 See Fig. P6.52; weight of beam is 80 lb/ft.

6.53 See Fig. P6.53.

6.54 See Fig. P6.54; weight of beam is 36 lb/ft.

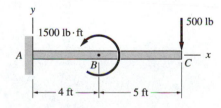

Fig. P6.51, P6.52

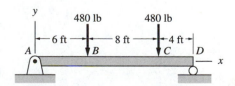

Fig. P6.53, P6.54

6.55 See Fig. P6.55.

6.56 See Fig. P6.56.

6.57 See Fig. P6.57.

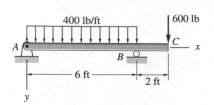

Fig. P6.55

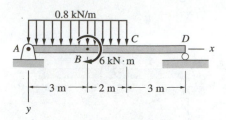

Fig. P6.56

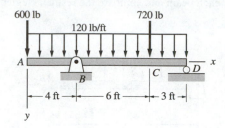

Fig. P6.57

6.58 See Fig. P6.58.

6.59 See Fig. P6.59.

6.60 Use w_0 = 80 lb/ft for the beam shown in the figure.

6.61 Use w_0 = 200 lb/ft for the beam shown in the figure.

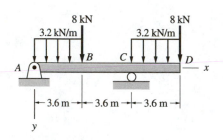

Fig. P6.58

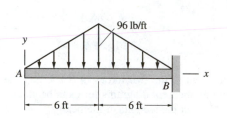

Fig. P6.59

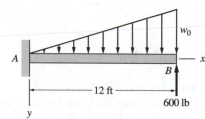

Fig. P6.60, P6.61

PART B: *Cables*

*6.5 *Cables under Distributed Loads*

a. *General discussion*

Flexible cables are used in numerous engineering applications. Common examples are power transmission lines and suspension bridges. The term *flexible* means that the cables are incapable of developing internal forces other than tension. In earlier chapters we treated cables as two-force members; that is, the weights of the cables were neglected, and the loading consisted of end forces only. Here we consider the effects of distributed forces, such as the weight of the cable or the weight of a structure that is suspended from the cable. Concentrated loads are covered in the next article.

Figure 6.10(a) shows a cable that is suspended from its endpoints A and B. In order to support the distributed loading of intensity w, the cable must assume a curved shape. It turns out that the equation describing this shape is simplified if we place the origin of the xy-coordinate system at the lowest point O of the cable. We let s be the distance measured along the cable from O. The shape of the cable and the location of point O are generally unknown at the beginning of the analysis.

The units of the load intensity w are lb/ft or N/m. The length can be measured in two ways: along the horizontal x-axis (w as a function of x) or along the cable (w as a function of s). Although these two cases must be treated separately, we first consider the elements of the analyses that are common to both.

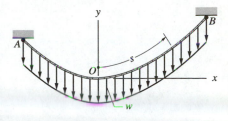

(a)

The free-body diagram (FBD) of a segment of the cable, extending from the lowest point O to an arbitrary point C, is shown in Fig. 6.10(b). The tensile forces in the cable at O and C are denoted by T_0 and T, respectively; W is the resultant of the distributed loading; and θ represents the slope angle of the cable at C. The force equilibrium equations of the cable segment are

$$\Sigma F_x = 0 \quad \xrightarrow{+} \quad T\cos\theta - T_0 = 0$$

$$\Sigma F_y = 0 \quad +\uparrow \quad T\sin\theta - W = 0$$

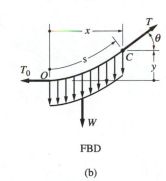

FBD

(b)

Fig. 6.10

from which we obtain

$$\boxed{T\cos\theta = T_0 \qquad T\sin\theta = W} \tag{6.7}$$

The first of Eqs. (6.7) shows that the horizontal component of the cable force, namely $T\cos\theta$, is constant throughout the cable. The solution of Eqs. (6.7) for θ and T yields

$$\boxed{\tan\theta = \frac{W}{T_0} \qquad T = \sqrt{T_0^2 + W^2}} \tag{6.8}$$

b. Parabolic cable

Here we analyze the special case in which the loading is distributed uniformly along the horizontal; that is, $w(x) = w_0$, where w_0 is the constant load intensity. This case arises, for example, in the main cables of a suspension bridge (see Fig. 6.11), where w_0 represents the weight of the roadway per unit length (it is assumed that the roadway is connected to the main cables by a large number of vertical cables and that the weights of all cables are negligible compared to the weight of the roadway).

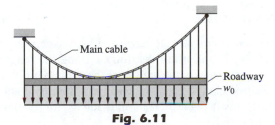

Fig. 6.11

Taking Eqs. (6.8) as the starting point, we now derive several useful equations that describe the geometry of the cable and the variation of the tensile force within the cable.

θ and T as functions of x and T_0   Because the resultant of the loading shown in Fig. 6.10(b) is $W = w_0 x$, Eqs. (6.8) become

$$\tan \theta = \frac{w_0 x}{T_0} \qquad T = \sqrt{T_0^2 + (w_0 x)^2} \qquad (6.9)$$

y as a function of x and T_0   Substituting $\tan \theta = dy/dx$, the first of Eqs. (6.9) can be written as $dy/dx = w_0 x/T_0$. Upon integration, we get

$$y = \frac{w_0 x^2}{2T_0} \qquad (6.10)$$

where the constant of integration was set equal to zero in order to satisfy the condition $y = 0$ when $x = 0$. Equation (6.10), which represents a parabola with its vertex at O, could also be obtained from a moment equilibrium equation using the FBD in Fig. 6.10(b).

s as a function of x and T_0   It is often necessary to compute the length s of the cable between points O and C in Fig. 6.10(b). The infinitesimal length of the cable is

$$ds = \sqrt{dx^2 + dy^2} = \sqrt{1 + \left(\frac{dy}{dx}\right)^2}\, dx \qquad (a)$$

Substituting $dy/dx = w_0 x/T_0$ and integrating, we obtain

$$s(x) = \int_0^x \sqrt{1 + \left(\frac{w_0 x}{T_0}\right)^2}\, dx \qquad (6.11)$$

Therefore, the length of the cable between points O and C is (see a table of integrals)

$$s(x) = \frac{x}{2}\sqrt{1 + \left(\frac{w_0 x}{T_0}\right)^2}$$
$$+ \frac{1}{2}\left(\frac{T_0}{w_0}\right)\ln\left[\left(\frac{w_0 x}{T_0}\right) + \sqrt{1 + \left(\frac{w_0 x}{T_0}\right)^2}\right] \qquad (6.12)$$

c.   Catenary cable

Consider a homogeneous cable that carries no load except its own weight. In this case, the loading is uniformly distributed along the length of the cable; that is, $w(s) = w_0$, where w_0 is the weight of the cable per unit length, and the distance s is measured along the cable. Therefore, the resultant of the loading shown in Fig. 6.10(b) is $W = w_0 s$. The following useful relationships can now be derived from Eqs. (6.8).

θ and T as functions of s and T_0   Substituting $W = w_0 s$ into Eq. (6.8) gives

$$\tan\theta = \frac{w_0 s}{T_0} \qquad T = \sqrt{T_0^2 + (w_0 s)^2} \qquad (6.13)$$

s as a function of x and T_0   We start with Eq. (a), which can be written as $(dy/dx)^2 = (ds/dx)^2 - 1$. Substituting $dy/dx = \tan\theta = w_0 s/T_0$, and solving for dx, yields

$$dx = \frac{ds}{\sqrt{1 + \left(\dfrac{w_0 s}{T_0}\right)^2}} \qquad (b)$$

Using a table of integrals, Eq. (b) yields

$$x(s) = \int_0^s dx = \frac{T_0}{w_0}\ln\left[\frac{w_0 s}{T_0} + \sqrt{1 + \left(\frac{w_0 s}{T_0}\right)^2}\right] \qquad (6.14)$$

Solving this equation for s gives

$$s(x) = \frac{T_0}{w_0}\sinh\frac{w_0 x}{T_0} \qquad (6.15)$$

The functions $\sinh u$ and $\cosh u$, called the *hyperbolic sine* and *hyperbolic cosine*, respectively, are defined as

$$\sinh u = \frac{1}{2}\left(e^u - e^{-u}\right) \qquad \cosh u = \frac{1}{2}\left(e^u + e^{-u}\right)$$

It can be seen that the rules for differentiation are

$$\frac{d}{du}\sinh u = \cosh u \qquad \frac{d}{du}\cosh u = \sinh u$$

y as a function of x and T_0   We substitute Eq. (6.15) into the first equation of Eqs. (6.13), which yields $\tan\theta = \sinh(w_0 x/T_0)$. Using $\tan\theta = dy/dx$, we obtain

$$dy = \tan\theta\, dx = \sinh\frac{w_0 x}{T_0}\, dx$$

which gives

$$y(x) = \int_0^x dy = \frac{T_0}{w_0}\left(\cosh\frac{w_0 x}{T_0} - 1\right) \qquad (6.16)$$

The curve represented by Eq. (6.16) is called a *catenary*.

If the slope of the catenary is small everywhere, then the curve differs very little from a parabola. As a proof of this statement, we note that if $\theta \ll 1$, then $dx = ds\cos\theta \approx ds$. Consequently, $w(s) \approx w(x)$, which means that the weight of the cable may be approximated by a uniformly distributed loading along the horizontal. This approximation usually simplifies the solution, because parabolic cables are generally easier to analyze than catenary cables.

T as a function of x and T_0   According to Eq. (6.7), the tension in the cable is $T = T_0 / \cos \theta$. Utilizing the geometrical relationship $\cos \theta = dx/ds$, this becomes $T = T_0 \, ds/dx$. On substituting for s from Eq. (6.15), we get

$$T = T_0 \cosh \frac{w_0 x}{T_0}$$ (6.17)

d.   Note on the solution of problems

There is no standard, step-by-step procedure for solving problems involving flexible cables. The reason is that the solution method for every problem is highly dependent on the information that is given in the problem statement. However, here are two guidelines that are applicable in most situations and that may be helpful.

1. It is not always wise to depend entirely on Eqs. (6.9)–(6.17). More often than not, a good starting point is the free-body diagram of the entire cable, or a portion of it, similar to Fig. 6.10(b). This FBD, in conjunction with Eqs. (6.9)–(6.17), should be used to formulate a method of analysis before proceeding to the actual computations.

2. Observe that point O, the origin of the coordinate system, and T_0 appear in all of the cable equations. If the location of point O and/or T_0 are not known, they should be determined first.

Sample Problem 6.8

The 36-m cable shown in Fig. (a) weighs 1.5 kN/m. Determine the sag H and the maximum tension in the cable.

Solution

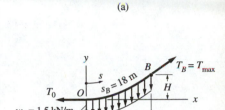

(a)

Method of Analysis

Because the loading is distributed along the cable, the shape of the cable is catenary. The cable is obviously symmetric about the midpoint of AB, which means that the location of the lowest point O of the cable is known. From the second of Eqs. (6.13), we note that the maximum cable tension occurs at the endpoints, where s is a maximum.

We now draw the free-body diagram of the right half of the cable, shown in Fig. (b). Although we could use this FBD in the solution of the problem, it is easier to use Eqs. (6.15)–(6.17). However, the FBD is convenient for identifying the various terms that arise in the solution.

By studying Eqs. (6.15)–(6.17), we conclude that the solution can be obtained by the following three steps.

(b) FBD of segment OB

Step 1: Equation (6.15)—Substitute $w_0 = 1.5$ kN/m and the coordinates of B ($s = 18$ m, $x = 15$ m); solve for T_0.

Step 2: Equation (6.16)—Substitute $w_0 = 1.5$ kN/m, the coordinates of B ($x = 15$ m, $y = H$), and the value found for T_0; solve for H.

Step 3: Equation (6.17)—Substitute $w_0 = 1.5$ kN/m, $x = 15$ m, $T = T_{max}$, and the value found for T_0; solve for T_{max}.

Mathematical Details

Step 1

Equation (6.15) is

$$s = \frac{T_0}{w_0} \sinh \frac{w_0 x}{T_0}$$

$$18 = \frac{T_0}{1.5} \sinh \frac{1.5(15)}{T_0}$$

This equation must be solved numerically. The result, which may be verified by substitution, is $T_0 = 21.13$ kN.

Step 2

Equation (6.16) is

$$y = \frac{T_0}{w_0} \left(\cosh \frac{w_0 x}{T_0} - 1 \right)$$

$$H = \frac{21.13}{1.5} \left[\cosh \frac{1.5(15)}{21.13} - 1 \right] = 8.77 \text{ m} \qquad \textit{Answer}$$

Step 3

Equation (6.17) is

$$T = T_0 \cosh \frac{w_0 x}{T_0}$$

$$T_{max} = 21.13 \cosh \frac{1.5(15)}{21.13} = 34.3 \text{ kN} \qquad \textit{Answer}$$

Sample Problem 6.9

Figure (a) shows a cable that carries the uniformly distributed load $w_0 = 80$ lb/ft, where the distance is measured along the horizontal. Determine the shortest cable for which the cable tension does not exceed 10 000 lb, and find the corresponding vertical distance H.

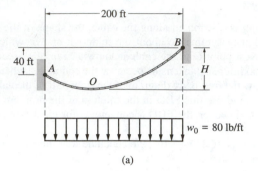

(a)

Solution

Method of Analysis

Because the loading is distributed uniformly over the horizontal distance, we know that the shape of the cable is parabolic. It is also apparent that the location of the lowest point O of the cable and the cable tension T_0 at that point are not known. Therefore, the computation of these unknowns is addressed first.

A good starting point is the free-body diagram of the entire cable, shown in Fig. (b). The forces appearing on this diagram are the cable tensions at the endpoints (T_A and T_B) and the resultant of the distributed load: $W = (80$ lb/ft$)(200$ ft$) = 16 000$ lb. According to the second equation of Eqs. (6.9), the tension in the cable is proportional to x^2 (x is measured from the vertex of the parabola). It follows that the maximum cable tension occurs at B; that is, $T_B = 10 000$ lb, as shown in the figure. The FBD in Fig. (b) now contains three unknowns: the slope angles θ_A and θ_B and the tension T_A, all of which could be computed from the three available equilibrium equations. It turns out that we need only θ_B, which can be obtained from the moment equation $\Sigma M_A = 0$.

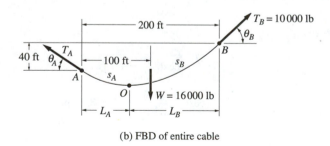

(b) FBD of entire cable

As the next step, we draw the FBD of the portion of the cable that lies to the right of point O, as shown in Fig. (c). Assuming that θ_B has already been computed, this FBD contains three unknowns: L_B (which locates point O), T_0, and H. Because there are also three equilibrium equations available, all the unknowns can now be calculated. The final step is to calculate the length of the cable from Eq. (6.12).

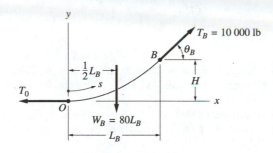

(c) FBD of segment OB

Mathematical Details

From the FBD of the entire cable in Fig. (b), we obtain

$$\Sigma M_A = 0 \quad \overset{+}{\curvearrowleft} \quad (10\,000 \sin \theta_B)(200) - (10\,000 \cos \theta_B)(40)$$
$$- (16\,000)(100) = 0$$

which reduces to

$$\sin \theta_B - 0.2 \cos \theta_B - 0.8 = 0$$

The smallest positive root of this equation can be found by numerical methods. The result, which can be verified by substitution, is $\theta_B = 62.98°$.

From the FBD of segment OB in Fig. (c) we obtain

$$\Sigma F_x = 0 \quad T_0 = T_B \cos \theta_B = 10\,000 \cos 62.98° = 4543 \text{ lb}$$

Note that this equation is identical to the first equation of Eqs. (6.7). Using the FBD in Fig. (c), we also get

$$\Sigma F_y = 0 \quad 80 L_B = T_B \sin \theta_B = 10\,000 \sin 62.98° = 8908 \text{ lb}$$

Therefore, $L_B = 8908/80 = 111.35$ ft and $L_A = 200 - 111.35 = 88.65$ ft. The FBD in Fig. (c) also gives

$$\Sigma M_O = 0 \quad \overset{+}{\curvearrowleft} \quad T_B \sin \theta_B (L_B) - T_B \cos \theta_B (H) - 80 L_B \frac{L_B}{2} = 0$$

which, on substituting the values for T_B, θ_B, and L_B, becomes

$$10\,000 \sin 62.98°(111.35) - 10\,000 \cos 62.98°(H) - 80 \frac{(111.35)^2}{2} = 0$$

Solving for H, we find

$$H = 109.2 \text{ ft} \qquad\qquad Answer$$

The length of each of the two cable segments can be computed from Eq. (6.12). For the segment AO, we substitute $x = -L_A = -88.65$ ft, $w_0 = 80$ lb/ft, and $T_0 = 4543$ lb. Therefore, $w_0 x / T_0 = 80(-88.65)/4543 = -1.5611$, and Eq. (6.12) becomes

$$s(-L_A) = \frac{-88.65}{2} \sqrt{1 + (-1.5611)^2}$$

$$+ \frac{1}{2} \frac{4543}{80} \ln\left[(-1.5611) + \sqrt{1 + (-1.5611)^2}\right]$$

$$= -117.0 \text{ ft}$$

The negative result is due to the sign convention: The positive direction of s points to the right of O, whereas point A is to the left of O. Therefore, the length of segment OA is $s_A = 117.0$ ft.

The length of segment OB is obtained by using $x = L_B = 111.35$ ft in Eq. (6.12). Omitting the details of this computation, the result is $s_B = 163.1$ ft. Hence the total length of the cable is

$$s_A + s_B = 117.0 + 163.1 = 280.1 \text{ ft} \qquad \textit{Answer}$$

If the length of the cable were smaller than 280.1 ft, the maximum cable tension would exceed the limiting value of 10 000 lb.

Problems

6.62 Show that the tension acting at a point in a parabolic cable varies with the *xy*-coordinates of the point as

$$T(x, y) = w_0 x \left[1 + (x/2y)^2 \right]^{1/2} \qquad (x > 0)$$

6.63 The cable *AB* spans the length $L = 60$ m with the sag $H = 10$ m. The cable supports a uniformly distributed load of w_0 N/m. If the maximum allowable force in the cable is 36 kN, determine the largest permissible value of w_0.

6.64 Cable *AB* of span $L = 210$ ft supports the uniformly distributed load $w_0 = 80$ lb/ft. If the horizontal component of the force exerted by the cable on the supports is limited to 30 kips, find the smallest allowable sag *H* and the corresponding angle θ.

6.65 Cable *AB* supports the uniformly distributed load of 2 kN/m. If the slope of the cable at *A* is zero, compute (a) the maximum tensile force in the cable; and (b) the length of the cable.

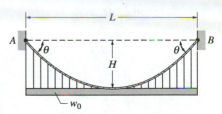

Fig. P6.63, P6.64

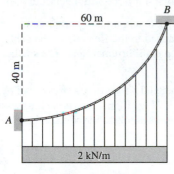

Fig. P6.65

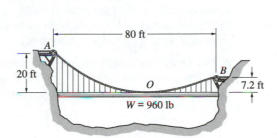

Fig. P6.66

6.66 A uniform 80-ft pipe that weighs 960 lb is supported entirely by a cable *AB* of negligible weight. Determine the length of the cable and the maximum force in the cable. (Hint: First locate the point *O* where the cable is tangent to the pipe.)

6.67 The cable *AB* supports a uniformly distributed load of 12 lb/ft. Determine the maximum force in the cable and the distance *h*.

6.68 The string attached to the kite weighs 0.4 oz/ft. If the force in the string at *O* (the point where the string leaves the ground) is 2.5 lb, determine the height *H* of the kite and the string tension at *B*.

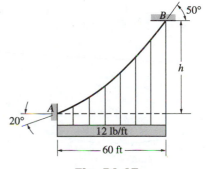

Fig. P6.67

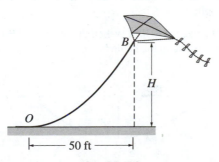

Fig. P6.68

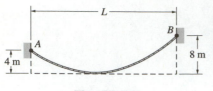

Fig. P6.70

6.69   Show that the tension acting at a point in a catenary cable varies with the y-coordinate of the point as $T(y) = T_0 + w_0 y$.

6.70   A uniform cable weighing 15 N/m is suspended from points A and B. The force in the cable at B is known to be 500 N. Using the result of Prob. 6.69, calculate (a) the force in the cable at A; and (b) the span L.

6.71   The span L and the sag H of the cable AB are 100 m and 10 m, respectively. If the cable weighs 50 N/m, determine the maximum force in the cable using (a) the equations of the catenary; and (b) the parabolic approximation. (c) Compute the percentage error in the parabolic approximation.

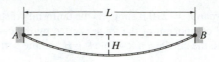

Fig. P6.71, P6.72

6.72   Determine the ratio H/L that minimizes the maximum force in the uniform cable AB of a given span L. (Hint: Minimize the maximum force with respect to T_0.)

6.73   The cable of mass 1.5 kg/m is attached to a rigid support at A and passes over a smooth pulley at B. If the mass $M = 35$ kg is attached to the free end of the cable, find the two values of H for which the cable will be in equilibrium. (Note: The smaller value of H represents *stable* equilibrium.)

6.74   One end of cable AB is fixed, whereas the other end passes over a smooth pulley at B. If the mass of the cable is 1.5 kg/m and the sag is $H = 1.8$ m, determine the mass M that is attached to the free end of the cable. Use the parabolic approximation.

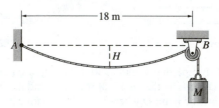

Fig. P6.73–P6.75

6.75   Solve Prob. 6.74 using the equations of the catenary.

6.76   The 50-ft steel tape AB that weighs 2.4 lb is used to measure the horizontal distance between points A and C. If the spring scale at B reads 7.5 lb when the length of tape between A and C is 36 ft, calculate the horizontal distance L_{AC} between A and C to four significant digits.

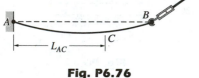

Fig. P6.76

6.77   The cable AOB weighs 4 N/m. When the horizontal 30-N force is applied to the roller support at B, the sag in the cable is 5 m. Find the span L of the cable.

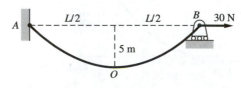

Fig. P6.77

6.78 The chain AB weighs 5 lb/ft. If the force in the chain at B is 800 lb, determine the length of the chain.

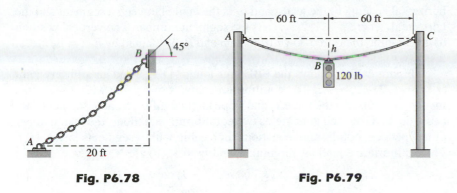

Fig. P6.78   Fig. P6.79

6.79 The 120-lb traffic light is suspended from two identical cables AB and BC, each weighing 0.75 lb/ft. If the maximum allowable horizontal force exerted by a cable on a vertical post is 180 lb, determine the shortest possible length of each cable and the corresponding vertical distance h.

*6.6 *Cables under Concentrated Loads*

a. *General discussion*

Sometimes a cable is called on to carry a number of concentrated vertical loads, such as in Fig. 6.12(a). If the weight of the cable is negligible compared to the applied loads, then each segment of the cable is a two-force member and the shape of the cable consists of a series of straight lines. The analysis of a cable loaded in this manner is similar to truss analysis, except that with cables the locations of the joints (i.e., points where the loads are applied) are sometimes unknown. As in the case of truss analysis, we can use the method of joints and/or the method of sections to determine the equilibrium equations. However, it is often necessary to include equations of geometric constraints in order to have enough equations to find all the unknowns.

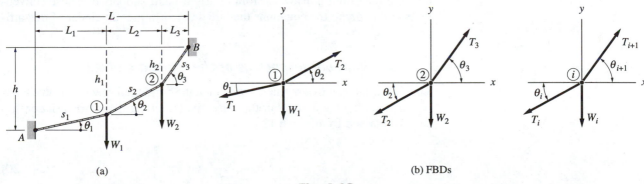

(a)   (b) FBDs

Fig. 6.12

If a cable has n segments, then there are $(n-1)$ joints. For example, the cable in Fig. 6.12(a) has $n = 3$ segments, and $(n-1) = 2$ joints, labeled ① and ②. We use the following notation: s_i is the segment length; L_i is the horizontal spacing of the loads; and θ_i is the angle between a segment and the horizontal, where $i = 1, 2, \ldots, n$ is the segment number. The vertical position of the ith joint, measured *downward* from end B, is denoted by h_i, $i = 1, 2, \ldots, n - 1$.

Figure 6.12(b) shows the FBDs for joints ①, ②, and an arbitrary joint (i), $i = 1, 2, \ldots, n - 1$. The equilibrium analysis of a cable with n segments involves calculating the force T_i and slope angle θ_i of each cable segment. Because the FBD of each joint yields two equilibrium equations, the total number of independent equilibrium equations for a cable with n segments is $2(n-1)$. The equilibrium equations for joint (i) in Fig. 6.12(b) are

$$\Sigma F_x = 0 \quad \xrightarrow{+} \quad T_{i+1}\cos\theta_{i+1} - T_i\cos\theta_i = 0 \tag{a}$$

$$\Sigma F_y = 0 \quad +\!\uparrow \quad T_{i+1}\sin\theta_{i+1} - T_i\sin\theta_i - W_i = 0 \tag{b}$$

where $i = 1, 2, \ldots, n - 1$. From Eq. (a) we see that the horizontal component $T_i\cos\theta_i$ is the same for each segment. Labeling this component as T_0, we can replace Eq. (a) with

$$\boxed{T_i\cos\theta_i = T_0 \qquad i = 1, 2, \ldots, n} \tag{6.18}$$

and Eq. (b) can be rewritten as

$$\boxed{T_0(\tan\theta_{i+1} - \tan\theta_i) = W_i \qquad i = 1, 2, \ldots, n - 1} \tag{6.19}$$

Observe that Eqs. (6.19) represent $(n-1)$ equations that contain the $(n+1)$ unknowns $T_0, \theta_1, \theta_2, \ldots, \theta_n$. Therefore, we must obtain two additional independent equations before we can calculate all of the unknowns.

The source of the additional equations depends primarily on the nature of the problem. It is convenient to divide problems into two categories depending on whether the horizontal spacings of the loads (L_i) or the lengths of the cable segments (s_i) are given and to discuss each category separately (we assume that the relative position of the supports—the distances h and L in Fig. 6.12(a)—are known).

Because Eqs. (6.18) and (6.19) have been derived from Fig. 6.12, the figure also defines the sign conventions that have been used in the derivations: *tensile forces* and *counterclockwise angles* are positive, and h is the vertical distance measured *downward* from the right-hand support B. These conventions also apply to the equations that are derived in the remainder of this article.

b. *Horizontal spacings of the loads are given*

Consider a cable with n segments for which the horizontal spacings of the loads ($L_1, L_2, \ldots, L_n$) are given. For this case, the following geometric relationship can be obtained from Fig. 6.12(a).

$$\boxed{h = \sum_{i=1}^{n} L_i\tan\theta_i} \tag{6.20}$$

However, the problem is still not solvable, unless one additional piece of information is given. This information may take several forms. For example, the horizontal pull T_0 or the maximum cable tension may be specified (both conditions are relevant from a design viewpoint), the vertical position of one of the joints (e.g., h_1) may be prescribed, or the total length of the cable may be known.

We should point out that, in general, the analysis involves the solution of simultaneous equations that are nonlinear in the angles θ_i. In many problems this difficulty can be avoided by considering an appropriate moment equation using the FBD of the entire cable (see Sample Problem 6.10) or the FBD of a section of the cable containing two or more joints. However, these moment equilibrium equations are not independent of Eqs. (6.18)–(6.20).

c. Lengths of the segments are given

Consider next a cable with n segments for which the lengths of the segments $s_1, s_2, \ldots, s_n$ are known. For this case, Fig. 6.12(a) yields two independent geometric relationships:

$$h = \sum_{i=1}^{n} s_i \sin \theta_i \qquad L = \sum_{i=1}^{n} s_i \cos \theta_i \qquad (6.21)$$

These two equations, coupled with the $(n - 1)$ equilibrium equations given in Eq. (6.19), can be solved for the $(n + 1)$ unknowns without the need for additional information. After $T_0, \theta_1, \theta_2, \ldots, \theta_n$ have been computed, the forces in the cables can be found from Eq. (6.18).

Unfortunately, in this case it is not always possible to avoid the solution of simultaneous, nonlinear equations (a very difficult task to perform analytically). Therefore, a computer program capable of solving simultaneous, nonlinear equations may be necessary for solving problems in this category.

Sample Problem 6.10

For the cable loaded as shown in Fig. (a), determine the angles β_1 and β_2, the force in each segment, and the length of the cable.

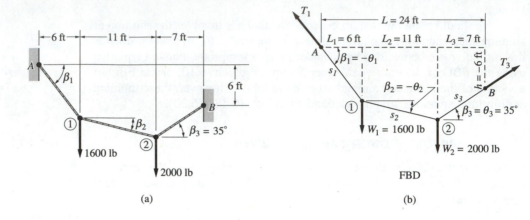

(a)

(b)

FBD

Solution

Method of Analysis

The free-body diagram of the entire cable is shown in Fig. (b), where the labeling of the variables is consistent with the notation used in Fig. 6.12 (recall that the positive direction for θ_1, θ_2, and θ_3 is counterclockwise from the horizontal, and positive h is measured downward from end B).

    We note that the information given in Fig. (a) includes the horizontal spacing of the loads and the angle β_3. Therefore, according to the discussion in Art. 6.6, the problem is statically determinate, and a solution can be obtained by writing and solving Eqs. (6.18)–(6.20). In this problem, however, the difficulty of solving these simultaneous, nonlinear equations can be avoided.

    Examination of the FBD in Fig. (b) reveals that T_3 can be calculated from the equation $\Sigma M_A = 0$. Equilibrium equations for joints ② and ① will then determine the other unknowns without having to solve the equations simultaneously.

Mathematical Details

From the FBD of the entire cable, Fig. (b), we obtain

$$\Sigma M_A = 0 \quad \overset{+}{\curvearrowleft} \quad T_3 \sin 35°(24) + T_3 \cos 35°(6) - 1600(6) - 2000(17) = 0$$

which gives

$$T_3 = 2334 \text{ lb} \qquad \qquad \textit{Answer}$$

The constant horizontal component T_0 of the cable tension can now be found by computing the horizontal component of T_3.

$$T_0 = T_3 \cos \theta_3 = 2334 \cos 35° = 1912 \text{ lb}$$

Substituting $i = 2$ in Eq. (6.19), we obtain the equilibrium equation for joint ②:

$$T_0(\tan \theta_3 - \tan \theta_2) = W_2$$
$$1912(\tan 35° - \tan \theta_2) = 2000$$

which gives

$$\theta_2 = -19.08° \quad \text{or} \quad \beta_2 = 19.08° \qquad \textit{Answer}$$

With $i = 1$, Eq. (6.19) represents the equilibrium equation for joint ①:

$$T_0(\tan\theta_2 - \tan\theta_1) = W_1$$
$$1912[\tan(-19.08°) - \tan\theta_1] = 1600$$

which gives

$$\theta_1 = -49.78° \quad \text{or} \quad \beta_1 = 49.78° \qquad \textit{Answer}$$

The tensions in the first and second segments can now be found from Eqs. (6.18):

$$T_1 = \frac{T_0}{\cos\theta_1} = \frac{1912}{\cos(-49.78°)} = 2961 \text{ lb} \qquad \textit{Answer}$$

$$T_2 = \frac{T_0}{\cos\theta_2} = \frac{1912}{\cos(-19.08°)} = 2023 \text{ lb} \qquad \textit{Answer}$$

The total length s of the cable is

$$s = s_1 + s_2 + s_3$$
$$= \frac{L_1}{\cos\beta_1} + \frac{L_2}{\cos\beta_2} + \frac{L_3}{\cos\beta_3}$$
$$= \frac{6}{\cos 49.78°} + \frac{11}{\cos 19.08°} + \frac{7}{\cos 35°}$$
$$= 9.29 + 11.64 + 8.55 = 29.48 \text{ ft} \qquad \textit{Answer}$$

Sample Problem 6.11

For the cable loaded as shown in Fig. (a), calculate the angles β_1, β_2, and β_3 and the force in each segment of the cable.

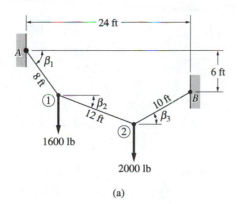

(a)

Solution

Method of Analysis

The free-body diagram of the entire cable is shown in Fig. (b). Its main function is to identify the variables and to enforce the sign conventions defined in Fig. 6.12 (recall that the positive directions for the θs are counterclockwise and that positive h is measured downward from end B).

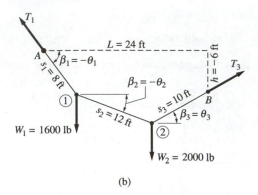

(b)

Observe that the length of each cable segment is given. As pointed out in Art. 6.6, in problems of this type (and this problem is no exception) it is seldom possible to obtain a solution without having to solve nonlinear, simultaneous equations. An inspection of Fig. (b) reveals that two or more unknown angles would appear in each equilibrium equation for the entire cable (this would also be the situation if the equilibrium of any portion of the cable is analyzed). Therefore, the most direct method of solution is to write down and solve Eqs. (6.19) and (6.21), giving us a total of four equations containing the unknowns T_0, θ_1, θ_2, and θ_3. Because the equations are nonlinear, the solution must be obtained numerically by a computer program capable of solving nonlinear, simultaneous equations. Once the solution has been found, the force in each segment can be calculated from Eq. (6.18).

Mathematical Details

On substituting the given values, Eqs. (6.19) yield

$$(i = 1) \qquad T_0(\tan\theta_2 - \tan\theta_1) = W_1$$
$$T_0(\tan\theta_2 - \tan\theta_1) = 1600 \qquad \text{(a)}$$

and

$$(i = 2) \qquad T_0(\tan\theta_3 - \tan\theta_2) = W_2$$
$$T_0(\tan\theta_3 - \tan\theta_2) = 2000 \qquad \text{(b)}$$

and Eqs. (6.21) become

$$s_1\sin\theta_1 + s_2\sin\theta_2 + s_3\sin\theta_3 = h$$
$$8\sin\theta_1 + 12\sin\theta_2 + 10\sin\theta_3 = -6 \qquad \text{(c)}$$

and

$$s_1\cos\theta_1 + s_2\cos\theta_2 + s_3\cos\theta_3 = L$$
$$8\cos\theta_1 + 12\cos\theta_2 + 10\cos\theta_3 = 24 \qquad \text{(d)}$$

The solution of Eqs. (a)–(d), which can be verified by substitution, is

$$T_0 = 1789 \text{ lb}$$
$$\theta_1 = -53.62°(= -\beta_1)$$
$$\theta_2 = -24.83°(= -\beta_2)$$
$$\theta_3 = 33.23°(= \beta_3)$$

Answer

Using Eq. (6.18), the tensions in the cable segments are

$$T_1 = \frac{T_0}{\cos \theta_1} = \frac{1789}{\cos(-53.62°)} = 3020 \text{ lb}$$

$$T_2 = \frac{T_0}{\cos \theta_2} = \frac{1789}{\cos(-24.83°)} = 1971 \text{ lb}$$

$$T_3 = \frac{T_0}{\cos \theta_3} = \frac{1789}{\cos 33.23°} = 2140 \text{ lb}$$

Answer

Problems

6.80   The cable carrying 40-lb loads at *B* and *C* is held in the position shown by the horizontal force *P* = 60 lb applied at *A*. Determine *h* and the forces in segments *BC* and *CD*.

6.81   The cable carrying 40-lb loads at *B* and *C* is held in the position shown by the horizontal force *P* applied at *A*. Find *P* and the maximum force in the cable if *h* = 7.5 ft.

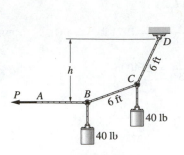

Fig. P6.80, P6.81

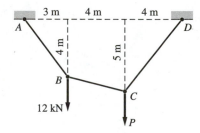

Fig. P6.82

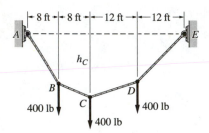

Fig. P6.83, P6.84

6.82   Determine the angles β_2 and β_3 and the force in each cable segment if $\beta_1 = 40°$.

6.83   The cable carrying three 400-lb loads has a sag at *C* of h_C = 16 ft. Calculate the force in each segment of the cable.

6.84   The cable supports three 400-lb loads as shown. If the maximum allowable tension in the cable is 900 lb, find the smallest possible sag h_C at *C*.

6.85   Cable *ABC* of length 5 m supports the force *W* at *B*. Determine (a) the angles β_1 and β_2; and (b) the force in each cable segment in terms of *W*.

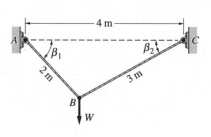

Fig. P6.85

Fig. P6.86

6.86   When the 12-kN load and the unknown force *P* are applied, the cable assumes the configuration shown. Determine *P* and the force in each segment of the cable.

6.87 The cable is subjected to a 150-lb horizontal force at B, and an 80-lb vertical force at C. Determine the force in segment CD of the cable, and the distance b.

6.88 The 15-m-long cable supports the loads W_1 and W_2 as shown. Find the ratio W_1/W_2 for which the segment BC will be horizontal; that is, $\beta_2 = 0$.

6.89 The cable of length 15 m supports the forces $W_1 = W_2 = W$ at B and C. (a) Derive the simultaneous equations for β_1, β_2, and β_3. (b) Show that the solution to these equations is $\beta_1 = 41.0°$, $\beta_2 = 9.8°$, and $\beta_3 = 50.5°$. (c) Compute the force in each segment in terms of W.

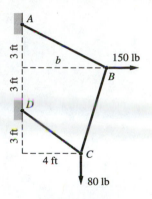

Fig. P6.87

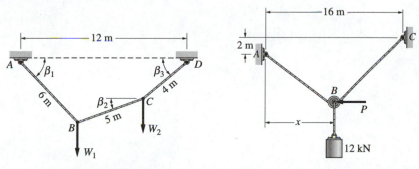

Fig. P6.88, P6.89   **Fig. P6.90**

6.90 The 12-kN weight is suspended from a small pulley that is free to roll on the cable. The length of the cable ABC is 18 m. Determine the horizontal force P that would hold the pulley in equilibrium in the position $x = 4$ m.

6.91 The cable of length 9 m supports a 25-kg mass at C. Knowing that $\beta_1 = 70°$, determine (a) the angles β_2 and β_3; and (b) the force in segment CD of the cable.

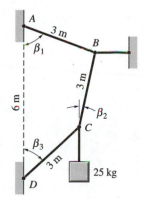

Fig. P6.91

7.1 Introduction

In most of the equilibrium problems that we have analyzed up to this point, the surfaces of contact have been frictionless. The reactive forces were, therefore, normal to the contact surfaces. The concept of a frictionless surface is, of course, an idealization. All real surfaces also provide a force component that is tangent to the surface, called the *friction force,* that resists sliding.

In many situations, friction forces are helpful. For example, friction enables you to walk without slipping, it holds nails and screws in place, and it allows us to transmit power by means of clutches and belts. On the other hand, friction can also be detrimental: It causes wear in machinery and reduces efficiency in the transmission of power by converting mechanical energy into heat.

Dry friction refers to the friction force that exists between two unlubricated solid surfaces. *Fluid friction* acts between moving surfaces that are separated by a layer of fluid. The friction in a lubricated journal bearing is classified as fluid friction, because the two halves of the bearing are not in direct contact but are separated by a thin layer of liquid lubricant. In this chapter, we consider only dry friction.* A study of fluid friction involves hydrodynamics, which is beyond the scope of this text.

7.2 Coulomb's Theory of Dry Friction

Dry friction is a complex phenomenon that is not yet completely understood. This article introduces a highly simplified theory, known as Coulomb's theory of dry friction, that has been found to give satisfactory results in many practical problems.

Coulomb's theory is best explained by considering two bodies that are in contact with each other, as shown in Fig. 7.1(a). Although a single point of contact is indicated in this figure, the following discussion also applies for a finite contact area. The *plane of contact* shown in Fig. 7.1(a) is tangent to both bodies at the point of contact. Figure 7.1(b) displays the free-body diagrams of the bodies, where N is the normal contact force and F is the friction force. The

---

*Dry friction is also known as Coulomb friction, after C.-A. de Coulomb (1736–1806), the first investigator to completely state the laws of dry friction.

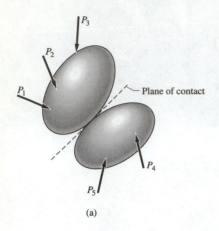

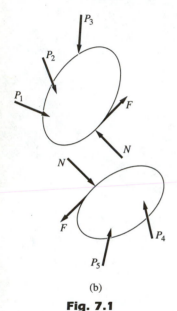

(a)

(b)

Fig. 7.1

force N is perpendicular to the plane of contact, whereas F lies in the plane of contact. Coulomb's theory consists of several postulates that are explained in the following.

a. Static case

Coulomb proposed the following law: If there is no relative motion between two surfaces that are in contact, the normal force N and the friction force F satisfy the following relationship.

$$F \leq F_{\max} = \mu_s N \qquad (7.1)$$

where $F_{\max}$ is the *maximum* static friction force that *can* exist between the contacting surfaces and μ_s is known as the *coefficient of static friction*. The coefficient of static friction is an experimental constant that depends on the composition and roughness of the contacting surfaces. Typical values of μ_s are listed in Table 7.1. Observe that Eq. (7.1) states simply that the friction force F that exists under static conditions (no relative motion) has an upper limit that is proportional to the normal force.

Materials in Contact	μ_s	μ_k
Hard steel on hard steel	0.78	0.42
Aluminum on mild steel	0.61	0.47
Teflon on steel	0.04	—
Nickel on nickel	1.10	0.53
Copper on cast iron	1.05	0.29

The data are extracted by permission from a more complete table in *Marks' Standard Handbook for Mechanical Engineers,* 10th ed., McGraw-Hill Book Co., New York, 1996.

Table 7.1 *Appoximate Coefficients of Friction for Dry Surfaces*

b. Case of impending sliding

Consider the static case in which the friction force equals its limiting value; that is,

$$F = F_{\max} = \mu_s N \qquad (7.2)$$

For this condition, the surfaces are on the verge of sliding, a condition known as *impending sliding*. When sliding impends, the surfaces are at rest relative to each other. However, any change that would require an increase in the friction force would cause sliding. The direction for $F_{\max}$ can be determined from the observation that friction always opposes relative motion; therefore,

$F_{\max}$ always opposes impending sliding

c. *Dynamic case*

If the two contact surfaces are sliding relative to each other, the friction force F is postulated to be

$$F = F_k = \mu_k N \qquad (7.3)$$

where N is the contact normal force; μ_k is an experimental constant called the *coefficient of kinetic friction*; and F_k is referred to as the *kinetic,* or *dynamic friction force*. As indicated in Table 7.1, the coefficient of kinetic friction is usually smaller than its static counterpart. As in the static case,

$$F_k \text{ always opposes sliding}$$

d. *Further discussion of Coulomb friction*

When applying Coulomb's theory, the difference between F_{max} and F_k must be clearly understood: F_{max} is the *maximum* friction force that *can* exist under static conditions; F_k is the friction force that *does* exist during sliding.

To illustrate Coulomb's laws of friction, consider the situation depicted in Fig. 7.2(a). The block of weight W is assumed to be at rest on a horizontal surface when it is subjected to the horizontal force P. (We limit our attention here to sliding motion; the possibility that the block may tip about its corner is considered later.) The free-body diagram of the block is shown in Fig. 7.2(b). Because the friction force F resists the tendency of the block to slide, F is directed opposite to P. We now examine the variation of F with P as the latter increases slowly from zero.

If P is relatively small, the block will remain at rest, and the force equations of equilibrium, $\Sigma F_x = 0$ and $\Sigma F_y = 0$, yield $F = P$ and $N = W$.* Therefore, as long as the block remains at rest, the friction force F equals the applied force P. Figure 7.3 shows the plot of F versus P. In the static region, $0 \le F \le F_{max}$; the variation is a straight line with unit slope. When

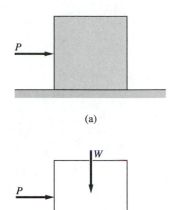

(a)

(b)

Fig. 7.2

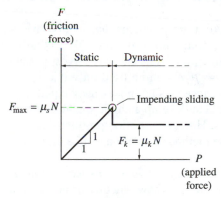

Fig. 7.3

---

*The moment equation of equilibrium would determine the line of action of the normal force N, an important consideration in the analysis of tipping. However, because we are delaying a discussion of tipping, this equation is not of interest at the present time.

$F = F_{\max}$, the block is still in static equilibrium, but sliding impends. However, the slightest increase in P would result in sliding. In Fig. 7.3, the point referring to impending sliding marks the end of the static region. When P exceeds $F_{\max}$, the block starts to slide, and the friction force F drops to its kinetic value F_k. If P is further increased, F remains constant at F_k. Consequently, the plot of F versus P is a horizontal line in the dynamic range.

e. Limitations

Because there is no theoretical explanation that accurately describes friction phenomena, engineers must rely on empirical constants, such as the coefficient of friction. Handbook values for the coefficients of friction should be treated as approximate values. Experimental results indicate that the coefficients may vary widely with environmental conditions, such as humidity, the cleanliness of the surfaces, and so on.

The theory of dry friction is applicable only to surfaces that are dry or that contain only a small amount of lubricant. If there is relative motion between the surfaces of contact, the theory is valid for low speeds only. If the surfaces are well lubricated and are moving with high relative speeds, the frictional characteristics are best described by the theories of fluid friction, which are beyond the scope of this text.

It is interesting to note that Coulomb's theory of dry friction does not depend on the area of contact. There are, of course, many situations where this is not the case. For example, the traction (friction force) between an automobile tire and the pavement can be increased under certain conditions by letting a small amount of air out of the tire, thus increasing the contact area. Obviously, Coulomb's theory of dry friction is not applicable in this situation. The maximum traction in this case is also influenced by factors, such as surface adhesion, that depend on the area of contact.

7.3 Problem Classification and Analysis

The analysis of equilibrium problems that involve friction can be somewhat complicated because Coulomb's law, Eq. (7.1), is an inequality. It does not tell us the friction force; it tells us only the largest possible friction force. The equality $F = F_{\max} = \mu_s N$ can be used only if it is known beforehand that sliding impends. Because F is not necessarily equal to $F_{\max}$ at a friction surface, it is not possible to develop a single method of analysis that is valid for all friction problems. However, friction problems can be classified into three types, and a separate method of solution can be outlined for each type.

Type I *The problem statement does not specify impending motion.* In problems of this type, we do not know whether or not the body is in equilibrium. Therefore, the analysis must begin with an assumption about equilibrium.

Method of Analysis

1. *Assume equilibrium* You are strongly advised to write down this assumption as a reminder that the solution will not be complete unless the assumption has been checked. The sense of each friction force can be assumed because the solution of the equilibrium equations will determine the correct sense.

2. *Solve the equilibrium equations* for the friction forces required for equilibrium.*

3. *Check the assumption*   If the friction forces required for equilibrium do not exceed their limits (i.e., if $F \leq \mu_s N$ at each friction surface), then the assumption is correct, and the remaining unknowns can be computed using equilibrium analysis. (Note that if $F = \mu_s N$ at a surface, which would imply impending sliding, then the assumption is still correct.) If equilibrium requires that $F > \mu_s N$ at any friction surface (which is physically impossible), the assumption of equilibrium is incorrect. Therefore, we have a dynamic problem in which the friction forces at the sliding surfaces are $F = F_k = \mu_k N$.

See Sample Problems 7.1 and 7.4 for examples of Type I problems.

Type II   *The problem statement implies impending sliding, and the surfaces where sliding impends are known.* Friction problems of this type have the most straightforward analyses, because no assumptions and, therefore, no checks are required. It is not necessary to assume equilibrium—a body known to be in a state of impending sliding is in equilibrium by definition.

Method of Analysis

1. *Set $F = F_{max} = \mu_s N$* at the surfaces where sliding impends. Make sure that the sense of each F_{max} is correctly shown on the FBD (opposing impending sliding), because the solution of the equilibrium equations may depend on the assumed directions of the friction forces.

2. *Solve for the unknowns* using the equilibrium equations.

See Sample Problems 7.2, 7.5, and 7.6 for examples of Type II problems.

Type III   *The problem statement implies impending sliding, but the surfaces at which sliding impends are not known.* Problems of this type are the most tedious to analyze, because the surfaces at which sliding impends must be identified by trial and error. Once an assumption has been made, the analysis is similar to that for Type II problems. Two methods of analysis can be used here, both of which are described in the following.

Method of Analysis 1

1. *Determine all possible ways* in which sliding can impend.

2. *For each case, set $F = F_{max}$* at the surfaces where sliding impends and solve the equilibrium equations. Again, the sense of each F_{max} should be

---

*This analysis presupposes that the friction forces are statically determinate. Statically indeterminate friction forces are omitted from the present discussion. Problems of this type are best solved using the principle of virtual work (see Chapter 10).

correct on the FBD. In general, a different solution is obtained for each mode of impending sliding.

3. *Choose the correct answer* by inspection of the solutions.

Method of Analysis 2

1. *Determine all possible ways* in which sliding can impend.

2. *For one of the cases,* set $F = F_{max}$ at the surfaces where sliding impends and solve the equilibrium equations.

3. *Check the solution* by comparing the friction force at each of the other surfaces with its limiting value. If all these forces are less than or equal to their maximum permissible values, then the solution is correct. If a friction force exceeds its limiting value $\mu_s N$, the solution is invalid and another mode of impending sliding must be analyzed. This procedure must be continued until the correct solution is found.

See Sample Problems 7.3 and 7.7 for examples of Type III problems.

Caution   Remember that the equation $F = \mu_s N$ is valid only in the special case of impending sliding. Many difficulties encountered by students can be traced to the incorrect assumption that the equation $F = \mu_s N$ is always true.

Sample Problem 7.1

The 100-lb block in Fig. (a) is initially at rest on a horizontal plane. Determine the friction force between the block and the surface after P was gradually increased from 0 to 30 lb.

Solution

From the problem statement we conclude that this is a Type I problem (impending motion is not specified). Furthermore, we do not know if the block will even remain at rest in static equilibrium when $P = 30$ lb.

Assume Equilibrium

Once we have assumed that the body remains at rest, the equilibrium equations for the free-body diagram in Fig. (b) can be used to calculate the two unknowns (N and F), as follows.

$$\Sigma F_y = 0 \quad +\uparrow \quad N - 100 = 0$$
$$N = 100 \text{ lb}$$

$$\Sigma F_x = 0 \quad \xrightarrow{+} \quad P - F = 0$$
$$F = P = 30 \text{ lb}$$

Before we can accept this solution, the assumption of equilibrium must be checked.

Check

The maximum static friction force is

$$F_{\max} = \mu_s N = 0.5(100) = 50 \text{ lb}$$

Because $F < F_{\max}$, we conclude that the block is in static equilibrium, and the correct value of the friction force is

$$F = 30 \text{ lb} \qquad\qquad \textit{Answer}$$

Comment

If the coefficient of static friction had been 0.25, instead of 0.5, the block would not be in equilibrium. The 30-lb friction force required for equilibrium would be greater than $F_{\max} = 0.25(100) = 25$ lb; therefore, the friction force would be

$$F = F_k = \mu_k N = 0.2(100) = 20 \text{ lb}$$

and the block would be accelerating to the right.

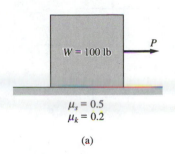

$\mu_s = 0.5$
$\mu_k = 0.2$

(a)

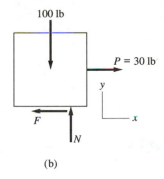

(b)

Sample Problem 7.2

The 100-lb block in Fig. (a) is at rest on a rough horizontal plane before the force P is applied. Determine the magnitude of P that would cause impending sliding to the right.

Solution

The problem statement clearly specifies that sliding impends. Because we know where it impends (there is only one friction surface), we conclude that this is a Type II problem.

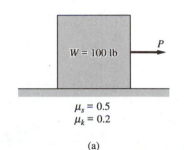

$\mu_s = 0.5$
$\mu_k = 0.2$

(a)

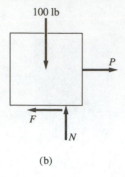

100 lb

F

N

(b)

The free-body diagram of the block is shown in Fig. (b), where the friction force is shown acting to the left, opposite the direction of impending sliding. There are three unknowns in this FBD: P, N, and F. There are also three independent equations: two equilibrium equations and Coulomb's law for impending sliding.

From the FBD we see that the equilibrium equations give $N = 100$ lb and $P = F$. Coulomb's law then yields

$$P = F = F_{max} = \mu_s N = 0.5(100) = 50 \text{ lb} \qquad \textit{Answer}$$

This completes the solution. Because there were no assumptions, no checks are necessary.

Comment

Note that Fig. (a) in both Sample Problems 7.1 and 7.2 is identical. The differences are revealed only in the problem statements. This shows that you must read each problem statement very carefully, because it determines the problem type.

A problem statement can *imply* impending sliding, without using the precise words. For example, the following are equivalent to the original statement of this problem: They both imply that P is to be calculated for impending sliding.

1. Determine the largest force P that can be applied without causing the block to slide to the right.
2. Determine the smallest force P that will cause the block to slide to the right.

Sample Problem 7.3

Determine the maximum force P that can be applied to block A in Fig. (a) without causing either block to move.

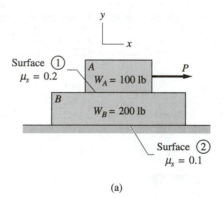

(a)

Solution

The problem statement indicates that we are to find P that would cause impending motion. However, there are two possible ways in which motion can impend: impending sliding at surface ①, or impending sliding at surface ②. Because impending sliding is specified but not its location, this is a Type III problem.

The free-body diagrams of the entire system and each block are shown in Figs. (b) and (c), respectively. Note that the equilibrium of each block yields $N_1 = 100$ lb and $N_2 = 300$ lb, as shown on the FBDs. Attention should be paid to the friction forces. The friction force F_2 on the bottom of block B is directed to the left, opposite the direction in which sliding would impend. At surface ①, block A would tend to slide to the right, across the top of block B. Therefore, F_1 is directed to the left on block A, and to the right on block B. The tendency of F_1 to slide B to the right is resisted by the friction force F_2. Note that F_1 and N_1 do not appear in the FBD in Fig. (b), because they are internal to the system of both blocks.

Two solutions are presented here to illustrate both methods of analysis described in Art. 7.3.

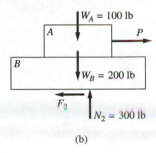

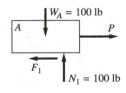

(b)

Method of Analysis 1

First, assume impending sliding at surface ①. Under this assumption we have

$$F_1 = (F_1)_{max} = (\mu_s)_1 N_1 = 0.2(100) = 20 \text{ lb}$$

The FBD of block A then gives

$$\Sigma F_x = 0 \quad \xrightarrow{+} \quad P - F_1 = 0$$
$$P = F_1 = 20 \text{ lb}$$

Next, assume impending sliding at surface ②, which gives

$$F_2 = (F_2)_{max} = (\mu_s)_2 N_2 = 0.1(300) = 30 \text{ lb}$$

From the FBD of the entire system, Fig. (b), we then obtain

$$\Sigma F_x = 0 \quad \xrightarrow{+} \quad P - F_2 = 0$$
$$P = F_2 = 30 \text{ lb}$$

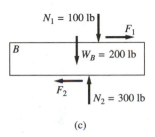

(c)

So far, we have determined that $P = 20$ lb will cause motion to impend at surface ① and that $P = 30$ lb will cause motion to impend at surface ②. Therefore, the largest force that can be applied without causing either block to move is

$$P = 20 \text{ lb} \qquad\qquad\qquad \textit{Answer}$$

with sliding impending at surface ①.

Be sure you understand that the largest force that can be applied is the smaller of the two values determined in the preceding calculations. If sliding impends when $P = 20$ lb, then the system would not be at rest when $P = 30$ lb.

Method of Analysis 2

Assume impending motion at surface ①. We would then obtain $P = (F_1)_{max} = 20$ lb, as determined in Method of Analysis 1. Next, we check the assumption.

Check

The assumption of impending motion at surface ① is checked by comparing the friction force F_2 with $(F_2)_{max}$, its maximum possible value. Using the FBD of block B, we obtain

$$\Sigma F_x = 0 \quad \xrightarrow{+} \quad F_1 - F_2 = 0$$
$$F_1 = F_2 = 20 \text{ lb}$$

Because $(F_2)_{max} = (\mu_s)_2 N_2 = 0.1(300) = 30$ lb, we have $F_2 < (F_2)_{max}$. Consequently, we conclude that impending motion at surface ① is the correct assumption, so that the answer is $P = 20$ lb.

311

Had F_2 turned out to be greater than $(F_2)_{max}$, we would know that sliding would first impend at surface ②, and the problem would have to be solved again making use of this fact.

Comment

There are five unknowns in this problem: P, N_1, F_1, N_2, and F_2. There are four independent equilibrium equations: two for each block. The assumption of impending motion at one surface provides the fifth equation, $F = \mu_s N$, making the problem statically determinate.

In our solution, we have considered two possible modes of impending motion—impending sliding at surface ① and impending sliding at surface ②. Impending sliding at both surfaces at the same time is obviously a third possibility, but it need not be examined independently. Both of the foregoing analyses would determine if simultaneous impending sliding is indeed the case. In Method of Analysis 1 the two computed values of P would be equal. In Method of Analysis 2 the check would reveal that $F = F_{max}$ at both surfaces.

Caution A mistake that is often made in the analysis of Type III problems is to assume that motion impends at the surface with the smallest coefficient of static friction. The solution to this problem illustrates that this need not be the case.

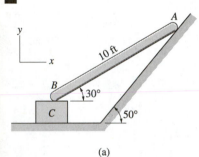

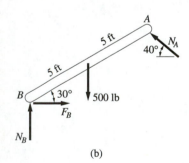

(a)

Sample Problem 7.4

Can the system in Fig. (a) be in static equilibrium in the position shown? The uniform bar AB weighs 500 lb, and the weight of block C is 300 lb. Friction at A is negligible, and the coefficient of static friction is 0.4 at the other two contact surfaces.

Solution

Because it is not known whether motion impends, we identify this as a Type I problem. Note that the FBDs of the bar and the block, Figs. (b) and (c), contain five unknowns: N_A, N_B, F_B, N_C, and F_C.

Assume Equilibrium

Under this assumption, there are five equilibrium equations: three for the bar AB and two for the block C. The unknowns may be computed by the following procedure.

FBD of AB [Fig. (b)]

$$\Sigma M_B = 0 \quad \overset{+}{\curvearrowleft} \quad N_A \sin 40°(10\cos 30°) + N_A \cos 40°(10 \sin 30°)$$
$$- 500(5\cos 30°) = 0$$
$$N_A = 230.4 \text{ lb}$$

$$\Sigma F_x = 0 \quad \underset{+}{\rightarrow} \quad F_B - N_A \cos 40° = 0$$
$$F_B = 230.4 \cos 40° = 176.50 \text{ lb}$$

$$\Sigma F_y = 0 \quad +\uparrow \quad N_B + N_A \sin 40° - 500 = 0$$
$$N_B = -230.4 \sin 40° + 500 = 351.9 \text{ lb}$$

FBD of Block C [Fig. (c)]

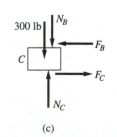

(c)

$$\Sigma F_y = 0 \quad +\uparrow \quad N_C - N_B - 300 = 0$$
$$N_C = 351.9 + 300 = 651.9 \text{ lb}$$

$$\Sigma F_x = 0 \quad \underset{+}{\rightarrow} \quad F_C - F_B = 0$$
$$F_C = F_B = 176.50 \text{ lb}$$

Check

To check the assumption of equilibrium, we must compare each of the friction forces against its maximum static value.

$$(F_B)_{max} = 0.4N_B = 0.4(351.9) = 140.76 \text{ lb} < F_B = 176.50 \text{ lb}$$
$$(F_C)_{max} = 0.4N_C = 0.4(651.9) = 260.8 \text{ lb} > F_C = 176.50 \text{ lb}$$

We conclude that the system cannot be in equilibrium. Although there is sufficient friction beneath B, the friction force under C exceeds its limiting value.

Sample Problem 7.5

Determine the largest and smallest values of the force P for which the system in Fig. (a) will be in static equilibrium. The homogeneous bars AB and BC are identical, each having a mass of 100 kg. The coefficient of static friction between the bar at C and the horizontal plane is 0.5.

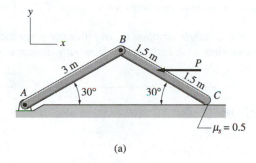

(a)

Solution

This is a Type II problem because impending sliding at C is implied. However, finding the largest and smallest values of P are two separate problems.

Note that the weights of the bars have a tendency to slide C to the right. Therefore, impending sliding of C to the right corresponds to the *smallest P*. The *largest P* occurs when sliding of C impends to the left; in this case, P must overcome both the friction and the tendency of the weights to slide C to the right. Consequently, the only difference between the two problems is the direction of the friction force at C.

The FBD of the system consisting of both bars is shown in Fig. (b); the two directions of F_C are indicated by dashed lines. The weight of each bar, $W = mg = 100(9.81) = 981$ N, is also shown on the diagram.

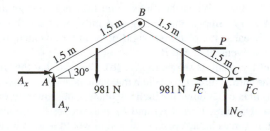

(b)

313

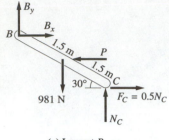

(c) Largest P

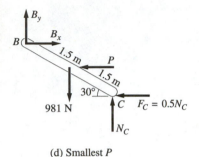

(d) Smallest P

An equation involving only N_C and P is obtained by summing moments about A in Fig. (a):

$$\Sigma M_A = 0 \quad \overset{+}{\frown} \quad N_C(6\cos 30°) + P(1.5\sin 30°)$$
$$- 981(1.5\cos 30°) - 981(4.5\cos 30°) = 0 \quad \text{(a)}$$

The FBDs of bar BC corresponding to the largest and smallest values of P are shown in Figs. (c) and (d), respectively. In both cases, F_C is set equal to $(F_C)_{\text{max}}$ because sliding impends. Summing moments about B yields another equation containing N_C and P.

$$\Sigma M_B = 0 \quad \overset{+}{\frown} \quad N_C(3\cos 30°) - 981(1.5\cos 30°)$$
$$- P(1.5\sin 30°) \pm 0.5N_C(3\sin 30°) = 0 \quad \text{(b)}$$

where the positive (negative) sign on the last term corresponds to the largest (smallest) value of P.

Solving Eqs. (a) and (b) gives

$$\text{largest } P = 1630 \text{ N} \qquad \qquad \textit{Answer}$$
$$\text{smallest } P = 530 \text{ N} \qquad \qquad \textit{Answer}$$

Therefore, the system is in static equilibrium for values of P in the range 530 N $\leq$ $P \leq$ 1630 N.

The solution of this sample problem clearly illustrates that the directions of the friction forces must be shown correctly on the free-body diagrams when sliding impends.

Sample Problem 7.6

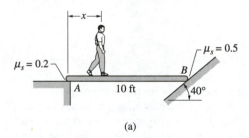

(a)

The uniform 100-lb plank in Fig. (a) is resting on friction surfaces at A and B. The coefficients of static friction are shown in the figure. If a 200-lb man starts walking from A toward B, determine the distance x when the plank will start to slide.

Solution

This is a Type II problem. When the plank is on the verge of moving, sliding must impend at both A and B. Impending sliding at A only, or at B only, would be physically impossible. Because the plank is a rigid body, any movement of end A must be accompanied by a movement of end B.

The FBD of the plank is shown in Fig. (b). Observe that the friction forces are shown acting in the correct directions. When the plank is ready to move, the direction of impending sliding of end B is down the inclined plane. Consequently, end A would tend to slide to the left. The directions of F_A and F_B must oppose these motions. Showing either of the friction forces in the opposite direction would lead to incorrect results.

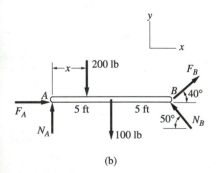

(b)

Inspection of the FBD in Fig. (b) reveals that there are five unknowns: N_A, F_A, N_B, F_B, and x. There are also five equations: three equilibrium equations and two friction equations. Therefore, all the unknowns can be computed from the FBD as follows.

$$\Sigma M_A = 0 \qquad \overset{+}{\curvearrowright} \qquad N_B \sin 50°(10) + F_B \sin 40°(10)$$
$$- 200x - 100(5) = 0 \tag{a}$$

$$\Sigma F_x = 0 \qquad \overset{+}{\longrightarrow} \qquad F_A - N_B \cos 50° + F_B \cos 40° = 0 \tag{b}$$

$$\Sigma F_y = 0 \qquad +\uparrow \qquad N_A - 200 - 100 + N_B \sin 50° + F_B \sin 40° = 0 \tag{c}$$

Substituting the friction equations, $F_A = 0.2N_A$ and $F_B = 0.5N_B$, and solving Eqs. (a)–(c) give $N_A = 163.3$ lb, $N_B = 125.7$ lb, and

$$x = 4.34 \text{ ft} \qquad\qquad\qquad Answer$$

Sample Problem 7.7

The spool in Fig. (a) weighs 25 N, and its center of gravity is located at the geometric center. The weight of block C is 50 N. The coefficients of static friction at the two points of contact are as shown. Determine the largest horizontal force P that can be applied without disturbing the equilibrium of the system.

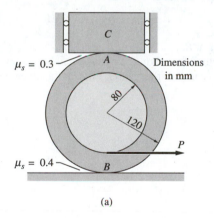

(a)

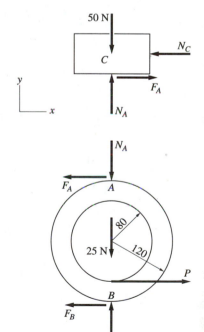

(b)

Solution

The key to the analysis of this problem is understanding that there are two possibilities for impending motion (both could occur simultaneously): (1) impending sliding at A with impending rolling (without sliding) at B, and (2) impending sliding at B with impending rolling (without sliding) at A. Because it is initially not known which of these possibilities represents the actual mode of impending motion, this is a Type III problem.

The free-body diagrams for the block and the spool are shown in Fig. (b). Observe that both friction forces have been shown in the correct directions. The force P tends to slide points A and B on the spool to the right. Therefore, both friction forces are shown acting to the left on the FBD of the spool.

Inspecting the FBDs in Fig. (b), we conclude from $\Sigma F_y = 0$ that $N_A = 50$ N and $N_B = 75$ N. At this stage three unknowns remain in the FBD of the spool: F_A, F_B, and P. Because only two equilibrium equations are left ($\Sigma F_y = 0$ has already been used), the remainder of the solution depends on the assumption regarding impending motion.

315

Assume Impending Sliding at A

This assumption gives us the additional equation $F_A = 0.3N_A = 0.3(50) = 15$ N. The FBD of the spool then yields

$$\Sigma M_B = 0 \qquad \overset{+}{\curvearrowright} \qquad F_A(240) - P(40) = 0$$

which gives

$$P = 6F_A = 6(15) = 90.0 \text{ N}$$

Assume Impending Sliding at B

This assumption gives $F_B = 0.4N_B = 0.4(75) = 30$ N. From the FBD of the spool, we now obtain

$$\Sigma M_A = 0 \qquad \overset{+}{\curvearrowright} \qquad -F_B(240) + P(200) = 0$$

which gives

$$P = 1.2F_B = 1.2(30) = 36.0 \text{ N}$$

Choose the Correct Answer

Up to this point, the analysis has determined that sliding impends at A if $P = 90.0$ N and at B if $P = 36.0$ N. Consequently, the largest force P that can be applied without disturbing the static equilibrium of the spool is

$$P = 36.0 \text{ N} \hspace{4cm} \textit{Answer}$$

with sliding impending at B.

   An alternate method for solving this problem is to assume impending sliding at one surface and then to compare the friction force at the other surface with its limiting static value.

Problems

7.1 Can the two blocks be in equilibrium in the position shown? Justify your answer. All surfaces are frictionless except the horizontal surface beneath block *B*.

7.2 Determine the range of *P* for which the system of two blocks will be in equilibrium. Friction is negligible except for the surface under block *B*.

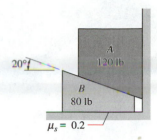

Fig. P7.1

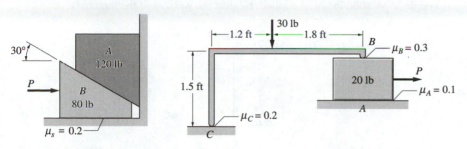

Fig. P7.2   **Fig. P7.3**

7.3 Find the maximum force *P* that can be applied to the 20-lb block without disturbing the equilibrium of the system. Neglect the weight of bar *BC*. The coefficients of static friction at the three contact surfaces (*A*, *B*, and *C*) are shown in the figure.

7.4 The sliding collars *A* and *B* weigh 20 lb and 14 lb, respectively. The weights of bars *AC* and *BC* are negligible. Determine the largest weight *W* that can be suspended from pin *C*. The coefficients of static friction between the sliding collars and the supporting rod are shown in the figure.

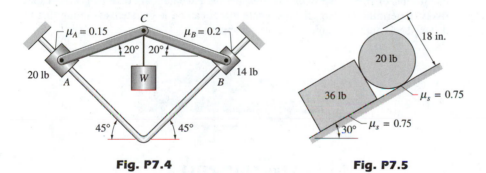

Fig. P7.4   **Fig. P7.5**

7.5 The contact surface between the 36-lb block and 20-lb homogenous cylinder is frictionless. Can the system be in static equilibrium on the rough inclined plane?

7.6 Determine the smallest angle θ at which the uniform triangular plate of weight *W* can remain at rest. The coefficient of static friction at *A* and *B* is 0.5.

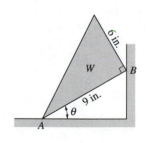

Fig. P7.6

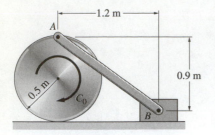

Fig. P7.7

7.7 The 50-kg uniform cylinder and the 30-kg block are connected by a weightless rod AB. The coefficient of static friction is 0.2 under both the cylinder and the block. Determine the largest clockwise couple C_0 that can be applied to the cylinder without disturbing the equilibrium of the system.

7.8 The brake pads at C and D are pressed against the cylinder by the spring BF. The coefficient of static friction between the pads and the cylinder is 0.2. Find the smallest tension in the spring that would prevent the cylinder from rotating when the clockwise couple $M = 3000$ lb · in. is applied. Neglect the weights of the members.

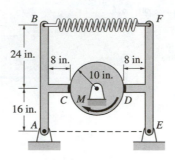

Fig. P7.8   **Fig. P7.9**

7.9 Can the three identical cylinders be in equilibrium if they are stacked as shown? The static coefficient of friction is 0.30 between the cylinders and 0.1 between the cylinders and the ground.

7.10 The rear-wheel-drive pickup truck, with its center of gravity at G, is to negotiate a bump from a standing start in the position shown. The static and kinetic coefficients of friction between the tires and the pavement are 0.18 and 0.15, respectively. Determine the largest slope angle θ that can be negotiated, assuming that the drive wheels are (a) spinning; and (b) not spinning.

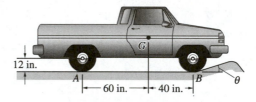

Fig. P7.10, P7.11

7.11 Solve Prob. 7.10 assuming that the pick-up truck has front-wheel drive.

7.12 The 0.8-kg bar is pinned at A and rests on the 1.6-kg spool at B. Both bodies are homogenous. If the coefficient of static friction is 0.25 at both B and C, calculate the largest force P that can be applied without disturbing the equilibrium of the system.

7.13 The homogeneous 0.8-kg bar, which is pin-supported at A, rests on the 1.6-kg spool. The coefficients of static friction at B and C are μ_B and μ_C, respectively. When a sufficiently large force P is applied, it is noted that sliding impends at B and C simultaneously. Determine the ratio μ_C/μ_B.

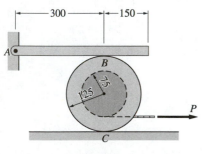

Dimensions in mm

Fig. P7.12, P7.13

7.14 The uniform bar and the homogeneous cylinder each have a mass of 24 kg. The static coefficient of friction is μ_s at *A*, *B*, and *C* (the three points of contact). (a) Assuming equilibrium, calculate the normal and friction forces at *A*, *B*, and *C*. (b) What is the smallest value of μ_s necessary for equilibrium?

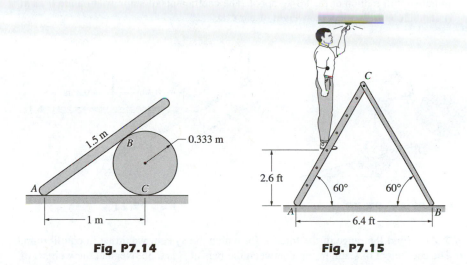

Fig. P7.14

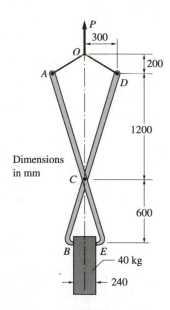

Fig. P7.15

7.15 A stepladder consisting of two legs pinned together at *C* is resting on a rough floor. Will a 160-lb worker be able to change the light bulb if he is required to climb to a height of 2.6 ft? The uniform legs *AC* and *BC* weigh 22 lb and 14 lb, respectively. The coefficient of static friction at *A* and *B* is 0.48.

7.16 The mass of the unbalanced disk is *m*, and its center of gravity is located at *G*. If the coefficient of static friction is 0.2 between the cylinder and the inclined surface, determine whether the cylinder can be at rest in the position shown. Note that the string *AB* is parallel to the incline.

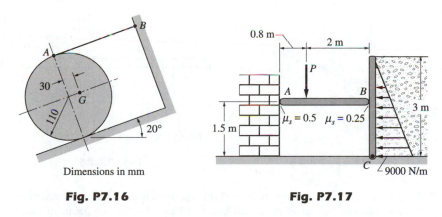

Fig. P7.16

Fig. P7.17

Fig. P7.18

7.17 The piling on the side of an excavation is hinged at *C* and propped up with the horizontal bar *AB*. The force of the earth on the piling has the triangular distribution shown. Neglecting the weights of the members, compute the force *P* required to move bar *AB*.

7.18 Find the smallest coefficient of static friction at *B* and *E* that would permit the tongs to lift the 40-kg block. Neglect the weight of the tongs.

7.19   Determine the smallest force P that the worker must apply to the bar CD to prevent the 80-kg spool from moving down the hill. The coefficients of static friction are 0.12 at A and 0.36 at B. Neglect the weight of bar CD.

7.20   Find the smallest force P that the worker must apply to the bar CD in order to initiate uphill motion of the 80-kg spool. The coefficients of static friction are 0.12 at A and 0.36 at B. Neglect the weight of bar CD.

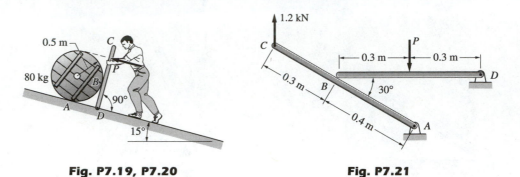

Fig. P7.19, P7.20   **Fig. P7.21**

7.21   Find the range of the force P for which the system would be in equilibrium. The coefficient of static friction between the bars at B is 0.35. Neglect the weights of the bars.

7.22   A 2.2-lb disk A is placed on the inclined surface. The coefficient of static friction between the disk and the surface is 0.4. Is the disk in equilibrium if $P = 1.2$ lb and $\theta = 30°$?

7.23   The coefficient of static friction between the 2.2-lb disk and the inclined surface is 0.4. The force P acts on the disk in the direction $\theta = 45°$. Determine (a) the magnitude of P that would cause impending sliding; and (b) the direction in which sliding impends.

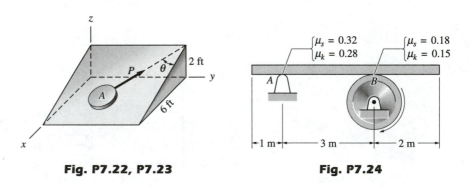

Fig. P7.22, P7.23   **Fig. P7.24**

7.24   A uniform plank is supported by a rigid support at A and a drum at B that rotates clockwise. The coefficients of static and kinetic friction for the two points of contact are as shown. Determine whether the plank moves from the position shown if (a) the plank is placed in position before the drum is set in motion; and (b) the plank is first placed on the support at A and then lowered onto the drum, which is already rotating.

7.25 The uniform bar of weight W is supported by a ball-and-socket joint at A and rests against a vertical wall at B. If sliding impends when the bar is in the position shown, determine the static coefficient of friction at B. [Hint: The direction of impending sliding is tangent to the dashed circle (the potential path of motion of point B).]

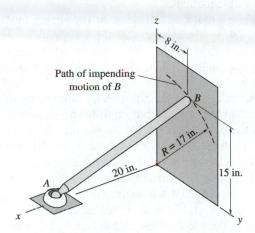

Fig. P7.25

*7.26 The uniform plank is initially at rest on the fixed support at A and the stationary drum at B. If the drum begins rotating slowly counterclockwise, determine how far the plank will travel before it comes to rest again. (Note: Because the drum rotates slowly, the inertia of the plank may be neglected.)

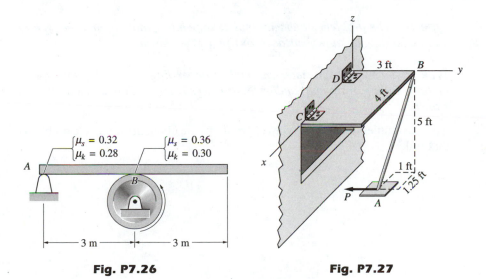

Fig. P7.26   **Fig. P7.27**

*7.27 The 20-lb homogeneous door is held in a horizontal position by the weightless strut AB. The strut is connected to the door by a ball-and-socket joint at B and rests on the ground at A, where the coefficient of static friction is 0.4. Determine the magnitude of the force P, applied parallel to the y-axis, that will move the strut.

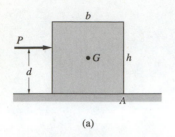

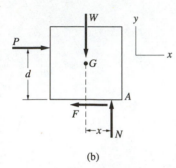

(a)

(b)

Fig. 7.4

7.4 *Impending Tipping*

In the preceding article, we restricted our attention to sliding; the possibility of tipping was neglected. We now discuss problems that include both sliding and tipping as possible motions.

Consider again a block on a friction surface being pushed by a force P, as shown in Fig. 7.4(a). We assume that the weight W of the block, and the dimensions b, h, and d are known. We wish to determine the magnitude of P that will cause impending motion of the block, either impending sliding or impending tipping.

We can gain insight into the solution by comparing the number of unknowns with the number of available equilibrium equations. From the free-body diagram of the block, Fig. 7.4(b), we see that there are four unknowns: the applied force P, the resultant normal force N, the friction force F, and the distance x that locates the line of action of N. Because there are only three independent equilibrium equations, an additional equation must be found before all unknowns can be calculated. If we assume impending sliding, the additional equation is $F = F_{max} = \mu_s N$. On the other hand, if impending tipping about corner A is assumed, the additional equation is $x = b/2$, because N acts at the corner of the block when tipping impends.

In the preceding article, three classes of friction problems were introduced for impending sliding. This classification can be easily reworded to include the possibility of impending tipping.

Type I *The problem statement does not specify impending motion (sliding or tipping).*

Type II *The problem statement implies impending motion, and the type of motion (sliding at known surfaces, or tipping) is known.*

Type III *The problem statement implies impending motion, but the type of motion (sliding or tipping) and/or the surfaces where sliding impends are not known.*

Examples of the three types of problems are given in the sample problems that follow.

Sample Problem 7.8

The man in Fig. (a) is trying to move a packing crate across the floor by applying a horizontal force P. The center of gravity of the 250-N crate is located at its geometric center. Does the crate move if $P = 60$ N? The coefficient of static friction between the crate and the floor is 0.3.

Solution

This is a Type I problem because the problem statement does not specify impending motion. To determine if the crate moves for the conditions stated, we first assume equilibrium and then check the assumption. However, the check must answer two questions—(1) does the crate slide and (2) does the crate tip?

The free-body diagram of the crate is shown in Fig. (b). If the block is assumed to remain in equilibrium, the three equilibrium equations can be used to calculate the three unknowns: the normal force N_1, the friction force F_1, and the distance x locating the line of action of N_1, as shown in the following.

Assume Equilibrium

$$\Sigma F_x = 0 \quad \xrightarrow{+} \quad P - F_1 = 0$$
$$F_1 = P = 60 \text{ N}$$

$$\Sigma F_y = 0 \quad +\uparrow \quad N_1 - 250 = 0$$
$$N_1 = 250 \text{ N}$$

$$\Sigma M_O = 0 \quad \overset{+}{\curvearrowleft} \quad N_1 x - P(0.9) = 0$$

which gives

$$x = P(0.9)/N_1 = 60(0.9)/250 = 0.216 \text{ m}$$

Check

The largest possible value for x is 0.3 m (half the width of the crate). Because $x = 0.216$ m, as obtained from equilibrium analysis, is smaller than that, we conclude that the block *will not tip*.

The limiting static friction force is $(F_1)_{max} = \mu_s N_1 = 0.3(250) = 75.0$ N, which is larger than the force $F_1 = 60$ N that is required for equilibrium. We therefore conclude that the crate *will not slide*.

Crate *will not move* when $P = 60$ N   *Answer*

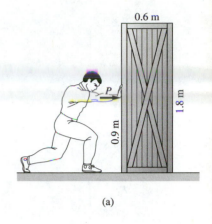

(a)

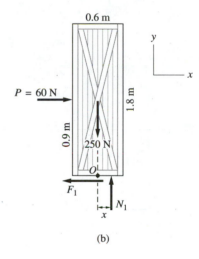

(b)

Sample Problem 7.9

Calculate the force P required to cause tipping of the packing crate in Sample Problem 7.8. Also determine the minimum coefficient of static friction that permits tipping.

Solution

This is a Type II problem because impending tipping is specified. The free-body diagram for the crate is shown in the figure. Note that when the crate is on the verge of tipping, the normal force N_A acts through corner A. There are three equilibrium equations that can be solved for the unknowns P, N_A, and F_A:

$$\Sigma M_A = 0 \quad \overset{+}{\curvearrowright} \quad 250(0.3) - P(0.9) = 0$$
$$P = 83.3 \text{ N} \qquad \qquad \textit{Answer}$$

$$\Sigma F_x = 0 \quad \xrightarrow{+} \quad P - F_A = 0$$
$$F_A = P = 83.3 \text{ N}$$

$$\Sigma F_y = 0 \quad +\!\uparrow \quad N_A - 250 = 0$$
$$N_A = 250 \text{ N}$$

The minimum coefficient of static friction that permits tipping is

$$\mu_s = F_A/N_A = 83.3/250 = 0.333 \qquad \qquad \textit{Answer}$$

Note that if the coefficient of static friction were exactly 0.333, then the force $P = 83.3$ N would result in simultaneous impending sliding and impending tipping.

Sample Problem 7.10

The winch in Fig. (a) is used to move the 300-lb uniform log AB. Compute the largest tension in the cable for which the log remains at rest. The coefficient of static friction between the log and the plane is 0.4.

Solution

Although we are asked to find the cable tension that would cause impending motion, we do not know whether sliding or tipping impends. Therefore, this is a Type III problem.

The free-body diagram of the log in Fig. (b) contains four unknowns: tension T, resultant normal force N, friction force F, and x (the distance from A to the line of action of N). Because there are only three independent equilibrium equations, all unknowns cannot be calculated unless an assumption is made concerning the type of impending motion.

Assume Impending Sliding

Under this assumption, we have $F = F_{max} = 0.4 N$, and the force equilibrium equations for the FBD in Fig. (b) are as follows.

$$\Sigma F_x = 0 \quad \xrightarrow{+} \quad 0.4N - T\cos 60° = 0 \qquad \text{(a)}$$

$$\Sigma F_y = 0 \quad +\!\uparrow \quad N - 300 + T\sin 60° = 0 \qquad \text{(b)}$$

Solving Eqs. (a) and (b) simultaneously, we obtain $T = 141.8$ lb and $N = 177.2$ lb.

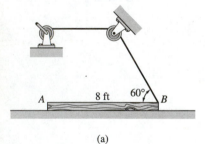

(a)

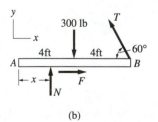

(b)

Assume Impending Tipping

Under this assumption, N will act at A, as shown in the FBD in Fig. (c). The cable tension T can be computed from the moment equation

$$\Sigma M_A = 0 \qquad \overset{+}{\curvearrowleft} \qquad T \sin 60°(8) - 300(4) = 0$$

$$T = 173.2 \text{ lb}$$

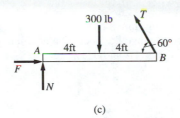

(c)

Because $T = 141.8$ lb for impending sliding and $T = 173.2$ lb for impending tipping, the maximum tension that can be applied without moving the log is

$$T = 141.8 \text{ lb} \hspace{3cm} \textit{Answer}$$

Alternate Solutions

As with most equilibrium problems, there are several equivalent methods of analysis that could be used. Two such methods are

1. Assume impending sliding, and solve for T. Continue the equilibrium analysis to find x. Then check to see if this value of x is physically possible.
2. Assume impending tipping, and solve for T. Continue the equilibrium analysis to find F. Then check to see if $F \leq F_{max}$.

Problems

7.28   Can the 200-lb homogeneous block be in static equilibrium when subjected to the 90-lb vertical force?

7.29   The 60-kg crate has its center of gravity at G. Determine the smallest force P that will initiate motion if $\theta = 30°$.

7.30   Re-solve Prob. 7.29 if $\theta = 0$.

7.31   The 120-lb door is hung from a horizontal track at A and B. Find the smallest horizontal force P required to move the door. The coefficient of static friction is 0.5 between the track and the sliders at A and B. The center of gravity of the door is located at G.

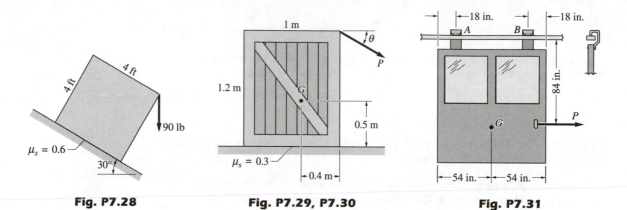

Fig. P7.28   Fig. P7.29, P7.30   Fig. P7.31

7.32   Determine the largest force P for which the 18-kg uniform bar remains in equilibrium.

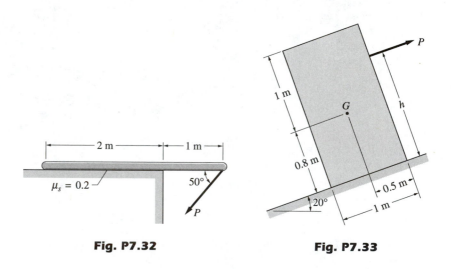

Fig. P7.32   Fig. P7.33

7.33   The center of gravity of the 60-kg block is at G. The static coefficient of friction between the block and the inclined surface is 0.3. Find the smallest force P and the largest distance h for which the block will move up the incline without tipping.

7.34   The hub of the 4-lb spool is resting on a horizontal rail. A string wound around the hub is connected to the 2-lb block. The coefficient of static friction is 0.4 at both contact surfaces. Determine the largest weight W that can be suspended from the rim of the spool without disturbing equilibrium of the system.

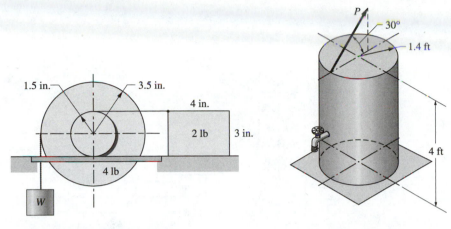

Fig. P7.34

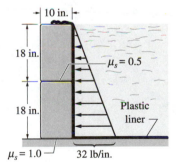

Fig. P7.35

7.35   The weight of the cylindrical tank is negligible in comparison to the weight of water it contains (water weighs 62.4 lb/ft^3). The coefficient of static friction between the tank and the horizontal surface is μ_s. (a) Assuming a full tank, find the smallest force P required to tip the tank, and the smallest μ_s that would allow tipping to take place. (b) If the force $P = 200$ lb initiates tipping, determine the depth of water in the tank.

7.36   Find the smallest angle θ for which a sufficiently large force P would cause the uniform log AB of weight W to tip about A.

7.37   The 40-lb ladder AC is leaning on a 10-lb block at B and a frictionless corner at C. Both bodies are homogeneous. Can the system remain at rest in the position shown? Be sure to consider all possibilities.

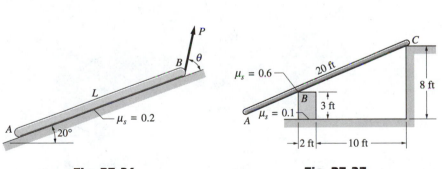

Fig. P7.36   **Fig. P7.37**   **Fig. P7.38**

7.38   Two concrete blocks weighing 320 lb each form part of the retaining wall of a swimming pool. Will the blocks be in equilibrium when the pool is filled and the water exerts the line loading shown?

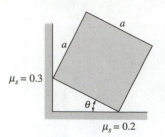

Fig. P7.39

7.39 Excluding the trivial cases $\theta = 0$ and $\theta = 90°$, determine the range of values of the angle θ for which the homogeneous block can be in equilibrium.

7.40 Find the weight of the lightest block *D* that can be used to support the 200-lb uniform pole *ABC* in the position shown.

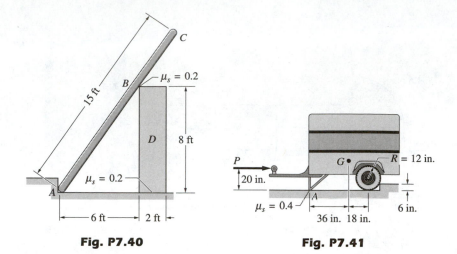

Fig. P7.40   **Fig. P7.41**

7.41 The 2000-lb weight of the trailer is distributed equally between its two wheels, one on each side of the trailer. The center of gravity is at *G*, and the wheels are free to rotate. Determine whether the trailer can be pushed over a 6-in. curb without tipping, and, if so, compute the required horizontal force *P*.

7.42 Determine the smallest force *P*, applied to the plunger *D*, that will prevent the couple $C = 250 \text{ N} \cdot \text{m}$ from moving the cylinder. Friction may be neglected at all surfaces, except between the plunger and cylinder. The masses of the bodies are negligible.

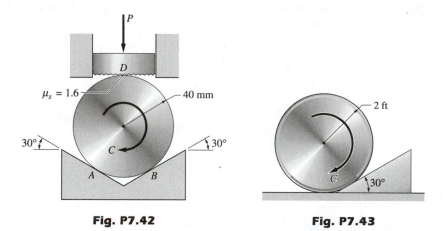

Fig. P7.42   **Fig. P7.43**

7.43 The weights of the homogeneous roller and the wedge are 100 lb and 40 lb, respectively. The coefficient of static friction is 0.2 between all contact surfaces. Find the smallest couple *C* that will move the cylinder.

7.5 *Angle of Friction; Wedges and Screws*

a. *Angle of friction*

Figure 7.5 shows a block on a friction surface subjected to the horizontal force P. As shown in the free-body diagram, we let ϕ be the angle between the contact force R and the normal n to the contact surface. We see that ϕ is given by $\tan \phi = F/N$, where N and F are the normal and friction forces, respectively. The upper limit of ϕ, denoted by ϕ_s, is reached at impending sliding when $F = F_{\max} = \mu_s N$. Therefore, we have

$$\boxed{\tan \phi_s = \mu_s} \qquad (7.4)$$

The angle ϕ_s is called the *angle of static friction*. Note that $\phi \le \phi_s$ signifies equilibrium and that $\phi = \phi_s$ indicates impending sliding. Therefore, the direction of the contact force R is known at all surfaces where sliding impends. This knowledge can be frequently utilized to gain insight into problems involving two- and three-force bodies.

In Fig. 7.5, the friction force F opposes the tendency of P to slide the block to the right. If the direction of P is reversed, the direction of F would also be reversed. This leads to the conclusion that the block can be in equilibrium only if the line of action of R stays within the sector AOB (bounded by $\pm\phi_s$), as shown in Fig. 7.6. For more general loadings, the line of action of R must lie within the cone, called the *cone of static friction,* that is formed by rotating sector AOB about the normal n. Observe that the vertex angle of the cone of static friction is $2\phi_s$.

When sliding occurs, the friction force is $F = \mu_k N$, and the value of ϕ that specifies the direction of R is given by

$$\boxed{\tan \phi_k = \mu_k} \qquad (7.5)$$

The angle ϕ_k is called the *angle of kinetic friction.* For this case, the cone of static friction is replaced by the smaller *cone of kinetic friction,* for which the vertex angle is $2\phi_k$.

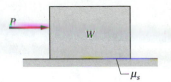

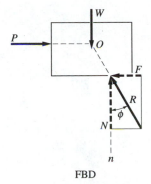

FBD

Fig. 7.5

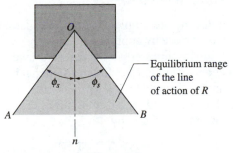

Equilibrium range
of the line
of action of R

Fig. 7.6

b. *Wedges*

A wedge is a simple device that is used for the same purpose as a lever—that is, to create a mechanical advantage. Consider, for example, the wedge shown in Fig. 7.7(a) that is being forced into a crack by the applied force P. The angle formed by the tip of the wedge is 2β, where β is called the *wedge angle*. Neglecting the weight, the free-body diagram of the wedge at impending sliding is shown in Fig. 7.7(b). As before, we let ϕ be the angle between the contact force R and the normal n to the contact surface. Because sliding impends, $\phi = \phi_s$, where $\phi_s = \tan^{-1} \mu_s$ is the angle of friction. From the force diagram in Fig. 7.7(c) we see that $R = P/[2\sin(\phi_s + \beta)]$, which is substantially larger than P if the wedge angle β is small and the sides of the wedge are lubricated (giving a small value for ϕ_s).

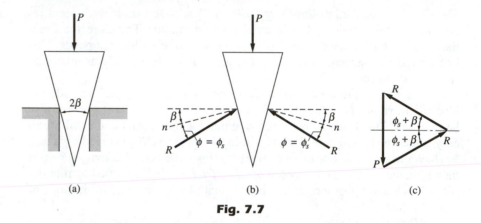

(a)   (b)   (c)

Fig. 7.7

Ideally, a wedge should be slippery enough to be easily driven into the crack, but have enough friction so that it stays in place when the driving force is removed. In the absence of P, the wedge becomes a two-force body. Therefore, the contact forces R must be collinear, as indicated in the free-body diagram in Fig. 7.8, where now $\phi = \beta$. Recalling that equilibrium can exist only if $\phi \le \phi_s$, we conclude that the wedge will stay in place provided that $\beta \le \phi_s$.

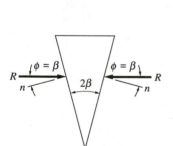

Fig. 7.8

c. *Square-threaded screws*

Screws with square threads are employed in jacks, vises, and other devices that produce a large axial force by applying a relatively small couple about the axis of the screw. A square-threaded screw can be viewed as a bar of rectangular cross section wrapped around a cylinder in a helical fashion, as shown in Fig. 7.9. The helix angle θ is called the *lead angle,* the distance p between the threads is known as the *pitch,* and the mean radius of the threads is denoted by r. These three parameters are related by

$$p = 2\pi r \tan \theta \qquad (7.6)$$

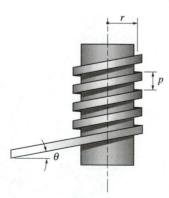

Fig. 7.9

Figure 7.10(a) shows a screw being employed as a jack. Assuming that the couple C_0, called the *torque,* is large enough, it will cause the screw to advance, thereby elevating the weight W. The analysis of this problem is simplified if we recall that in Coulomb's theory the friction force is independent of the contact area. Therefore, we can assume the contact area to be very small, as illustrated in Fig. 7.10(b). Note that the entire weight W is carried by the contact area and that the horizontal force $Q = C_0/r$ is supplied by the applied torque C_0. We can now see that this problem is identical to the one shown in Fig. 7.11(a)—namely, a block of weight W being pushed up an incline by the horizontal force Q.

The smallest torque required to start the weight W moving *upward* can now be obtained from the FBD in Fig. 7.11(b). Note that at impending sliding the angle between R and the normal n to the contact surface is $\phi = \phi_s$, and that the direction of ϕ_s relative to the normal n indicates that the impending motion is directed up the incline. For equilibrium of the block, we have

$$\Sigma F_x = 0 \qquad \xrightarrow{+} \qquad \frac{C_0}{r} - R\sin(\phi_s + \theta) = 0 \qquad \text{(a)}$$

$$\Sigma F_y = 0 \qquad +\uparrow \qquad R\cos(\phi_s + \theta) - W = 0 \qquad \text{(b)}$$

Solving Eqs. (a) and (b), we find that the smallest torque that will cause the weight W to move *upward* is

$$\boxed{(C_0)_{\text{up}} = Wr\tan(\phi_s + \theta)} \qquad \text{(7.7a)}$$

If we reverse the direction of C_0 and assume impending motion down the incline, the FBD in Fig. 7.11(c) must be used. It is seen from the equilibrium equations that the torque required to cause the weight W to move *downward* is

$$\boxed{(C_0)_{\text{down}} = Wr\tan(\phi_s - \theta)} \qquad \text{(7.7b)}$$

If $\phi_s \geq \theta$, the torque C_0 in Eq. (7.7b) is positive, which means that the weight W remains at rest if C_0 is removed. In this case, the screw is said to be *self-locking*. On the other hand, if $\phi_s < \theta$, the torque C_0 in Eq. (7.7b) is negative, indicating that the weight W would come down by itself in the absence of C_0.

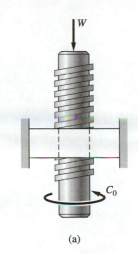

(a)

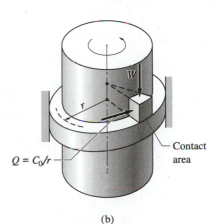

(b)

Fig. 7.10

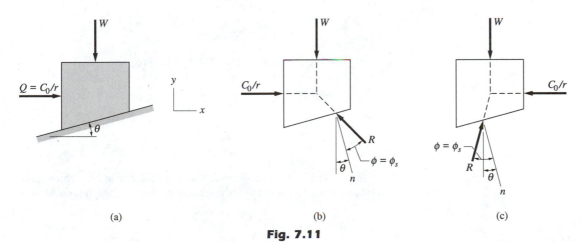

(a) (b) (c)

Fig. 7.11

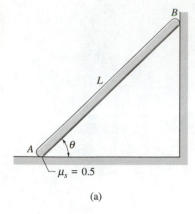

(a)

Sample Problem 7.11

Using the angle of friction, determine the smallest positive angle θ for which the slender bar shown in Fig. (a) can remain at rest. The bar is homogeneous of weight W and length L. Neglect friction between the bar and wall at B.

Solution

As shown in the FBD, Fig. (b), the bar is acted on by three forces: the weight W acting at the midpoint of the bar, the horizontal normal force N_B, and the reaction R_A at the horizontal surface. Because impending motion is specified, the angle ϕ between R_A and the normal to the contact surface is equal to its limiting value: $\phi = \phi_s = \tan^{-1}\mu_s = \tan^{-1} 0.5 = 26.57°$. Because the bar is a three-force member, the forces must intersect at point C.

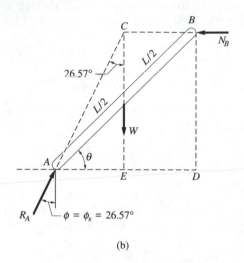

(b)

From triangle ABD we see that

$$\tan\theta = \frac{\overline{BD}}{\overline{AD}} \tag{a}$$

From triangle ACE we obtain $\overline{AE} = \overline{CE}\tan 26.57°$. Because $\overline{AD} = 2\overline{AE}$, this becomes

$$\overline{AD} = 2\overline{CE}\tan 26.57° \tag{b}$$

Substituting Eq. (b) into Eq. (a) together with $\overline{BD} = \overline{CE}$, we get

$$\tan\theta = \frac{\overline{CE}}{2\overline{CE}\tan 26.57°} = \frac{1}{2\tan 26.57°}$$

which yields

$$\theta = 45.0° \qquad \textit{Answer}$$

By using the angle of friction and by recognizing the bar to be a three-force member, we were able to find θ from geometry, without having to write the equilibrium equations.

Sample Problem 7.12

The screw press shown is used in bookbinding. The screw has a mean radius of 10 mm and its pitch is 5 mm. The static coefficient of friction between the threads is 0.18. If a clamping force of 1000 N is applied to the book, determine (1) the torque that was applied to the handle of the press; and (2) the torque required to loosen the press.

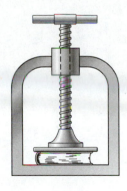

Solution

The lead angle of the screw is computed from Eq. (7.6).

$$\theta = \tan^{-1} \frac{p}{2\pi r} = \tan^{-1} \frac{5}{2\pi(10)} = 4.550°$$

The friction angle is

$$\phi_s = \tan^{-1} \mu_s = \tan^{-1} 0.18 = 10.204°$$

Part 1

The torque required to apply the force $W = 1000$ N can be calculated from Eq. (7.7a).

$$C_0 = Wr\tan(\phi_s + \theta) = 1000(0.01)\tan(10.204° + 4.550°)$$

$$= 2.63 \text{ N} \cdot \text{m} \qquad \qquad \textit{Answer}$$

Part 2

The torque needed to loosen the press is obtained from Eq. (7.7b).

$$C_0 = Wr\tan(\phi_s - \theta) = 1000(0.01)\tan(10.204° - 4.550°)$$

$$= 0.990 \text{ N} \cdot \text{m} \qquad \qquad \textit{Answer}$$

Problems

The following problems are to be solved using the angle of friction. Utilize the characteristics of two and/or three-force bodies wherever applicable.

7.44 The coefficient of static friction between the uniform bar AB of weight W and the supporting surfaces is $\mu_s = 0.8$. Determine the smallest angle θ for which the bar would be in equilibrium.

7.45 The movable bracket of negligible weight is mounted on a vertical post. The coefficient of static friction between the bracket and the post is 0.2. Determine the smallest ratio b/a for which the bracket can support the vertical force P. Assume that the diameter of the post is negligible. (Note: Because the result is independent of P, the bracket is said to be self-locking.)

7.46 The 200-lb man walks up the inclined plank of negligible weight. The coefficients of static friction at A and B are 0.3 and 0.2, respectively. Determine the distance x at which the plank would begin to slide.

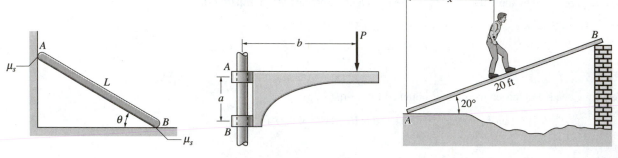

Fig. P7.44   Fig. P7.45   Fig. P7.46

7.47 The four-wheel drive vehicle of weight W attempts to climb a vertical obstruction at A. The center of gravity of the vehicle is at G, and the coefficient of static friction is μ_s at A and B. Find the smallest μ_s necessary to initiate the climb. (Hint: Slipping must impend at A and B simultaneously.)

7.48 Find the smallest distance d for which the hook will remain at rest when acted on by the force P. Neglect the weight of the hook, and assume that the vertical wall is frictionless.

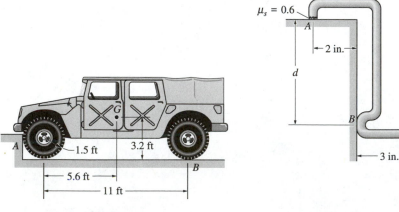

Fig. P7.47   Fig. P7.48

7.49 The figure shows a locking device that is used in some belt buckles. When the belt is pulled to the right by the force P, the roller A becomes jammed between the belt and the upper surface of the buckle. If the coefficient of static friction between all surfaces is 0.24, determine the largest angle θ for which the buckle is self-locking (motion of the belt in either direction is prevented even after the force P is removed). Neglect the weight of roller A.

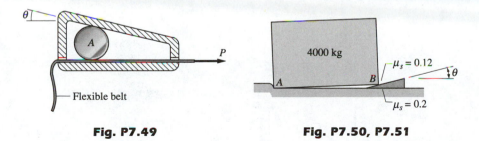

Fig. P7.49   **Fig. P7.50, P7.51**

7.50 A small wedge is placed beneath corner B of the 4000-kg block of marble. Determine the largest angle θ for which the wedge is self-locking; that is, the wedge will not slide out from under the block. Neglect the mass of the wedge and the small angle between surface AB and the horizontal.

7.51 A small wedge is placed under corner B of the 4000-kg block of marble. If the wedge angle is $\theta = 15°$, determine the horizontal force that must be applied to the wedge in order to move it (a) to the left; and (b) to the right.

7.52 The cord carrying the weight W is wrapped around a 10-lb pulley. If the coefficient of static friction between the pulley and the shaft is 0.3, determine the largest weight W that can be supported without rotating the pulley.

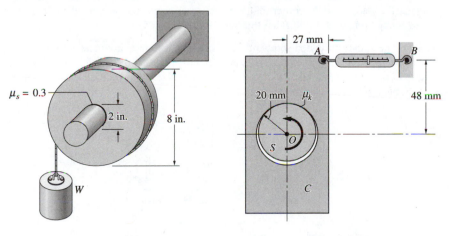

Fig. P7.52   **Fig. P7.53**

7.53 The device shown is used to measure the kinetic coefficient of friction between the rotating shaft S and the homogeneous stationary collar C. The entire 840-N weight of the collar is supported by the shaft. The spring scale attached to the collar at A measures the tension in AB caused by the counterclockwise rotation of the shaft. What is the coefficient of kinetic friction if the reading on the scale is 150 N?

7.54 End *B* of the L-shaped rod is supported by a frictionless slider bearing, and end *A* rests on a friction plane inclined at the angle θ. The rod is subjected to the force *P* and couple C_0, as shown. Show that the couple required to move end *A* up the incline is $C_0 = Pr \tan(\phi_s + \theta)$, where ϕ_s is the angle of static friction at *A*. Neglect the weight of the rod.

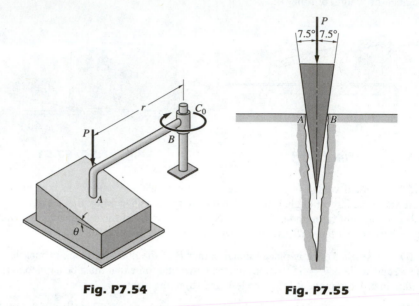

Fig. P7.54   **Fig. P7.55**

7.55 The force $P = 900$ lb is required to push the wedge into the crack, starting from the position shown. It is also known that a force of 250 lb is required to pull the wedge out of the crack from this position. Determine the coefficient of static friction between the wedge and the corners of the crack. Assume that the horizontal components of the contact forces at *A* and *B* are the same for both cases.

7.56 The square-threaded screw of the C-clamp has a mean diameter of 9 mm and a pitch of 1.5 mm. The coefficient of static friction between the threads is 0.2. If the torque $C = 1.25$ N · m is used to tighten the clamp, determine (a) the clamping force; and (b) the torque required to loosen the clamp.

Fig. P7.56

7.57   The square-threaded screw with a pitch of 10 mm and a mean radius of 18 mm drives a gear that has a mean radius of 75 mm. The static and kinetic coefficients of friction between the gear and the screw are 0.12 and 0.06, respectively. The input torque applied to the screw is $C_0 = 10$ N · m. Assuming constant speed operation, determine the output torque C_1 acting on the gear.

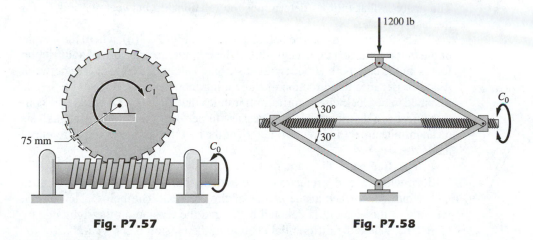

Fig. P7.57   **Fig. P7.58**

7.58   The screw of the car jack has a pitch of 0.1 in. and a mean radius of 0.175 in. Note that the ends of the screw are threaded in opposite directions (right- and left-handed threads). The coefficient of static friction between the threads is 0.08. Calculate the torque that must be applied to the screw in order to start the 1200-lb load moving (a) upward; and (b) downward.

*7.6   *Ropes and Flat Belts*

The theory of Coulomb friction can also be used to analyze the forces acting between a flexible body, such as a rope or belt, and a friction surface.

Figure 7.12 shows a weight W that is held in static equilibrium by a rope that passes over a peg. If the peg is frictionless, then $P = W$; that is, the peg simply reverses the direction of the rope without changing its tension. If the contact surface between the peg and the rope has friction, the friction force will help to keep the weight from falling. In this case, it is possible to have $P < W$ and still maintain equilibrium. A good example of this principle in action is the capstan—a device for fastening a ship to the dock. Other applications are belt drives and band brakes. In a belt drive, the friction between the belt and the pulleys enables power to be transmitted between rotating shafts. Band brakes use friction between a band (belt) and a cylindrical drum to reduce the speed of rotating machinery.

Fig. 7.12

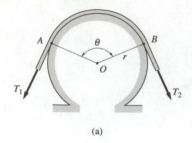

(a)

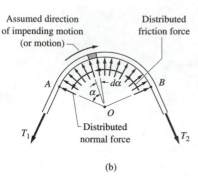

Assumed direction of impending motion (or motion)

Distributed friction force

Distributed normal force

(b)

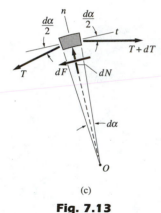

(c)

Fig. 7.13

Figure 7.13(a) shows a thin, flat belt that passes over a cylinder of radius r. Because of the friction between the belt and the cylinder, the tensions T_1 and T_2 are not equal. According to the theory of dry friction, the belt is in one of the following states depending on the values of T_1 and T_2: equilibrium without impending slipping, equilibrium with impending slipping, or slipping. The analysis that follows determines the relationship between T_1 and T_2 for the last two cases.

The forces acting on the belt are shown in Fig. 7.13(b), where the weight of the belt is assumed to be negligible. The cylinder exerts normal and friction forces on the belt, both of which are distributed along the contact area between A and B. Because the direction of impending motion (or motion) of the belt is assumed to be clockwise, equilibrium implies that $T_2 > T_1$.* Because the contact forces are distributed, it is necessary to perform the equilibrium analysis on an infinitesimal (differential) length of the belt that subtends the differential angle $d\alpha$.

The free-body diagram of the differential element is shown in Fig. 7.13(c), where n and t refer to the directions that are normal and tangent to the cylindrical surface at the center of the element. The belt tension on the left side of the element is denoted by T, and the tension on the right side by $(T + dT)$, where the differential change in the tension is $dT = (dT/d\alpha) d\alpha$. The angle between each tension and the t-direction is $d\alpha/2$. The element is also subjected to the normal force dN and to the friction force dF, acting in the n- and t-directions, respectively.

Equilibrium of forces in the tangential direction yields

$$\Sigma F_t = 0 \qquad \xrightarrow{+} \quad (T + dT)\cos\frac{d\alpha}{2} - T\cos\frac{d\alpha}{2} - dF = 0 \qquad \text{(a)}$$

Because the cosine of an infinitesimal angle equals 1, Eq. (a) reduces to

$$dF = dT \qquad \text{(b)}$$

The balance of forces in the normal direction gives

$$\Sigma F_n = 0 \qquad +\uparrow \quad dN - (T + dT)\sin\frac{d\alpha}{2} - T\sin\frac{d\alpha}{2} = 0 \qquad \text{(c)}$$

Assuming that α is measured in radians, $\sin(d\alpha/2)$ can be replaced by $d\alpha/2$, an identity that is valid for infinitesimal angles. Making this substitution gives

$$dN - (T + dT)\frac{d\alpha}{2} - T\frac{d\alpha}{2} = 0 \qquad \text{(d)}$$

Neglecting the product of differentials $(dT d\alpha)$ compared to $T d\alpha$ (this is not an approximation), we have

$$dN = T d\alpha \qquad \text{(e)}$$

If the belt is slipping or if motion impends, we have the additional equation $dF = \mu dN$, where $\mu = \mu_k$ (slipping), or $\mu = \mu_s$ (impending slipping).

---

*The equilibrium equations are applicable even if the belt is moving. Because the weight of the belt is assumed to be negligible, inertial effects can be omitted except for very high speeds.

Substituting this for dF in Eq. (b) and eliminating dN between Eqs. (b) and (e), we obtain

$$\frac{dT}{T} = \mu \, d\alpha \qquad \text{(f)}$$

Integrating both sides of Eq. (f) over the contact angle θ, shown in Fig. 7.13(a), gives

$$\int_0^\theta \frac{dT}{T} = \mu \int_0^\theta d\alpha$$

Noting that $T = T_1$ when $\alpha = 0$, and $T = T_2$ when $\alpha = \theta$, integration yields

$$\ln(T_2/T_1) = \mu\theta$$

which can be written as

$$\boxed{T_2 = T_1 e^{\mu\theta}} \qquad \text{(7.8)}$$

where $e = 2.718\ldots$ is the base of natural (Naperian) logarithms. If the tension in one side of the belt is known, Eq. (7.8) can be used to calculate the belt tension in the other side.

The following points should be kept in mind when using Eq. (7.8).

- T_2 is the belt tension that is directed opposite the belt friction. Thus, T_2 must always refer to the larger of the two tensions.
- For impending motion, use $\mu = \mu_s$. If there is relative motion between the belt and cylinder, use $\mu = \mu_k$.
- The angle of contact θ must be expressed in radians.
- Because Eq. (7.8) is independent of r, its use is not restricted to circular contact surfaces; it may also be used for a surface of arbitrary shape.

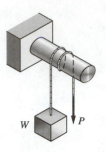

Sample Problem 7.13

The block of weight W is supported by a rope that is wrapped one-and-one-half times around the circular peg. Determine the range of values of P for which the block remains at rest. The coefficient of static friction between the rope and the peg is 0.2.

Solution

The tension in the portion of the rope that is attached to the block is obviously equal to W. Because motion impends, Eq. (7.8) can be used to relate this tension to P. Because the angle of contact is $\theta = 1.5(2\pi) = 3\pi$ rad, Eq. (7.8) becomes

$$T_2 = T_1 e^{\mu_s \theta} = T_1 e^{0.2(3\pi)} = 6.59 T_1$$

Recall that in this equation, T_2 refers to the larger of the two tensions.

The largest value of P for equilibrium occurs when the block is on the verge of moving upward. For this case we must substitute $T_1 = W$ and $T_2 = P$ into the preceding equation. The result is $P = 6.59W$.

The smallest value of P corresponds to impending motion of the block downward, when W will be larger than P. Substituting $T_1 = P$ and $T_2 = W$, we have $W = 6.59P$, or $P = W/6.59 = 0.152W$.

Therefore, the block is at rest if P is in the range

$$0.152W \le P \le 6.59W \qquad\qquad \textit{Answer}$$

Sample Problem 7.14

As shown in Fig. (a), a flexible belt placed around a rotating drum of 4-inch radius acts as a brake when the arm $ABCD$ is pulled down. The coefficient of kinetic friction between the belt and the drum is 0.2. Determine the force P that would result in a braking torque of 400-lb·in., assuming that the drum is rotating counterclockwise. Neglect the weight of the brake arm.

Solution

The free-body diagram of the *belt* is shown in Fig. (b). The distributed contact forces exerted by the drum have been replaced by the equivalent force-couple system at O. The resultant force R is not of interest to us, but the couple C^R represents the braking torque; that is, $C^R = 400$ lb·in. Note that C^R has the same sense as the rotation of the drum—namely, counterclockwise. The moment equation of equilibrium, with O as the moment center, is

$$\Sigma M_O = 0 \quad \underset{+}{\curvearrowleft} \quad (T_C - T_B)4 - 400 = 0 \qquad (a)$$

Equation (7.8) provides us with another relationship between the belt tensions. Substituting $\mu = \mu_k = 0.2$, $T_1 = T_B$, $T_2 = T_C$ (note that $T_C > T_B$), and $\theta = 240(\pi/180) = 1.333\pi$ rad, Eq. (7.8) becomes

$$T_C = T_B e^{0.2(1.333\pi)} = 2.311 T_B \qquad (b)$$

The solution of Eqs. (a) and (b) is $T_B = 76.3$ lb and $T_C = 176.3$ lb.

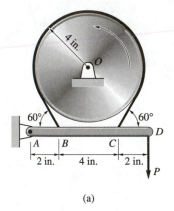

(a)

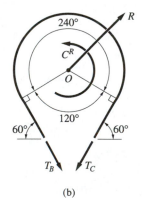

(b)

The force P can now be found by the balance of moments about A on the FBD of the brake arm, shown in Fig. (c):

$$\Sigma M_A = 0 \quad \overset{\curvearrowleft}{+} \quad T_B \sin 60°(2) + T_C \sin 60°(6) - P(8) = 0$$

Substituting the values for T_B and T_C, and solving for P, gives

$$P = 131.0 \text{ lb} \qquad\qquad Answer$$

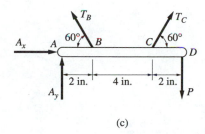

(c)

Problems

7.59 How many turns of rope around the capstan are needed for the 60-lb force to resist the 9000-lb pull of a docked ship? The static coefficient of friction between the capstan and the rope is 0.2.

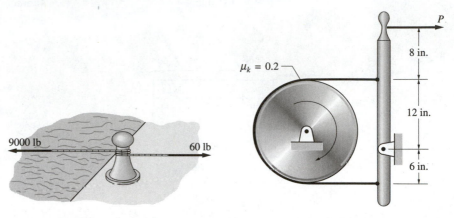

Fig. P7.59   Fig. P7.60, P7.61

7.60 The force P applied to the brake handle enables the band brake to reduce the angular speed of a rotating drum. If the tensile strength of the band is 3800 lb, find the maximum safe value of P and the corresponding braking torque acting on the drum. Assume that the drum is rotating clockwise.

7.61 Solve Prob. 7.60 if the drum is rotating counterclockwise.

7.62 The two blocks are joined with a cord that passes over a fixed drum. The co-efficient of static friction is 0.2 between block B and the inclined plane, as well as between the cord and the drum. Friction between the blocks is negligible. Find the smallest weight W of block A that would prevent the blocks from sliding.

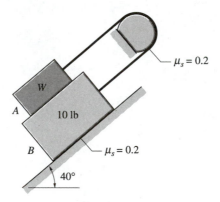

Fig. P7.62

7.63 The leather rein used to fasten the horse to the hitching rail weighs 4 oz per foot. The coefficient of static friction between the rail and the rein is 0.6. If a 30-lb force acting on the bridle is sufficient to restrain the horse, determine the smallest safe length L for the free end of the rein.

7.64 The uniform semi-cylinder of radius $R = 5$ in. weighs 6 lb. It is supported by a pin at O and the cable AB. The weight $W = 1.5$ lb is attached to the free end of the cable. Determine the smallest coefficient of static friction between the cable and the cylinder that is necessary to maintain equilibrium. The center of gravity of the cylinder is at G.

7.65 The rail AB of negligible weight is suspended from a rope that runs around two fixed pegs. The coefficient of static friction between the rope and the pegs is 0.5. As the weight W moves along the rail toward end B, determine its position x when the rope is about to slip on the pegs.

7.66 The sling with a sliding hook is used to hoist a homogeneous drum. If the static coefficient of friction between the cable and the eye of the hook is 0.6, determine the smallest possible value for the angle θ.

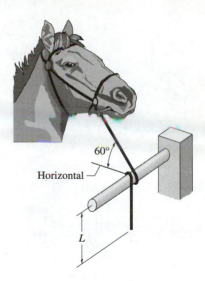

Fig. P7.63

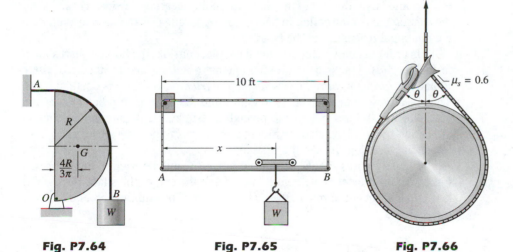

Fig. P7.64   **Fig. P7.65**   **Fig. P7.66**

7.67 The rope, attached to the wall at A and passing over a peg at C, supports two identical weights. One of the weights runs on a small pulley, so that it is always positioned under the midpoint of AC. Determine the range of the angle θ for which the weights will be in equilibrium if the coefficient of static friction between the rope and peg at C is 0.2.

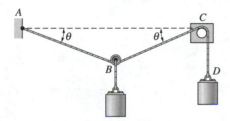

Fig. P7.67

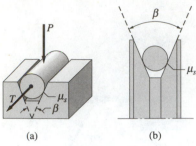

(a)   (b)

Fig. P7.68

(a) Friction clutch

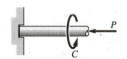

(b) Pivot bearing

(c) Collar bearing

Fig. 7.14

7.68   The largest pull that can be applied without causing the cylinder in Fig. (a) to slide is $T = \mu_s' P$, where μ_s' is a constant. (a) Determine μ_s' in terms of μ_s (the static coefficient of friction) and the groove angle β. (b) By making appropriate changes in derivation of Eq. (7.8), show that the friction equation for a circular belt in a grooved pulley, shown in Fig. (b), is $T_2 = T_1 e^{\mu_s' \theta}$ when sliding impends.

*7.7   Disk Friction

When a disk or the end of a shaft is pressed against a flat surface, its rotation is resisted by a frictional couple, known as *disk friction*. Some examples of disk friction are illustrated in Fig. 7.14. The friction clutch consists of two disks that are coated with special high-friction materials. When the disks are pressed together by an axial force P, as shown in Fig. 7.14(a), they are capable of transmitting a large torque C without slipping. Axial loads carried by rotating shafts are sometimes supported by pivot bearings and collar bearings, shown in Fig. 7.14(b) and (c), respectively. In a pivot bearing, the axial force is distributed over the end of the shaft; in a collar bearing, the load is carried by the annular area of the collar. In both cases the torque C is required to overcome the rotational resistance of the bearing.

In order to analyze the frictional couple, consider the hollow shaft shown in Fig. 7.15(a). The shaft is pressed against a flat surface by the *axial force P*, and the *torque required to overcome the frictional couple* is denoted by C. The objective of our analysis is to determine the relationship between P and C. Following the practice of the preceding article, we denote the coefficient of friction by μ. If the shaft is rotating, then μ is to be interpreted as μ_k; for impending rotation, μ_s should be used.

Figure 7.15(b) shows the normal force dN and the friction force dF acting on the infinitesimal element of area dA at the end of the shaft. If the shaft is rotating, or about to rotate, then $dF = \mu\, dN$. The equilibrium equations of the shaft are

$$\Sigma F_{\text{axial}} = 0 \qquad P - \int_{\mathcal{A}} dN = 0$$

$$\Sigma M_{\text{axis}} = 0 \qquad C - \int_{\mathcal{A}} r\, dF = C - \mu \int_{\mathcal{A}} r\, dN = 0$$

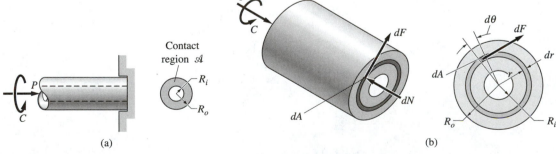

(a)   (b)

Fig. 7.15

where the integrals are to be computed over the contact region $\mathscr{A}$, which is an annular ring of area $A = \pi(R_o^2 - R_i^2)$. Denoting the normal contact pressure by p, we have $dN = p\,dA$, and the equilibrium equations become

$$P = \int_{\mathscr{A}} p\,dA \qquad (7.9)$$

$$C = \mu \int_{\mathscr{A}} pr\,dA \qquad (7.10)$$

In order to perform the integrations, the variation of the normal pressure p over the contact region must be known. This variation depends on whether the contact surfaces are new or worn.

New Surfaces For new, flat contact surfaces, it is reasonable to assume that the pressure p is uniformly distributed. Therefore, Eq. (7.9) becomes $P = pA$, and the contact pressure is given by

$$p = \frac{P}{A} = \frac{P}{\pi(R_o^2 - R_i^2)}$$

Taking p outside the integral in Eq. (7.10) gives $C = \mu p \int_{\mathscr{A}} r\,dA$. As shown in Fig. 7.15(b), dA can be expressed in terms of polar coordinates as $dA = r\,d\theta\,dr$. Therefore, the torque required to overcome the friction couple becomes

$$C = \frac{\mu P}{\pi(R_o^2 - R_i^2)} \int_{R_i}^{R_o} \int_0^{2\pi} r^2\,d\theta\,dr$$

which, after evaluating the integrals, becomes:

$$\boxed{C = \frac{2\mu P}{3} \frac{\left(R_o^3 - R_i^3\right)}{\left(R_o^2 - R_i^2\right)}} \qquad (7.11)$$

If the cross section is a solid circle of radius $R_o (R_i = 0)$, the above expression reduces to

$$\boxed{C = \frac{2\mu P R_o}{3}} \qquad (7.12)$$

Worn Surfaces Although the normal pressure p may be initially uniform between two new, flat surfaces, the wear will not be uniform. The wear at a given point on the cross section will depend on both the pressure and the distance traveled by the point during slipping. Because the distance traveled is proportional to r (a point at a radial distance r travels the distance $2\pi r$ in one revolution of the shaft), greater wear will occur at points further from the axis of the shaft. Once the contact surfaces have been broken in, it is reasonable to assume that the cross section will have worn to a shape for which the rate of wear is constant. In this situation, we would have $pr = K$, where K is a constant. For a hollow cross section, Eq. (7.10) then becomes

$$C = \mu K \int_{\mathcal{A}} dA = \mu K \pi \left(R_o^2 - R_i^2\right) \tag{7.13}$$

The constant K can be calculated by substituting $p = K/r$ into Eq. (7.9), resulting in

$$P = \int_{\mathcal{A}} p \, dA = \int_{R_i}^{R_o} \int_0^{2\pi} \frac{K}{r} r \, d\theta \, dr = 2\pi K(R_o - R_i)$$

from which

$$K = \frac{P}{2\pi (R_o - R_i)}$$

Substituting this expression for K into Eq. (7.13), the torque required to overcome the friction couple is

$$C = \frac{\mu P}{2} (R_o + R_i) \tag{7.14}$$

For a solid shaft of radius R_o, we have $R_i = 0$ and the torque reduces to

$$C = \frac{\mu P R_o}{2} \tag{7.15}$$

Sample Problem 7.15

Figure (a) shows a disk clutch that transmits torque from the input shaft on the left to the output shaft on the right. The disk is splined to the input shaft, thereby forcing the disk and the shaft to rotate together but allowing the disk to slide along the shaft. The normal force between the two halves of the clutch is provided by the compression spring. The force F applied to the clutch pedal can disengage the clutch by sliding the throw-out bearing to the left. Determine the largest torque that can be transmitted if the value of F necessary to disengage the clutch is 20 lb. Solve for both new and worn friction surfaces.

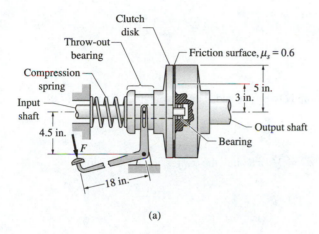

(a)

Solution

Utilizing the free-body diagram of the clutch pedal assembly, Fig. (b), we get

$$\Sigma M_A = 0 \quad \overset{+}{\curvearrowleft} \quad 20(18) - P(4.5) = 0$$

$$P = 80 \text{ lb}$$

This is the normal force on the friction surface when the clutch is engaged. The largest torque that can be transmitted equals the friction couple for this value of P.

For a new friction surface, Eq. (7.11) gives

$$C = \frac{2\mu_s P}{3} \frac{\left(R_o^3 - R_i^3\right)}{\left(R_o^2 - R_i^2\right)} = \frac{2(0.6)(80)}{3} \frac{\left(5^3 - 3^3\right)}{\left(5^2 - 3^2\right)}$$

$$= 196.0 \text{ lb} \cdot \text{in.} \qquad \qquad \textit{Answer}$$

After the friction surfaces have become worn, we have from Eq. (7.14)

$$C = \frac{\mu_s P}{2}(R_o + R_i) = \frac{0.6(80)}{2}(5 + 3) = 192.0 \text{ lb} \cdot \text{in.} \qquad \textit{Answer}$$

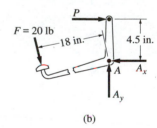

(b)

Sample Problem 7.16

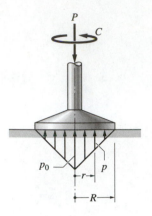

The normal pressure under a circular industrial glass polisher is axially symmetric and varies linearly from p_0 at $r = 0$ to zero at $r = R$, as shown in the figure. Derive the expression for the torque required to rotate the polisher in terms of the axial load P. The coefficient of kinetic friction between the polisher and the glass is μ_k.

Solution

The pressure p at the radial distance r is given by $p = p_0(R - r)/R$. Substituting this expression into Eq. (7.10), the torque C required to rotate the polisher is

$$C = \mu_k \int_{\mathcal{A}} p r \, dA = \frac{\mu_k p_0}{R} \int_0^R \int_0^{2\pi} (R - r) r^2 \, d\theta \, dr$$

$$= \frac{\pi \mu_k p_0 R^3}{6}$$

The relationship between p_0 and P can be obtained from Eq. (7.9).

$$P = \int_{\mathcal{A}} p \, dA = \frac{p_0}{R} \int_0^R \int_0^{2\pi} (R - r) r \, d\theta \, dr = \frac{\pi p_0 R^2}{3}$$

Substituting $p_0 = 3P/\pi R^2$ into the expression for C yields

$$C = \frac{\mu_k P R}{2} \qquad\qquad \textit{Answer}$$

Problems

7.69   The collar bearing carries the axial load $P = 400$ N. Assuming uniform pressure between the collar and the horizontal surface, determine the couple C required to start the shaft turning. Use $\mu_s = 0.15$ for the coefficient of static friction.

7.70   Solve Sample Problem 7.16 if the contact pressure under the polisher varies linearly from p_0 at $r = 0$ to $p_0/2$ at $r = R$.

7.71   The 500-lb cable spool is placed on a frictionless spindle that has been driven into the ground. If the force required to start the spool rotating is $F = 110$ lb, determine the coefficient of friction between the ground and the spool. Neglect the diameter of the spindle compared to the diameter of the spool.

7.72   Determine the braking torque acting on the rotating disk when a force $P = 200$ N is applied to each brake pad. Assume that the brakes are (a) new; and (b) worn.

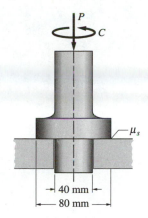

Fig. P7.69

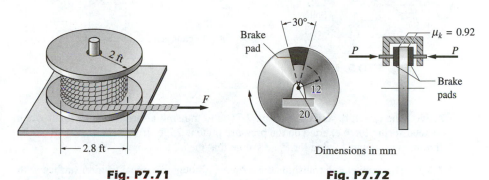

Fig. P7.71

Fig. P7.72

7.73   The normal pressure acting on the disk of the sander is given by $p = (4/3) + (r^2/6)$, where p is the pressure in pounds per square inch (psi) and r represents the radial distance in inches. Determine the torque C required to operate the sander at constant speed if the kinetic coefficient of friction for the surface being sanded is 0.86.

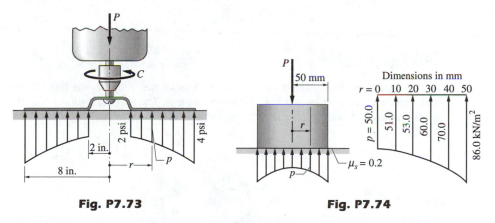

Fig. P7.73

Fig. P7.74

7.74   When the hard cylinder is pressed against a much softer substrate, the contact pressure is considerably higher near the edge of the cylinder than in the middle. If the force P causes the contact pressure shown on the diagram, determine the couple required to initiate rotation of the cylinder about its axis. Use the trapezoidal rule to evaluate the resulting integral.

7.75 The single-plate clutch transmits the torque C from the input shaft on the left to the output shaft on the right. Compression springs between the clutch housing and the pressure plate provide the necessary pressure on the friction surface. The splines prevent the clutch plate from rotating relative to the output shaft. If $R_i = 4$ in. and $R_o = 9$ in., determine the total force that must be applied to the pressure plate by the springs if the clutch is to transmit a torque of $C = 56$ lb · ft when it is new.

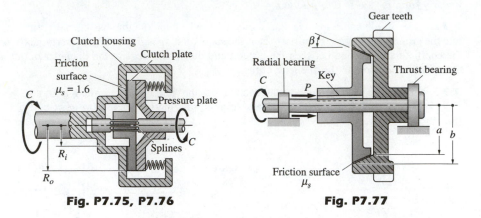

Fig. P7.75, P7.76   **Fig. P7.77**

7.76 The clutch described in Prob. 7.75 is to transmit a torque of 120 lb · ft when the total spring force exerted on the pressure plate is 75 lb. If $R_i = 4$ in., calculate the minimum acceptable value for R_o. Assume that the clutch is new.

*7.77 The cone clutch transmits the torque C through a conical friction surface with cone angle β. The inner and outer radii of the friction surface are a and b, respectively. The left half of the clutch is keyed to the shaft, and the right half drives a machine (not shown) through a gear attached to its outer rim. Assuming uniform pressure on the friction surface, show that the maximum torque that can be transmitted by the clutch is

$$C = \frac{2\mu_s P}{3\sin\beta}\frac{\left(b^3 - a^3\right)}{\left(b^2 - a^2\right)}$$

Review Problems

7.78 Determine the smallest force P that will move the wedge to the right if the coefficient of static friction is 0.5 at all contact surfaces. The uniform cylinder weighs W, and the weight of the wedge may be neglected.

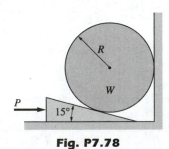

Fig. P7.78

7.79 Find the range of *P* for which the 120-lb block will be in equilibrium. Neglect the possibility of tipping.

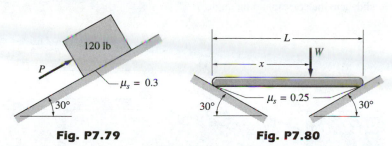

Fig. P7.79   **Fig. P7.80**

7.80 Determine the smallest ratio *x*/*L* for which the bar can remain at rest in the position shown. Neglect the weight of the bar.

7.81 Find the smallest angle *β* for which the uniform crate can be tipped about corner *A*. Also compute the corresponding value of *P*.

7.82 The belt is placed between two rollers, which are free to rotate about *A* and *B*. Determine the smallest coefficient of static friction between the belt and the rollers for which the device is self-locking; that is, the belt cannot be pulled down for any value of *P*.

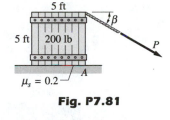

Fig. P7.81

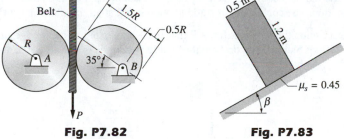

Fig. P7.82   **Fig. P7.83**

7.83 Determine the largest angle *β* for which the uniform box can be in equilibrium.

7.84 Can the uniform bar of weight *W* remain at rest in the position shown?

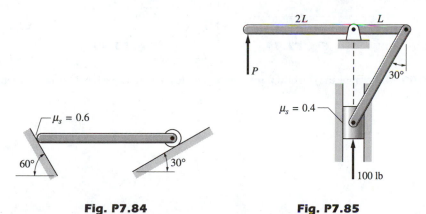

Fig. P7.84   **Fig. P7.85**

7.85 Determine the largest force *P* for which the system can remain at rest. Neglect the weights of the members.

7.86 The person is trying to move the crate of weight W by pulling on the rope at the angle θ to the horizontal. Find the smallest possible tension that would cause the crate to slide and the corresponding angle θ.

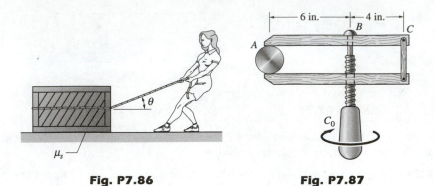

Fig. P7.86   **Fig. P7.87**

7.87 The screw of the clamp has a square thread of pitch 0.15 in. and a mean diameter of 0.5 in. The coefficient of static friction between the threads is 0.5. Determine (a) the torque C_0 that must be applied to the screw in order to produce a 30-lb clamping force at A; and (b) the torque required to loosen the clamp.

7.88 Find the largest clockwise couple C that can be applied to cylinder A without causing motion. The coefficient of static friction is 0.2 at all three contact surfaces.

7.89 Determine the smallest force P that must be applied to the wedge in order to slide the 200-kg block. The coefficient of static friction is 0.3 at all contact surfaces. Neglect the weight of the wedge.

40 kg

0.3 m

A C

60 kg

B

Fig. P7.88

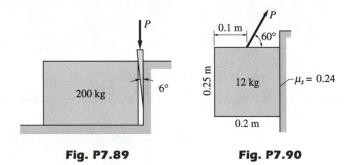

Fig. P7.89   **Fig. P7.90**

7.90 Determine the smallest force P necessary to hold the 12-kg block in the position shown.

8

Centroids and Distributed Loads

8.1 Introduction

In this chapter we investigate centroids, centers of gravity, and centers of mass. Centroids were first discussed in Art. 3.6 in conjunction with distributed normal loads. The more rigorous treatment of centroids presented here will enable us to analyze normal loads that are distributed in a complex manner. We also discuss the theorems of Pappus-Guldinus, which utilize centroids for calculating areas and volumes of revolution.

8.2 Centroids of Plane Areas and Curves

a. Definitions

Consider the plane region $\mathcal{A}$ shown in Fig. 8.1. Let dA be a differential (infinitesimal) area element of $\mathcal{A}$, located at (x, y). There are certain properties of $\mathcal{A}$ that occur frequently in various branches of the physical sciences. One of these is, of course, the area

$$A = \int_{\mathcal{A}} dA \qquad (8.1)$$

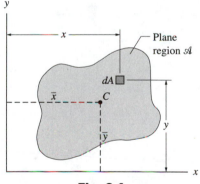

Fig. 8.1

Other properties are the *first moments of the area* about the x- and y-axes, defined as

$$Q_x = \int_{\mathcal{A}} y\, dA \qquad Q_y = \int_{\mathcal{A}} x\, dA \qquad (8.2)$$

and the *second moments of the area* (also called the *moments of inertia*), which are treated separately in Chapter 9.

The importance of the *centroid of the area* stems from its close association with the first moments of areas. The centroid C of a plane area is defined as the point that has the coordinates (see Fig. 8.1)

$$\bar{x} = \frac{Q_y}{A} = \frac{\displaystyle\int_{\mathcal{A}} x\, dA}{\displaystyle\int_{\mathcal{A}} dA} \qquad \bar{y} = \frac{Q_x}{A} = \frac{\displaystyle\int_{\mathcal{A}} y\, dA}{\displaystyle\int_{\mathcal{A}} dA} \qquad (8.3)$$

It can be seen that if A and $(\bar{x}, \bar{y})$ of an area are known, its first moments can be computed by $Q_x = A\bar{y}$ and $Q_y = A\bar{x}$, thereby avoiding the evaluation of the integrals in Eqs. (8.2).

The centroid is sometimes referred to as the geometric center of the region. The centroid is not to be confused with the mass center, which is a property of the mass distribution within the region. Centroids and mass centers coincide only if the distribution of mass is uniform—that is, if the body is homogeneous.

The following characteristics of centroids and first moments of areas should be noted.

- The dimension of Q_x and Q_y is $[L^3]$; hence the units are mm³, in.³, and so on.
- Q_x and Q_y may be positive, negative, or zero, depending on the location of the coordinate axes relative to the centroid of the region. If the x-axis passes through the centroid, then $Q_x = 0$. Similarly, $Q_y = 0$ if the centroid lies on the y-axis.
- If the region is symmetric, then its centroid is located on the axis of symmetry. This can be demonstrated by considering the region in Fig. 8.2, which is symmetric about the y-axis. Clearly, the integral $\int x\, dA$ over the left half (where x is negative) cancels the integral $\int x\, dA$ over the right half (where x is positive). Consequently, $Q_y = 0$, and it follows that $\bar{x} = 0$.

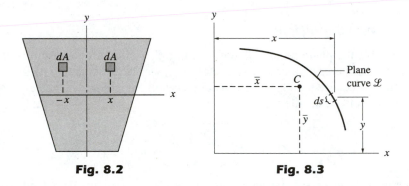

Fig. 8.2 **Fig. 8.3**

The definitions of centroids and first moments of plane curves are analogous to those of plane areas. Letting ds be the differential length of the plane curve $\mathcal{L}$ in Fig. 8.3, the length of the curve is

$$L = \int_{\mathcal{L}} ds \tag{8.4}$$

and the first moments of the curve about the coordinate axes are defined as

$$Q_x = \int_{\mathcal{L}} y\, ds \qquad Q_y = \int_{\mathcal{L}} x\, ds \tag{8.5}$$

The dimension of the first moment of a curve is $[L^3]$. The coordinates of the *centroid of the curve* are, by definition,

$$\bar{x} = \frac{Q_y}{L} = \frac{\displaystyle\int_{\mathcal{L}} x\,ds}{\displaystyle\int_{\mathcal{L}} ds} \qquad \bar{y} = \frac{Q_x}{L} = \frac{\displaystyle\int_{\mathcal{L}} y\,ds}{\displaystyle\int_{\mathcal{L}} ds} \qquad (8.6)$$

b. Integration techniques

The details of the integration for plane areas in Eqs. (8.1) and (8.2) depend on the choice of the area element dA. There are two basic choices for dA: the double differential elements shown in Fig. 8.4(a) and (b); and the single differential elements in Fig. 8.4(c)–(e). In the latter case, the coordinates x and y of the differential element must be interpreted as the coordinates of the centroid of the element. These coordinates, denoted by $\bar{x}_{el}$ and $\bar{y}_{el}$, are shown in Fig. 8.4(c)–(e).

The expressions for dA, $\bar{x}_{el}$, and $\bar{y}_{el}$ also depend on the choice of the coordinate system. Figure 8.4 illustrates elements using both rectangular and polar coordinates. The most convenient coordinate system for a given problem

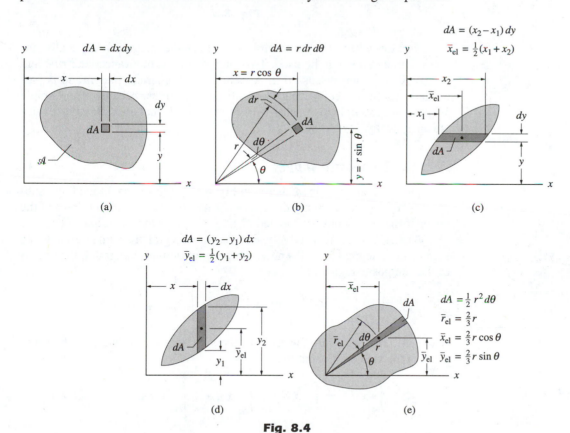

Fig. 8.4

is determined primarily by the shape of the region $\mathscr{A}$. Obviously, rectangular regions are best handled by rectangular coordinates, whereas polar coordinates should be chosen for circular regions.

The properties of curves always involve a single integration, carried out along the length of the curve. The expressions for the differential length ds for rectangular and polar coordinates are shown in Fig. 8.5.

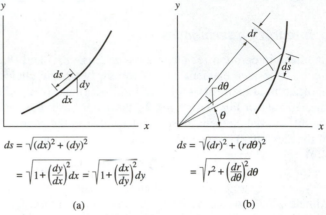

$$ds = \sqrt{(dx)^2 + (dy)^2}$$
$$= \sqrt{1 + \left(\frac{dy}{dx}\right)^2}\,dx = \sqrt{1 + \left(\frac{dx}{dy}\right)^2}\,dy$$

(a)

$$ds = \sqrt{(dr)^2 + (rd\theta)^2}$$
$$= \sqrt{r^2 + \left(\frac{dr}{d\theta}\right)^2}\,d\theta$$

(b)

Fig. 8.5

In cases where it is not possible to evaluate the integrals analytically, numerical integration can be used. Two such methods, the trapezoidal rule and Simpson's rule, are discussed in Appendix A. Numerical integration is particularly useful for computing the centroids of curves, because even simple shapes—for example, parabolas—result in integrals that are difficult to evaluate analytically.

c. Composite shapes

Consider the plane region $\mathscr{A}$ shown in Fig. 8.6 that has been divided into subregions $\mathscr{A}_1, \mathscr{A}_2, \mathscr{A}_3, \ldots$ (only three subregions are shown). The centroids of the areas of the subregions are labeled $C_1, C_2, C_3, \ldots$, with coordinates $(\bar{x}_1, \bar{y}_1)$, $(\bar{x}_2, \bar{y}_2), (\bar{x}_3, \bar{y}_3), \ldots$, respectively. Because the integral of a sum is equal to the sum of the integrals (a well-known property of definite integrals), the area A of the composite region $\mathscr{A}$ is

$$A = \int_{\mathscr{A}} dA = \int_{\mathscr{A}_1} dA + \int_{\mathscr{A}_2} dA + \int_{\mathscr{A}_3} dA + \cdots = \sum_i A_i$$

where $A_1, A_2, A_3, \ldots$ are the areas of the subregions. Similarly, the first moment of the area of $\mathscr{A}$ about the y-axis is

$$Q_y = \int_{\mathscr{A}} x\,dA = \int_{\mathscr{A}_1} x\,dA + \int_{\mathscr{A}_2} x\,dA + \int_{\mathscr{A}_3} x\,dA + \cdots = \sum_i (Q_y)_i$$

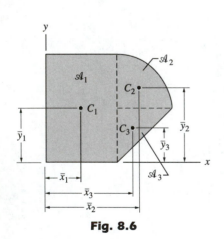

Fig. 8.6

where $(Q_y)_i$ refers to the first moment of the area of $\mathcal{A}_i$ about the y-axis. A similar analysis may be used to determine Q_x, the first moment of the area of $\mathcal{A}$ about the x-axis.

Therefore, the centroidal coordinates of the area of $\mathcal{A}$ can be written as

$$\bar{x} = \frac{Q_y}{A} = \frac{\Sigma_i (Q_y)_i}{\Sigma_i A_i} \qquad \bar{y} = \frac{Q_x}{A} = \frac{\Sigma_i (Q_x)_i}{\Sigma_i A_i} \qquad (8.7)$$

Determining the centroid of an area by this technique is called the *method of composite areas*. Substituting $(Q_y)_i = A_i \bar{x}_i$ and $(Q_x)_i = A_i \bar{y}_i$, the preceding equations become

$$\bar{x} = \frac{Q_y}{A} = \frac{\Sigma_i A_i \bar{x}_i}{\Sigma_i A_i} \qquad \bar{y} = \frac{Q_x}{A} = \frac{\Sigma_i A_i \bar{y}_i}{\Sigma_i A_i} \qquad (8.8)$$

Caution   The centroid of the composite area is not equal to the sum of the centroids of its subregions; that is, $\bar{x} \neq \Sigma_i (A_i \bar{x}_i / A_i)$ and $\bar{y} \neq \Sigma_i (A_i \bar{y}_i / A_i)$.

The method of composite curves is analogous to the method of composite areas. The centroidal coordinates of the curve $\mathcal{L}$ of length L that has been subdivided into the segments $\mathcal{L}_1, \mathcal{L}_2, \mathcal{L}_3, \ldots$ are given by

$$\bar{x} = \frac{Q_y}{L} = \frac{\Sigma_i L_i \bar{x}_i}{\Sigma_i L_i} \qquad \bar{y} = \frac{Q_x}{L} = \frac{\Sigma_i L_i \bar{y}_i}{\Sigma_i L_i} \qquad (8.9)$$

where L_i is the length of the segment $\mathcal{L}_i$ with its centroid located at $(\bar{x}_i, \bar{y}_i)$.

You will discover that tables that list the properties of common plane figures, such as Tables 8.1 and 8.2, are very useful when applying the methods of composite areas and composite lines.

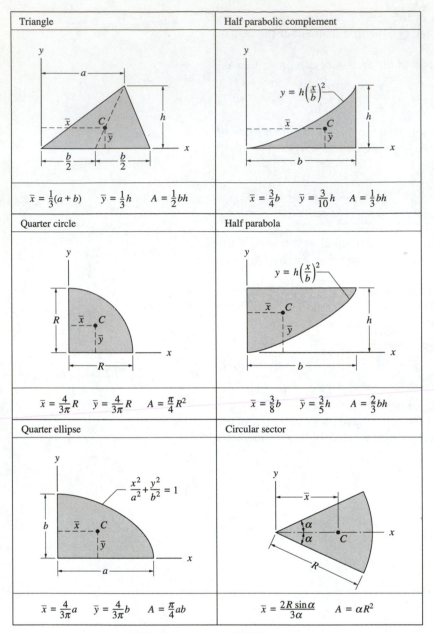

Triangle

$$\bar{x} = \tfrac{1}{3}(a + b) \qquad \bar{y} = \tfrac{1}{3}h \qquad A = \tfrac{1}{2}bh$$

Half parabolic complement

$$y = h\left(\frac{x}{b}\right)^2$$

$$\bar{x} = \tfrac{3}{4}b \qquad \bar{y} = \tfrac{3}{10}h \qquad A = \tfrac{1}{3}bh$$

Quarter circle

$$\bar{x} = \frac{4}{3\pi}R \qquad \bar{y} = \frac{4}{3\pi}R \qquad A = \frac{\pi}{4}R^2$$

Half parabola

$$y = h\left(\frac{x}{b}\right)^2$$

$$\bar{x} = \tfrac{3}{8}b \qquad \bar{y} = \tfrac{3}{5}h \qquad A = \tfrac{2}{3}bh$$

Quarter ellipse

$$\frac{x^2}{a^2} + \frac{y^2}{b^2} = 1$$

$$\bar{x} = \frac{4}{3\pi}a \qquad \bar{y} = \frac{4}{3\pi}b \qquad A = \frac{\pi}{4}ab$$

Circular sector

$$\bar{x} = \frac{2R\sin\alpha}{3\alpha} \qquad A = \alpha R^2$$

Table 8.1 *Centroids of Plane Areas*

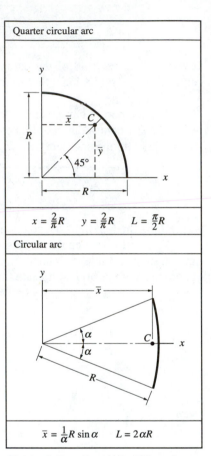

Quarter circular arc

$$x = \frac{2}{\pi}R \qquad y = \frac{2}{\pi}R \qquad L = \frac{\pi}{2}R$$

Circular arc

$$\bar{x} = \frac{1}{\alpha}R\sin\alpha \qquad L = 2\alpha R$$

Table 8.2 *Centroids of Plane Curves*

Sample Problem 8.1

Determine the coordinates of the centroid of the area that lies between the straight line $x = 2y/3$ and the parabola $x^2 = 4y$, where x and y are measured in inches— see Fig. (a). Use the following methods: (1) double integration; (2) single integration using a horizontal differential area element; and (3) single integration using a vertical differential area element.

Solution

Part 1   Double Integration

The double differential area element is shown in Fig. (b). Note that dA can be written as $dx\,dy$, or $dy\,dx$, depending on whether you choose to integrate on x or y first. Choosing to integrate over y first, the area A of the region $\mathcal{A}$ is

$$A = \int_{\mathcal{A}} dA = \int_0^6 \left(\int_{x^2/4}^{3x/2} dy \right) dx$$

$$= \int_0^6 \left(\frac{3x}{2} - \frac{x^2}{4} \right) dx$$

$$= \left[\frac{3x^2}{4} - \frac{x^3}{12} \right]_0^6 = \frac{3(6)^2}{4} - \frac{(6)^3}{12}$$

$$= 27 - 18 = 9 \text{ in.}^2$$

The first moment of the area about the y-axis is

$$Q_y = \int_{\mathcal{A}} x\, dA = \int_0^6 \left(\int_{x^2/4}^{3x/2} x\, dy \right) dx$$

$$= \int_0^6 \left(\frac{3x}{2} - \frac{x^2}{4} \right) x\, dx$$

$$= \left[\frac{3x^3}{6} - \frac{x^4}{16} \right]_0^6 = \frac{3(6)^3}{6} - \frac{(6)^4}{16}$$

$$= 108 - 81 = 27 \text{ in.}^3$$

The first moment of the area about the x-axis is

$$Q_x = \int_{\mathcal{A}} y\, dA = \int_0^6 \left(\int_{x^2/4}^{3x/2} y\, dy \right) dx$$

$$= \int_0^6 \frac{1}{2} \left(\frac{9x^2}{4} - \frac{x^4}{16} \right) dx$$

$$= \frac{1}{2} \left[\frac{9x^3}{12} - \frac{x^5}{80} \right]_0^6 = \frac{1}{2} \left[\frac{9(6)^3}{12} - \frac{(6)^5}{80} \right]$$

$$= \frac{1}{2}(162 - 97.2) = 32.4 \text{ in.}^3$$

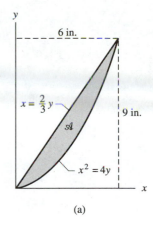

(a)

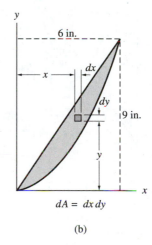

$dA = dx\,dy$

(b)

359

Therefore, the coordinates of the centroid of the area are

$$\bar{x} = \frac{Q_y}{A} = \frac{27}{9} = 3 \text{ in.} \qquad\qquad \textit{Answer}$$

$$\bar{y} = \frac{Q_x}{A} = \frac{32.4}{9} = 3.6 \text{ in.} \qquad\qquad \textit{Answer}$$

If you choose to integrate over x first, the respective integrals are as follows:

$$A = \int_{s\!A} dA = \int_0^9 \left(\int_{2y/3}^{2\sqrt{y}} dx \right) dy$$

$$Q_y = \int_{s\!A} x \, dA = \int_0^9 \left(\int_{2y/3}^{2\sqrt{y}} x \, dx \right) dy$$

$$Q_x = \int_{s\!A} y \, dA = \int_0^9 \left(\int_{2y/3}^{2\sqrt{y}} y \, dx \right) dy$$

You may wish to verify that the evaluation of these integrals yields the same centroidal coordinates as determined previously.

Part 2 Single Integration: Horizontal Differential Area Element

The horizontal differential area element is shown in Fig. (c), together with the expressions for dA and $\bar{x}_{el}$. For the area we have

$$A = \int_{s\!A} dA = \int_0^9 \left(2\sqrt{y} - \frac{2y}{3} \right) dy = \left[\frac{4y^{3/2}}{3} - \frac{y^2}{3} \right]_0^9$$

$$= \frac{4}{3}(3)^3 - \frac{(9)^2}{3} = 36 - 27 = 9 \text{ in.}^2$$

Using $dQ_y = \bar{x}_{el} \, dA$ we obtain

$$dQ_y = \frac{1}{2}\left(2\sqrt{y} + \frac{2y}{3} \right)\left(2\sqrt{y} - \frac{2y}{3} \right) dy = \left(2y - \frac{2y^2}{9} \right) dy$$

The first moment of the area about the y-axis becomes

$$Q_y = \int_0^9 \left(2y - \frac{2y^2}{9} \right) dy = \left[y^2 - \frac{2y^3}{27} \right]_0^9$$

$$= (9)^2 - \frac{2(9)^3}{27} = 81 - 54 = 27 \text{ in.}^3$$

Similarly, $dQ_x = y \, dA$ gives

$$dQ_x = y\left(2\sqrt{y} - \frac{2y}{3} \right) dy = \left(2y^{3/2} - \frac{2y^2}{3} \right) dy$$

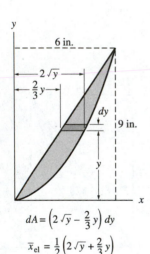

$$dA = \left(2\sqrt{y} - \frac{2}{3}y \right) dy$$

$$\bar{x}_{el} = \frac{1}{2}\left(2\sqrt{y} + \frac{2}{3}y \right)$$

(c)

The first moment about the x-axis is

$$Q_x = \int_0^9 \left(2y^{3/2} - \frac{2y^2}{3}\right) dy = \left[\frac{4}{5}y^{5/2} - \frac{2y^3}{9}\right]_0^9$$

$$= \frac{4}{5}(3)^5 - \frac{2}{9}(9)^3 = 194.4 - 162 = 32.4 \text{ in.}^3$$

Note that A, Q_x, and Q_y are identical to the values computed in Part 1. Therefore, the coordinates of the centroid of the area are also the same.

Part 3 Single Integration: Vertical Differential Area Element

The vertical differential area element is shown in Fig. (d), which also gives the expressions for dA and $\bar{y}_{el}$. The area of the region is

$$A = \int_{\mathcal{A}} dA = \int_0^6 \left(\frac{3x}{2} - \frac{x^2}{4}\right) dx = \left[\frac{3x^2}{4} - \frac{x^3}{12}\right]_0^6$$

$$= \frac{3(6)^2}{4} - \frac{(6)^3}{12} = 27 - 18 = 9 \text{ in.}^2$$

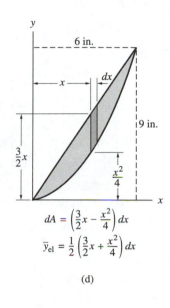

$$dA = \left(\frac{3}{2}x - \frac{x^2}{4}\right) dx$$

$$\bar{y}_{el} = \frac{1}{2}\left(\frac{3}{2}x + \frac{x^2}{4}\right) dx$$

(d)

Using the information in Fig. (d), we obtain

$$dQ_y = x\,dA = x\left(\frac{3x}{2} - \frac{x^2}{4}\right) dx$$

Therefore, the first moment about the y-axis is

$$Q_y = \int_{\mathcal{A}} dQ_y = \int_0^6 \left(\frac{3x^2}{2} - \frac{x^3}{4}\right) dx = \left[\frac{x^3}{2} - \frac{x^4}{16}\right]_0^6$$

$$= \frac{(6)^3}{2} - \frac{(6)^4}{16} = 108 - 81 = 27 \text{ in.}^3$$

For $dQ_x = \bar{y}_{el}\,dA$, we get

$$dQ_x = \frac{1}{2}\left(\frac{3x}{2} + \frac{x^2}{4}\right)\left(\frac{3x}{2} - \frac{x^2}{4}\right) dx$$

$$= \frac{1}{2}\left(\frac{9x^2}{4} - \frac{x^4}{16}\right) dx$$

Integration of this expression yields

$$Q_x = \int_{\mathcal{A}} dQ_x = \int_0^6 \frac{1}{2}\left(\frac{9x^2}{4} - \frac{x^4}{16}\right) dx = \frac{1}{2}\left[\frac{9x^3}{12} - \frac{x^5}{80}\right]_0^6$$

$$= \frac{1}{2}\left[\frac{9(6)^3}{12} - \frac{(6)^5}{80}\right]$$

$$= \frac{1}{2}(162 - 97.2) = 32.4 \text{ in.}^3$$

Again, the same values of A, Q_x, and Q_y have been obtained as in Parts 1 and 2. Thus $\bar{x}$ and $\bar{y}$ would also be identical.

Sample Problem 8.2

Using the method of composite areas, determine the location of the centroid of the shaded area shown in Fig. (a).

Solution

The area can be viewed as a rectangle, from which a semicircle and a triangle have been removed. The areas and centroidal coordinates for each of these shapes can be determined using Table 8.1. The results are shown in Figs. (b)–(d).

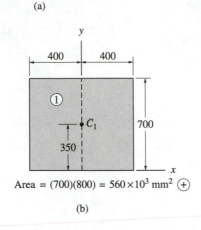

Dimensions in mm

(a)

Area = $(700)(800) = 560 \times 10^3$ mm^2 $\oplus$

(b)

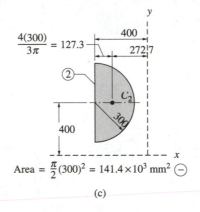

Area = $\frac{\pi}{2}(300)^2 = 141.4 \times 10^3$ mm^2 $\ominus$

(c)

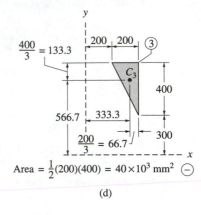

Area = $\frac{1}{2}(200)(400) = 40 \times 10^3$ mm^2 $\ominus$

(d)

When applying the method of composite areas, it is convenient to tabulate the data in the following manner.

Shape	Area A (mm)^2	$\bar{x}$ (mm)	$A\bar{x}$ (mm)^3	$\bar{y}$ (mm)	$A\bar{y}$ (mm)^3
1 (Rectangle)	$+560.0 \times 10^3$	0	0	$+350$	196.0×10^6
2 (Semicircle)	-141.4×10^3	-272.7	$+38.56 \times 10^6$	$+400$	-56.56×10^6
3 (Triangle)	-40.0×10^3	$+333.3$	-13.33×10^6	$+566.7$	-22.67×10^6
Σ	$+378.6 \times 10^3$	$\cdots$	$+25.23 \times 10^6$	$\cdots$	$+116.77 \times 10^6$

Be certain that you understand each of the entries in this table, paying particular attention to signs. For the rectangle, A is positive, $A\bar{x}$ is zero, and $A\bar{y}$ is positive. The area of the semicircle is assigned a negative value because it must be subtracted from the area of the rectangle. Because $\bar{x}$ for the semicircle is also negative, its $A\bar{x}$ is positive; however, $\bar{y}$ is positive, so that its $A\bar{y}$ is negative. The area of the triangle is also assigned a negative value, but $\bar{x}$ and $\bar{y}$ are both positive, resulting in negative values for both $A\bar{x}$ and $A\bar{y}$.

According to the tabulated results, the coordinates of the centroid of the composite area are

$$\bar{x} = \frac{\Sigma A\bar{x}}{\Sigma A} = \frac{+25.23 \times 10^6}{+378.6 \times 10^3} = +66.6 \text{ mm} \qquad \textit{Answer}$$

$$\bar{y} = \frac{\Sigma A\bar{y}}{\Sigma A} = \frac{+116.77 \times 10^6}{+378.6 \times 10^3} = +308 \text{ mm} \qquad \textit{Answer}$$

Because $\bar{x}$ and $\bar{y}$ are both positive, the centroid of the composite area lies in the first quadrant of the coordinate plane.

Sample Problem 8.3

Using the method of composite curves, determine the centroidal coordinates of the line in Fig. (a) that consists of the circular arc ①, and the straight lines ② and ③.

Solution

The length and centroidal coordinates of the circular arc can be calculated using Table 8.2; the results are shown in Fig. (b). Figure (c) displays the properties of the two straight line segments; the centroidal coordinates are at the midpoints of the segments.

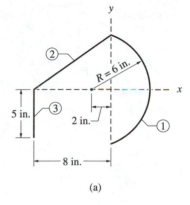

(a)

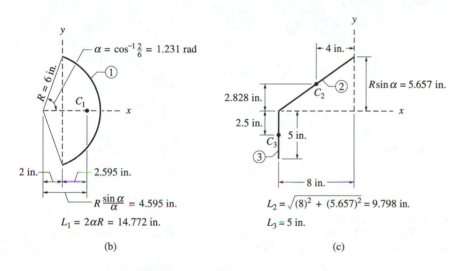

(b)

(c)

It is convenient to organize the analysis in tabular form, as follows:

Segment	Length L (in.)	$\bar{x}$ (in.)	$L\bar{x}$ (in.^{2})	$\bar{y}$ (in.)	$L\bar{y}$ (in.^{2})
1	14.772	+2.595	+38.33	0	0
2	9.798	−4.0	−39.19	+2.828	+27.71
3	5.0	−8.0	−40.0	−2.5	−12.50
Σ	29.570	$\cdots$	−40.86	$\cdots$	+15.21

363

Therefore, the coordinates of the centroid of the composite curve are

$$\bar{x} = \frac{\Sigma L\bar{x}}{\Sigma L} = \frac{-40.86}{29.570} = -1.382 \text{ in.} \qquad Answer$$

$$\bar{y} = \frac{\Sigma L\bar{y}}{\Sigma L} = \frac{+15.21}{29.570} = +0.514 \text{ in.} \qquad Answer$$

Because $\bar{x}$ is negative and $\bar{y}$ is positive, the centroid of the composite curve lies in the third quadrant of the coordinate plane.

Sample Problem 8.4

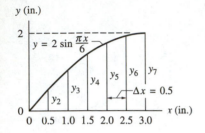

Using numerical integration (Simpson's rule), determine the centroidal coordinates of the sine curve shown in the figure. Use six panels, each of width $\Delta x = 0.5$ in.

Solution

Simpson's rule can be summarized as follows (see Appendix A).

$$\int_a^b f(x)\, dx \approx \sum_{i=1}^{n+1} W_i f_i \qquad \text{(a)}$$

where n is the number of panels of width Δx (n must be an even number) and W_i are the weights, given by

$$W_1 = W_{n+1} = \frac{\Delta x}{3}$$

$$\left.\begin{array}{ll} W_i = \dfrac{4\Delta x}{3} & i \text{ even} \\[2mm] W_i = \dfrac{2\Delta x}{3} & i \text{ odd} \end{array}\right\} \quad 2 \le i \le n$$

In Eq. (a) the expression

$$\sum_{i=1}^{n+1} W_i f_i$$

is called the *weighted summation*.

The integrals to be evaluated in the calculation of the centroidal coordinates of a plane curve are given by Eqs. (8.4) and (8.5):

$$L = \int_{\mathcal{L}} ds \qquad Q_y = \int_{\mathcal{L}} x\, ds \qquad Q_x = \int_{\mathcal{L}} y\, ds \qquad \text{(b)}$$

Substituting $ds = (ds/dx)\, dx$ and applying Simpson's rule, the integrals in Eqs. (b) become

$$L = \int_{\mathcal{L}} \left(\frac{ds}{dx}\right) dx \approx \sum_{i=1}^{n+1} W_i \left(\frac{ds}{dx}\right)_i$$

$$Q_y = \int_{\mathcal{L}} \left(x\frac{ds}{dx}\right) dx \approx \sum_{i=1}^{n+1} W_i x_i \left(\frac{ds}{dx}\right)_i \qquad \text{(c)}$$

$$Q_x = \int_{\mathcal{L}} \left(y\frac{ds}{dx}\right) dx \approx \sum_{i=1}^{n+1} W_i y_i \left(\frac{ds}{dx}\right)_i$$

For our problem, the weighted summations in Eqs. (c) are to compute using $n = 6$ and $\Delta x = 0.5$ in. The values of y_i and $(ds/dx)_i$ can be obtained from the following sequence of computations (see Fig. 8.5).

$$y_i = 2 \sin \frac{\pi x_i}{6} \qquad \left(\frac{dy}{dx}\right)_i = \frac{\pi}{3} \cos \frac{\pi x_i}{6} \qquad \left(\frac{ds}{dx}\right)_i = \sqrt{1 + \left(\frac{dy}{dx}\right)_i^2}$$

The numerical computations giving the weighted summations are contained in the following table.

i	x (in.)	y (in.)	dy/dx	ds/dx	$x(ds/dx)$ (in.)	$y(ds/dx)$ (in.)	W (in.)
1	0.0	0.0000	1.0472	1.4480	0.0000	0.0000	0.5/3
2	0.5	0.5176	1.0115	1.4224	0.7112	0.7363	4(0.5/3)
3	1.0	1.0000	0.9069	1.3500	1.3500	1.3500	2(0.5/3)
4	1.5	1.4142	0.7405	1.2443	1.8665	1.7597	4(0.5/3)
5	2.0	1.7321	0.5236	1.1288	2.2576	1.9551	2(0.5/3)
6	2.5	1.9319	0.2710	1.0361	2.5902	2.0016	4(0.5/3)
7	3.0	2.0000	0.0000	1.0000	3.0000	2.0000	0.5/3

Substituting the values from this table into Eqs. (c) yields

$$L \approx \frac{0.5}{3}[1(1.4480) + 4(1.4224) + 2(1.3500) + 4(1.2443) + 2(1.1288)$$

$$+ 4(1.0361) + 1(1.0000)] = 3.7028 \text{ in.}$$

$$Q_y \approx \frac{0.5}{3}[1(0) + 4(0.7112) + 2(1.3500) + 4(1.8665) + 2(2.2576)$$

$$+ 4(2.5902) + 1(3.0000)] = 5.1478 \text{ in.}^2$$

$$Q_x \approx \frac{0.5}{3}[1(0) + 4(0.7363) + 2(1.3500) + 4(1.7597) + 2(1.9551)$$

$$+ 4(2.0016) + 1(2.0000)] = 4.4334 \text{ in.}^2$$

from which we obtain

$$\bar{x} = \frac{Q_y}{L} = \frac{5.1478}{3.7028} = 1.390 \text{ in.} \qquad\qquad \textit{Answer}$$

$$\bar{y} = \frac{Q_x}{L} = \frac{4.4334}{3.7028} = 1.197 \text{ in.} \qquad\qquad \textit{Answer}$$

Problems

8.1–8.6   Use integration to determine the coordinates of the centroid of the plane region shown.

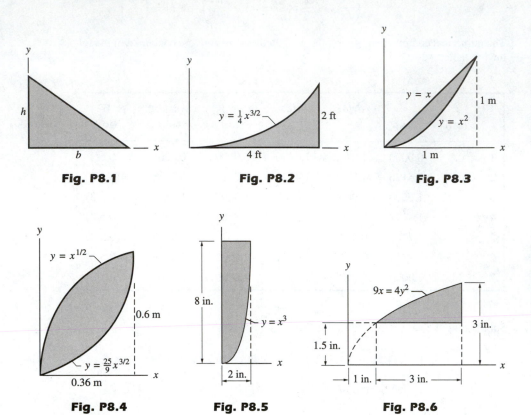

Fig. P8.1   **Fig. P8.2**   **Fig. P8.3**

Fig. P8.4   **Fig. P8.5**   **Fig. P8.6**

8.7   (a) Using integration, locate the centroid of the area under the nth order parabola in terms of b, h, and n (n is a positive integer). (b) Check the result of part (a) with Table 8.1 for the case $n = 2$.

8.8   Use integration to compute the coordinates of the centroid of the triangle. Check your results with Table 8.1.

8.9   Determine the y-coordinate of the centroid of the semicircular segment, given that $a = 18$ in. and $\alpha = 45°$.

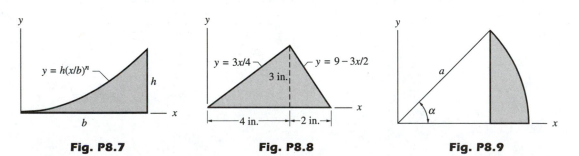

Fig. P8.7   **Fig. P8.8**   **Fig. P8.9**

8.10 (a) Use integration to locate the centroid of the shaded region in terms of R and t. (b) Show that when $t \to 0$ the result of part (a) agrees with that given in Table 8.2 for a quarter circular arc.

*8.11 The plane region lies between the x-axis and the cardioid $r = a(1 + \cos \theta)$, $0 \le \theta \le \pi$ rad. Calculate the centroidal y-coordinate by integration.

8.12 Use integration to locate the centroid of the quarter circular arc shown in Table 8.2.

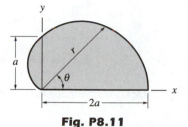

Fig. P8.11

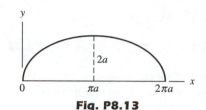

Fig. P8.13

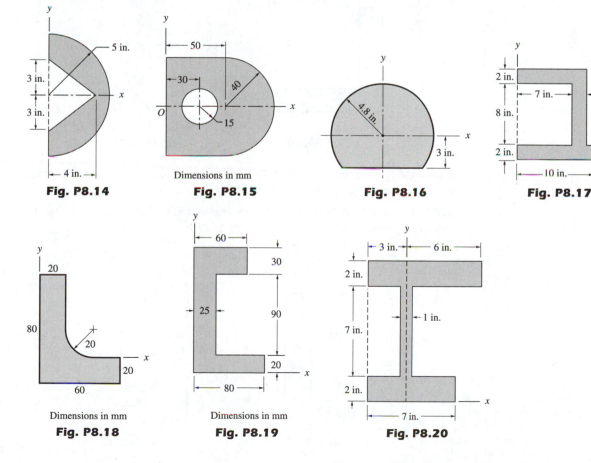

Fig. P8.10

*8.13 The parametric equations of the plane curve known as a cycloid are $x = a(\theta - \sin \theta)$ and $y = a(1 - \cos \theta)$. Use integration to find the coordinates of the centroid of the cycloid obtained by varying θ from 0 to 2π rad.

8.14–8.23 Use the method of composite areas to calculate the centroidal coordinates of the plane regions shown.

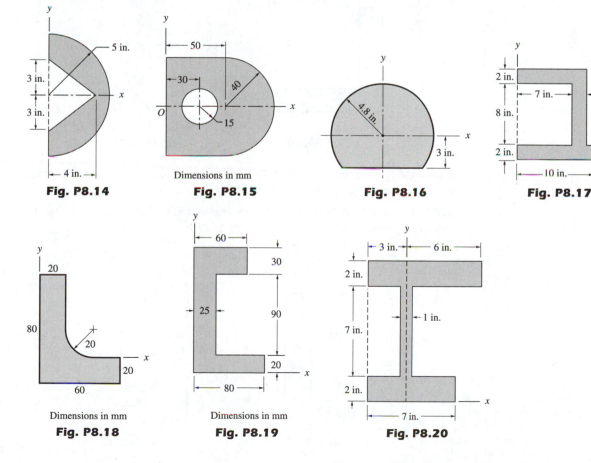

Fig. P8.14

Fig. P8.15
Dimensions in mm

Fig. P8.16

Fig. P8.17

Fig. P8.18
Dimensions in mm

Fig. P8.19
Dimensions in mm

Fig. P8.20

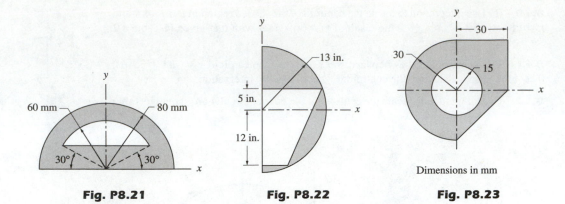

Fig. P8.21   **Fig. P8.22**   **Fig. P8.23**

8.24 Compute the centroidal coordinates of the L-shaped region in terms of b and t using the method of composite areas.

8.25 By the method of composite areas, derive the expression for the centroidal x-coordinate of the circular segment in terms of R and α.

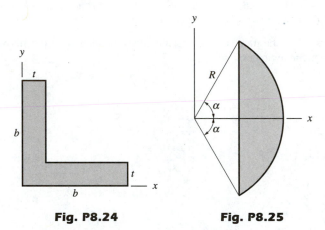

Fig. P8.24   **Fig. P8.25**

8.26 Using composite areas, find the dimension h for which the centroid C of the plane region is located at the position shown.

8.27 Given that the centroid of the plane region is at C, find the radius R. Use the method of composite areas.

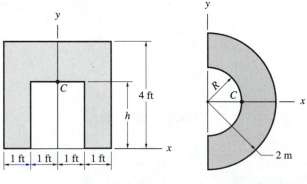

Fig. P8.26   **Fig. P8.27**

8.28–8.33   Using the method of composite curves, locate the centroids of the plane curves shown.

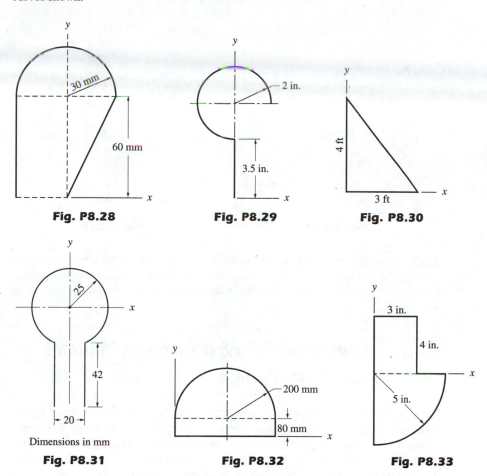

Fig. P8.28

Fig. P8.29

Fig. P8.30

Fig. P8.31

Dimensions in mm

Fig. P8.32

Fig. P8.33

8.34   Determine the ratio a/b for which the centroid of the composite curve will be located at point O.

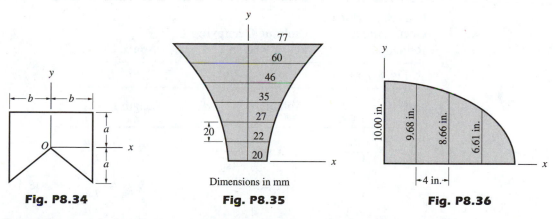

Fig. P8.34

Dimensions in mm

Fig. P8.35

Fig. P8.36

8.35   Use numerical integration to locate the centroid of the symmetric plane region.

8.36   Determine the centroidal coordinates of the plane region by numerical integration.

8.37   Compute the y-coordinate of the centroid of the parabola shown, the equation of which is $y = 40(1 - x^2/3600)$, where x and y are in inches. Use numerical integration with $\Delta x = 15$ in.

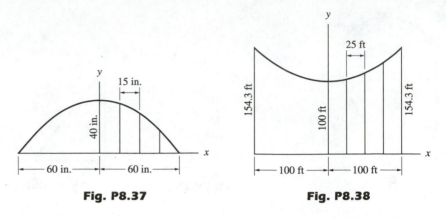

Fig. P8.37   **Fig. P8.38**

8.38   The equation of the catenary shown is $y = 100\cosh(x/100)$ where x and y are measured in feet (the catenary is the shape of a cable suspended between two points). Locate the y-coordinate of the centroid of the catenary by numerical integration using $\Delta x = 25$ ft.

8.3   *Centroids of Curved Surfaces, Volumes, and Space Curves*

The centroids of curved surfaces, volumes, and space curves are defined by expressions that are analogous to those used for plane figures. The only difference is that three coordinates, instead of two, are required to locate the centroids for three-dimensional shapes. The following table lists the expressions that define the centroidal coordinates for various three-dimensional shapes.

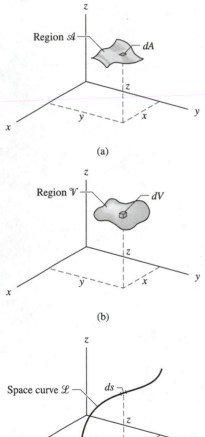

Region $\mathcal{A}$ — dA

(a)

Region $\mathcal{V}$ — dV

(b)

Space curve $\mathcal{L}$ — ds

(c)

Fig. 8.7

Curved Surface Occupying a Region $\mathcal{A}$	Volume Occupying a Region $\mathcal{V}$	Space Curve $\mathcal{L}$	
Fig. 8.7(a)	Fig. 8.7(b)	Fig. 8.7(c)	
Area $A = \displaystyle\int_{\mathcal{A}} dA$	Volume $V = \displaystyle\int_{\mathcal{V}} dV$	Length $L = \displaystyle\int_{\mathcal{L}} ds$	
$\bar{x} = \dfrac{\int_{\mathcal{A}} x\, dA}{A}$	$\bar{x} = \dfrac{\int_{\mathcal{V}} x\, dV}{V}$	$\bar{x} = \dfrac{\int_{\mathcal{L}} x\, ds}{L}$	
$\bar{y} = \dfrac{\int_{\mathcal{A}} y\, dA}{A}$	$\bar{y} = \dfrac{\int_{\mathcal{V}} y\, dV}{V}$	$\bar{y} = \dfrac{\int_{\mathcal{L}} y\, ds}{L}$	(8.10)
$\bar{z} = \dfrac{\int_{\mathcal{A}} z\, dA}{A}$	$\bar{z} = \dfrac{\int_{\mathcal{V}} z\, dV}{V}$	$\bar{z} = \dfrac{\int_{\mathcal{L}} z\, ds}{L}$	

The term $\int_{\mathcal{A}} x\,dA$ is sometimes labeled Q_{yz} and is referred to as the *first moment of the area relative to the yz-plane*. Similarly, $Q_{xz} = \int_{\mathcal{A}} y\,dA$ and $Q_{xy} = \int_{\mathcal{A}} z\,dA$ are called the *first moments of the area relative to the xz- and xy-planes,* respectively. Extensions of this notation and terminology to volumes and space curves are obvious.

The definitions in Eqs. (8.10) assume that the differential element (dA, dV, or ds) is located at the point that has coordinates x, y, and z. For other choices of elements, such as those occurring in single or double integration, it may be necessary to replace x, y, or z in Eqs. (8.10) with the centroidal coordinates of the element: $\bar{x}_{el}$, $\bar{y}_{el}$, $\bar{z}_{el}$. Because this integration procedure is similar to that described in Art. 8.2, it is not repeated here.

Symmetry of a body can play an important role in the determination of its centroid, as explained in the following:

- *If a volume has a plane of symmetry, its centroid lies in that plane.* (The analogous statement for plane areas has been proven in Art. 8.2; the proof for volumes is essentially the same.)
- *If a volume has two planes of symmetry that intersect along a line, its centroid lies on that line.* (The proof of this statement follows directly from the preceding symmetry argument.)

  This symmetry argument is particularly useful when determining the centroids of volumes of revolution. For example, consider the volume shown in Fig. 8.8 that is generated by rotating a plane area about the *y*-axis. Because any plane that contains the *y*-axis is a plane of symmetry, the centroid C of the volume must lie on the *y*-axis; that is, $\bar{x} = 0$ and $\bar{z} = 0$.

- *The centroid of the volume of a prismatic body is located at the centroid of the cross-sectional area that forms the middle plane of the volume.*

  To prove this statement, consider Fig. 8.9, which shows a prismatic body of thickness h that occupies the region $\mathcal{V}$. The xy-coordinate plane coincides with the middle plane of the body, and the z-axis is the centroidal axis (passes through the centroids of the cross-sectional areas). The foregoing statement will be proven if we can show that the origin of the coordinate system is the centroid C of the body—that is, if we can show that $\bar{x} = \bar{y} = \bar{z} = 0$. There is no question that C lies on the xy-plane ($\bar{z} = 0$), because it is the plane of symmetry. To show that

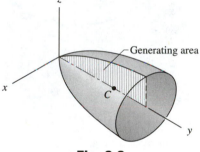

Generating area

Fig. 8.8

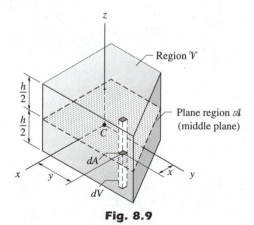

Region $\mathcal{V}$

Plane region $\mathcal{A}$ (middle plane)

dA

dV

Fig. 8.9

$\bar{x} = \bar{y} = 0$, it is sufficient to demonstrate that $\int_{\mathcal{V}} y \, dV = \int_{\mathcal{V}} x \, dV = 0$. Using the differential volume element $dV = h \, dA$ shown in Fig. 8.9, we get $\int_{\mathcal{V}} y \, dV = \int_{\mathcal{A}} y \, (h \, dA) = h \int_{\mathcal{A}} y \, dA = 0$, where $\mathcal{A}$ is the plane region of the cross section. The last equality follows from the knowledge that $\int_{\mathcal{A}} y \, dA = 0$ if y is measured from the centroid of the cross-sectional area. It can be proven in a similar manner that $\int_{\mathcal{A}} x \, dA = 0$.

Identical symmetry arguments can be used to locate the centroids of surfaces and space curves. For example, knowing the centroidal coordinates of a semicircular arc, shown in Fig. 8.10(a), we can immediately deduce the centroidal coordinates of the half cylindrical surface in Fig. 8.10(b).

The method of composite shapes also applies to curved surfaces, volumes, and space curves. The expressions for the centroidal coordinates of composite surfaces and curves can be obtained by extending Eqs. (8.8) and (8.9) to three dimensions. The equations for composite volumes can be written by analogy with composite areas. The results are

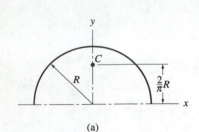

(a)

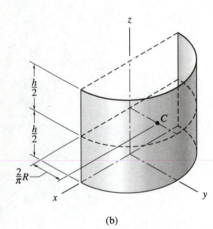

(b)

Fig. 8.10

Composite Areas

$$\bar{x} = \frac{\Sigma_i A_i \bar{x}_i}{\Sigma_i A_i} \qquad \bar{y} = \frac{\Sigma_i A_i \bar{y}_i}{\Sigma_i A_i} \qquad \bar{z} = \frac{\Sigma_i A_i \bar{z}_i}{\Sigma_i A_i} \qquad (8.11)$$

Composite Volumes

$$\bar{x} = \frac{\Sigma_i V_i \bar{x}_i}{\Sigma_i V_i} \qquad \bar{y} = \frac{\Sigma_i V_i \bar{y}_i}{\Sigma_i V_i} \qquad \bar{z} = \frac{\Sigma_i V_i \bar{z}_i}{\Sigma_i V_i} \qquad (8.12)$$

Composite Curves

$$\bar{x} = \frac{\Sigma_i L_i \bar{x}_i}{\Sigma_i L_i} \qquad \bar{y} = \frac{\Sigma_i L_i \bar{y}_i}{\Sigma_i L_i} \qquad \bar{z} = \frac{\Sigma_i L_i \bar{z}_i}{\Sigma_i L_i} \qquad (8.13)$$

Note that these expressions are identical to Eqs. (8.10) except that the integrations have been replaced by summations.

In order to facilitate the application of Eqs. (8.11) and (8.12), the properties of some basic surfaces and volumes are shown in Tables 8.3 and 8.4.

Right tetrahedron	Pyramid
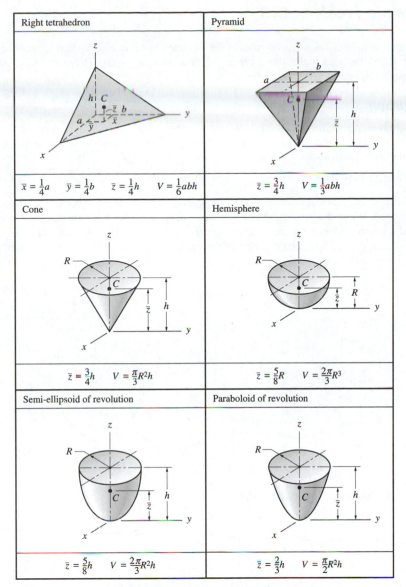	
$\bar{x} = \frac{1}{4}a$ $\bar{y} = \frac{1}{4}b$ $\bar{z} = \frac{1}{4}h$ $V = \frac{1}{6}abh$	$\bar{z} = \frac{3}{4}h$ $V = \frac{1}{3}abh$
Cone	Hemisphere
$\bar{z} = \frac{3}{4}h$ $V = \frac{\pi}{3}R^2h$	$\bar{z} = \frac{5}{8}R$ $V = \frac{2\pi}{3}R^3$
Semi-ellipsoid of revolution	Paraboloid of revolution
$\bar{z} = \frac{5}{8}h$ $V = \frac{2\pi}{3}R^2h$	$\bar{z} = \frac{2}{3}h$ $V = \frac{\pi}{2}R^2h$

Table 8.3   *Centroids of Volumes*

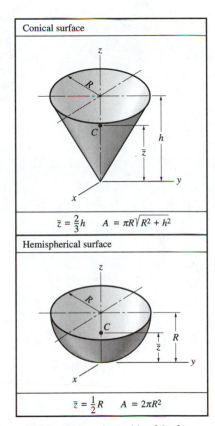

Conical surface
$\bar{z} = \frac{2}{3}h$ $A = \pi R\sqrt{R^2 + h^2}$
Hemispherical surface
$\bar{z} = \frac{1}{2}R$ $A = 2\pi R^2$

Table 8.4   *Centroids of Surfaces*

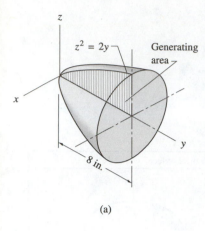

$z^2 = 2y$

Generating area

x

z

y

8 in.

(a)

Sample Problem 8.5

Determine the centroidal coordinates of the volume shown in Fig. (a) that is generated by rotating the area under the curve $z^2 = 2y$ about the y-axis. The coordinates are measured in inches.

Solution

By symmetry, $\bar{x} = \bar{z} = 0$. Integration must be used to find $\bar{y}$. There are two convenient single-integration techniques for volumes of revolution; the method of thin disks and the method of thin shells.

Method I: Thin Disks

In this method, a differential element of the generating area is the vertical strip shown in Fig. (b). When the generating area is rotated about the y-axis to form the volume of revolution, the differential area element generates the thin disk of thickness dy that is shown in Fig. (c). Thus, the properties of the volume may be determined by integrating the properties of the disk.

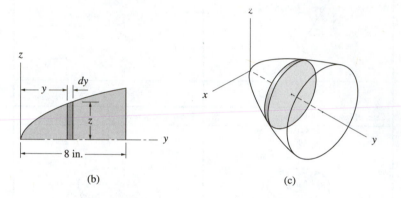

(b)　　　　　　　　　　　　(c)

The volume of the disk is

$$dV = \pi z^2 \, dy = 2\pi y \, dy$$

Integrating to determine the volume V, we obtain

$$V = 2\pi \int_0^8 y \, dy = 2\pi \left[\frac{y^2}{2} \right]_0^8 = 64\pi \text{ in.}^3$$

Because the distance of the disk from the xz-plane is y, the corresponding first moment is $dQ_{xz} = y \, dV$, which on substituting the expression for dV becomes

$$dQ_{xz} = y(2\pi y \, dy) = 2\pi y^2 \, dy$$

Integration yields for the first moment of the volume

$$Q_{xz} = 2\pi \int_0^8 y^2 \, dy = 2\pi \left[\frac{y^3}{3} \right]_0^8 = 341\pi \text{ in.}^4$$

Therefore, the centroidal coordinates of the volume are

$$\bar{y} = \frac{Q_{xz}}{V} = \frac{341\pi}{64\pi} = 5.33 \text{ in.} \qquad \bar{x} = \bar{z} = 0 \qquad \textit{Answer}$$

Method II: Thin Shells

In this method, a differential element of the generating area is the horizontal strip shown in Fig. (d). Rotation about the y-axis generates a shell of infinitesimal thickness dz, as shown in Fig. (e).

The volume of the thin shell (circumference × thickness × length) is

$$dV = 2\pi z \, dz \, (8 - y) = 2\pi z \left(8 - \frac{z^2}{2}\right) dz = \pi(16z - z^3) \, dz$$

Noting that the range of z is from 0 to 4 in., the volume is given by

$$V = \pi \int_0^4 (16z - z^3) \, dz = \pi \left[8z^2 - \frac{z^4}{4}\right]_0^4 = 64\pi \text{ in.}^3$$

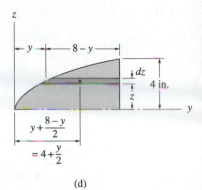

(d)

Referring to Fig. (d), we see that the y-coordinate of the centroid of the thin shell is $\bar{y}_{el} = 4 + (y/2) = 4 + (z^2/4)$, and its first moment with respect to the xz-plane becomes

$$dQ_{xz} = \bar{y}_{el} \, dV = \left(4 + \frac{z^2}{4}\right)\pi(16z - z^3) \, dz = \pi\left(64z - \frac{z^5}{4}\right) dz$$

Therefore, we obtain

$$Q_{xz} = \pi \int_0^4 \left(64z - \frac{z^5}{4}\right) dz = \pi \left[32z^2 - \frac{z^6}{24}\right]_0^4 = 341\pi \text{ in.}^4$$

and the centroidal coordinates of the volume become

$$\bar{y} = \frac{Q_{xz}}{V} = \frac{341\pi}{64\pi} = 5.33 \text{ in.} \qquad \bar{x} = \bar{z} = 0 \qquad \textit{Answer}$$

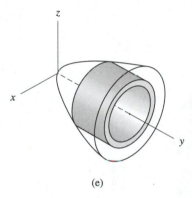

(e)

Of course, these answers are identical to those obtained by the method of thin disks.

Sample Problem 8.6

Locate the centroid of the hyperbolic paraboloid shown in Fig. (a) using (1) single integration with a differential volume element parallel to the xz-plane; and (2) double integration. The equation of the surface that bounds the volume is $(y^2/b^2) - (x^2/a^2) = z/c$.

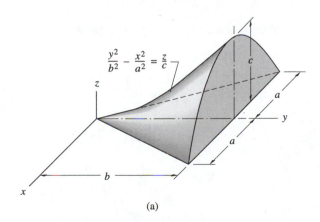

(a)

Solution

Noting that $\bar{x} = 0$ by symmetry, we see that integration is required to find $\bar{y}$ and $\bar{z}$ only.

Part 1: Single Integration

The dimensions of a differential volume element dV, parallel to the xz-plane, are shown in Fig. (b). Because the cross section of the element is a parabola, we may use Table 8.1 to determine dV and $\bar{z}_{el}$ (the volume and centroidal coordinate of the element).

$$dV = \frac{2}{3}\left(2\frac{a}{b}y\right)\left(\frac{cy^2}{b^2}\right)dy = \frac{4}{3}\frac{ac}{b^3}y^3\,dy$$

$$\bar{z}_{el} = \frac{2}{5}\frac{cy^2}{b^2}$$

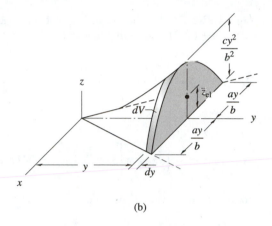

(b)

Because the limits of integration are $y = 0$ to $y = b$, the volume becomes

$$V = \int_{\mathcal{V}} dV = \frac{4}{3}\frac{ac}{b^3}\int_0^b y^3\,dy = \frac{abc}{3}$$

and the first moments with respect to the xz- and xy-planes are

$$Q_{xz} = \int_{\mathcal{V}} y\,dV = \frac{4}{3}\frac{ac}{b^3}\int_0^b y^4\,dy = \frac{4ab^2c}{15}$$

$$Q_{xy} = \int_{\mathcal{V}} \bar{z}_{el}\,dV = \left(\frac{4}{3}\frac{ac}{b^3}\right)\left(\frac{2}{5}\frac{c}{b^2}\right)\int_0^b y^5\,dy = \frac{4abc^2}{45}$$

Therefore, the centroidal coordinates of the hyperbolic paraboloid are

$$\bar{x} = 0$$

$$\bar{y} = \frac{Q_{xz}}{V} = \frac{4ab^2c/15}{abc/3} = \frac{4b}{5} \qquad \textit{Answer}$$

$$\bar{z} = \frac{Q_{xy}}{V} = \frac{4abc^2/45}{abc/3} = \frac{4c}{15}$$

Part 2: Double Integration

Using the double differential element of volume shown in Fig. (c), we have

$$dV = z\,dx\,dy \qquad \bar{z}_{el} = \frac{z}{2}$$

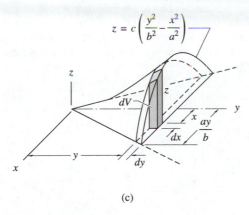

$$z = c\left(\frac{y^2}{b^2} - \frac{x^2}{a^2}\right)$$

(c)

Choosing to integrate first on x, then on y, and using the fact that the volume is symmetric with respect to the yz-plane, we have

$$V = \int_{\mathcal{V}} dV = 2\int_0^b \left(\int_0^{ay/b} z\,dx\right) dy$$

$$Q_{xz} = \int_{\mathcal{V}} y\,dV = 2\int_0^b \left(\int_0^{ay/b} yz\,dx\right) dy$$

$$Q_{xy} = \int_{\mathcal{V}} \bar{z}_{el}\,dV = 2\int_0^b \left(\int_0^{ay/b} \frac{z}{2}z\,dx\right) dy$$

Substituting $z = c[(y^2/b^2) - (x^2/a^2)]$ into the preceding expressions and performing the integrations yield the same results as found in Part 1.

Sample Problem 8.7

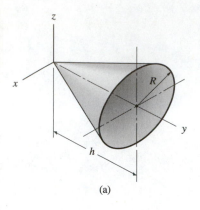

(a)

Locate the centroid of the conical surface shown in Fig. (a).

Solution

By symmetry, we see that $\bar{x} = \bar{z} = 0$. Because the conical surface is a surface of revolution, single integration can be used to calculate $\bar{y}$.

As shown in Fig. (b), the differential area element is taken to be the area of the "ring," technically known as a *frustum,* that is generated by rotating the line segment of differential length ds about the y-axis. The area of this differential element (circumference $\times$ slant height) is

$$dA = 2\pi z\, ds$$

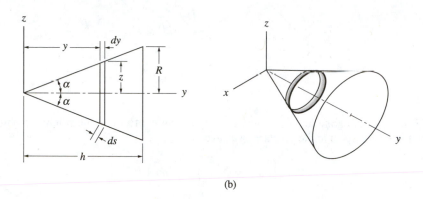

(b)

Letting 2α be the cone angle as shown in Fig. (b), we have $z = y\tan\alpha$ and $ds = dy/\cos\alpha$. Substituting these identities, together with $\tan\alpha = R/h$ and $\cos\alpha = h/\sqrt{R^2 + h^2}$, into the expressions for the differential area, we obtain

$$dA = \frac{2\pi R\sqrt{R^2 + h^2}}{h^2}\, y\, dy$$

Integrating to find the area of the conical surface, we obtain

$$A = \frac{2\pi R\sqrt{R^2 + h^2}}{h^2}\int_0^h y\, dy = \frac{2\pi R\sqrt{R^2 + h^2}}{h^2}\left[\frac{y^2}{2}\right]_0^h$$

$$= \pi R\sqrt{R^2 + h^2}$$

From Fig. (b), we see that y is the distance from the xz-plane to the differential area element. Therefore, its first moment relative to that plane is $dQ_{xz} = y\, dA$. Substituting the expression previously determined for dA, and integrating, the first moment of the conical surface becomes

$$Q_{xz} = \frac{2\pi R\sqrt{R^2 + h^2}}{h^2}\int_0^h y^2\, dy = \frac{2\pi R\sqrt{R^2 + h^2}}{h^2}\left[\frac{y^3}{3}\right]_0^h$$

$$= \frac{2}{3}\pi Rh\sqrt{R^2 + h^2}$$

Therefore, the centroidal coordinates of the conical surface are

$$\bar{x} = \bar{z} = 0 \qquad\qquad\qquad Answer$$

$$\bar{y} = \frac{Q_{xz}}{A} = \frac{(2/3)\pi Rh\sqrt{R^2 + h^2}}{\pi R\sqrt{R^2 + h^2}} = \frac{2h}{3}$$

Observe that these results agree with data given for a conical surface in Table 8.4.

Sample Problem 8.8

Use the method of composite volumes to determine the location of the centroid of the volume for the machine part shown in Fig. (a).

Solution

We note that $\bar{x} = 0$ because the yz-plane is a plane of symmetry of the volume.

To calculate $\bar{y}$ and $\bar{z}$, the machine part can be considered to be composed of the four volumes shown in Fig. (b): the rectangular solid ①, plus the semicylinder ②, plus the rectangular solid ③, minus the cylinder ④. Most centroidal coordinates of these volumes can be determined by symmetry; only $\bar{z}$ of volume ② must be found from Table 8.1.

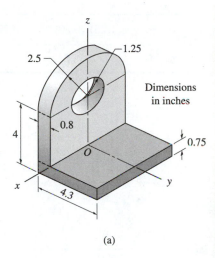

(a)

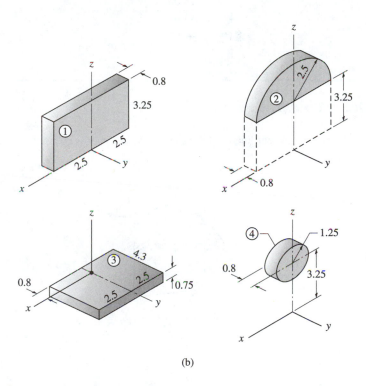

(b)

The computations of the volumes and first moments relative to the xy- and xz-planes are shown in the following table.

Part	Volume V (in.^{3})	$\bar{y}$ (in.)	$V\bar{y}$ (in.^{4})	$\bar{z}$ (in.)	$V\bar{z}$ (in.^{4})
1	$5(3.25)(0.8) = 13.000$	-0.40	-5.20	$+\dfrac{3.25}{2} = +1.625$	$+21.13$
2	$\dfrac{\pi(2.5)^2}{2}(0.8) = 7.854$	-0.40	-3.14	$+3.25 + \dfrac{4(2.5)}{3\pi} = +4.311$	$+33.86$
3	$5(4.30)(0.75) = 16.125$	$+\dfrac{4.30}{2} - 0.80 = +1.35$	$+21.77$	$-\dfrac{0.75}{2} = -0.375$	-6.05
4	$-\pi(1.25)^2(0.80) = -3.927$	-0.40	$+1.57$	$+3.250$	-12.76
Σ	33.052	$\cdots$	$+15.00$	$\cdots$	$+36.18$

Using the results displayed in the table, the centroidal coordinates of the machine part are

$$\bar{x} = 0$$

$$\bar{y} = \frac{\Sigma V\bar{y}}{V} = \frac{15.00}{33.05} = 0.454 \text{ in.} \qquad \textit{Answer}$$

$$\bar{z} = \frac{\Sigma V\bar{z}}{V} = \frac{36.18}{33.05} = 1.095 \text{ in.}$$

Sample Problem 8.9

Calculate the centroidal coordinates of the shaded surface shown in Fig. (a).

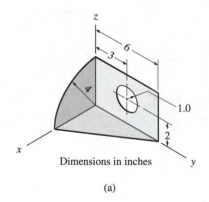

Dimensions in inches

(a)

Solution

The surface in Fig. (a) can be decomposed into the four plane areas in Fig. (b): the rectangle ①, plus the quarter circle ②, plus the triangle ③, minus the circle ④. The location of the centroid of each composite area can be found by symmetry or from Table 8.1.

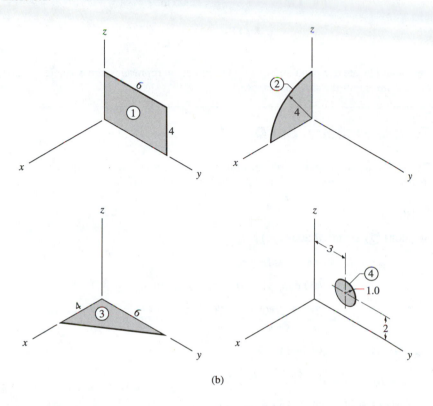

(b)

The following table lists the computations for the areas and first moments of the areas.

Part	Area A (in.^{2})	$\bar{x}$ (in.)	$A\bar{x}$ (in.^{3})	$\bar{y}$ (in.)	$A\bar{y}$ (in.^{3})	$\bar{z}$ (in.)	$A\bar{z}$ (in.^{3})
1	$4(6) = 24$	0	0	$+3$	$+72.00$	$+2$	$+48.00$
2	$\dfrac{\pi(4)^2}{4} = 4\pi$	$+\dfrac{4R}{3\pi} = +\dfrac{4(4)}{3\pi}$	$+21.33$	0	0	$+\dfrac{4R}{3\pi} = +\dfrac{4(4)}{3\pi}$	$+21.33$
3	$\dfrac{1}{2}(4)(6) = 12$	$+\dfrac{1}{3}(4)$	$+16.00$	$+\dfrac{1}{3}(6)$	$+24.00$	0	0
4	$-\pi(1)^2 = -\pi$	0	0	$+3$	-3π	$+2$	-2π
Σ	45.42	$\cdots$	$+37.33$	$\cdots$	$+86.58$	$\cdots$	$+63.05$

Therefore, the centroidal coordinates of the shaded surface shown in Fig. (a) are

$$\bar{x} = \frac{37.33}{45.42} = 0.822 \text{ in.}$$

$$\bar{y} = \frac{86.58}{45.42} = 1.906 \text{ in.} \qquad \textit{Answer}$$

$$\bar{z} = \frac{63.05}{45.42} = 1.388 \text{ in.}$$

You should locate this point on Fig. (a) to verify that it represents a reasonable location of the centroid for the shaded surface.

Sample Problem 8.10

Determine the centroidal coordinates for the composite curve made up of three segments: the semicircular arc ①, and the straight lines ② and ③.

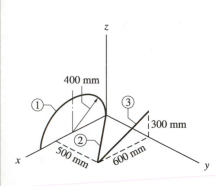

Solution

Segment ① (semicircular arc)

$$L_1 = \pi R = 400\pi \text{ mm}$$

$$\bar{x}_1 = 400 \text{ mm}, \quad \bar{y}_1 = 0 \qquad \text{(by inspection)}$$

$$\bar{z}_1 = \frac{2R}{\pi} = \frac{2(400)}{\pi} = \frac{800}{\pi} \text{ mm} \qquad \text{(from Table 8.2)}$$

Segment ② (straight line)

$$L_2 = \sqrt{500^2 + 600^2} = 781.0 \text{ mm}$$

$$\bar{x}_2 = 300 \text{ mm}, \qquad \bar{y}_2 = 250 \text{ mm}, \qquad \bar{z}_2 = 0 \qquad \text{(by inspection)}$$

Segment ③ (straight line)

$$L_3 = \sqrt{600^2 + 300^2} = 670.8 \text{ mm}$$

$$\bar{x}_3 = 300 \text{ mm}, \qquad \bar{y}_3 = 500 \text{ mm}, \qquad \bar{z}_3 = 150 \text{ mm} \qquad \text{(by inspection)}$$

The remaining computations are carried out in the following table.

Segment	Length L (mm)	$\bar{x}$ (mm)	$L\bar{x}$ (mm²)	$\bar{y}$ (mm)	$L\bar{y}$ (mm²)	$\bar{z}$ (mm)	$L\bar{z}$ (mm²)
1	400π	400	502.7×10^3	0	0	$\frac{800}{\pi}$	320.0×10^3
2	781.0	300	234.3×10^3	250	195.3×10^3	0	0
3	670.8	300	201.2×10^3	500	335.4×10^3	150	100.6×10^3
Σ	2708.4	$\cdots$	938.2×10^3	$\cdots$	530.7×10^3	$\cdots$	420.6×10^3

Therefore, the centroidal coordinates of the composite curve are

$$\bar{x} = \frac{\Sigma L \bar{x}}{\Sigma L} = \frac{938.2 \times 10^3}{2708.4} = 346 \text{ mm}$$

$$\bar{y} = \frac{\Sigma L \bar{y}}{\Sigma L} = \frac{530.7 \times 10^3}{2708.4} = 196 \text{ mm} \qquad \textit{Answer}$$

$$\bar{z} = \frac{\Sigma L \bar{z}}{\Sigma L} = \frac{420.6 \times 10^3}{2708.4} = 155 \text{ mm}$$

You should locate this point on the figure to verify that it represents a reasonable location for the centroid of the composite curve.

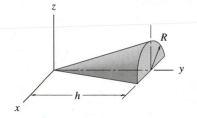

Fig. P8.39, P8.40

Problems

8.39 Use integration to locate the centroid of the volume of the hemisphere. Compare your results with Table 8.3.

8.40 By integration, find the centroid of the surface of the hemisphere. Compare your result with Table 8.4.

8.41 Locate the centroid of the volume obtained by revolving the triangle about the x-axis. Use integration.

8.42 Repeat Prob. 8.41 assuming that the triangle is revolved about the y-axis.

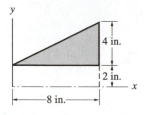

Fig. P8.41, P8.42

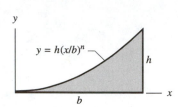

Fig. P8.43, P8.44

8.43 Use integration to find the centroidal coordinates for the volume obtained by revolving the area shown about the x-axis.

8.44 Repeat Prob. 8.43 assuming that the area is revolved about the y-axis.

8.45 Verify the centroidal z-coordinate of the pyramid shown in Table 8.3 by integration.

8.46 Use integration to compute the z-coordinate of the centroid of the half cone.

8.47 Determine the centroidal z-coordinate of the curved surface of the half cone by integration.

8.48 By integration, determine the x- and y-centroidal coordinates for the volume shown.

8.49 Locate the centroid of the volume between the parabolic surface and the xy-plane using integration.

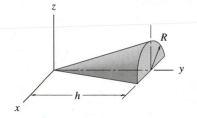

Fig. P8.46, P8.47

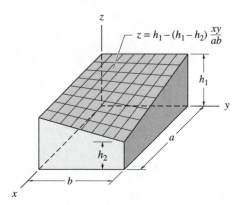

Fig. P8.48

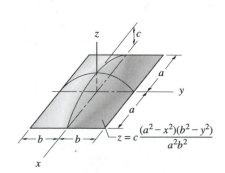

Fig. P8.49

8.50 Use integration to locate the centroid of the curved surface.

8.51 By integration, determine the centroidal coordinates of the curve connecting points A and B.

8.52–8.57 By the method of composite volumes, determine the centroidal coordinates of the volume.

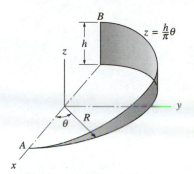

Fig. P8.50, P8.51

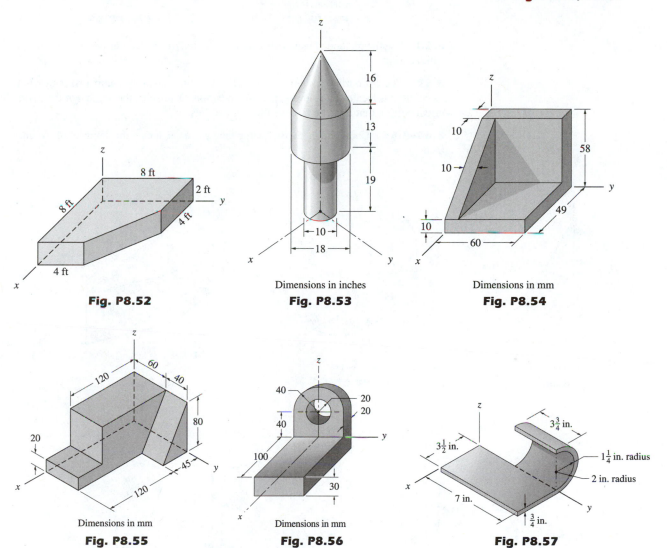

Fig. P8.52

Dimensions in inches
Fig. P8.53

Dimensions in mm
Fig. P8.54

Dimensions in mm
Fig. P8.55

Dimensions in mm
Fig. P8.56

Fig. P8.57

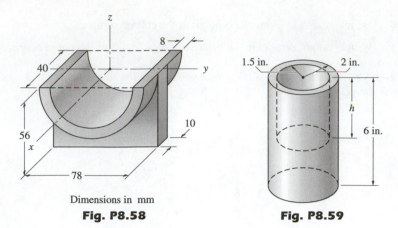

Dimensions in mm

Fig. P8.58

Fig. P8.59

8.58 Use the method of composite volumes to find the centroidal z-coordinate of the split bearing.

8.59 The cylindrical container will have maximum stability against tipping when its centroid is located at its lowest possible position. Determine the depth h of the cylindrical portion that must be removed to achieve this.

8.60–8.65 Using the method of composite surfaces, locate the centroid of the surface.

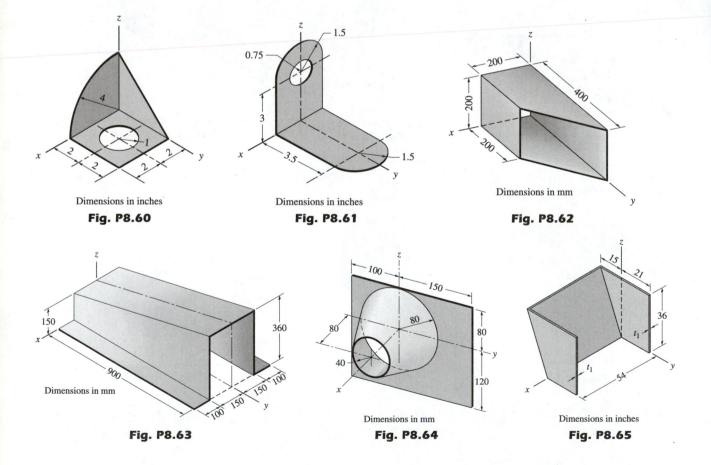

Dimensions in inches

Fig. P8.60

Dimensions in inches

Fig. P8.61

Dimensions in mm

Fig. P8.62

Dimensions in mm

Fig. P8.63

Dimensions in mm

Fig. P8.64

Dimensions in inches

Fig. P8.65

8.66   The picture board and its triangular supporting bracket form a composite surface. Calculate the height h of the support that minimizes the centroidal z-coordinate of the assembly.

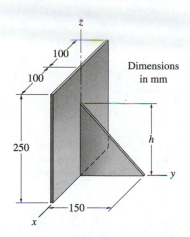

Fig. P8.66

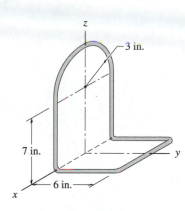

Fig. P8.67

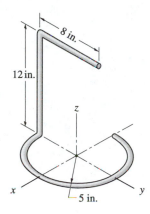

Fig. P8.68

8.67–8.69   By the method of composite curves, locate the centroid of the wire figure.

8.70   Use numerical integration to find the centroid of the volume generated by revolving the area shown about the x-axis.

8.71   Repeat Prob. 8.70 assuming that the area is revolved about the y-axis.

8.72   Locate the centroid of the volume generated by revolving the area shown about the line AB. Use numerical integration.

8.73   (a) Repeat Prob. 8.72 assuming that the area is revolved about the x-axis. (b) Check your result in part (a) with Table 8.3 knowing that the curve OB is a parabola.

8.74   Use numerical integration with $\Delta x = 1.0$ m to locate the centroid of the surface generated by revolving the parabola about the y-axis.

8.75   Repeat Prob. 8.74 assuming that the parabola OA is revolved about the x-axis.

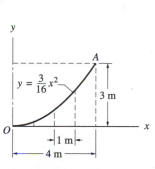

Fig. P8.69

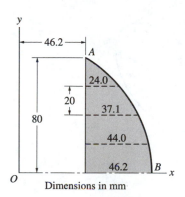

Fig. P8.70, P8.71

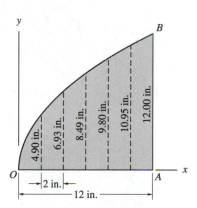

Fig. P8.72, P8.73

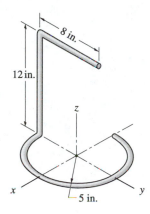

Fig. P8.74, P8.75

8.4 *Theorems of Pappus-Guldinus*

The theorems of Pappus-Guldinus* provide relatively simple methods for calculating surface areas and volumes of bodies of revolution, utilizing first moments of curves and areas.

Theorem I *The surface area generated by revolving a plane curve through 360° about a nonintersecting axis in the plane of the curve is equal to 2π times the first moment of the curve about the axis of revolution.*

Proof

Consider the curve $\mathcal{L}$ with length L, shown in Fig. 8.11, that lies in the xy-plane. When this curve is rotated through 360° about the x-axis, the area of the ring generated by the differential curve length ds is $dA = 2\pi y\,ds$. Therefore, the surface area generated by the entire curve $\mathcal{L}$ becomes

$$A = 2\pi \int_{\mathcal{L}} y\,ds = 2\pi Q_x \qquad (8.14)$$

where $Q_x = \int_{\mathcal{L}} y\,ds$ is the first moment of the curve about the x-axis. This completes the proof.

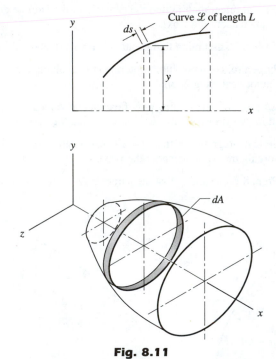

Fig. 8.11

---

*Named after the Greek geometrician Pappus (fourth century A.D.) and the Swiss mathematician Paul Guldinus (1577–1643).

Theorem II   *The volume generated by revolving a plane area through 360° about a nonintersecting axis in the plane of the area is equal to 2π times the first moment of the area about the axis of revolution.*

Proof

Consider the region $\mathcal{A}$ with area A, shown in Fig. 8.12, that lies in the xy-plane. When this area is rotated through 360° about the x-axis, the volume generated by the differential area dA is $dV = 2\pi y\, dA$. Therefore, the volume generated by the entire area is

$$V = 2\pi \int_{\mathcal{A}} y\, dA = 2\pi Q_x \qquad (8.15)$$

where $Q_x = \int_{\mathcal{A}} y\, dA$ is the first moment of the area about the x-axis. This completes the proof.

Note that if the generating curve or area is revolved through an angle less than 360°, this angle, measured in radians, should replace 2π in Eqs. (8.14) or (8.15).

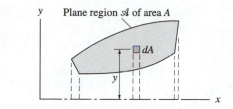

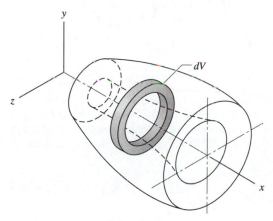

Fig. 8.12

Sample Problem 8.11

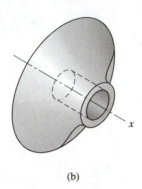

(a)

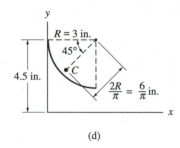

(b)

The area shown in Fig. (a) is revolved around the x-axis to form the body shown in Fig. (b). Using the theorems of Pappus-Guldinus, calculate (1) the volume of the body; and (2) the area of the curved surface of the body.

Solution

Part 1

By Theorem II, the volume of the body is $V = 2\pi Q_x$, where Q_x is the first moment of the area in Fig. (a) about the x-axis. By the method of composite areas, Q_x equals the first moment of a rectangle minus the first moment of a quarter circle, both taken about the x-axis—see Fig. (c). Table 8.1 has been used to locate the centroid for the quarter circle, and the centroid for the rectangle has been determined from symmetry. Therefore, we obtain

$$Q_x = \Sigma A\bar{y} = (3 \times 3.5)(2.75) - \frac{\pi(3)^2}{4}\left(4.5 - \frac{4}{\pi}\right) = 6.066 \text{ in.}^3$$

Hence the volume of the body is

$$V = 2\pi Q_x = 2\pi(6.066) = 38.1 \text{ in.}^3 \qquad \qquad \text{Answer}$$

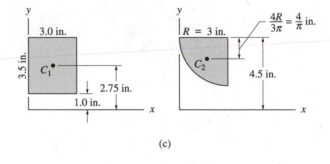

(c)

Part 2

The curved surface of the body in Fig. (b) is generated by revolving the quarter circular arc shown in Fig. (d) about the x-axis. By Theorem I, this area equals $2\pi Q_x$, where Q_x is the first moment of the arc about the x-axis. The location of the centroid for the arc can be found in Table 8.3. Using $L = \pi R/2$ for the length of the arc and referring to Fig. (d), we obtain

$$Q_x = L\bar{y} = \frac{\pi(3)}{2}\left(4.5 - \frac{6}{\pi}\sin 45°\right) = 14.842 \text{ in.}^2$$

The area of the curved surface is

$$A = 2\pi Q_x = 2\pi(14.842) = 93.3 \text{ in.}^2 \qquad \qquad \text{Answer}$$

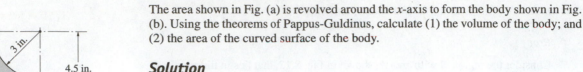

(d)

Problems

8.76 A 4-in. diameter hole is drilled in the conical frustum. Calculate the volume and the surface area of the resulting body.

8.77 Find the volume of the tapered washer that is generated when the rectangular area is revolved about the axis AB.

8.78 A solid of revolution is formed by rotating the plane area shown about the y-axis. Determine the surface area and the volume of the solid.

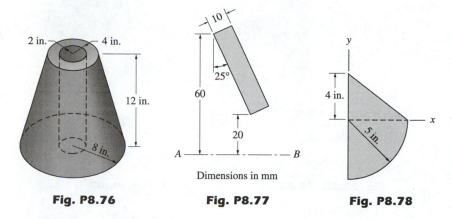

| Fig. P8.76 | Fig. P8.77 | Fig. P8.78 |

8.79 Compute the volume of the spherical cap that is formed when the circular segment is revolved about the y-axis.

8.80 Calculate the surface area of the truncated sphere that is formed by rotating the circular arc AB about the y-axis.

8.81 The rim of a steel V-belt pulley is formed by rotating the plane area shown about the axis AB. Find the mass of the rim, given that the mass density of steel is 7850 kg/m³.

8.82 Compute the volume of the bowl.

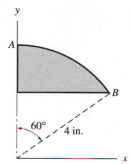

Fig. P8.79, P8.80

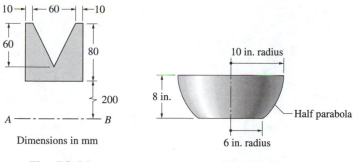

| Fig. P8.81 | Fig. P8.82 |

8.83   The sheet metal bellows is formed by revolving the generating curve about the z-axis. Find the surface area of the bellows.

8.84   Compute the volume of the sheet metal bellows described in Prob. 8.83.

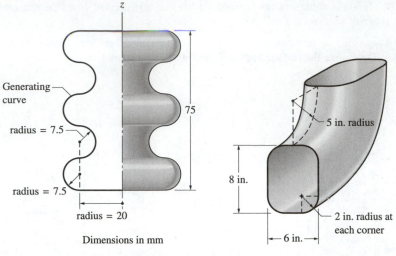

Dimensions in mm

Fig. P8.83, P8.84

Fig. P8.85

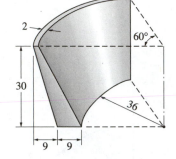

Dimensions in meters

Fig. P8.86

8.85   Find the surface area of the 90° duct elbow.

8.86   Determine the volume of the concrete arch dam.

8.87   (a) Find the volume of liquid contained in the flask when it is filled to the "full" mark. (b) Determine the elevation h of the "half full" mark.

8.88   Determine the ratio b/a for which the volume of the fill equals the volume of the material removed from the conical excavation. Assume that the fill is axisymmetric about the excavation.

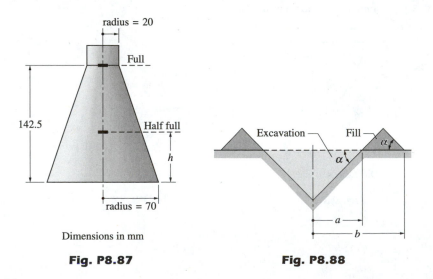

Fig. P8.87

Fig. P8.88

8.5   *Center of Gravity and Center of Mass*

The resultant of gravity forces acting on a body, which we know as the *weight* of the body, acts through a point called the *center of gravity* of the body. The center of gravity is thus determined by the distribution of weight within the body.

The *center of mass* of a body is a very important concept in dynamics (it is the point through which the resultant inertia force acts); it is a property of the distribution of mass within the body. However, because weight and mass differ only by a constant factor (provided that the gravitational field is uniform), we find that the centers of mass and gravity coincide in most applications. Differences arise only in problems in which the gravitational field is not uniform. Therefore, it is not surprising that engineers frequently use the terms *center of gravity* and *center of mass* interchangeably.

a.   *Center of gravity*

The weight of a body is the most common example of a distributed force. For a body occupying the region $\mathcal{V}$, as in Fig. 8.13, the weight of a differential volume element is $dW = \gamma\, dV$, where γ is the weight density (weight per unit volume). The total weight W of the body is thus the resultant of an infinite number of parallel forces dW; that is,

$$W = \int_{\mathcal{V}} dW = \int_{\mathcal{V}} \gamma\, dV$$

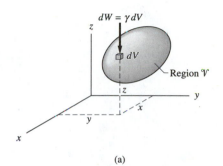

(a)

The point G through which the weight W acts is called the *center of weight*—or, more commonly—the *center of gravity of the body*—see Fig. 8.13(b).

The coordinates of G can be determined by equating the resultant moment of the distributed weight to the moment of W about the coordinate axes. Referring to Fig. 8.13(a) and (b), we obtain

$$\Sigma M_x = -W\bar{y} = -\int_{\mathcal{V}} y\, dW \qquad \Sigma M_y = W\bar{x} = \int_{\mathcal{V}} x\, dW$$

from which we obtain

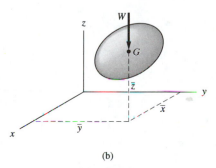

(b)

Fig. 8.13

$$\bar{x} = \frac{\displaystyle\int_{\mathcal{V}} x\, dW}{\displaystyle\int_{\mathcal{V}} dW} = \frac{\displaystyle\int_{\mathcal{V}} x\gamma\, dV}{\displaystyle\int_{\mathcal{V}} \gamma\, dV} \qquad (8.16a)$$

$$\bar{y} = \frac{\displaystyle\int_{\mathcal{V}} y\, dW}{\displaystyle\int_{\mathcal{V}} dW} = \frac{\displaystyle\int_{\mathcal{V}} y\gamma\, dV}{\displaystyle\int_{\mathcal{V}} \gamma\, dV} \qquad (8.16b)$$

If we next imagine that the body and the coordinate axes are rotated so that either the x- or y-axis is vertical, the equality of moments would then yield

$$
\bar{z} = \frac{\displaystyle\int_{\mathcal{V}} z \, dW}{\displaystyle\int_{\mathcal{V}} dW} = \frac{\displaystyle\int_{\mathcal{V}} z\gamma \, dV}{\displaystyle\int_{\mathcal{V}} \gamma \, dV}
\tag{8.16c}
$$

The integrals in Eqs. (8.16) can be evaluated by the same techniques that are used to locate the centroids of volumes in Art. 8.3.

For homogeneous bodies, the weight density γ is constant and thus cancels in Eqs. (8.16), giving

$$
\bar{x} = \frac{\int_{\mathcal{V}} x \, dV}{\int_{\mathcal{V}} dV} \qquad \bar{y} = \frac{\int_{\mathcal{V}} y \, dV}{\int_{\mathcal{V}} dV} \qquad \bar{z} = \frac{\int_{\mathcal{V}} z \, dV}{\int_{\mathcal{V}} dV}
$$

Comparing these equations with Eqs. (8.10), we can make the following important observation.

> The center of gravity of a homogeneous body coincides with the centroid of its volume.

Therefore, tables listing the centroids of volumes, such as Table 8.3, can be used to determine the location of the centers of gravity of homogeneous bodies.

b. Center of mass

The center of mass of a body is defined as the point that has the following coordinates.

$$
\begin{aligned}
\bar{x} &= \frac{\displaystyle\int_{\mathcal{V}} x \, dm}{\displaystyle\int_{\mathcal{V}} dm} = \frac{\displaystyle\int_{\mathcal{V}} x\rho \, dV}{\displaystyle\int_{\mathcal{V}} \rho \, dV} \\[2mm]
\bar{y} &= \frac{\displaystyle\int_{\mathcal{V}} y \, dm}{\displaystyle\int_{\mathcal{V}} dm} = \frac{\displaystyle\int_{\mathcal{V}} y\rho \, dV}{\displaystyle\int_{\mathcal{V}} \rho \, dV} \\[2mm]
\bar{z} &= \frac{\displaystyle\int_{\mathcal{V}} z \, dm}{\displaystyle\int_{\mathcal{V}} dm} = \frac{\displaystyle\int_{\mathcal{V}} z\rho \, dV}{\displaystyle\int_{\mathcal{V}} \rho \, dV}
\end{aligned}
\tag{8.17}
$$

In Eqs. (8.17), dm is the mass of the differential element dV in Fig. 8.13(a), given by $dm = \rho \, dV$, where ρ is the mass density (mass per unit volume). If the gravitational field is constant, which is a valid assumption in most engineering problems, the weight density γ and the mass density ρ are related by $\gamma = \rho g$, where g is the gravitational acceleration (constant). In this case, the center of gravity and the center of mass for a given body coincide.

c. Composite bodies

If a body is composed of several simple shapes, we can use the method of composite bodies to find its center of gravity of mass center. By replacing the integrals in Eqs. (8.16) by summations, we obtain for the center of gravity

$$\bar{x} = \frac{\Sigma W_i \bar{x}_i}{\Sigma W_i} \qquad \bar{y} = \frac{\Sigma W_i \bar{y}_i}{\Sigma W_i} \qquad \bar{z} = \frac{\Sigma W_i \bar{z}_i}{\Sigma W_i} \qquad (8.18)$$

where W_i is the weight of the ith component of the body and $(\bar{x}_i, \ \bar{y}_i, \ \bar{z}_i)$ are the coordinates of its center of gravity.

Similarly, location of the mass center of a composite body can be obtained from Eq. (8.17):

$$\bar{x} = \frac{\Sigma m_i \bar{x}_i}{\Sigma m_i} \qquad \bar{y} = \frac{\Sigma m_i \bar{y}_i}{\Sigma m_i} \qquad \bar{z} = \frac{\Sigma m_i \bar{z}_i}{\Sigma m_i} \qquad (8.19)$$

where now m_i is the mass of the ith component and $(\bar{x}_i, \ \bar{y}_i, \ \bar{z}_i)$ is the location of its mass center.

Sample Problem 8.12

The machine part in Fig. (a) consists of a steel hemisphere joined to an aluminum cylinder into which a hole has been drilled. Determine the location of the center of the mass. The mass densities for aluminum and steel are 2700 kg/m³ and 7850 kg/m³, respectively.

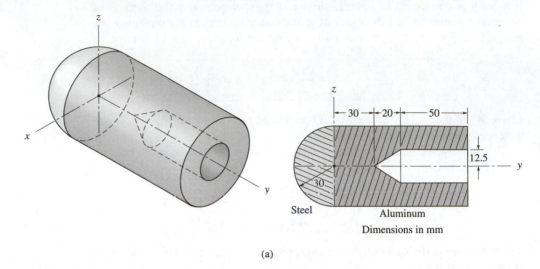

Dimensions in mm

(a)

Solution

By symmetry, we note that $\bar{x} = \bar{z} = 0$. If the machine part were homogeneous, its center of mass would coincide with the centroid of the enclosing volume, and $\bar{y}$ could be determined using the method of composite volumes described in Art. 8.3. Because the machine part is not homogeneous, $\bar{y}$ must be determined by the method of composite bodies.

The part is composed of the four masses shown in Fig. (b): the aluminum cylinder ①, plus the steel hemisphere ②, minus the aluminum cylinder ③, minus the

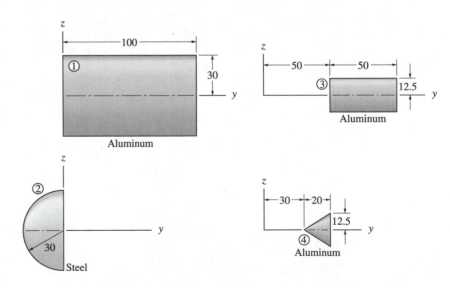

(b)

aluminum cone ④. Because each of these masses is homogeneous, each center of mass coincides with the centroid of the enclosing volume.

Aluminum Cylinder ①

$$m = \rho V = \rho \pi R^2 h = 2700\pi(0.030)^2(0.100) = 0.7634 \text{ kg}$$

$$\bar{y} = \frac{100}{2} = 50 \text{ mm} \qquad \text{(by symmetry)}$$

$$m\bar{y} = (0.7634)(50) = 38.17 \text{ kg} \cdot \text{mm}$$

Steel Hemisphere ②

$$m = \rho V = \rho \frac{2\pi}{3} R^3 = 7850 \frac{2\pi}{3}(0.030)^3 = 0.4439 \text{ kg}$$

$$\bar{y} = -\frac{3}{8}R = -\frac{3}{8}(30) = -11.25 \text{ mm} \qquad \text{(using Table 8.3)}$$

$$m\bar{y} = (0.4439)(-11.25) = -4.994 \text{ kg} \cdot \text{mm}$$

Aluminum Cylinder ③ (to be subtracted)

$$m = -\rho V = -\rho \pi R^2 h = -2700\pi(0.0125)^2(0.050) = -0.066\,27 \text{ kg}$$

$$\bar{y} = 75 \text{ mm} \qquad \text{(by symmetry)}$$

$$m\bar{y} = (-0.066\,27)(75) = -4.970 \text{ kg} \cdot \text{mm}$$

Aluminum Cone ④ (to be subtracted)

$$m = -\rho V = -\rho \frac{\pi}{3} R^2 h = -2700 \frac{\pi}{3}(0.0125)^2(0.020) = -0.008\,836 \text{ kg}$$

$$\bar{y} = 30 + \frac{3}{4}(20) = 45 \text{ mm} \qquad \text{(using Table 8.3)}$$

$$m\bar{y} = (-0.008\,836)(45) = -0.3976 \text{ kg} \cdot \text{mm}$$

Totals

$$\Sigma m = 0.7634 + 0.4439 - 0.066\,27 - 0.008\,836 = 1.1322 \text{ kg}$$

$$\Sigma m\bar{y} = 38.17 - 4.994 - 4.970 - 0.3976 = 27.81 \text{ kg} \cdot \text{mm}$$

Therefore, the coordinates of the mass center of the machine part are

$$\bar{x} = \bar{z} = 0$$

$$\bar{y} = \frac{\Sigma m\bar{y}}{\Sigma m} = \frac{27.81}{1.1322} = 24.6 \text{ mm} \qquad\qquad \textit{Answer}$$

Problems

8.89 The steel cylinder with a cylindrical hole is connected to the copper cone. Find the center of gravity of the assembly. The weight densities of steel and copper are 489 lb/ft^3 and 556 lb/ft^3, respectively.

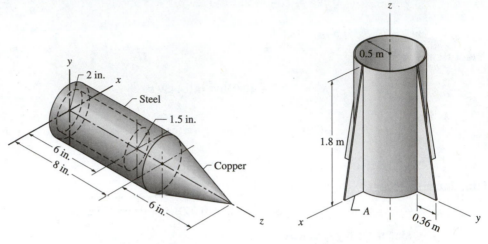

Fig. P8.89   **Fig. P8.90**

8.90 The steel rocket casing consists of a 120-kg cylindrical shell and four triangular fins of mass 15 kg each. All components are thin and of uniform thickness. Determine the coordinates of the mass center after fin *A* has been removed.

8.91 What is the ratio *L/R* for which the uniform wire figure can be balanced in the position shown?

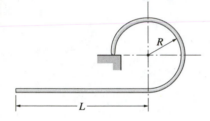

Fig. P8.91

8.92 Small screws are used to fasten a piece of hardwood to the bracket that is formed from 1/20-in.-thick steel sheet metal. For steel, $\gamma = 0.283$ lb/in.^{3}, and for hardwood, $\gamma = 0.029$ lb/in.^{3} Locate the center of gravity of the assembly.

8.93 Plywood with two different thicknesses is used to fabricate the partition shown. Find the largest allowable ratio t_2/t_1 if the partition is not to tip over.

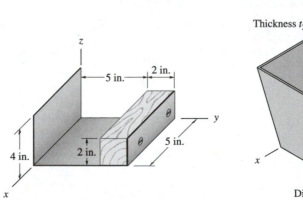

Fig. P8.92

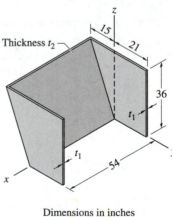

Dimensions in inches

Fig. P8.93

8.94 The hemispherical ceramic crucible ($\rho = 2000$ kg/m³) containing molten iron ($\rho = 7200$ kg/m³) is suspended from the rod AB. To pour the iron, a cable attached to the crucible at C is pulled with the vertical force F. Determine F as a function of the tilt angle α.

8.95 Two uniform bars of different diameters are joined together as shown. Bar AB is made of aluminum ($\rho = 2700$ kg/m³) and bar CD is made of copper ($\rho = 8910$ kg/m³). Find the coordinates of the mass center of the assembly.

8.96 The assembly is formed by joining a semicircular steel ($\gamma = 490$ lb/ft³) plate to a triangular aluminum ($\gamma = 166$ lb/ft³) plate. Both plates are homogeneous and of thickness 1/8 inch. Find the coordinates of the center of gravity of the assembly.

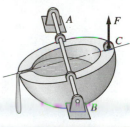

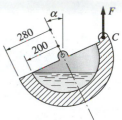

Dimensions in mm

Fig. P8.94

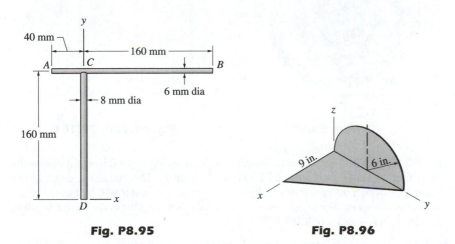

Fig. P8.95   **Fig. P8.96**

8.97 The dish antenna, shown in cross section, is composed of a parabolic reflector and a boom-receiver assembly. Assuming that each part is uniform and using the weights shown on the figure, locate the center of gravity of the antenna.

8.98 Locate the center of gravity of the hammer if the mass of the steel head is 0.919 kg, and the mass of the hardwood handle is 0.0990 kg.

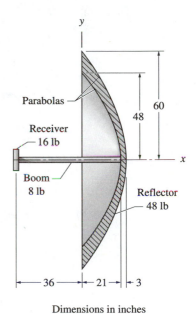

Dimensions in inches

Fig. P8.97

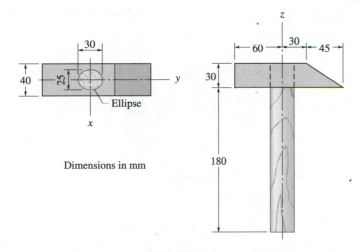

Dimensions in mm

Fig. P8.98

8.99 The total weight of the car wheel and tire is 24 lb. To statically balance the wheel-tire assembly (to move its center of gravity to point O), 2-oz lead weights are attached to the rim at A and B. What was the location of the center of gravity of the assembly before the balancing weights were added?

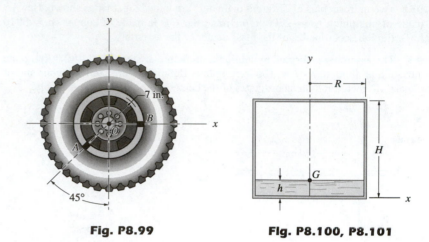

Fig. P8.99   **Fig. P8.100, P8.101**

8.100 The cylindrical water tank with $R = 12$ ft and $H = 18$ ft has thin steel walls of uniform thickness and weighs 23 000 lb when empty. Determine the depth of the water h for which the center of gravity G of the tank plus water will be located at the surface of the water. For water, use $\gamma = 62.4$ lb/ft³. (Note: This is the lowest possible location for G—see Prob. 8.101.)

*8.101 A cylindrical tank, such as the one described in Prob. 8.100, contains liquid at a depth h. Without differentiating, show that the lowest location for G (the center of gravity of the vessel-fluid combination) occurs when G lies on the surface of the liquid. (Hint: First assume that G lies on the surface of the liquid; then show that it always moves up if liquid is added or removed.)

8.102 Five $\frac{3}{4}$-in. diameter holes are to be drilled in a uniform plate. Determine the angle θ for the circular segment that must also be removed if the center of gravity is to remain at point O.

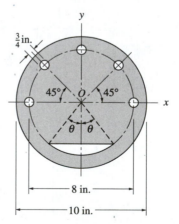

Fig. P8.102

8.6   *Distributed Normal Loads*

Our discussion of distributed loads in Art. 3.6 was limited to simple cases because the treatment of centroids has been postponed until the present chapter. It is now possible to discuss distributed normal loads in a general context.

a.   *General case*

Consider the general case of a distributed normal loading shown in Fig. 8.14(a). The loading is assumed to act normal to the load area $\mathcal{A}$, and its magnitude is characterized by $p(x, y, z)$, the *load intensity* (force per unit area). Letting $d\mathbf{R}$ be the force acting on the differential area dA, we have $d\mathbf{R} = \mathbf{n}p\,dA$, where $\mathbf{n}$ is a unit vector normal to dA (in the direction of p). Then the resultant of the distributed load is the force $\mathbf{R}$ shown in Fig. 8.14(b), where

$$\mathbf{R} = \int_{\mathcal{A}} d\mathbf{R} = \int_{\mathcal{A}} \mathbf{n}p\,dA \qquad (8.20)$$

Moment equations can be used to determine the line of action of $\mathbf{R}$—that is, the coordinates $\bar{x}$, $\bar{y}$, and $\bar{z}$ (of course, if $\mathbf{R} = \mathbf{0}$ and the sum of the moments is not zero, the resultant is a couple). In general, $\mathbf{n}$, p, and dA are functions of x, y, and z, and evaluation of the integrals may be complicated. The following special cases occur frequently enough to warrant special attention: flat surfaces, line loads, uniform pressure on curved surfaces, and fluid pressure.

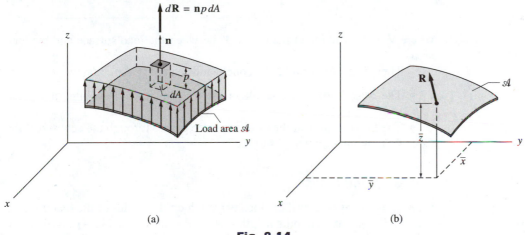

(a)   (b)

Fig. 8.14

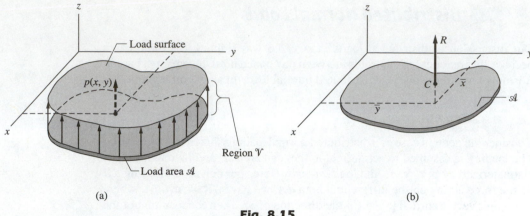

Fig. 8.15

b. Normal loads on flat surfaces

The load shown in Fig. 8.15(a) is parallel to the z-axis and is assumed to be distributed across the load area $\mathcal{A}$, which lies in the xy-plane. In this case, the load intensity p is a function only of x and y. The resultant force is shown in Fig. 8.15(b). According to Eqs. (3.18) and (3.19) of Art. 3.6, the magnitude of the resultant force and its line of action are determined by

$$R = \int_{\mathcal{V}} dV = V \qquad \bar{x} = \frac{\int_{\mathcal{V}} x\,dV}{V} \qquad \bar{y} = \frac{\int_{\mathcal{V}} y\,dV}{V} \qquad (8.21)$$

where V is the volume of the region $\mathcal{V}$ between the load surface and the load area.

Therefore, we arrive at the conclusions stated in Art. 3.6:

- The magnitude of the resultant force is equal to the volume under the load surface.
- The line of action of the resultant force passes through the centroid of the volume under the load surface.

c. Line loads

Line loads—that is, distributed loads for which the width of the loading area is negligible compared to its length—were also introduced in Art. 3.6. In that article, emphasis was given to loads distributed along straight lines. Here, we determine the resultants of loadings that are distributed along plane curves.

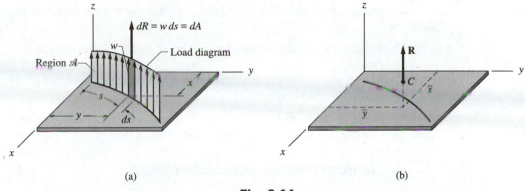

Fig. 8.16

Fig. 8.16(a) shows a loading, parallel to the z-axis, that acts along a curve lying in the xy-plane. We let s be the length measured along the curve. The loading is characterized by the *load intensity* $w(s)$ with units lb/ft, N/m, etc. The plot of $w(s)$ is called the *load diagram*. We let A be the area of the region (curved surface) $\mathcal{A}$ under the load diagram. The resultant $\mathbf{R}$ of the line load is shown in Fig. 8.16(b), with its line of action located by $\bar{x}$ and $\bar{y}$. Because the line load may be thought of as an infinite number of parallel, differential forces, integration is used to find the magnitude and line of action of $\mathbf{R}$.

As shown in Fig. 8.16(a), we let dR represent the infinitesimal force that acts on the differential line length ds. Because $dR = w\,ds = dA$, where dA is the differential area under the loading diagram, we see that the magnitude of the resultant force equals the area of the curved surface $\mathcal{A}$; that is,

$$R = \int_{\mathcal{A}} dR = \int_{\mathcal{A}} w\,ds = \int_{\mathcal{A}} dA = A \qquad (8.22)$$

Equating moments of the load system in Fig. 8.16(a) and (b) about the coordinate axes yields

$$\Sigma M_x = R\bar{y} = \int_{\mathcal{A}} y\,dR \qquad \Sigma M_y = -R\bar{x} = -\int_{\mathcal{A}} x\,dR$$

After substituting $R = A$ and $dR = dA$, we get

$$A\bar{y} = \int_{\mathcal{A}} y\,dA \qquad A\bar{x} = \int_{\mathcal{A}} x\,dA$$

from which we obtain

$$\bar{x} = \frac{\int_{\mathcal{A}} x\,dA}{A} \qquad \bar{y} = \frac{\int_{\mathcal{A}} y\,dA}{A} \qquad (8.23)$$

Comparing Eqs. (8.23) and (8.10), we conclude that $\mathbf{R}$ acts at the centroid of the region under the load diagram. This point is labeled C in Fig. 8.16(b). The z-coordinate of the centroid is irrelevant, because $\mathbf{R}$ is a sliding vector.

In summary, line loads distributed along a plane curve have the following properties:

- The magnitude of the resultant force is equal to the area under the load diagram.
- The line of action of the resultant force passes through the centroid of the area under the load diagram.

These conclusions were made in Art. 3.6 for loads distributed along straight lines; now we see that they are also valid for curved lines.

d.   *Uniform pressure on curved surfaces*

Uniform pressure refers to the special case in which the magnitude of the load intensity is constant. If p is constant, Eq. (8.20) becomes

$$\mathbf{R} = \int_{\mathcal{A}} p\mathbf{n}\, dA = p \int_{\mathcal{A}} \mathbf{n}\, dA \tag{8.24}$$

It can be shown by vector analysis (see Prob. 8.122) that

$$\mathbf{n}\, dA = dA_x\, \mathbf{i} + dA_y\, \mathbf{j} + dA_z\, \mathbf{k}$$

where dA_x, dA_y, and dA_z are the projections of dA on the three coordinate planes, as shown in Fig. 8.17. Therefore, Eq. (8.24) can be written as

$$\mathbf{R} = p \left(\mathbf{i} \int_{\mathcal{A}} dA_x + \mathbf{j} \int_{\mathcal{A}} dA_y + \mathbf{k} \int_{\mathcal{A}} dA_z \right) = p(A_x \mathbf{i} + A_y \mathbf{j} + A_z \mathbf{k})$$

or

$$\boxed{R_x = pA_x \qquad R_y = pA_y \qquad R_z = pA_z} \tag{8.25}$$

where A_x, A_y, and A_z are the projections of the load area on the coordinate planes. In principle, moment equations could be used to determine the line of action of $\mathbf{R}$. However, in many instances the line of action can be determined by symmetry.

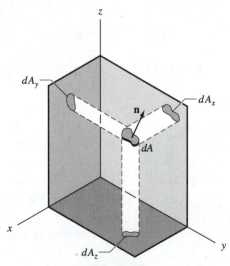

Fig. 8.17

There are many practical applications in which the results of Eqs. (8.25) can be used. For example, consider the cylindrical vessel with inside diameter D, shown in Fig. 8.18(a), that is subjected to a uniform internal pressure p. Half of the vessel and the pressure acting on it are shown in Fig. 8.18(b). It is clear that the pressure p has a resultant force P_b that tends to rupture the vessel (to split it lengthwise). This bursting force can be calculated using integration as indicated in Eq. (8.24). However, from Eq. (8.25), we immediately obtain $P_b = pA_y = pDL$, as shown in Fig. 8.18(c).

Although the formula "force equals pressure times projected area" is very useful, you must remember that it is valid only for uniform pressure.

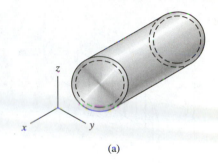

(a)

e. *Fluid pressure*

If a surface is submerged in a fluid of weight density γ, the pressure exerted by the fluid is $p = \gamma h$, where h is the depth measured from the free surface of the fluid. The resultant of this pressure could be obtained by integration, but if the surface is curved, the analysis may become complicated because both the magnitude and direction of p vary. It is usually easier to find the resultant by equilibrium analysis of a volume of water that is bounded by the curved surface, as illustrated below.

Let us find the resultant of the water pressure acting on the curved surface AB of the dam in Fig. 8.19(a) by considering equilibrium of the body of water contained in the region ABC shown in Fig. 8.19(b). The FBD of the water in Fig. 8.19(c) contains the following forces:

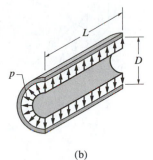

(b)

R_1: resultant of the uniform pressure exerted by the water above the region.
R_2: resultant of the linearly varying pressure due to water lying to the right of the region.
R_{AB}: resultant of the pressure exerted by the dam; it is equal and opposite to the force of the water acting on portion AB of the dam.
W: weight of the water contained in the region; it acts at the centroid G of region ABC.

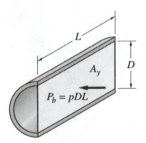

(c)

Fig. 8.18

The forces R_1, R_2, and W can be found by the methods described previously in this article. The horizontal and vertical components of R_{AB} can then be found relatively simply by force equations of equilibrium. A moment equation would be required to determine the line of action of R_{AB}.

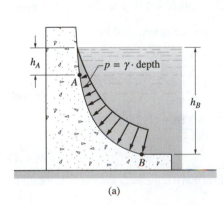

(a)

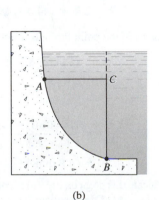

(b)

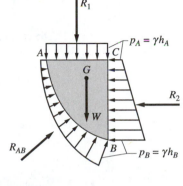

(c)

Fig. 8.19

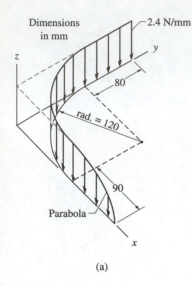

Dimensions in mm

2.4 N/mm

y

80

rad. = 120

90

Parabola

x

(a)

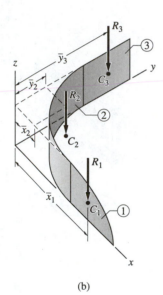

R_3

③

$\bar{y}_3$

$\bar{y}_2$

R_2

C_3

y

②

$\bar{x}_2$

C_2

R_1

$\bar{x}_1$

C_1

①

x

(b)

Sample Problem 8.13

Determine the resultant of the line load shown in Fig. (a).

Solution

As shown in Fig. (b), we represent the original line loading as the sum of the following three components: ①—the parabolic load distributed along a straight line on the x-axis, ②—the uniform load distributed along the quarter circle lying in the xy-plane, and ③ —the rectangular load along the y-axis. The resultant force of each load component equals the area under the corresponding load diagram, acting through the centroid of the diagram.

Load Component ①

Using Table 8.1 for the parabola, we have

$$R_1 = A_1 = \frac{2}{3}(90)(2.4) = 144.0 \text{ N}$$

$$\bar{x}_1 = 120 + \frac{3}{8}(90) = 153.75 \text{ mm} \qquad \bar{y}_1 = 0$$

Load Component ②

$$R_2 = A_2 = \frac{\pi R}{2}(2.4) = \frac{\pi(120)}{2}(2.4) = 452.4 \text{ N}$$

Noting that the x- and y-centroidal coordinates for the curved surface are the same as the centroidal coordinates for the quarter circle lying in the xy-plane, Table 8.2 yields

$$\bar{x}_2 = \bar{y}_2 = R - \frac{2R}{\pi} = 0.3634R = 0.3634(120) = 43.6 \text{ mm}$$

Load Component ③

$$R_3 = A_3 = (80)(2.4) = 192.0 \text{ N}$$

$$\bar{x}_3 = 0 \qquad \bar{y}_3 = 120.0 + 40.0 = 160.0 \text{ mm}$$

Resultant Load

From the foregoing data, we find that the magnitude of the resultant is

$$R = \Sigma R = 144.0 + 452.4 + 192.0 = 788.4 \text{ N} \qquad \textit{Answer}$$

The coordinates of the point through which **R** acts are determined by

$$\bar{x} = \frac{\Sigma R\bar{x}}{R} = \frac{(144.0)(153.75) + (452.4)(43.6) + (192.0)(0)}{788.4}$$

$$= 53.1 \text{ mm} \qquad \textit{Answer}$$

$$\bar{y} = \frac{\Sigma R\bar{y}}{R} = \frac{(144.0)(0) + (452.4)(43.6) + (192.0)(160.0)}{788.4}$$

$$= 64.0 \text{ mm} \qquad \textit{Answer}$$

You should locate this point on Fig. (a) to confirm that it represents a reasonable location for the line of action of the resultant force.

Sample Problem 8.14

The undersides of the 36-in. × 24-in. corrugated sheets in Figs. (a) and (b) carry a uniform normal pressure of $p_0 = 0.6$ lb/in.^{2} Calculate the rectangular components of the resultant force acting on the underside of each sheet.

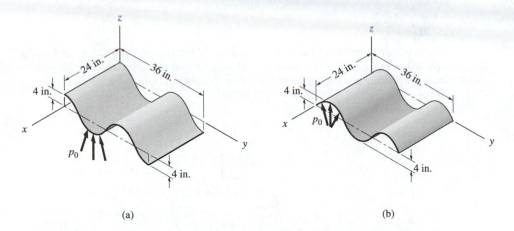

(a)   (b)

Solution

Because each sheet is subjected to uniform normal pressure it is possible to determine the components of the resultant force by using the pressure times the projected area: $R_x = p_0 A_x$, $R_y = p_0 A_y$, and $R_z = p_0 A_z$, where A_x, A_y, and A_z are the projected areas of the loading surface on the three coordinate planes.

The construction in Fig. (c) shows that the projected areas are

Sheet a

$$A_x = 0 \qquad A_y = 8 \times 24 = 192 \text{ in.}^2$$
$$A_z = 36 \times 24 = 864 \text{ in.}^2$$

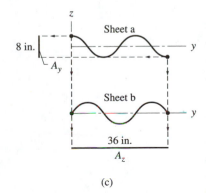

(c)

Sheet b

$$A_x = 0 \qquad A_y = 0$$
$$A_z = 36 \times 24 = 864 \text{ in.}^2$$

Therefore, for sheet (a), the rectangular components of the resultant force are

$$R_x = p_0 A_x = 0 \qquad R_y = p_0 A_y = 0.6(192) = 115.2 \text{ lb}$$
$$R_z = p_0 A_z = 0.6(864) = 518 \text{ lb} \qquad\qquad \textit{Answer}$$

For sheet (b), we have

$$R_x = p_0 A_x = 0 \qquad R_y = p_0 A_y = 0$$
$$R_z = p_0 A_z = 0.6(864) = 518 \text{ lb} \qquad\qquad \textit{Answer}$$

Sample Problem 8.15

A swimming pool is filled with water to a depth of 3 m, as shown in Fig. (a). Determine the magnitude and line of action of the resultant force that acts on the circular portion AB of the wall. The length of the wall (dimension perpendicular to the paper) is 25 m, and the mass density of water is 1000 kg/m^3.

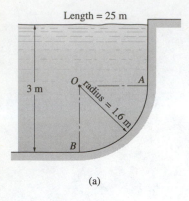

Length = 25 m

3 m

O radius = 1.6 m

A

B

(a)

Solution

The free-body diagram of the water contained in the region OAB (25 m in length) is shown in Fig. (b). The pressures p_1 and p_2 are equal to the products of the weight density of water γ and the depths at A and B, respectively. Letting ρ be the mass density, we have $\gamma = \rho g = 1000(9.81) = 9.81$ kN/m^3.

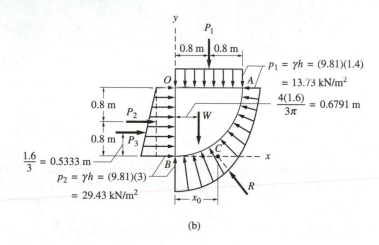

y

P_1

0.8 m | 0.8 m

O A

$p_1 = \gamma h = (9.81)(1.4)$
$= 13.73$ kN/m^2

0.8 m P_2 W

$\dfrac{4(1.6)}{3\pi} = 0.6791$ m

0.8 m P_3

C x

$\dfrac{1.6}{3} = 0.5333$ m

B

R

$p_2 = \gamma h = (9.81)(3)$
$= 29.43$ kN/m^2

x_0

(b)

The resultant forces P_1 and P_2 are caused by the uniformly distributed loadings shown, and P_3 is due to the triangular portion of the loading. Each of these forces equals the product of the pressure (average pressure is used for P_3) and the area on which the pressure acts.

$$P_1 = p_1 A = (13.73)(1.6)(25) = 549.2 \text{ kN}$$

$$P_2 = p_1 A = (13.73)(1.6)(25) = 549.2 \text{ kN}$$

$$P_3 = \frac{p_2 - p_1}{2} A = \frac{29.43 - 13.73}{2}(1.6)(25) = 314.0 \text{ kN}$$

The lines of action of P_1, P_2, and P_3 pass through the centroids of the corresponding load diagrams. The weight W of the water contained in the region OAB is

$$W = \gamma \times \text{volume} = (9.81)\frac{\pi(1.6)^2}{4}(25) = 493.1 \text{ kN}$$

Letting **R** be the resultant force exerted on the water by the curved portion of the wall, we can find its components from the force equilibrium equations. Referring to the FBD in Fig. (b), we obtain

$$\Sigma F_x = 0 \quad \xrightarrow{+} \quad P_2 + P_3 - R_x = 0$$
$$R_x = P_2 + P_3 = 549.2 + 314.0 = 863.2 \text{ kN}$$

$$\Sigma F_y = 0 \quad +\uparrow \quad R_y - W - P_1 = 0$$
$$R_y = W + P_1 = 493.1 + 549.2 = 1042.3 \text{ kN}$$

Therefore, the magnitude and direction of **R** are

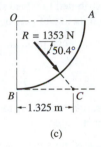

$$R = \sqrt{(863.2)^2 + (1042.3)^2}$$
$$= 1353 \text{ kN}$$

$$\theta = \tan^{-1} \frac{1042.3}{863.2} = 50.4°$$

To find the line of action of **R**, we need the moment equation of equilibrium. Choosing point B in Fig. (b) as the moment center and letting x_0 be the distance between B and C (the point where R intersects the x-axis), we obtain

$$\Sigma M_B = 0 \quad \overset{\curvearrowleft}{+} \quad R_y x_0 - P_1(0.8) - P_2(0.8)$$
$$- P_3(0.5333) - W(0.6791) = 0$$

from which we have

$$x_0 = \frac{549.2(0.8) + 549.2(0.8) + 314.0(0.5333) + 493.1(0.6791)}{1042.3} = 1.325 \text{ m}$$

The force exerted by the water on the wall is equal and opposite to the force determined above, as shown in Fig. (c).

(c)

409

Problems

Unless integration is specified, the following problems are to be analyzed using the information in Tables 8.1–8.3.

8.103 Wind pressure acting on a cylinder can be approximated by $p = p_0 \cos \theta$, where p_0 is a constant (note that on the lee side the pressure is negative). Determine the resultant force of the wind pressure on a cylinder of radius R and length L by integration.

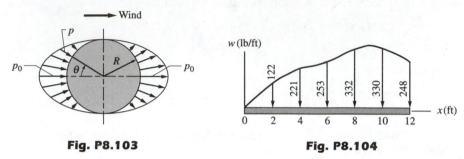

Fig. P8.103   **Fig. P8.104**

8.104 The beam carries the distributed line load shown. Use numerical integration to determine the resultant force and its line of action.

8.105 The pressure acting on the square plate varies as

$$p = p_0 \left(2 - \frac{x}{L} - \frac{y}{L} + \frac{xy}{L^2} \right)$$

where p_0 is a constant. Use integration to find the resultant force of the pressure, and the x- and y-coordinates of its line of action.

8.106 The intensity of the line loading acting on the rim of the semicircular plate varies as $w = w_0 y/a$, where w_0 is a constant and a represents the radius of the plate. Use integration to determine the resultant force and to locate its line of action.

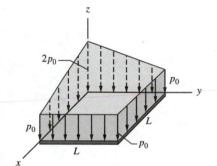

Fig. P8.105

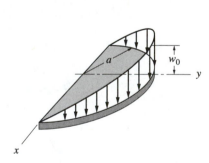

Fig. P8.106

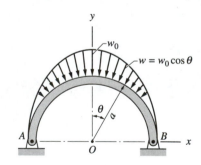

Fig. P8.107

8.107 Use integration to determine the resultant of the normal line loading acting on the circular arch.

8.108 If the intensity of the line loading is $w = [(40x - x^2)/40]$ lb/in., where x is measured in inches, use integration to find the resultant.

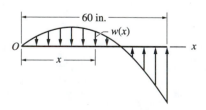

Fig. P8.108

8.109 Determine the resultant of the line loading, given that $w_0 = 1.5$ lb/in.

8.110 Find the resultant of the line load shown.

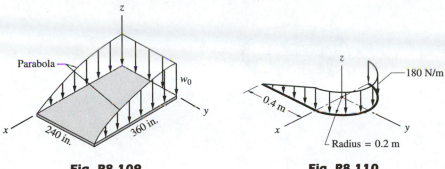

Fig. P8.109   **Fig. P8.110**

8.111 Determine the resultant force or resultant couple for each of the line loads shown. In each case the loading is normal to the line and has constant intensity w_0.

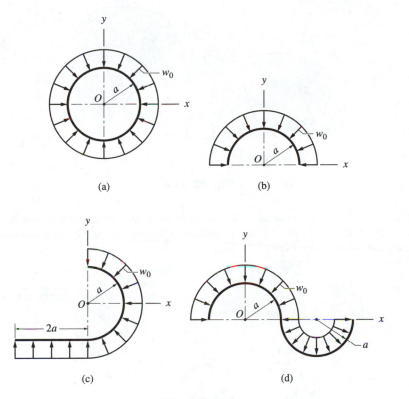

Fig. P8.111

8.112   The inside surface of each thin shell carries a uniform normal pressure of intensity p_0. Compute R, the magnitude of the resultant force for each case.

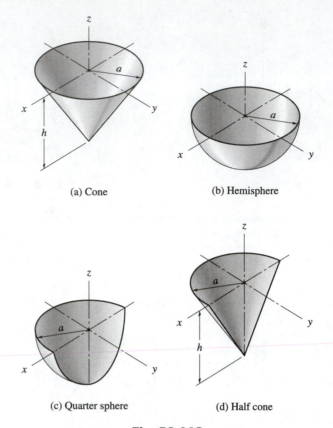

(a) Cone

(b) Hemisphere

(c) Quarter sphere

(d) Half cone

Fig. P8.112

8.113   Calculate the resultant force caused by the water acting on the parabolic arch dam. The water levels are 16 ft on the upstream side and zero on the downstream side. For water, use $\gamma = 62.4$ lb/ft^3.

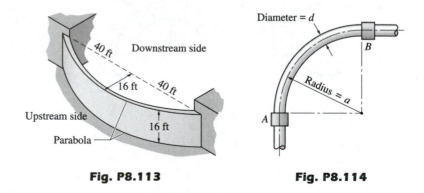

Fig. P8.113

Fig. P8.114

8.114   Determine the resultant force acting on the elbow of the thin-walled pipe when the pipe carries a uniform internal pressure p_0.

8.115 Consider the equilibrium of the liquid contained in the region $\mathcal{V}$ shown in Fig. (a). The forces acting on $\mathcal{V}$ are the weight W of the water and the resultant R of the pressure acting on the boundary of $\mathcal{V}$. Clearly, the liquid in $\mathcal{V}$ can be in equilibrium only if $W = R$ (and W and R are collinear). Use this result to derive the *principle of buoyancy:* A body floating on the surface of a liquid, as shown in Fig. (b), displaces a volume of liquid whose weight equals the weight of the body.

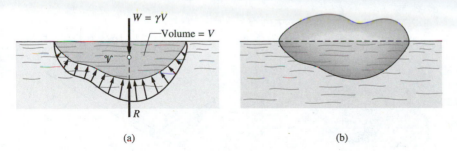

(a)   (b)

Fig. P8.115

8.116 Each of the three gates has a constant width c (perpendicular to the paper). Calculate the force P required to maintain each gate in the position shown. Express your answers in terms of b, c, h, and γ (the weight density of the fluid).

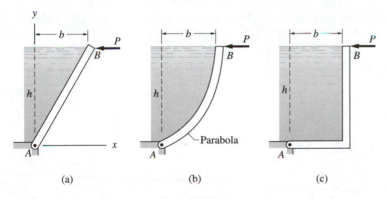

(a)   (b)   (c)

Fig. P8.116

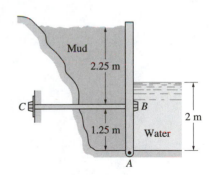

Fig. P8.117

8.117 The piling that restrains a liquid mud wall is supported by equally spaced anchor bolts BC that are embedded in bedrock at C. The attachment of the piling to the bedrock at A is equivalent to a pin support. What is the minimum safe spacing of the anchor bolts if the allowable tension in each bolt is 40 kN? Use $\rho = 1.0 \text{ Mg/m}^3$ for water and $\rho = 1.76 \text{ Mg/m}^3$ for mud.

8.118 A concrete seawater dam is shown in cross section. Is the dam safe against tipping about edge A? The mass densities are 2400 kg/m^3 for concrete and 1030 kg/m^3 for seawater.

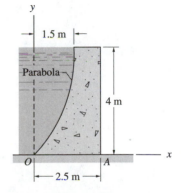

Fig. P8.118

8.119 Determine the force F required to pull up the 2.5-oz stopper from the drain of a sink if the depth of water is 9 in. Use $\gamma = 0.036$ lb/in.^{3} for water and neglect the weight of the chain.

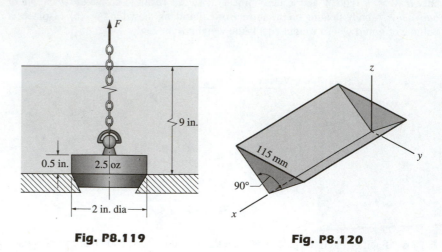

Fig. P8.119

Fig. P8.120

8.120 If the trough is filled with water ($\rho = 1000$ kg/m^3), determine the resultant of the water pressure acting on one of the equilateral triangular end plates.

8.121 The pressure acting on the circular plate of radius a varies as $p = p_0\left[1 + (r/a)\cos\theta\right]$, where p_0 is a constant. Use integration to find the resultant force and to locate its line of action.

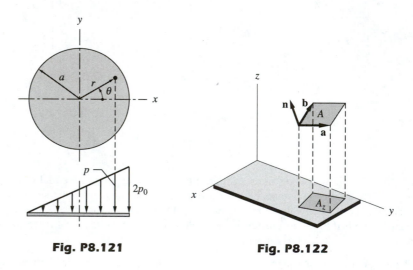

Fig. P8.121

Fig. P8.122

8.122 The parallelogram of area A is formed by the vectors $\mathbf{a}$ and $\mathbf{b}$; consequently, $A\mathbf{n} = \mathbf{a} \times \mathbf{b}$, where $\mathbf{n}$ is the unit normal to A. Show that the area of the projection of the parallelogram on the xy-plane is $A_z = An_z$.

Review Problems

8.123   Locate the center of gravity of the plane wire figure.

8.124   The 10-m wide gate restrains water at a depth of 6 ft. Calculate the magnitude of the hinge reaction at A, and the contact force between the gate and the smooth surface at B. Neglect the weight of the gate.

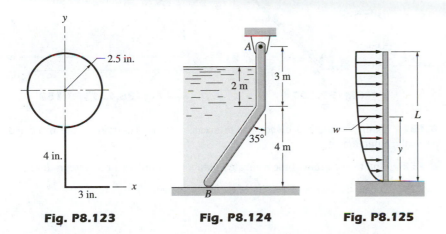

| **Fig. P8.123** | **Fig. P8.124** | **Fig. P8.125** |

8.125   The pressure of wind acting on a tall pole results in the line loading $w = w_0 \left[1 - \exp(-5y/L) \right]$, where w_0 is a constant. Use integration to determine the reactive couple at the base of the pole.

8.126   The bin, open at the top and closed at the bottom, is made from sheet metal of uniform thickness. Locate its center of gravity.

8.127   Find the volume of the solid that is generated by rotating the plane area shown about the y-axis.

8.128   Determine the surface area of the solid that is generated by rotating the plane area shown about the y-axis.

8.129   Determine the resultant of the distributed loading shown by integration.

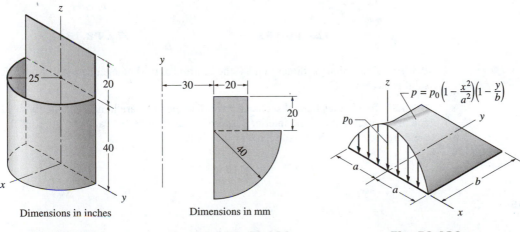

Dimensions in inches   Dimensions in mm

| **Fig. P8.126** | **Fig. P8.127, P8.128** | **Fig. P8.129** |

8.130 Calculate the tension in each of the three ropes, which support the uniform steel plate weighing 0.284 lb/in.³

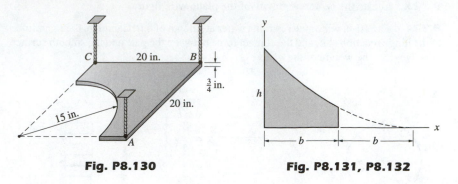

Fig. P8.130   Fig. P8.131, P8.132

8.131 Using the method of composite areas, find the centroid of the truncated parabolic complement.

8.132 Find the centroid of the truncated parabolic complement by integration.

8.133 Locate the center of gravity of the body shown.

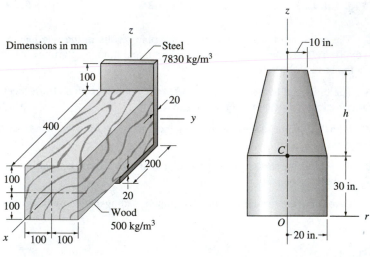

Fig. P8.133   Fig. P8.134

8.134 Determine the dimension h if the centroid of the homogeneous axisymmetric solid is located at C.

8.135 Two hemispherical shells of inner diameter 1 m are joined together with 12 equally spaced bolts. If the interior pressure is raised to 300 kPa above atmospheric pressure, determine the tensile force in each bolt.

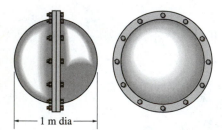

Fig. P8.135

8.136 Calculate the area of the surface generated when the plane Z-curve is rotated about (a) the x-axis; and (b) the y-axis.

8.137 Determine the resultant of the line loading, given that $w_0 = 36$ lb/ft.

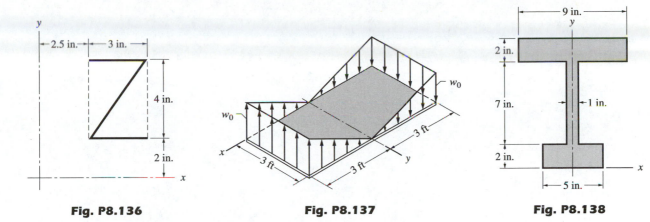

Fig. P8.136   Fig. P8.137   Fig. P8.138

8.138 Find the coordinates of the centroid of the plane region.

8.139 The sheet metal trough has a uniform wall thickness. Determine the coordinates of its center of gravity.

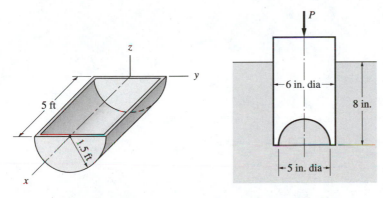

Fig. P8.139   Fig. P8.140

8.140 The thin-walled cylindrical can with a spherical dimple weighs 0.2 lb. Determine the force P required to push the can into water to a depth of 8 in. Use $\gamma = 0.036$ lb/in.^{3} for water.

8.141 Find the location of the centroid of the plane region.

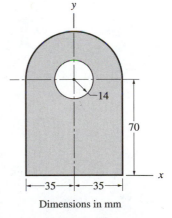

Dimensions in mm

Fig. P8.141

9

Moments and Products of Inertia of Areas

9.1 *Introduction*

First moments of areas, as presented in Chapter 8, dealt with the integrals $\int x \, dA$ and $\int y \, dA$. In this chapter, we discuss the second moments of plane areas, also known as moments of inertia, $\int x^2 \, dA$ and $\int y^2 \, dA$. We also introduce the product of inertia $\int xy \, dA$.

Moments and products of inertia arise in the analysis of linear load distributions acting on plane areas. Such distributions occur in members subjected to bending (beams), and in circular shafts carrying twisting couples. In addition, moments and products of inertia are encountered in the determination of resultants acting on submerged surfaces.

In this chapter we also discuss the dependence of moments and products of inertia on the orientation of the coordinate system. This dependence results in the transformation equations for moments and products of inertia, which are used to determine the maximum and minimum moments of inertia at a point. This chapter concludes with a discussion of Mohr's circle, a graphical method for representing the transformation equations.

9.2 *Moments of Inertia of Areas and Polar Moments of Inertia*

a. *Moment of inertia of area*

In Art. 8.2, the first moments of the area of a plane region $\mathcal{A}$ about the x- and y-axes were defined as

$$Q_x = \int_{\mathcal{A}} y \, dA \qquad Q_y = \int_{\mathcal{A}} x \, dA \qquad (9.1)$$

where A is the area of the region and x and y are the coordinates of the differential area element dA, as shown in Fig. 9.1.

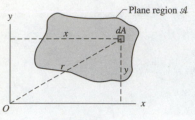

Fig. 9.1

The *moments of inertia of the area* about the x- and y-axes, respectively, are defined by*

$$
I_x = \int_{\mathcal{A}} y^2 \, dA \qquad I_y = \int_{\mathcal{A}} x^2 \, dA
\tag{9.2}
$$

Because the distances x and y are squared, I_x and I_y are sometimes called the *second moments of the area* about the axes.

The dimension for moment of inertia of area is $[L^4]$. Therefore, the units are in.^{4}, mm^4, and so forth. Although the first moment of an area can be positive, negative, or zero, its moment of inertia is always positive, because both x and y in Eqs. (9.2) are squared.

Caution   Recall that the first moment of an area can be obtained from $Q_x = A\bar{y}$, where $\bar{y}$ is the centroidal coordinate of the area. A mistake frequently made is to assume that $I_x = A\bar{y}^2$. Although the first moment of an area equals the area times the centroidal distance, the second moment of an area is *not* equal to the area times the centroidal distance squared.

b.   *Polar moment of inertia*

Referring again to Fig. 9.1, the polar moment of inertia of the area about point O (strictly speaking, about an axis through O, perpendicular to the plane of the area) is defined by

$$
J_O = \int_{\mathcal{A}} r^2 \, dA
\tag{9.3}
$$

where r is the distance from O to the differential area element dA. Note that the polar moment of an area is always positive and its dimension is $[L^4]$.

From Fig. 9.1, we note that $r^2 = y^2 + x^2$, which gives the following relationship between polar moment of inertia and moment of inertia:

$$
J_O = \int_{\mathcal{A}} r^2 \, dA = \int_{\mathcal{A}} (y^2 + x^2) \, dA = \int_{\mathcal{A}} y^2 \, dA + \int_{\mathcal{A}} x^2 \, dA
$$

or

$$
J_O = I_x + I_y
\tag{9.4}
$$

This relationship states that the polar moment of inertia of an area about a point O equals the sum of the moments of inertia of the area about two perpendicular axes that intersect at O.

---

*The term *moment of inertia of an area* should not be confused with *moment of inertia of a body*, which occurs in the study of dynamics. The latter refers to the ability of a body to resist a change in its rotation and is a property of mass. Because an area does not have mass, it does not possess inertia. However, the term *moment of inertia* is used because the integrals in Eqs. (9.2) are similar to the expression $\int r^2 \, dm$ that defines the moment of inertia of a body.

c. *Parallel-axis theorems*

There is a simple relationship between the moments of inertia about two parallel axes, provided that one of the axes passes through the centroid of the area. Referring to Fig. 9.2(a), let C be the centroid of the area contained in the plane region $\mathcal{A}$ and let the x'-axis be the centroidal axis that is parallel to the x-axis. We denote the moment of inertia about the x'-axis with $\bar{I}_x$, which is to be read as the "moment of inertia about the centroidal x-axis" (about the axis that is parallel to the x-axis and passes through the centroid of the area). Observe that the y-coordinate of the differential area dA can be written as $y = \bar{y} + y'$, where $\bar{y}$ (the centroidal coordinate of the area) is the distance between the two axes. Equations (9.2) yield the following expression for the moment of inertia of the area about the x-axis (note that $\bar{y}$ is constant).

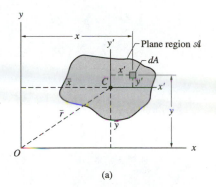

(a)

$$I_x = \int_{\mathcal{A}} y^2\, dA = \int_{\mathcal{A}} (\bar{y} + y')^2\, dA$$

$$= \bar{y}^2 \int_{\mathcal{A}} dA + 2\bar{y} \int_{\mathcal{A}} y'\, dA + \int_{\mathcal{A}} (y')^2 dA \tag{a}$$

Noting that $\int_{\mathcal{A}} dA = A$ (the area of the region), $\int_{\mathcal{A}} y'\, dA = 0$ (the first moment of the area about a centroidal axis vanishes), and $\int_{\mathcal{A}} (y')^2 dA = \bar{I}_x$ (the second moment of the area about the x'-axis), Eq. (a) simplifies to

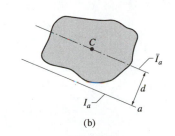

(b)

Fig. 9.2

$$I_x = \bar{I}_x + A\bar{y}^2 \tag{9.5a}$$

This relationship is known as the *parallel-axis theorem* for moment of inertia of an area. The distance $\bar{y}$ is sometimes called the *transfer distance* (the distance through which the moment of inertia is to be "transferred"). It is important to remember that the theorem is valid only if $\bar{I}_x$ is the moment of inertia about the centroidal x-axis. If this is not the case, the term $2\bar{y} \int_{\mathcal{A}} y'\, dA$ in Eq. (a) would not vanish, giving rise to an additional term in Eq. (9.5a).

Because the direction of the x-axis in Fig. 9.2(a) can be chosen arbitrarily, the parallel-axis theorem applies to axes of any orientation. For example, applying the theorem to the y-axis yields

$$I_y = \bar{I}_y + A\bar{x}^2 \tag{9.5b}$$

where $\bar{I}_y$ is the moment of inertia of the area about the centroidal y-axis—that is, the y'-axis in Fig. 9.2(a), and $\bar{x}$ is the x-coordinate of the centroid.

In general, the parallel-axis theorem can be written as

$$\boxed{I_a = \bar{I}_a + Ad^2} \tag{9.6}$$

As illustrated in Fig. 9.2(b), I_a is the moment of inertia about an arbitrarily oriented a-axis, $\bar{I}_a$ represents the moment of inertia about the parallel axis that passes through the centroid C, and d is the distance between the axes (transfer distance).

By inspection of Eq. (9.6), we see that, given the direction of the axis, the moment of inertia of an area is smallest about the axis that passes through the centroid of the area. In other words, $\bar{I}_a$ is smaller than the moment of inertia about any other axis that is parallel to the a-axis.

The parallel-axis theorem also applies to the polar moment of inertia. Denoting the polar moment of inertia of the area about the origin O by J_O, and about the centroid C by $\bar{J}_C$, we have from Eqs. (9.4) and (9.5)

$$J_O = I_x + I_y = (\bar{I}_x + A\bar{y}^2) + (\bar{I}_y + A\bar{x}^2)$$

Using $\bar{I}_x + \bar{I}_y = \bar{J}_C$, this equation becomes

$$\boxed{J_O = \bar{J}_C + A\bar{r}^2} \tag{9.7}$$

where $\bar{r} = \sqrt{\bar{x}^2 + \bar{y}^2}$ is the distance between O and C, as shown in Fig. 9.2(a).

d.   Radius of gyration

In some structural engineering applications, it is common practice to introduce the *radius of gyration of area*. The radii of gyration of an area about the x-axis, the y-axis, and the point O are defined as

$$\boxed{k_x = \sqrt{\frac{I_x}{A}} \qquad k_y = \sqrt{\frac{I_y}{A}} \qquad k_O = \sqrt{\frac{J_O}{A}}} \tag{9.8}$$

The dimension of the radius of gyration is $[L]$. However, the radius of gyration is not a distance that has a clear-cut physical meaning, nor can it be determined by direct measurement; its value can be determined only by computation using Eqs. (9.8). The radii of gyration are related by the equation

$$\boxed{k_O^2 = k_x^2 + k_y^2} \tag{9.9}$$

which can be obtained by substituting Eqs. (9.8) into Eq. (9.4).

e.   Integration techniques

When computing the moment of inertia of an area about a given axis by integration, we must choose a coordinate system and decide whether to use single or double integration. The differential area elements dA associated with various coordinate systems were discussed in Art 8.2 and illustrated in Fig. 8.4.

If double integration is used, the moments of inertia can be calculated from Eqs. (9.2) in a straightforward manner. However, in single integration we must view Eqs. (9.2) in the form

$$I_x = \int_{\mathcal{A}} dI_x \qquad I_y = \int_{\mathcal{A}} dI_y$$

where dI_x and dI_y are the moments of inertia of the area element dA about the x- and y-axes. In general, $dI_x = y^2\,dA$ only if all parts of the area element are the same distance y from the x-axis. To satisfy this condition, the area element must be either a double differential element ($dA = dx\,dy$), or a strip of width dy that is parallel to the x-axis, as shown in Fig. 8.4(c). A similar argument applies to dI_y.

f.   *Method of composite areas*

Consider a plane region $\mathcal{A}$ that has been divided into the subregions $\mathcal{A}_1$, $\mathcal{A}_2$, $\mathcal{A}_3$, ... The moment of inertia of the area of $\mathcal{A}$ about an axis can be computed by summing the moments of inertia of the subregions about the same axis. This technique, known as the *method of composite areas*, follows directly from the property of definite integrals: the integral of a sum is equal to the sum of the integrals. For example, I_x, the moment of inertia about the x-axis, becomes

$$I_x = \int_{\mathcal{A}} y^2 \, dA = \int_{\mathcal{A}_1} y^2 \, dA + \int_{\mathcal{A}_2} y^2 \, dA + \int_{\mathcal{A}_3} y^2 \, dA + \cdots$$

which can be written as

$$I_x = (I_x)_1 + (I_x)_2 + (I_x)_3 + \cdots \qquad \text{(9.10a)}$$

where $(I_x)_i$ is the moment of the inertia of the area of the subregion $\mathcal{A}_i$ with respect to the x-axis. Obviously, the method of composite areas also applies to the computation of polar moments of areas:

$$J_O = (J_O)_1 + (J_O)_2 + (J_O)_3 + \cdots \qquad \text{(9.10b)}$$

where $(J_O)_i$ is the polar moment of inertia of the subregion $\mathcal{A}_i$ with respect to point O.

The moments of inertia of the component areas about *their centroidal axes* can be found in tables, such as Tables 9.1 and 9.2. The parallel-axis theorem must then be used to convert these moments of inertia to the *common axis* before they can be summed.

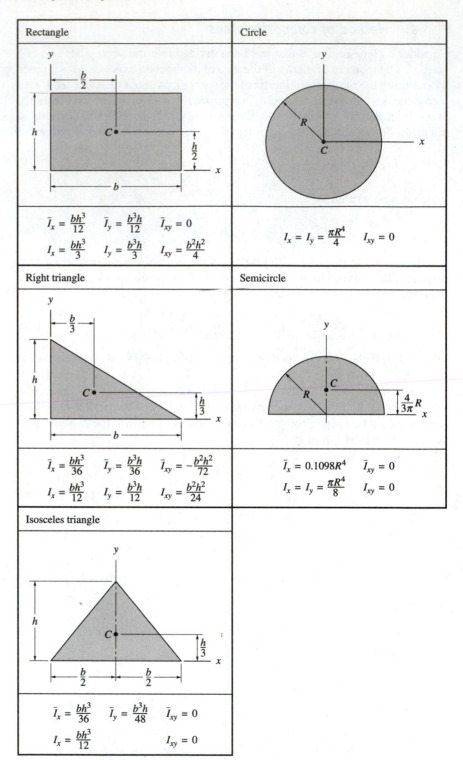

Table 9.1 *Inertial Properties of Plane Areas: Part 1*

Triangle		Half parabolic complement	
$\overline{I}_x = \dfrac{bh^3}{36}$	$I_x = \dfrac{bh^3}{12}$	$\overline{I}_x = \dfrac{37bh^3}{2100}$	$I_x = \dfrac{bh^3}{21}$
$\overline{I}_y = \dfrac{bh}{36}(a^2 - ab + b^2)$	$I_y = \dfrac{bh}{12}(a^2 + ab + b^2)$	$\overline{I}_y = \dfrac{b^3h}{80}$	$I_y = \dfrac{b^3h}{5}$
$\overline{I}_{xy} = \dfrac{bh^2}{72}(2a - b)$	$I_{xy} = \dfrac{bh^2}{24}(2a + b)$	$\overline{I}_{xy} = \dfrac{b^2h^2}{120}$	$I_{xy} = \dfrac{b^2h^2}{12}$

Quarter circle		Half parabola	
		$\overline{I}_x = \dfrac{8bh^3}{175}$	$I_x = \dfrac{2bh^3}{7}$
$\overline{I}_x = \overline{I}_y = 0.05488R^4$	$I_x = I_y = \dfrac{\pi R^4}{16}$	$\overline{I}_y = \dfrac{19b^3h}{480}$	$I_y = \dfrac{2b^3h}{15}$
$\overline{I}_{xy} = -0.01647R^4$	$I_{xy} = \dfrac{R^4}{8}$	$\overline{I}_{xy} = \dfrac{b^2h^2}{60}$	$I_{xy} = \dfrac{b^2h^2}{6}$

Quarter ellipse		Circular sector	
$\overline{I}_x = 0.05488ab^3$	$I_x = \dfrac{\pi ab^3}{16}$	$I_x = \dfrac{R^4}{8}(2\alpha - \sin 2\alpha)$	
$\overline{I}_y = 0.05488a^3b$	$I_y = \dfrac{\pi a^3b}{16}$	$I_y = \dfrac{R^4}{8}(2\alpha + \sin 2\alpha)$	
$\overline{I}_{xy} = -0.01647a^2b^2$	$I_{xy} = \dfrac{a^2b^2}{8}$	$I_{xy} = 0$	

Table 9.2   *Inertial Properties of Plane Areas: Part 2*

Sample Problem 9.1

The centroid of the plane region is located at C. If the area of the region is 2000 mm^2 and its moment of inertia about the x-axis is $I_x = 40 \times 10^6$ mm^4, determine I_u.

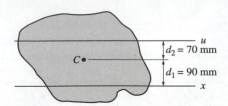

Solution

Note that we are required to transfer the moment of inertia from the x-axis to the u-axis, neither of which is a centroidal axis. Therefore, we must first calculate $\bar{I}_x$, the moment of inertia about the centroidal axis that is parallel to the x-axis.

From the parallel-axis theorem we have $I_x = \bar{I}_x + Ad_1^2$, which gives

$$\bar{I}_x = I_x - Ad_1^2 = (40 \times 10^6) - (2000)(90)^2 = 23.8 \times 10^6 \text{ mm}^4$$

After $\bar{I}_x$ has been found, the parallel-axis theorem enables us to compute the moment of inertia about any axis that is parallel to the centroidal axis. For I_u we have

$$I_u = \bar{I}_x + Ad_2^2 = (23.8 \times 10^6) + (2000)(70)^2$$
$$= 33.6 \times 10^6 \text{ mm}^4 \qquad \textit{Answer}$$

A common error is to use the parallel-axis theorem to transfer the moment of inertia between two axes, neither of which is a centroidal axis. In this problem, for example, it is tempting to write $I_u = I_x + A(d_1 + d_2)^2$, which would result in an incorrect answer for I_u.

Sample Problem 9.2

For the rectangle, compute the following: (1) the moment of inertia about the x-axis by integration; (2) the moment of inertia about the centroidal axis that is parallel to the x-axis; and (3) the polar moment of inertia about the centroid.

Solution

Part 1

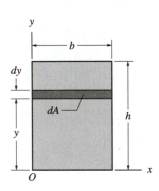

The area of the differential element shown in the figure is $dA = b\,dy$. Because all parts of the element are a distance y from the x-axis, we can use Eq. (9.2):

$$I_x = \int_{\mathscr{A}} y^2 \, dA = b \int_0^h y^2 \, dy = \frac{by^3}{3} \bigg]_0^h = \frac{bh^3}{3} \qquad \textit{Answer}$$

This result agrees with the information listed for a rectangle in Table 9.1.

If we had chosen to use double integration with $dA = dx\,dy$, the analysis would yield

$$I_x = \int_{\mathscr{A}} y^2 \, dA = \int_0^h \int_0^b y^2 \, dx \, dy = \frac{bh^3}{3}$$

which is identical to the previous result.

Part 2

We can calculate $\bar{I}_x$ either by integration, or from the parallel-axis theorem and the result of Part 1.

Substituting $I_x = bh^3/3$ into the parallel-axis theorem, and recognizing that the transfer distance d (the distance between the x-axis and the centroidal x-axis) is $h/2$, we find that

$$\bar{I}_x = I_x - Ad^2 = \frac{bh^3}{3} - bh\left(\frac{h}{2}\right)^2 = \frac{bh^3}{12} \qquad \textit{Answer}$$

This answer also agrees with the results in Table 9.1.

Part 3

One method of computing $\bar{J}_C$ is to use $\bar{J}_C = \bar{I}_x + \bar{I}_y$. From the results of Part 2, or Table 9.1, we have

$$\bar{J}_C = \bar{I}_x + \bar{I}_y = \frac{bh^3}{12} + \frac{hb^3}{12} = \frac{bh}{12}(h^2 + b^2) \qquad \textit{Answer}$$

Another method of computing $\bar{J}_C$ is to first compute $J_O = I_x + I_y$ and then transfer this result to the centroid. From the results of Part 1, we have

$$J_O = I_x + I_y = \frac{bh^3}{3} + \frac{hb^3}{3} = \frac{bh}{3}(h^2 + b^2)$$

The transfer distance is the distance between point O and the centroid of the rectangle; that is, $d = \sqrt{(b/2)^2 + (h/2)^2}$. From the parallel-axis theorem, we obtain

$$\bar{J}_C = J_O - Ad^2 = \frac{bh}{3}(h^2 + b^2) - bh\left(\frac{b^2}{4} + \frac{h^2}{4}\right)$$

or

$$\bar{J}_C = \frac{bh}{12}(h^2 + b^2) \qquad \textit{Answer}$$

which agrees with the previous result.

Sample Problem 9.3

By integration, calculate the moment of inertia about the y-axis of the area shown in Fig. (a) by the following methods: (1) single integration using a vertical differential area element; (2) double integration; and (3) single integration using a horizontal differential area element.

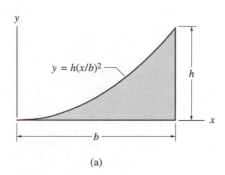

(a)

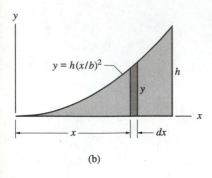

(b)

Solution

Part 1

The vertical differential area element is shown in Fig. (b). Because all parts of the element are the same distance x from the y-axis, we may use Eq. (9.2) directly. With $dA = y\,dx = h(x/b)^2\,dx$, we have

$$I_y = \int_{\mathcal{A}} x^2\,dA = \frac{h}{b^2}\int_0^b x^4\,dx = \frac{h}{b^2}\frac{b^5}{5} = \frac{b^3 h}{5} \qquad \text{Answer}$$

which agrees with the information in Table 9.2 for the half parabolic complement.

Part 2

Equation (9.2) can also be used with the double differential area element $dA = dy\,dx$. Choosing to integrate on y first, we obtain

$$I_y = \int_{\mathcal{A}} x^2\,dA = \int_0^b \int_0^{h(x/b)^2} x^2\,dy\,dx$$

Integrating over x first would yield

$$I_y = \int_{\mathcal{A}} x^2\,dA = \int_0^h \int_0^{b(y/h)^{1/2}} x^2\,dx\,dy$$

Performing either of the foregoing integrations yields the same expression for I_y as found in Part 1.

Part 3

The horizontal differential area element is shown in Fig. (c). Because all parts of the differential area element are not the same distance from the y-axis, Eq. (9.2) cannot be applied directly. To find I_y for the entire area, we must integrate dI_y, the moment of inertia of the differential area element about the y-axis.

Table 9.1 lists $\bar{I}_y = hb^3/12$ for a rectangle. Therefore, the moment of inertia of the differential element about its vertical centroidal axis (axis parallel to the y-axis passing through the centroid C_{el} of the element) is $d\bar{I}_y = dy\,(b-x)^3/12$. According to the parallel-axis theorem, $dI_y = d\bar{I}_y + dA\,(d_{el}^2)$, where d_{el} is the distance between the y-axis and the vertical centroidal axis of the element. Using $d_{el} = (b+x)/2$, as shown in Fig. (c), and integrating, we obtain I_y for the entire area:

$$I_y = \int_{\mathcal{A}} dI_y = \int_0^h \left[\frac{dy\,(b-x)^3}{12} + (b-x)\,dy\left(\frac{b+x}{2}\right)^2\right]$$

Substituting $x = b(y/h)^{1/2}$ and completing the integration gives the same result as found in Part 1.

Obviously, the horizontal differential area element is not as convenient as the other choices in this particular problem.

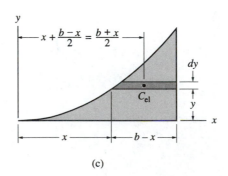

(c)

Sample Problem 9.4

For the area shown in Fig. (a), calculate the radii of gyration about the x- and y-axes.

Solution

We consider the area to be composed of the three parts shown in Figs. (b)–(d): a triangle, plus a semicircle, minus a circle. The moments of inertia of each part are obtained in two steps. First, the moments of inertia about the centroidal axes of the part are found from Table 9.1. The parallel-axis theorem is then used to calculate the moments of inertia about the x- and y-axes.

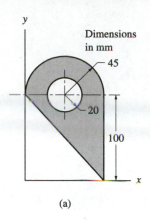

(a)

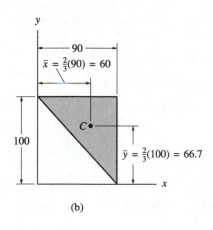

(b)

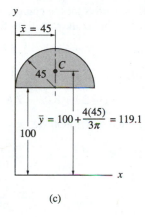

(c)

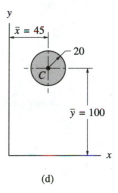

(d)

Triangle

$$A = \frac{bh}{2} = \frac{90(100)}{2} = 4500 \text{ mm}^2$$

$$\bar{I}_x = \frac{bh^3}{36} = \frac{90(100)^3}{36} = 2.50 \times 10^6 \text{ mm}^4$$

$$I_x = \bar{I}_x + A\bar{y}^2 = (2.50 \times 10^6) + (4500)(66.7)^2 = 22.52 \times 10^6 \text{ mm}^4$$

$$\bar{I}_y = \frac{hb^3}{36} = \frac{100(90)^3}{36} = 2.025 \times 10^6 \text{ mm}^4$$

$$I_y = \bar{I}_y + A\bar{x}^2 = (2.025 \times 10^6) + (4500)(60)^2 = 18.23 \times 10^6 \text{ mm}^4$$

Semicircle

$$A = \frac{\pi R^2}{2} = \frac{\pi (45)^2}{2} = 3181 \text{ mm}^2$$

$$\bar{I}_x = 0.1098R^4 = 0.1098(45)^4 = 0.450 \times 10^6 \text{ mm}^4$$

$$I_x = \bar{I}_x + A\bar{y}^2 = (0.450 \times 10^6) + (3181)(119.1)^2 = 45.57 \times 10^6 \text{ mm}^4$$

$$\bar{I}_y = \frac{\pi R^4}{8} = \frac{\pi (45)^4}{8} = 1.61 \times 10^6 \text{ mm}^4$$

$$I_y = \bar{I}_y + A\bar{x}^2 = (1.61 \times 10^6) + (3181)(45)^2 = 8.05 \times 10^6 \text{ mm}^4$$

Circle

$$A = \pi R^2 = \pi(20)^2 = 1257 \text{ mm}^2$$

$$\bar{I}_x = \frac{\pi R^4}{4} = \frac{\pi(20)^4}{4} = 0.1257 \times 10^6 \text{ mm}^4$$

$$I_x = \bar{I}_x + A\bar{y}^2 = (0.1257 \times 10^6) + (1257)(100)^2 = 12.70 \times 10^6 \text{ mm}^4$$

$$\bar{I}_y = \frac{\pi R^4}{4} = \frac{\pi(20)^4}{4} = 0.1257 \times 10^6 \text{ mm}^4$$

$$I_y = \bar{I}_y + A\bar{x}^2 = (0.1257 \times 10^6) + (1257)(45)^2 = 2.67 \times 10^6 \text{ mm}^4$$

Composite Area

To determine the properties for the composite area, we superimpose the foregoing results (taking care to subtract the quantities for the circle) and obtain

$$A = \Sigma A = 4500 + 3181 - 1257 = 6424 \text{ mm}^2$$

$$I_x = \Sigma I_x = (22.52 + 45.57 - 12.70) \times 10^6 = 55.39 \times 10^6 \text{ mm}^4$$

$$I_y = \Sigma I_y = (18.23 + 8.05 - 2.67) \times 10^6 = 23.61 \times 10^6 \text{ mm}^4$$

Therefore, for the radii of gyration we have

$$k_x = \sqrt{\frac{I_x}{A}} = \sqrt{\frac{55.39 \times 10^6}{6424}} = 92.9 \text{ mm} \qquad \textit{Answer}$$

$$k_y = \sqrt{\frac{I_y}{A}} = \sqrt{\frac{23.61 \times 10^6}{6424}} = 60.6 \text{ mm} \qquad \textit{Answer}$$

Sample Problem 9.5

By numerical integration, calculate the moments of inertia about the x- and y-axes for the half parabolic complement. Use Simpson's rule with four panels. Compare your results with the exact values in Table 9.2.

Solution

Numerical integration by Simpson's rule can be summarized as follows (see Appendix A).

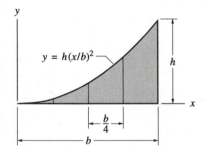

$$\int_a^b f(x)\, dx \approx \sum_{i=1}^{n+1} W_i f_i \tag{a}$$

where the range of integration is assumed to be divided into n panels of width Δx (n must be an even number) and where W_i are the weights, given by

$$W_1 = W_{n+1} = \frac{\Delta x}{3}$$

$$\left. \begin{array}{l} W_i = \dfrac{4\,\Delta x}{3} \quad i \text{ even} \\[2mm] W_i = \dfrac{2\,\Delta x}{3} \quad i \text{ odd} \end{array} \right\} \quad 2 \le i \le n$$

In Eq. (a), the expression $\sum_{i=1}^{n+1} W_i f_i$ is called the weighted summation.

Choosing a vertical strip of width dx and height y as the differential area element, its inertial properties are $dI_x = y^3\, dx/3$ (moment of inertia of a rectangle about its base, as listed in Table 9.1) and $dI_y = x^2\, dA$ (note that all parts of the area element are the same distance x from the y-axis). Therefore, we have

$$I_x = \int_{sA} dI_x = \int_0^b \frac{y^3\, dx}{3} \approx \sum_{i=1}^{n+1} W_i \frac{y_i^3}{3} \qquad \text{(b)}$$

$$I_y = \int_{sA} dI_y = \int_{sA} x^2\, dA = \int_0^b x^2 y\, dx \approx \sum_{i=1}^{n+1} W_i x_i^2 y_i \qquad \text{(c)}$$

For our problem, $n = 4$ and $\Delta x = b/4$. The numerical computations indicated in Eqs. (b) and (c) are shown in the following table.

i	x	y	$y^3/3$	$x^2 y$	W
1	0.00	0.0000	0.000 00	0.000 00	$b/12$
2	$0.25b$	$0.0625h$	$0.000\,08h^3$	$0.003\,91b^2 h$	$4(b/12)$
3	$0.50b$	$0.2500h$	$0.005\,21h^3$	$0.062\,50b^2 h$	$2(b/12)$
4	$0.75b$	$0.5625h$	$0.059\,33h^3$	$0.316\,41b^2 h$	$4(b/12)$
5	$1.00b$	$1.0000h$	$0.333\,33h^3$	$1.000\,00b^2 h$	$b/12$

Substituting the values from this table into Eqs. (b) and (c) gives

$$I_x = \frac{bh^3}{12}[1(0) + 4(0.000\,08) + 2(0.005\,21) + 4(0.059\,33)]$$

$$+ 1(0.333\,33)] = 0.0485bh^3 \qquad \textit{Answer}$$

$$I_y = \frac{b^3 h}{12}[1(0) + 4(0.003\,91) + 2(0.062\,50) + 4(0.316\,41)]$$

$$+ 1(1.000\,00)] = 0.2005b^3 h \qquad \textit{Answer}$$

According to Table 9.2, the half parabolic complement has $I_x = bh^3/21 = 0.0476bh^3$ and $I_y = b^3 h/5 = 0.2000b^3 h$. Therefore, the errors introduced by our numerical integration are 1.9% for I_x and 0.25% for I_y.

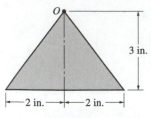

Fig. P9.1

Problems

9.1 Using the parallel-axis theorem and the information in Table 9.1, compute for the triangle shown (a) the polar moment of inertia about the centroid; (b) the polar moment of inertia about point O; and (c) the polar radius of gyration about point O.

9.2 The properties of the plane region are $\bar{J}_C = 50 \times 10^3$ mm^4, $I_x = 600 \times 10^3$ mm^4, and $I_y = 350 \times 10^3$ mm^4. Calculate A, $\bar{I}_x$, and $\bar{I}_y$ for the region.

9.3 The moments of inertia about the x- and u-axes of the plane region are $I_x = 14 \times 10^9$ mm^4 and $I_u = 38 \times 10^9$ mm^4, respectively. If $h = 200$ mm, determine the area of the region, and the radius of gyration about the centroidal axis parallel to the x-axis.

9.4 Find the distance h for which the moment of inertia of the plane region about the u-axis equals 120×10^9 mm^4, given that $A = 90 \times 10^3$ mm^2 and $I_x = 14 \times 10^9$ mm^4.

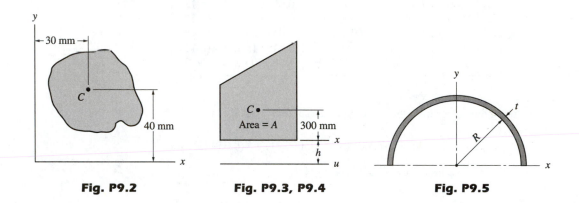

Fig. P9.2   **Fig. P9.3, P9.4**   **Fig. P9.5**

9.5 Using integration, find the moment of inertia and the radius of gyration about the x-axis for the thin ring ($t \ll R$).

9.6 By integration, derive the expression for the moment of inertia of the thin rectangle ($t \ll L$) about the x-axis.

9.7 Compute I_x for the shaded region using integration.

9.8 Using integration, compute the polar moment of inertia about point O for the circular sector. Check your result with Table 9.2.

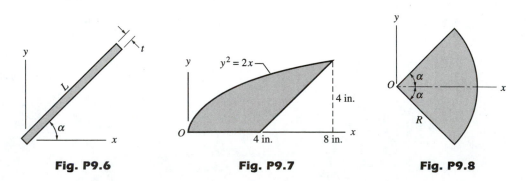

Fig. P9.6   **Fig. P9.7**   **Fig. P9.8**

9.9 Use integration to compute I_x and I_y for the *n*th order parabola. Check your answers with the results for the half parabolic complement in Table 9.2.

9.10 By integration, determine the moments of inertia about the *x*- and *y*-axes for the region shown.

9.11 Compute the moment of inertia about the *x*-axis for the region shown using integration.

9.12 By integration, find the moment of inertia about the *y*-axis for the region shown.

9.13 Figure (a) shows the cross section of a column that uses a structural shape known as W8 × 67 (wide-flange beam, nominally 8 in. deep, weighing 67 lb/ft). The American Institute of Steel Construction *Structural Steel Handbook* lists the following cross-sectional properties: $A = 19.7$ in.^{2}, $\bar{I}_x = 272$ in.^{4}, and $\bar{I}_y = 88.6$ in.^{4} Determine the dimensions of the rectangle in Fig. (b) that has the same $\bar{I}_x$ and $\bar{I}_y$ as a W8 × 67 section.

9.14 Compute the dimensions of the rectangle shown in Fig. (b) that has the same $\bar{k}_x$ and $\bar{k}_y$ as the W8 × 67 section in Fig. (a). (See Prob. 9.13 for properties of W8 × 67.)

Problems 9.15–9.29 are to be solved using the method of composite areas.

9.15 Compute $\bar{I}_x$ and $\bar{I}_y$ for the W8 × 67 shape dimensioned in the figure. Assume that the section is composed of rectangles, neglecting the effects due to rounding of the corners. Compare your results with the handbook values listed in Prob. 9.13.

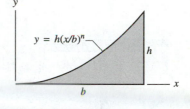

Fig. P9.9

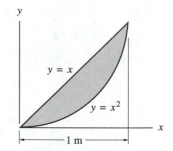

Fig. P9.10

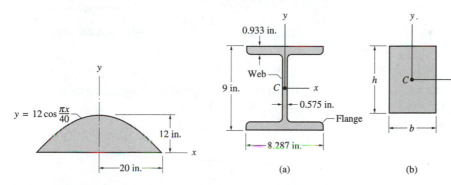

Fig. P9.11, P9.12   Fig. P9.13–P9.15

9.16 Figure (a) shows the cross-sectional dimensions for the structural steel section known as C10 × 20 (channel with a nominal depth of 10 in., weighing 20 lb/ft). The American Institute of Steel Construction *Structural Steel Handbook* lists the following properties for the cross section: $A = 5.88$ in.^{2}, $\bar{I}_x = 78.9$ in.^{4}, and $\bar{I}_y = 2.81$ in.^{4} If two of these channels are welded together as shown in Fig. (b), find $\bar{I}_x$ and $\bar{I}_y$ for the resulting cross section.

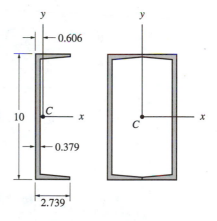

Dimensions in inches

(a)   (b)

Fig. P9.16

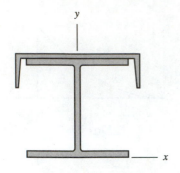

Fig. P9.17

9.17   A W8 × 67 section is joined to a C10 × 20 section to form a structural member that has the cross section shown. Calculate $\bar{I}_x$ and $\bar{I}_y$ for this cross section. (See Probs. 9.13 and 9.16 for the properties of the structural sections.)

9.18   Compute the polar moment of inertia about point O for the region shown.

9.19   Find $\bar{I}_x$ and $\bar{I}_y$ for the region shown.

9.20   Calculate $\bar{I}_x$ for the shaded region, knowing that $\bar{y} = 68.54$ mm.

9.21   Compute $\bar{I}_y$ for the region shown, given that $\bar{x} = 25.86$ mm.

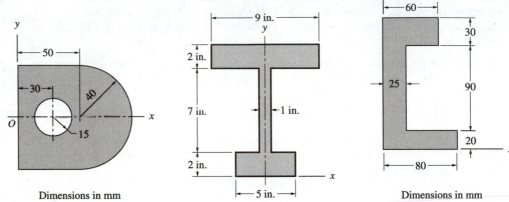

Dimensions in mm

Fig. P9.18   **Fig. P9.19**   **Fig. P9.20, P9.21**

9.22   Calculate $\bar{I}_x$ and $\bar{I}_y$ for the shaded region.

9.23   For the plane region with a circular cutout, (a) find I_x; and (b) compute $\bar{I}_x$ using the result of part (a) and the parallel-axis theorem.

9.24   Determine $\bar{I}_x$ for the triangular region shown.

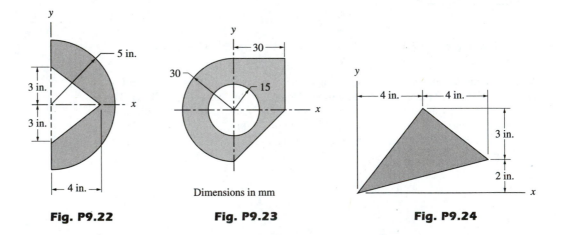

Dimensions in mm

Fig. P9.22   **Fig. P9.23**   **Fig. P9.24**

9.25 Determine the distance h for which the moment of inertia of the region shown about the x-axis will be as small as possible.

9.26 A circular region of radius $R/2$ is cut out from the circular region of radius R as shown. For what distance d will k_x for the new region be the same as k_x for the region before the cutout was removed?

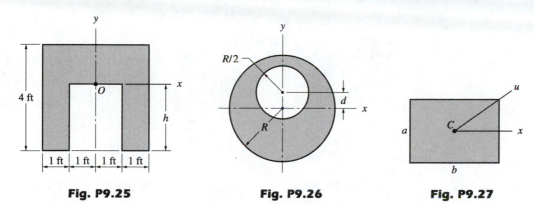

Fig. P9.25	Fig. P9.26	Fig. P9.27

9.27 Determine the ratio a/b for which $I_x = I_u$ for the rectangle.

9.28 Determine the ratio a/b for which $\bar{I}_x = \bar{I}_y$ for the isosceles triangle.

***9.29** As a round log passes through a sawmill, two slabs are cut off, resulting in the cross section shown. Calculate the distance h if it is known that the moment of inertia about the x-axis for the original cross section is reduced by 50 percent in the sawing process.

9.30 For the circular sector shown in Table 9.2, determine the angle α, other than $\alpha = 0$ and $\alpha = \pi$, for which $\bar{I}_x = \bar{I}_y$.

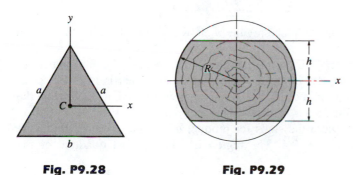

Fig. P9.28	Fig. P9.29

9.31 By numerical integration, compute the moments of inertia about the x- and y-axes for the region shown. The region is symmetric with respect to each coordinate axis.

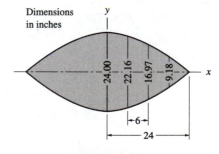

Dimensions in inches

Fig. P9.31

9.32 Use numerical integration to compute the moments of inertia about the *x*- and *y*-axes for the symmetric region shown.

9.33 Calculate the moments of inertia about the *x*- and *y*-axes for the quarter ellipse. Use numerical integration with $\Delta x = 2$ in. Compare your results with Table 9.2.

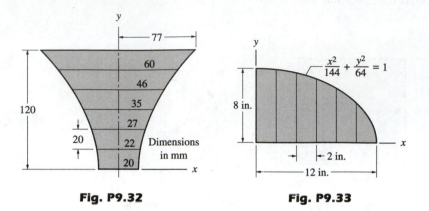

Fig. P9.32   Fig. P9.33

9.3 *Products of Inertia of Areas*

a. *Definition*

The *product of inertia of a plane area* (also called the *product of area*) about the *x*- and *y*-coordinate axes is defined by

Fig. 9.3

$$I_{xy} = \int_{\mathcal{A}} xy \, dA \qquad (9.11)$$

where *A* is the area of the plane region $\mathcal{A}$ shown in Fig. 9.3, and *x* and *y* are the coordinates of *dA*.

The dimension of product of inertia is $[L^4]$, the same as for moment of inertia and polar moment of area. Whereas moment of inertia is always positive, the product of inertia can be positive, negative, or zero, depending on the manner in which the area is distributed in the *xy*-plane.

To further explore the signs for product of inertia, consider the plane region $\mathcal{A}$ shown in Fig. 9.4. The region lies in the first quadrant of the

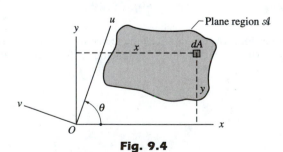

Fig. 9.4

xy-coordinate system. Because both x and y are positive for every differential area element dA, $I_{xy} = \int_{\mathcal{A}} xy\, dA$ is clearly positive. However, relative to the uv-coordinate system, the region $\mathcal{A}$ lies in the fourth quadrant, so that the u-coordinate of each dA is positive and the v-coordinate is negative. Therefore, $I_{uv} = \int_{\mathcal{A}} uv\, dA$ is negative.

Note that the uv-axes in Fig. 9.4 are rotated counterclockwise through the angle θ relative to the xy-axes. Because I_{xy} is positive and I_{uv} is negative, there must be an orientation of the axes for which the product of inertia is zero. As we see in the next article, the axes of zero product of inertia play a fundamental role in the calculation of the maximum and minimum moments of inertia.

Next, consider a region that has an axis of symmetry, such as that shown in Fig. 9.5. Because the y-axis is the axis of symmetry, for every dA with coordinates (x, y), there is a dA with coordinates $(x, -y)$. Therefore, $\int_{\mathcal{A}} xy\, dA = 0$ when the integration is performed over the entire region. Consequently, we have the following property:

> If an area has an axis of symmetry, that axis and the axis perpendicular to it constitute a set of axes for which the product of inertia is zero.

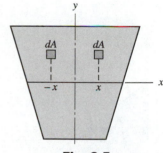

Fig. 9.5

b.   *Parallel-axis theorem*

The parallel-axis theorem for products of inertia can be derived by considering the plane region shown in Fig. 9.6. We let x' and y' be axes through the centroid C and parallel to the x- and y-axes. The coordinates of C relative to the xy-axes are $\bar{x}$ and $\bar{y}$. Using $x = x' + \bar{x}$ and $y = y' + \bar{y}$ in Eq. (9.11), we obtain

$$I_{xy} = \int_{\mathcal{A}} xy\, dA = \int_{\mathcal{A}} (x' + \bar{x})(y' + \bar{y})\, dA$$

$$= \int_{\mathcal{A}} x'y'\, dA + \bar{x} \int_{\mathcal{A}} y'\, dA + \bar{y} \int_{\mathcal{A}} x'\, dA + \bar{x}\bar{y} \int_{\mathcal{A}} dA$$

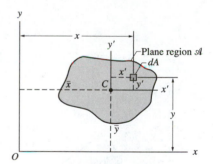

Fig. 9.6

The first term on the right-hand side is the product of inertia with respect to the centroidal axes, which we denote with $\bar{I}_{xy}$. The middle two terms are zero, because each integral represents the first moment of the area about a centroidal axis. The integral in the last term is simply the area A. Therefore, the parallel-axis theorem for products of inertia can be written as

$$\boxed{I_{xy} = \bar{I}_{xy} + A\bar{x}\bar{y}} \qquad (9.12)$$

To reiterate, the symbol $\bar{I}_{xy}$ is to be read as "the product of inertia relative to centroidal x- and y-axes" (axes through the centroid and parallel to the x- and y-axes).

It should be evident that the method of composite areas is also valid for products of inertia. Tables 9.1 and 9.2 list the products of inertia for common shapes, which can be utilized in the method of composite areas.

Sample Problem 9.6

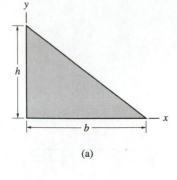

(a)

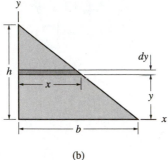

(b)

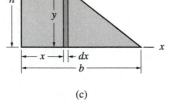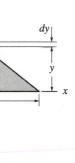

(c)

Calculate the product of inertia of the triangle shown in Fig. (a) about the x- and y-axes using (1) single integration; and (2) double integration.

Solution

Part 1

By definition, $I_{xy} = \int_{\mathcal{A}} xy\, dA$, where x and y are the coordinates of the differential area element $dA = dy\, dx$. However, this formula does not apply to single integration, where we must integrate dI_{xy}, the product of inertia of the differential area element. To find dI_{xy}, the parallel-axis theorem for products of inertia, $I_{xy} = \bar{I}_{xy} + A\bar{x}\bar{y}$, must be interpreted as $dI_{xy} = d\bar{I}_{xy} + dA\,\bar{x}_{el}\bar{y}_{el}$, where $d\bar{I}_{xy}$ is the product of inertia of dA about its centroidal axes, and $\bar{x}_{el}$ and $\bar{y}_{el}$ are the centroidal coordinates of the area element.

The analysis then proceeds as follows:

Horizontal element shown in Fig. (b)

$$dA = x\, dy$$

$$\bar{x}_{el} = \frac{x}{2}$$

$$\bar{y}_{el} = y$$

$$d\bar{I}_{xy} = 0 \quad \text{(by symmetry)}$$

$$dI_{xy} = dA\,\bar{x}_{el}\bar{y}_{el}$$

$$= (x\, dy)\left(\frac{x}{2}\right)(y)$$

$$= \frac{x^2 y}{2}\, dy$$

Substitute $x = \dfrac{b}{h}(h - y)$ and integrate

$$I_{xy} = \frac{b^2}{2h^2}\int_0^h (h - y)^2\, y\, dy$$

$$I_{xy} = \frac{b^2 h^2}{24} \qquad \textit{Answer}$$

Vertical element shown in Fig. (c)

$$dA = y\, dx$$

$$\bar{x}_{el} = x$$

$$\bar{y}_{el} = \frac{y}{2}$$

$$d\bar{I}_{xy} = 0 \quad \text{(by symmetry)}$$

$$dI_{xy} = dA\,\bar{x}_{el}\bar{y}_{el}$$

$$= (y\, dx)(x)\left(\frac{y}{2}\right)$$

$$= \frac{xy^2}{2}\, dx$$

Substitute $y = \dfrac{h}{b}(b - x)$ and integrate

$$I_{xy} = \frac{h^2}{2b^2}\int_0^b x(b - x)^2\, dx$$

$$I_{xy} = \frac{b^2 h^2}{24} \qquad \textit{Answer}$$

These results agree with the information listed for the triangle in Table 9.1.

Part 2

Using double integration with $dA = dx\, dy$, we use $I_{xy} = \int_{\mathcal{A}} xy\, dA$. Choosing to integrate on y first, we have

$$I_{xy} = \int_0^b \left(\int_0^{(h/b)(b-x)} xy\, dy \right) dx$$

Integrating on x first would yield

$$I_{xy} = \int_0^h \left(\int_0^{(b/h)(h-y)} xy\, dx \right) dy$$

Evaluating either of the above integrals yields $I_{xy} = b^2 h^2/24$, as in Part 1.

Sample Problem 9.7

Using the results of Sample Problem 9.6, calculate $\bar{I}_{xy}$, the product of inertia of the triangle shown about centroidal axes parallel to the x- and y-axes.

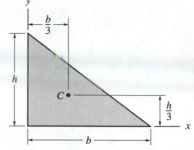

Solution

From the solution to Sample Problem 9.6, we have $I_{xy} = b^2 h^2/24$. The parallel-axis theorem yields

$$\bar{I}_{xy} = I_{xy} - A\bar{x}\bar{y} = \frac{b^2 h^2}{24} - \frac{bh}{2}\left(\frac{b}{3}\right)\left(\frac{h}{3}\right)$$

which simplifies to

$$\bar{I}_{xy} = -\frac{b^2 h^2}{72} \qquad \qquad \textit{Answer}$$

The above result agrees with the information in Table 9.1 for the right triangle.

Sample Problem 9.8

Calculate the product of inertia I_{xy} for the angle shown in Fig. (a) by the method of composite areas.

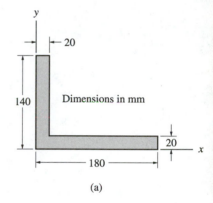

Solution

We may view the angle as the composite of the two rectangles shown in Fig. (b). For each rectangle, I_{xy} can be computed using the parallel-axis theorem for products of inertia: $I_{xy} = \bar{I}_{xy} + A\bar{x}\bar{y}$. Note that $\bar{I}_{xy} = 0$ for each rectangle, by symmetry. For the 20 mm $\times$ 140 mm rectangle,

$$I_{xy} = 0 + (140 \times 20)(10)(70) = 1.96 \times 10^6 \text{ mm}^4$$

For the 160 mm $\times$ 20 mm rectangle,

$$I_{xy} = 0 + (160 \times 20)(100)(10) = 3.20 \times 10^6 \text{ mm}^4$$

Therefore, the product of inertia for the angle is

$$I_{xy} = \Sigma I_{xy} = (1.96 + 3.20) \times 10^6 = 5.16 \times 10^6 \text{ mm}^4 \qquad \textit{Answer}$$

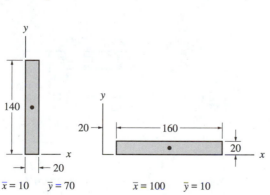

(b)

Problems

9.34 Use integration to verify the formula given in Table 9.2 for I_{xy} of a half parabola.

9.35 For the quarter ellipse in Table 9.2, verify the following formulas: (a) I_{xy} by integration; and (b) $\bar{I}_{xy}$ using the formula for I_{xy} and the parallel-axis theorem.

9.36 Determine the product of inertia with respect to the x- and y-axes for the quarter circular, thin ring ($t \ll R$) by integration.

9.37 By integration, determine I_{xy} for the triangle. Check your result with Table 9.2.

9.38 Calculate I_{uv} for the region shown, given that $\bar{x} = 30$ mm, $d = 40$ mm, and $I_{xy} = 520 \times 10^3$ mm^4.

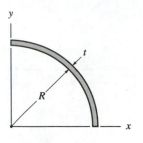

Fig. P9.36

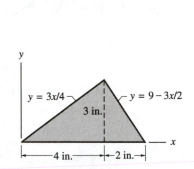

Fig. P9.37

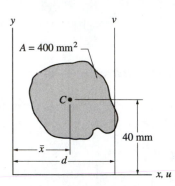

Fig. P9.38, P9.39

9.39 For the region shown, $I_{xy} = 320 \times 10^3$ mm^4 and $I_{uv} = 0$. Compute the distance d between the y- and v-axes. (Note: The result is independent of $\bar{x}$.)

Problems 9.40–9.46 are to be solved using the method of composite areas.

9.40 Compute the product of inertia with respect to the x- and y-axes.

9.41 Calculate the product of inertia with respect to the x- and y-axes.

9.42 Find $\bar{I}_{xy}$ for the region shown.

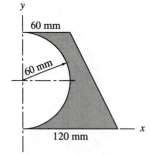

Fig. P9.40

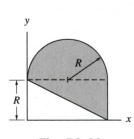

Fig. P9.41

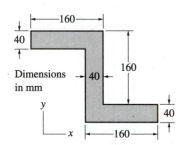

Fig. P9.42

9.43 Determine $\bar{I}_{xy}$ for the plate with parabolic cutouts.

9.44 The figure shows the cross section of a standard L80 × 60 × 10-mm structural steel, unequal angle section. Neglecting the effects of the small corner fillets, compute I_{xy} of the cross-sectional area.

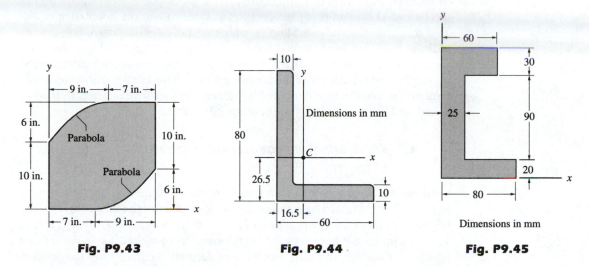

Fig. P9.43 **Fig. P9.44** **Fig. P9.45**

9.45 Calculate $\bar{I}_{xy}$ for the region shown, knowing that $\bar{x} = 25.86$ mm and $\bar{y} = 68.54$ mm.

9.46 Compute $\bar{I}_{xy}$ for the region shown.

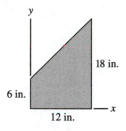

9.47 The plane region is subjected to a normal pressure of intensity $p = cy$, where c is a constant. The resultant force due to the pressure can be represented by a force $\mathbf{R}$ acting at O and a couple $\mathbf{C}^R$. Show that $\mathbf{R} = cQ_x\mathbf{k}$, where Q_x is the first moment of the area of the region about the x-axis and $\mathbf{C}^R = c(I_x\mathbf{i} - I_y\mathbf{j})$.

9.48 Use numerical integration to compute the product of inertia of the region shown with respect to the x- and y-axes.

9.49 Use numerical integration to calculate I_{xy} for the quarter ellipse. Compare your result with the formula given in Table 9.2.

Fig. P9.46

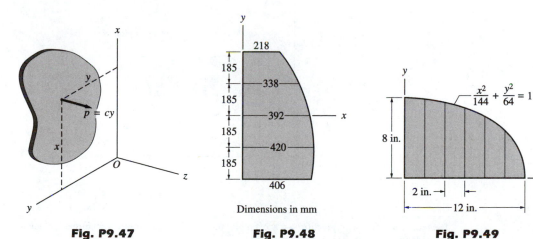

Fig. P9.47 **Fig. P9.48** **Fig. P9.49**

9.4   *Transformation Equations and Principal Moments of Inertia of Areas*

In general, the values of I_x, I_y, and I_{xy} of a given plane area depend on the location of O (the origin of the coordinate system) and the orientation of the xy-axes. The effect of relocating O, which is equivalent to translating the coordinate axes, has been studied and has resulted in the parallel-axis theorem. Here we investigate the changes in the moments and product of inertia caused by varying the orientation of the coordinate axes. This in turn enables us to determine the maximum and minimum moments of inertia associated with point O and find the orientation of the corresponding axes.

a.   *Transformation equations for moments and products of inertia*

Consider the plane region $\mathcal{A}$ with area A shown in Fig. 9.7, where the uv-axes at point O are obtained by rotating the xy-axes counterclockwise through the angle θ. We now derive formulas for I_u, I_v, and I_{uv} in terms of I_x, I_y, I_{xy}, and θ. These formulas are known as the *transformation equations for moments and products of inertia.** We start with the transformation equations of the position coordinates, which can be derived from Fig. 9.7:

$$u = y\sin\theta + x\cos\theta$$
$$v = y\cos\theta - x\sin\theta \tag{9.13}$$

Substituting these equations into the defining equation for I_u, we have

$$I_u = \int_{\mathcal{A}} v^2\, dA = \int_{\mathcal{A}} (y\cos\theta - x\sin\theta)^2\, dA$$

$$= \cos^2\theta \int_{\mathcal{A}} y^2\, dA - 2\sin\theta\cos\theta \int_{\mathcal{A}} xy\, dA + \sin^2\theta \int_{\mathcal{A}} x^2\, dA$$

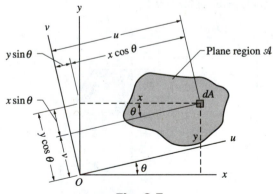

Fig. 9.7

---

*Several other physical quantities—for example, stress and strain—obey transformation equations identical to those for moment of inertia.

Identifying the moments and products of inertia, this equation becomes

$$I_u = I_x \cos^2 \theta - 2I_{xy} \sin \theta \cos \theta + I_y \sin^2 \theta \tag{9.14}$$

The equations for I_v and I_{uv} may be derived in a similar manner, the results being

$$I_v = I_x \sin^2 \theta + 2I_{xy} \sin \theta \cos \theta + I_y \cos^2 \theta \tag{9.15}$$

$$I_{uv} = (I_x - I_y) \sin \theta \cos \theta + I_{xy}(\cos^2 \theta - \sin^2 \theta) \tag{9.16}$$

The equation for I_v could also be derived by replacing θ with $(\theta + 90°)$ in Eq. (9.14).

Using the trigonometric identities

$$\sin 2\theta = 2 \sin \theta \cos \theta \qquad \cos 2\theta = \cos^2 \theta - \sin^2 \theta$$

$$\cos^2 \theta = \frac{1}{2}(1 + \cos 2\theta) \qquad \sin^2 \theta = \frac{1}{2}(1 - \cos 2\theta)$$

Eqs. (9.14)–(9.16) can also be written in the form

$$I_u = \tfrac{1}{2}(I_x + I_y) + \tfrac{1}{2}(I_x - I_y) \cos 2\theta - I_{xy} \sin 2\theta \tag{9.17}$$

$$I_v = \tfrac{1}{2}(I_x + I_y) - \tfrac{1}{2}(I_x - I_y) \cos 2\theta + I_{xy} \sin 2\theta \tag{9.18}$$

$$I_{uv} = \tfrac{1}{2}(I_x - I_y) \sin 2\theta + I_{xy} \cos 2\theta \tag{9.19}$$

From Eqs. (9.17)–(9.19) we see that $I_u + I_v = I_x + I_y$, a result that we expected, because both sides of the equation are equal to J_O, the polar moment of the area about O.

b. Principal moments of inertia

The maximum and minimum moments of inertia at a point are called the *principal moments of inertia* at that point. The axes about which the moments of inertia are maximum or minimum are called the *principal axes*, and the corresponding directions are referred to as *principal directions*. To find the maximum and minimum moments of inertia, we set the derivative of I_u in Eq. (9.17) equal to zero:

$$\frac{dI_u}{d\theta} = -(I_x - I_y) \sin 2\theta - 2I_{xy} \cos 2\theta = 0$$

Solving for 2θ, we obtain

$$\tan 2\theta = -\frac{2I_{xy}}{I_x - I_y} \tag{9.20}$$

Note that there are two solutions for the angle 2θ that differ by $180°$ or, equivalently, two solutions for θ that differ by $90°$. We denote these solutions by θ_1 and θ_2. From the graphical representations shown in Fig. 9.8 we find that

$$\sin 2\theta_{1,2} = \mp \frac{I_{xy}}{R}$$

$$\cos 2\theta_{1,2} = \pm \frac{I_x - I_y}{2R}$$

(9.21)

where

$$R = \sqrt{\left(\frac{I_x - I_y}{2}\right)^2 + I_{xy}^2}$$

(9.22)

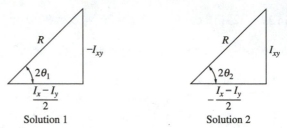

Solution 1   Solution 2

Fig. 9.8

The angles θ_1 and θ_2, measured counterclockwise from the x-axis, define the principal directions. Substituting Eq. (9.22) into Eq. (9.17) and simplifying, we obtain the principal moments of inertia

$$I_{1,2} = \frac{I_x + I_y}{2} \pm R$$

(9.23)

where I_1 and I_2 correspond to the axes defined by θ_1 and θ_2, respectively.* In Eqs. (9.21) and (9.23), the upper sign of either $\pm$ or $\mp$ is to be used with θ_1 and the lower sign with θ_2 (refer to Fig. 9.8).

To determine the product of inertia with respect to the principal axes, we substitute Eqs. (9.21) into Eq. (9.19), which yields

$$I_{uv}\Big|_{\theta=\theta_{1,2}} = \frac{I_x - I_y}{2}\left(\mp\frac{I_{xy}}{R}\right) + I_{xy}\left(\pm\frac{I_x - I_y}{2R}\right) = 0$$

Therefore, *the product of inertia with respect to the principal axes is zero.*

The properties of an area, in general, depend on the location of the origin O of the xy-coordinate system. Therefore, the principal moments of inertia and principal directions vary with the location of point O. However, most practical applications, such as those found in structural engineering, are concerned with moments of inertia with respect to centroidal axes.

---

*It can be shown that I_1 and I_2 are the two values of I that are the roots of the following quadratic equation.

$$\begin{vmatrix} I_x - I & -I_{xy} \\ -I_{xy} & I_y - I \end{vmatrix} = 0$$

Sample Problem 9.9

For the region shown in Fig. (a), calculate (1) the centroidal principal moments of inertia and the principal directions; and (2) the moments and product of inertia about the *uv*-axes through the centroid *C*.

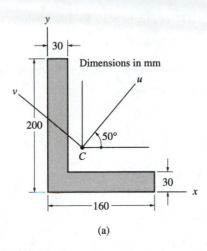

Dimensions in mm

(a)

Solution

Centroidal Properties

The table below lists the computations that have been used to determine the centroidal coordinates and the inertial properties with respect to the *x*- and *y*-axes. The region is considered as a composite of the two rectangles shown in Fig. (b). Their moments of inertia have been calculated using Table 9.1.

The centroidal coordinates, shown in Fig. (b), are computed from the results in Table 9.1 as follows:

$$\bar{x} = \frac{\Sigma A \bar{x}}{\Sigma A} = \frac{460.5 \times 10^3}{9900} = 46.52 \text{ mm}$$

$$\bar{y} = \frac{\Sigma A \bar{y}}{\Sigma A} = \frac{685.5 \times 10^3}{9900} = 66.52 \text{ mm}$$

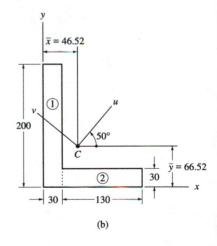

(b)

Part	A (mm^2)	$\bar{x}$ (mm)	$A\bar{x}$ (mm^3)	$\bar{y}$ (mm)	$A\bar{y}$ (mm^3)	I_x (mm^4)	I_y (mm^4)	$I_{xy} = \bar{I}_{xy} + A\bar{x}\bar{y}$ (mm^4)
1	$200(30)$ $= 6000$	15	90×10^3	100	600×10^3	$\dfrac{30(200)^3}{3}$ $= 80.00 \times 10^6$	$\dfrac{200(30)^3}{3}$ $= 1.800 \times 10^6$	$0 + 6000(15)(100)$ $= 9.00 \times 10^6$
2	$130(30)$ $= 3900$	95	370.5×10^3	15	58.5×10^3	$\dfrac{130(30)^3}{3}$ $= 1.17 \times 10^6$	$\dfrac{30(130)^3}{12}$ $+3900(95)^2$ $= 40.69 \times 10^6$	$0 + 3900(95)(15)$ $= 5.56 \times 10^6$
Σ	9900	$\cdots$	460.5×10^3	$\cdots$	658.5×10^3	81.17×10^6	42.49×10^6	14.56×10^6

445

The parallel-axis theorem is then used to calculate the inertial properties about the centroidal axes.

$$\bar{I}_x = I_x - A\bar{y}^2 = (81.17 \times 10^6) - (9900)(66.52)^2$$

$$= 37.36 \times 10^6 \text{ mm}^4$$

$$\bar{I}_y = I_y - A\bar{x}^2 = (42.49 \times 10^6) - (9900)(46.52)^2$$

$$= 21.07 \times 10^6 \text{ mm}^4$$

$$\bar{I}_{xy} = I_{xy} - A\bar{x}\bar{y} = (14.56 \times 10^6) - (9900)(46.52)(66.52)$$

$$= -16.08 \times 10^6 \text{ mm}^4$$

Part 1

Substituting the values for $\bar{I}_x$, $\bar{I}_y$, and $\bar{I}_{xy}$ into Eq. (9.22) yields

$$R = 10^6 \sqrt{\left(\frac{37.36 - 21.07}{2}\right)^2 + (-16.08)^2} = 18.03 \times 10^6 \text{ mm}^4$$

Therefore Eq. (9.23) becomes

$$I_{1,2} = \left(\frac{37.36 + 21.07}{2} \pm 18.03\right) \times 10^6 = (29.22 \pm 18.03) \times 10^6 \text{ mm}^4$$

from which we obtain the principal moments of inertia

$$I_1 = 47.3 \times 10^6 \text{ mm}^4 \qquad I_2 = 11.2 \times 10^6 \text{ mm}^4 \qquad \textit{Answer}$$

For the principal directions, Eqs. (9.21) yield

$$\sin 2\theta_{1,2} = \mp\frac{\bar{I}_{xy}}{R} = \mp\frac{(-16.08)}{18.03} = \pm 0.8919$$

Because the upper sign goes with θ_1, we have $2\theta_1 = 63.11°$ or $116.9°$, and $2\theta_2 = 243.11°$ or $-63.11°$. To determine the correct choices, we investigate the sign of $\cos 2\theta_1$. From Eqs. (9.21) we obtain $\cos 2\theta_1 = (\bar{I}_x - \bar{I}_y)/(2R) = (37.36 - 21.07)/[2(18.03)]$, which is positive. Therefore, $2\theta_1 = 63.11°$ and $2\theta_2 = 243.11°$ are the correct choices, which give

$$\theta_1 = 31.6° \qquad \theta_2 = 121.6° \qquad \textit{Answer}$$

The principal axes, labeled 1 and 2 in Fig. (c), correspond to the axes of I_1 and I_2, respectively.

Part 2

To compute the moments and product of inertia relative to the uv-axes in Fig. (a), we need only substitute $\bar{I}_x = 37.36 \times 10^6 \text{ mm}^4$, $\bar{I}_y = 21.07 \times 10^6 \text{ mm}^4$, $\bar{I}_{xy} = -16.08 \times 10^6 \text{ mm}^4$, and $\theta = 50°$ into the transformation equations. From Eq. (9.17) we obtain

$$I_u \times 10^{-6} = \frac{37.36 + 21.07}{2} + \frac{37.36 - 21.07}{2} \cos 100° - (-16.08) \sin 100°$$

$$I_u = 43.6 \times 10^6 \text{ mm}^4 \qquad \textit{Answer}$$

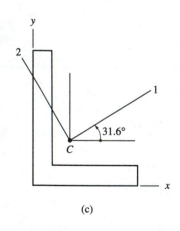

(c)

Equation (9.18) yields

$$I_v \times 10^{-6} = \frac{37.36 + 21.07}{2} - \frac{37.36 - 21.07}{2} \cos 100° + (-16.08) \sin 100°$$

$$I_v = 14.8 \times 10^6 \text{ mm}^4 \qquad\qquad\qquad\qquad\qquad\qquad \textit{Answer}$$

From Eq. (9.19) we have

$$I_{uv} \times 10^{-6} = \frac{37.36 - 21.07}{2} \sin 100° + (-16.08) \cos 100°$$

$$I_{uv} = 10.8 \times 10^6 \text{ mm}^4 \qquad\qquad\qquad\qquad\qquad\qquad \textit{Answer}$$

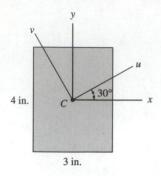

Fig. P9.50

Problems

9.50 For the rectangular region, determine (a) the principal moments of inertia and the principal directions at the centroid C; and (b) the moments and products of inertia about the u-v axes.

9.51 For the semicircular region, calculate (a) the principal moments of inertia and the principal directions at the centroid C; and (b) the moments and products of inertia about the u-v axes.

9.52 Find the principal moments of inertia and the principal directions at the centroid C of the triangle.

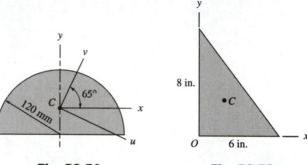

Fig. P9.51   **Fig. P9.52**

9.53 Compute I_u, I_v, and I_{uv} for the triangle shown.

9.54 Given that the properties of the region shown are $I_x = 4000$ in.^{4}, $I_y = 1000$ in.^{4}, and $I_{xy} = -800$ in.^{4}, determine I_u, I_v, and I_{uv} for $\theta = 120°$.

9.55 The properties of the region shown are $I_x = 10 \times 10^6$ mm^4, $I_y = 20 \times 10^6$ mm^4, and $I_{xy} = 12 \times 10^6$ mm^4. Compute I_u, I_v, and I_{uv} if $\theta = 33.7°$.

9.56 The u- and v-axes are the principal axes of the region shown. Given that $I_u = 7600$ in.^{4}, $I_v = 5000$ in.^{4}, and $\theta = 33.7°$, determine I_x, I_y, and I_{xy}.

9.57 The x- and y-axes are the principal axes for the region shown with $I_x = 6 \times 10^6$ mm^4 and $I_y = 2 \times 10^6$ mm^4. (a) Calculate the angle θ for which I_{uv} is maximum. (b) Determine I_u, I_v, and I_{uv} for the angle θ found in part (a).

9.58 Compute I_v for the region shown, given that $I_u = 160 \times 10^6$ mm^4, $I_{xy} = -30 \times 10^6$ mm^4, and $\theta = 18.44°$. The u- and v-axes are principal axes for the region.

9.59 For the region shown, determine the principal moments of inertia at point O and the corresponding principal directions.

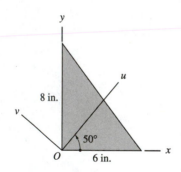

Fig. P9.53

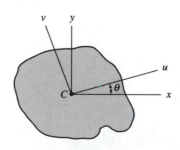   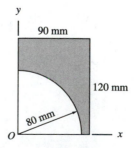

Fig. P9.54–P9.58   **Fig. P9.59**

9.60 Compute I_u for the rectangle by the following methods: (a) substitute the expressions for I_x, I_y, and I_{xy} into the transformation equation; (b) substitute the expressions for $\bar{I}_x$, $\bar{I}_y$, and $\bar{I}_{xy}$ into the transformation equation; and (c) use the method of composite areas, considering the rectangle to be the sum of two triangles.

9.61 Using I_x and I_u from Table 9.2, determine the moment of inertia of the circular sector about the OB-axis. Check your result for $\alpha = 45°$ with that given for a quarter circle in Table 9.2.

9.62 Show that every axis passing through the centroid of the equilateral triangle is a principal axis.

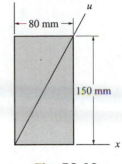

Fig. P9.60

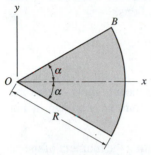

Fig. P9.61

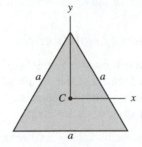

Fig. P9.62

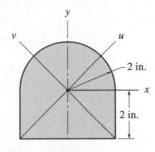

Fig. P9.63

9.63 Calculate I_u, I_v, and I_{uv} for the region shown.

9.64 The L80×60×10-mm structural angle has the following cross-sectional properties: $I_x = 0.808 \times 10^6$ mm⁴, $I_y = 0.388 \times 10^6$ mm⁴, and $I_2 = 0.213 \times 10^6$ mm⁴, where I_2 is a principal centroidal moment of inertia. Assuming I_{xy} is negative, compute (a) I_1 (the other principal centroidal moment of inertia); and (b) the principal directions.

9.65 The dimensions of the half parabolic complement are $h = 60$ mm and $b = 80$ mm. Determine (a) the principal centroidal moments of inertia; and (b) the corresponding principal directions. Use the data given in Table 9.2.

9.66 Substitute I_x, I_y, and I_{xy} from Table 9.2 into the transformation equation to calculate I_{OB} for the half parabolic complement. Use $b = 21$ in. and $h = 20$ in.

9.67 Determine the principal centroidal moments of inertia and the corresponding principal directions for the region shown.

9.68 Compute the principal centroidal moments of inertia and the corresponding principal directions for the region shown.

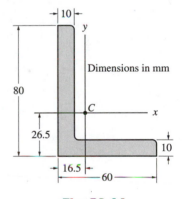

Fig. P9.64

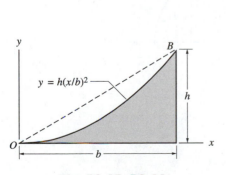

Fig. P9.65, P9.66

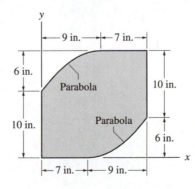

Fig. P9.67

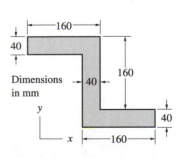

Fig. P9.68

*9.5 *Mohr's Circle for Moments and Products of Inertia*

Mohr's circle is a graphical representation of the transformation equations for moments and products of inertia. Developed by Otto Mohr, a German engineer in 1882, it is a popular alternative to the transformation equations. There are two advantages to using Mohr's circle. First, the circle gives a clear visual representation of how the inertial properties vary with the orientation of the axes. Second, by referring to the circle, you can obtain the numerical values without having to memorize the transformation equations.

a. *Construction of Mohr's circle*

Consider the plane region shown in Fig. 9.9(a). Let I_x, I_y, and I_{xy} be the moments and the product of inertia of the region with respect to the x-y axes that intersect at point O. Mohr's circle associated with point O is shown in Fig. 9.9(b). The circle is constructed as follows:

1. Draw a set of axes, the horizontal axis representing the moment of inertia (M.I.), and the vertical axis representing the product of inertia (P.I.).
2. Plot the point $\widehat{x}$ with coordinates (I_x, I_{xy}), and the point $\widehat{y}$ with coordinates $(I_y, -I_{xy})$.
3. Join $\widehat{x}$ and $\widehat{y}$ with a line, and draw a circle with this line as its diameter.

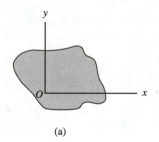

(a)

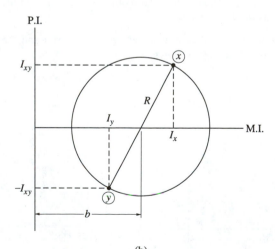

(b)

Fig. 9.9

Mohr's circle is now complete. Note that the radius of the circle is

$$R = \sqrt{\left(\frac{I_x - I_y}{2}\right)^2 + I_{xy}^2}$$

and its center is located at

$$b = \frac{1}{2}(I_x + I_y)$$

b. Properties of Mohr's circle

The key properties of Mohr's circle are

- The end points of every diameter of the circle represent the moments and the product of inertia associated with a set of perpendicular axes passing through point O.
- Angles between diameters on the circle are *twice* the angles between axes at point O, and these angles are measured in the *same sense* (CW or CCW).

The procedure for determining the inertial properties with respect to particular axes, such as the u-v axes shown in Fig. 9.10(a), is as follows:

1. Note the magnitude and sense of the angle θ between the x-y and the u-v coordinate axes. (The sense of the θ is the direction in which the x-y axes must be rotated so that they coincide with the u-v axes.)

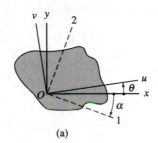

(a)

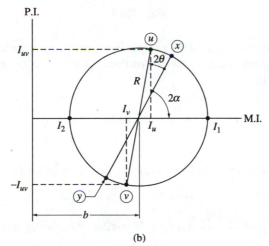

(b)

Fig. 9.10

2. Rotate the diameter $\widehat{x}$ $\widehat{y}$ of Mohr's circle through the angle 2θ in the same sense as θ. Label the end points of this diameter $\widehat{u}$ and $\widehat{v}$, as shown in Fig. 9.10(b). The coordinates of u are (I_u, I_{uv}), and the coordinates of v are $(I_v, -I_{uv})$.

Mohr's circle can also be used to find the principal moments of inertia and the principal directions. Referring to Fig. 9.10(b), we see that the maximum and the minimum moments of inertia are $I_1 = b + R$ and $I_2 = b - R$, respectively. The orientation of the principal axes, labeled "1" and "2" in Fig. 9.10(a), is obtained by rotating the x-y axes though the angle α. The magnitude and sense of α is determined from the Mohr's circle. In particular, note that α in Fig. 9.10(a) and 2α in Fig. 9.10(b) must have the same sense.

c. Verification of Mohr's circle

Figure 9.11 shows the circle that was constructed following the steps outlined in the preceding section. Because $\widehat{x}$ and $\widehat{u}$ are points located above the abscissa, we have assumed that both I_{xy} and I_{uv} are positive.

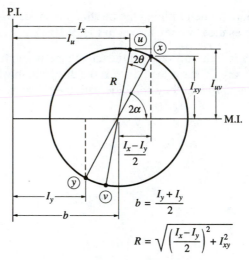

Fig. 9.11

In order to prove that Mohr's circle is a valid representation of the transformation equations, we must show that the coordinates of $\widehat{u}$ agree with Eqs. (9.17) and (9.19).

From Fig. 9.11, we have

$$I_u = b + R\cos(2\theta + 2\alpha)$$

Using the identity

$$\cos(2\theta + 2\alpha) = \cos 2\theta \cos 2\alpha - \sin 2\theta \sin 2\alpha$$

and substituting $b = (I_x + I_y)/2$, we get

$$I_u = \frac{I_x + I_y}{2} + R(\cos 2\theta \cos 2\alpha - \sin 2\theta \sin 2\alpha)$$

From Fig. 9.11, we see that $\sin 2\alpha = I_{xy}/R$ and $\cos 2\alpha = (I_x - I_y)/(2R)$. Substituting these relations into the last equation yields

$$I_u = \frac{I_x + I_y}{2} + R\left(\frac{I_x - I_y}{2R} \cos 2\theta - \frac{I_{xy}}{R} \sin 2\theta\right)$$

or

$$I_u = \frac{I_x + I_y}{2} + \frac{I_x - I_y}{2} \cos 2\theta - I_{xy} \sin 2\theta \qquad (9.24)$$

From Fig. 9.11, we also obtain

$$I_{uv} = R\sin(2\theta + 2\alpha)$$

Using the identity

$$\sin(2\theta + 2\alpha) = \sin 2\theta \cos 2\alpha + \cos 2\theta \sin 2\alpha$$

and the previously derived expressions for $\sin 2\alpha$ and $\cos 2\alpha$, we have

$$I_{uv} = R\left(\frac{I_x - I_y}{2R} \sin 2\theta + \frac{I_{xy}}{R} \cos 2\theta\right)$$

which becomes

$$I_{uv} = \frac{I_x - I_y}{2} \sin 2\theta + I_{xy} \cos 2\theta \qquad (9.25)$$

Because Eqs. (9.24) and (9.25) are identical to the transformation equations, Eqs. (9.17) and (9.19), we conclude that Mohr's circle is a valid representation of the transformation equations.

Sample Problem 9.10

For the region shown in Fig. (a), calculate (1) the centroidal principal moments of inertia and principal directions; and (2) the moments and product of inertia about the *uv*-axes through the centroid C. Note that this is the same region as in Sample Problem 9.9.

Solution

Construction of Mohr's Circle

From the solution to Sample Problem 9.9, we have $\bar{I}_x = 37.36 \times 10^6$ mm^4, $\bar{I}_y = 21.07 \times 10^6$ mm^4, and $\bar{I}_{xy} = -16.08 \times 10^6$ mm^4. Using these values, Mohr's circle is plotted as shown in Fig. (b), following the procedure outlined in Art. 9.5. Note that

1. The points on the circle that correspond to the centroidal axes that are parallel to the *x*- and *y*-axes are labeled $\widehat{x_C}$ and $\widehat{y_C}$, respectively.
2. Because $\bar{I}_{xy}$ is negative, $\widehat{x_C}$ is plotted below the abscissa and $\widehat{y_C}$ is plotted above.

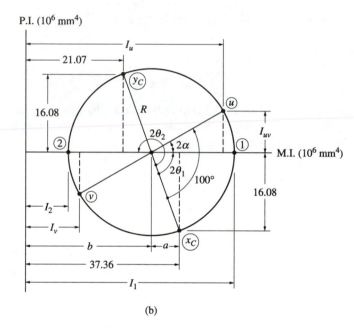

(b)

If the circle were drawn to scale, all unknown values could be determined by direct measurements. However, we will compute the requested values by trigonometry. Of course, all results in the analyses should agree with those found previously in the solution to Sample Problem 9.9.

The following values, computed directly from the circle, are shown in the figure.

$$b = \frac{37.36 + 21.07}{2} \times 10^6 = 29.22 \times 10^6 \text{ mm}^4$$

$$a = \frac{37.36 - 21.07}{2} \times 10^6 = 8.145 \times 10^6 \text{ mm}^4$$

$$R = \sqrt{(8.145)^2 + (16.08)^2} \times 10^6 = 18.03 \times 10^6 \text{ mm}^4$$

Part 1

In Fig. (b), ①, and ② correspond to the maximum and minimum moments of inertia, respectively. Therefore, we have $I_{1,2} = b \pm R = (29.22 \pm 18.03) \times 10^6$ mm⁴, from which we obtain

$$I_1 = 47.3 \times 10^6 \text{ mm}^4 \qquad I_2 = 11.2 \times 10^6 \text{ mm}^4 \qquad \textit{Answer}$$

The principal directions are found by calculating the angles θ_1 and θ_2. From the circle we find that $2\theta_1 = \sin^{-1}(16.08/18.03) = 63.11°$ and $2\theta_2 = 180 + 2\theta_1 = 243.11°$, which gives

$$\theta_1 = 31.6° \qquad \theta_2 = 121.6° \qquad \textit{Answer}$$

Note that on the circle the central angle from $\widehat{x_C}$ to ① is $2\theta_1$, counterclockwise. Therefore, the principal direction corresponding to I_1 is $\theta_1 = 31.6°$, measured counterclockwise from the centroidal x-axis. (Remember that angles on the circle are twice the angles between axes, measured in the same direction). Therefore, the centroidal principal axes are oriented as shown in Fig. (c).

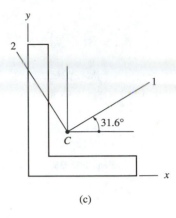

(c)

Part 2

To determine I_u, I_v, and I_{uv}, the points $\widehat{u}$ and $\widehat{v}$—corresponding to the u- and v-axes, respectively—must be identified on Mohr's circle. Because the u-axis is located at 50° counterclockwise from the centroidal x-axis, $\widehat{u}$ on the circle is 100° counterclockwise from $\widehat{x_C}$. Of course, $\widehat{v}$ is located at the opposite end of the diameter from $\widehat{u}$. To facilitate our computations, we have introduced the central angle 2α between points ① and $\widehat{u}$, given by $2\alpha = 100° - 2\theta_1 = 100° - 63.11° = 36.89°$. Referring to the circle, we find that

$$\begin{aligned} I_{u,v} &= b \pm R\cos 2\alpha \\ &= (29.22 \pm 18.03 \cos 36.89°) \times 10^6 \text{ mm}^4 \end{aligned}$$

or

$$I_u = 43.6 \times 10^6 \text{ mm}^4 \qquad I_v = 14.8 \times 10^6 \text{ mm}^4 \qquad \textit{Answer}$$

Additionally, the circle yields $|I_{uv}| = R\sin 2\alpha = (18.03 \sin 36.89°) \times 10^6 = 10.8 \times 10^6$ mm⁴. Because $\widehat{u}$ is above the abscissa, I_{uv} is positive. Therefore, we have

$$I_{uv} = +10.8 \times 10^6 \text{ mm}^4 \qquad \textit{Answer}$$

Recall that in the transformation equations, Eqs. (9.17)–(9.19), 2θ represents the angle measured counterclockwise from the x-axis to the u-axis. However, after Mohr's circle has been drawn, any convenient angle—clockwise or counterclockwise, and measured from any point on the circle—can be used to locate $\widehat{u}$. For example, on the circle in Fig. (b) we see that $\widehat{u}$ is located at 80° in the clockwise direction from $\widehat{y_C}$. This is consistent with Fig. (a), where the u-axis is reached from the centroidal y-axis by a 40° clockwise rotation.

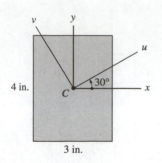

Fig. P9.69

Problems

The following problems are to be solved using Mohr's circle.

9.69   Find the moments and the product of inertia of the rectangle about the *u-v* axes at the centroid *C*.

9.70   Determine the moments and the product of inertia of the semicircle about the *u-v* axes that pass through the centroid *C*.

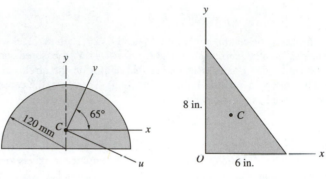

Fig. P9.70   **Fig. P9.71**

9.71   Find the principal moments of inertia and the principal directions at the centroid *C* of the triangle.

9.72   Determine the moments and the product of inertia of the triangle about the *u-v* axes.

9.73   Given that the properties of the region shown are $I_x = 4000$ in.^{4}, $I_y = 1000$ in.^{4}, and $I_{xy} = -800$ in.^{4}, determine I_u, I_v, and I_{uv} for $\theta = 120°$.

9.74   The properties of the region shown are $I_x = 10 \times 10^6$ mm^4, $I_y = 20 \times 10^6$ mm^4, and $I_{xy} = 12 \times 10^6$ mm^4. Determine I_u, I_v, and I_{uv} if $\theta = 33.7°$.

9.75   The *u*- and *v*-axes are the principal axes of the region shown. Given that $I_u = 8400$ in.^{4}, $I_v = 5000$ in.^{4}, and $\theta = 25°$, calculate I_x, I_y, and I_{xy}.

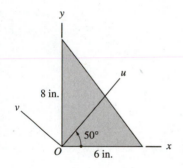

Fig. P9.72

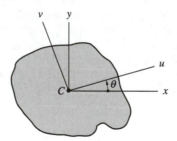

Fig. P9.73–P9.77

9.76   The *x*- and *y*-axes are the principal axes for the region shown, with $I_x = 8 \times 10^6$ mm^4 and $I_y = 2 \times 10^6$ mm^4. (a) Calculate the angle θ for which I_{uv} is maximum. (b) Determine I_u, I_v, and I_{uv} for the angle θ found in part (a).

9.77   Compute I_v for the region shown, given that $I_u = 140 \times 10^6$ mm^4, $I_{xy} = -30 \times 10^6$ mm^4, and $\theta = 18°$. The *u*- and *v*-axes are principal axes for the region.

9.78   The L80 × 60 × 10-mm structural angle has the following cross-sectional properties: $I_x = 0.808 \times 10^6$ mm^4, $I_y = 0.388 \times 10^6$ mm^4, and $I_2 = 0.213 \times 10^6$ mm^4, where I_2 is a centroidal principal moment of inertia. Assuming that I_{xy} is negative, compute (a) I_1 (the other centroidal principal moment of inertia); and (b) the principal directions at the centroid.

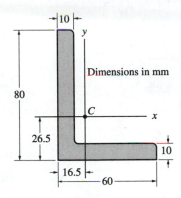

Fig. P9.78

Review Problems

9.79   Compute $\bar{I}_x$, $\bar{I}_y$, and $\bar{I}_{xy}$ for the rectangular region.

9.80   The principal moments of inertia at point O for the shaded region are 60×10^6 mm^4 and 30×10^6 mm^4. In addition, the product of inertia with respect to the x- and y-axes is 10×10^6 mm^4. Find (a) I_x and I_y; and (b) I_u and I_v.

9.81   By integration, show that the product of inertia with respect to the x- and y-axes for the quarter circular region is $R^4/8$.

9.82   Compute the $\bar{I}_x$ and $\bar{I}_y$ for the annular region.

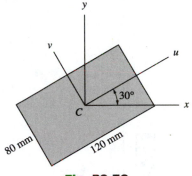

Fig. P9.79

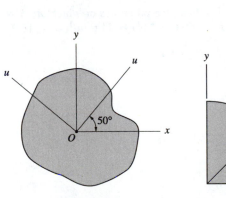

Fig. P9.80

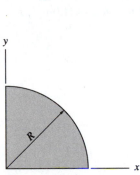

Fig. P9.81

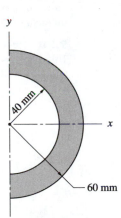

Fig. P9.82

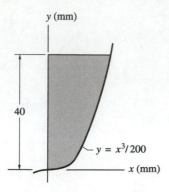

Fig. P9.83

9.83 Using integration, evaluate the moments of inertia about the *x*- and *y*-axes for the shaded region.

9.84 The inertial properties at point O for a plane region are $I_x = 200 \times 10^6$ mm^4, $I_y = 300 \times 10^6$ mm^4, and $I_{xy} = -120 \times 10^6$ mm^4. Determine the principal moments of inertia and principal directions at point O.

9.85 Compute $\bar{I}_x$ and $\bar{I}_y$ for the shaded region.

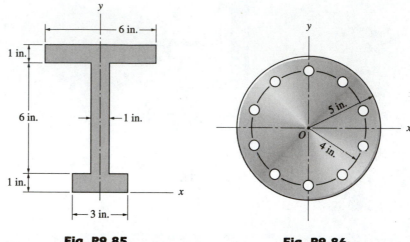

Fig. P9.85

Fig. P9.86

9.86 The flanged bolt coupling is fabricated by drilling 10 evenly spaced 0.5-in. diameter bolt holes in a steel plate. The radii of the plate and bolt circle are 5 in. and 4 in., respectively. Determine the percent reduction in the polar moment of the area about point O due to the drilling operation.

9.87 The figure shows a structural shape known as an unequal angle (L) section. From a table of structural shapes, the inertial properties of an L150 × 100 × 10-mm are $\bar{x} = 23.8$ mm, $\bar{y} = 48.8$ mm, $A = 2400$ mm^2, $\bar{I}_x = 5.58 \times 10^6$ mm^4, $\bar{I}_y = 2.03 \times 10^6$ mm^4. In addition, the angle α locating the axis of minimum centroidal moment of inertia (labeled as the 2-axis in the figure) is listed as 24.0°, with the corresponding radius of gyration being $\bar{k}_2 = 21.9$ mm. Compute (a) the other principal centroidal moment of inertia; and (b) $\bar{I}_{xy}$.

9.88 Re-solve Prob. 9.87 for an L8 × 4 × 1-in., the properties of which are $\bar{x} = 1.05$ in., $\bar{y} = 3.05$ in., area $= 11.0$ in.^{2}, $\bar{I}_x = 69.6$ in.^{4}, $\bar{I}_y = 11.6$ in.^{4}, $\alpha = 13.9°$, and $\bar{k}_2 = 0.846$ in.

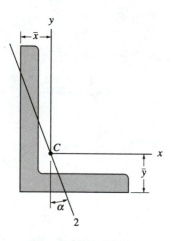

Fig. P9.87, P9.88

9.89 For the shaded region, evaluate (a) I_{xy}; and (b) $\bar{I}_x$.

9.90 Calculate $\bar{I}_x$, $\bar{I}_y$, and $\bar{I}_{xy}$ for the plane region shown.

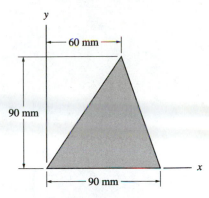

Fig. P9.89

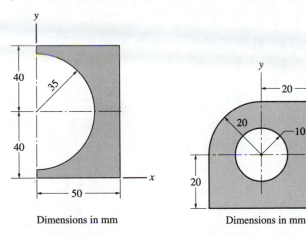

Dimensions in mm

Fig. P9.90

Dimensions in mm

Fig. P9.91

9.91 For the plane region shown, determine (a) I_x and I_y; and (b) $\bar{I}_x$ and $\bar{I}_y$ using the parallel axis theorem and the results of part (a).

9.92 Use integration to find I_x, I_y, and I_{xy} for the region shown.

9.93 Determine the principal moments of inertia and the principal directions at point O for the region shown.

9.94 The inertial properties of the region shown are $I_x = 140$ in.^{4}, $I_y = 264$ in.^{4}, and $I_{xy} = -116$ in.^{4} Determine I_u, I_v, and I_{uv}. Note that the u-axis passes through point B.

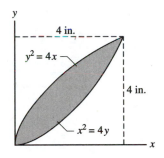

Fig. P9.92

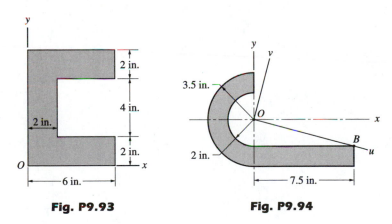

Fig. P9.93

Fig. P9.94

10
Virtual Work and Potential Energy

10.1 Introduction

Methods based on the concepts of virtual work and potential energy can be used as alternatives to Newton's laws in equilibrium analysis. These methods are best suited for the analysis of systems made up of several interconnected rigid bodies. The primary advantage of work and energy methods is that the reactions at certain connections, such as pins or inextensible cables, do not enter into the analysis. Therefore, the number of unknowns (and equations) is often considerably reduced.

The disadvantage of work and energy methods is that they require the use of kinematics (geometry of motion), which is a branch of dynamics. To keep the kinematics relatively simple, we confine our discussion to two-dimensional problems. In addition to kinematics, the concept of work must also be introduced.

10.2 Planar Kinematics of a Rigid Body

Any plane motion of a rigid body can be viewed as a superposition of two simple displacements: translation and rotation about a point. We will first consider finite motions and then specialize the results for virtual (infinitesimal) displacements.

a. *Finite displacements*

Translation   Translation of a rigid body is illustrated in Fig. 10.1. Two characteristics of translation are

- Any straight line embedded in the body, such as line AB, remains parallel to its original position. That is, the body (or any line embedded in it) does not rotate.
- All points of the body have the same displacement. Denoting the position vectors of A and B (measured from the fixed point O) by $\mathbf{r}_A$ and $\mathbf{r}_B$, then during translation we have

$$\Delta\mathbf{r}_A = \Delta\mathbf{r}_B \qquad (10.1)$$

where $\Delta\mathbf{r}_A$ and $\Delta\mathbf{r}_B$ are the displacement vectors (changes in the position vectors) of the two points.

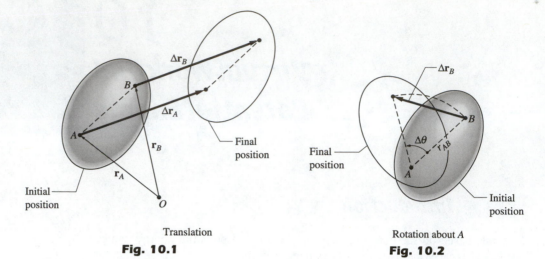

Translation
Fig. 10.1

Rotation about *A*
Fig. 10.2

Rotation about a Fixed Point Figure 10.2 shows rotation of a rigid body about point *A* through the angle $\Delta\theta$. The characteristics of this motion are

- Any point of the body, such as *B*, moves along a circular arc centered at *A*. From geometry we find that the magnitude of the displacement vector of *B* is

$$\Delta r_B = 2r_{AB} \sin \frac{\Delta\theta}{2} \qquad (10.2)$$

where r_{AB} is the distance between *A* and *B*.
- Every line embedded in the body undergoes the same rotation $\Delta\theta$.

General Plane Motion Consider the general plane displacement of the rigid body shown in Fig. 10.3. We can see that the final position of the body can be obtained by combining the following two displacements:

- The translation $\Delta\mathbf{r}_A$, which moves point *A* to its final position without changing the orientation of the body.
- The rotation $\Delta\theta$ about *A* in order to give the body its final orientation.

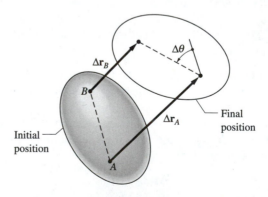

General plane motion
Fig. 10.3

The reference point A can be chosen arbitrarily. Its location will affect the translation but not the rotation. The order in which the two motions (translation and rotation) are carried out is irrelevant.

b. *Virtual displacements*

Definition and Notation   A virtual displacement is defined to be a fictitious displacement of *infinitesimal* magnitude. *Fictitious* means that the displacement may not be real; that is, it may not actually take place.

The usual practice is to precede an infinitesimal quantity with the letter d. Thus, an infinitesimal displacement of point A would be denoted by $d\mathbf{r}_A$. To draw attention to its fictitious nature, a virtual displacement is preceded by δ (lowercase delta). Hence the virtual displacement of point A would be written as $\delta\mathbf{r}_A$. However, $d\mathbf{r}_A$ and $\delta\mathbf{r}_A$ are mathematically identical. For example, if $\mathbf{r}_A$ is a function of some parameter θ, we can write $\delta\mathbf{r}_A = (d\mathbf{r}_A/d\theta)\delta\theta$, where $\delta\theta$ is the virtual change in θ.

Virtual Translation   As mentioned previously, all points of a rigid body have the same displacement during translation. Therefore,

$$\delta\mathbf{r}_A = \delta\mathbf{r}_B \tag{10.3}$$

where A and B are any two points of the body.

Virtual Rotation about a Fixed Point   If the virtual rotation takes place about point A, then we can use Eq. (10.2) to determine the virtual displacement of any other point B in the body. Replacing each Δ with δ, we get $\delta r_B = 2r_{AB}\sin(\delta\theta/2)$. Because the virtual rotation $\delta\theta$ of the body is infinitesimal, we have $\sin(\delta\theta/2) = \delta\theta/2$, yielding

$$\delta r_B = r_{AB}\,\delta\theta \tag{10.4}$$

Note that the direction of $\delta\mathbf{r}_B$ is perpendicular to r_{AB}, and its sense is determined by the sense of $\delta\theta$, as illustrated in Fig. 10.4.

It is sometimes convenient to use the vector form of Eq. (10.4):

$$\delta\mathbf{r}_B = \delta\boldsymbol{\theta} \times \mathbf{r}_{AB} \tag{10.5}$$

where $\delta\boldsymbol{\theta}$ is the virtual rotation vector* illustrated in Fig. 10.4. It can be seen that Eq. (10.4) produces the correct magnitude and direction of $\delta\mathbf{r}_B$.

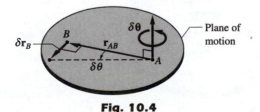

Fig. 10.4

---

*It can be shown that an infinitesimal rotation of a rigid body has all the characteristics of a vector.

General Virtual Plane Motion   To obtain the general case of plane motion, we superimpose the virtual translation $\delta\mathbf{r}_A$ of the body and the virtual rotation $\delta\boldsymbol{\theta}$ about point A. With Eqs. (10.3) and (10.5), this method yields for the virtual displacement of point B

$$\delta\mathbf{r}_B = \delta\mathbf{r}_A + \delta\boldsymbol{\theta} \times \mathbf{r}_{AB} \tag{10.6}$$

*10.3   Virtual Work

a.   Virtual work of a force

If the point of application of a force $\mathbf{F}$ undergoes a virtual displacement $\delta\mathbf{r}$, as shown in Fig. 10.5(a), the virtual work δU done by the force is defined to be

$$\delta U = \mathbf{F} \cdot \delta\mathbf{r} = F \cos\alpha\, \delta r \tag{10.7}$$

where α is the angle between $\mathbf{F}$ and $\delta\mathbf{r}$. Note that the virtual work is a scalar that can be positive, negative, or zero, depending upon the angle α. The dimension of virtual work is $[FL]$; hence the units are lb · ft, N · m, and so forth.

Referring to Figs. 10.5(b) and 10.5(c), we see that the virtual work can be viewed in two ways:

- $\delta U = (F \cos\alpha)\, \delta r$, where $F \cos\alpha$ (the component of $\mathbf{F}$ in the direction of $\delta\mathbf{r}$) is called the *working component* of the force.
- $\delta U = F(\delta r \cos\alpha)$, where $\delta r \cos\alpha$ (the component of $\delta\mathbf{r}$ in the direction of $\mathbf{F}$) is known as the *work-absorbing component* of the virtual displacement.

b.   Virtual work of a couple

Figure 10.6(a) shows a couple formed by the forces $-\mathbf{F}$ and $\mathbf{F}$ acting at points A and B of a rigid body. The corresponding couple-vector $\mathbf{C} = \mathbf{r}_{AB} \times \mathbf{F}$ is perpendicular to the plane of the couple, as indicated in Fig. 10.6(b). If the body undergoes a virtual motion in the plane of the couple, the virtual work of the couple is

$$\delta U = -\mathbf{F} \cdot \delta\mathbf{r}_A + \mathbf{F} \cdot \delta\mathbf{r}_B$$

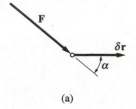

(a)

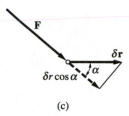

(b)

(c)

Fig. 10.5

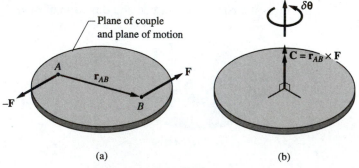

(a)

(b)

Fig. 10.6

Substituting for $\delta\mathbf{r}_B$ from Eq. (10.6), we get

$$\delta U = -\mathbf{F}\cdot\delta\mathbf{r}_A + \mathbf{F}\cdot(\delta\mathbf{r}_A + \delta\boldsymbol{\theta}\times\mathbf{r}_{AB}) = \mathbf{F}\cdot\delta\boldsymbol{\theta}\times\mathbf{r}_{AB} = \mathbf{r}_{AB}\times\mathbf{F}\cdot\delta\boldsymbol{\theta}$$

or

$$\delta U = \mathbf{C}\cdot\delta\boldsymbol{\theta} \qquad\qquad (10.8a)$$

Because $\mathbf{C}$ and $\delta\boldsymbol{\theta}$ are collinear (recall that we consider only two-dimensional problems), the virtual work of the couple can also be written as

$$\delta U = C\,\delta\theta \qquad\qquad (10.8b)$$

Note that δU is positive if $\mathbf{C}$ and $\delta\boldsymbol{\theta}$ have the same sense, and negative if they have opposite sense.

c.   *Virtual work performed on a rigid body*

The following theorem is sometimes useful in the computation of virtual work (this theorem is also needed to derive the principle of virtual work):

> *The virtual work of all forces that act on a rigid body is equal to the virtual work of their resultant.*

Proof

Consider a rigid body that is subjected to the coplanar forces $\mathbf{F}_1, \mathbf{F}_2, \ldots, \mathbf{F}_i \ldots$, as in Fig. 10.7(a). Figure 10.7(b) shows the resultant of this force system consisting of the force $\mathbf{R} = \Sigma_i\mathbf{F}_i$ acting at A, and the couple $\mathbf{C}^R = \Sigma_i\mathbf{r}_{Ai}\times\mathbf{F}_i$, where $\mathbf{r}_{Ai}$ is the vector drawn from A to the point of application of $\mathbf{F}_i$. If the body undergoes a virtual displacement, the virtual work of all the forces is

$$\delta U = \sum_i \mathbf{F}_i\cdot\delta\mathbf{r}_i$$

Using Eq. (10.6), we substitute $\delta\mathbf{r}_i = \delta\mathbf{r}_A + \delta\boldsymbol{\theta}\times\mathbf{r}_{Ai}$, which results in

$$\delta U = \sum_i \mathbf{F}_i\cdot\delta\mathbf{r}_A + \sum_i \mathbf{F}_i\cdot\delta\boldsymbol{\theta}\times\mathbf{r}_{Ai}$$

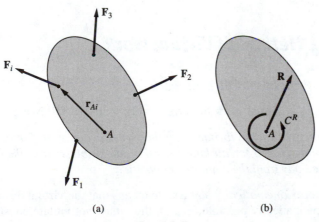

(a)   (b)

Fig. 10.7

The first term in this expression is

$$\sum_i \mathbf{F}_i \cdot \delta\mathbf{r}_A = \left(\sum_i \mathbf{F}_i\right) \cdot \delta\mathbf{r}_A = \mathbf{R} \cdot \delta\mathbf{r}_A$$

The second term can be written as

$$\sum_i \mathbf{F}_i \cdot \delta\boldsymbol{\theta} \times \mathbf{r}_{Ai} = \sum_i \mathbf{r}_{Ai} \times \mathbf{F}_i \cdot \delta\boldsymbol{\theta} = \left(\sum_i \mathbf{r}_{Ai} \times \mathbf{F}_i\right) \cdot \delta\boldsymbol{\theta} = \mathbf{C}^R \cdot \delta\boldsymbol{\theta}$$

Therefore, the virtual work performed on the body is

$$\delta U = \mathbf{R} \cdot \delta\mathbf{r}_A + \mathbf{C}^R \cdot \delta\boldsymbol{\theta} \qquad (10.9)$$

which completes the proof.

d. Virtual work for a system of rigid bodies

Consider a system of interconnected rigid bodies, where the frictional forces at the connections and the supports are negligible. Virtual work of friction forces would introduce complications that we wish to avoid at this level. The connections that we consider are thus limited to pins, rollers, inextensible and extensible cables, ideal springs, and so forth.

If a system of interconnected rigid bodies is given a virtual displacement, the virtual work done on the system equals the virtual work of the external forces, plus the virtual work of the internal forces. At connections that do not deform, the net work done by the internal forces is zero. For example, if a pin joins two rigid bodies, the positive work of the pin reaction acting on one body cancels the negative work of the pin reaction acting on the other body. The reason for this cancellation is that the pin reactions acting on the two bodies are equal in magnitude, opposite in sense, and undergo identical displacements. The forces provided by a deformable connector may also be equal and opposite; however, because of deformation, they do not necessarily undergo the same displacement. Consequently, a deformable connection is capable of doing virtual work on a system. Springs are the only deformable connections that we consider in this text.

*10.4   Method of Virtual Work

a. Principle of virtual work

The principle of virtual work for a rigid body states the following:

> *If a body is in equilibrium, then the virtual work of all forces acting on the body is zero for all kinematically admissible virtual displacements of the body from the equilibrium position.*

The term *kinematically admissible* means that the virtual displacements must be kinematically possible; that is, they must not violate constraints imposed by the supports.

Proof of the principle follows directly from Eq. (10.9). If a body is in equilibrium, the resultant of the forces acting on it vanishes; that is, $\mathbf{R} = \mathbf{0}$ and $\mathbf{C}^R = \mathbf{0}$, and Eq. (10.9) becomes

$$\boxed{\delta U = 0} \qquad\qquad (10.10)$$

The principle of virtual work also applies to systems of rigid bodies. Because a system can be in equilibrium only if each of its members (constituent bodies) is in equilibrium, we conclude that $\delta U = 0$ for each member. It follows that the virtual work done on the system also vanishes. In other words:

If a system of rigid bodies is in equilibrium, then the virtual work of all forces acting on the system is zero for all kinematically admissible virtual displacements of the system from the equilibrium position.

b.   *Kinematic constraints and independent coordinates*

The following terms are frequently used in kinematics:

- *Kinematic constraints* are geometric restrictions imposed on the configuration of a system.
- *Kinematically independent coordinates* of a system are parameters that define the configuration of the system and can be varied independently without violating kinematic constraints.
- *Number of degrees of freedom* (number of DOF) of a system is the number of kinematically independent coordinates required to completely define the configuration of the system.
- *Equations of constraint* are mathematical relations between position coordinates that describe the kinematic constraints.

To illustrate these terms, consider first the bar shown in Fig. 10.8. This bar has a single DOF, because it takes only one coordinate, such as the angle θ to define the position of every point in the bar. Because there are no kinematic restrictions on θ, it is a kinematically independent coordinate.

The system of two bars in Fig. 10.9 also has one DOF. For the kinematically independent coordinate we may choose either θ_1 or θ_2. Whichever of the two angles is selected, the other one is determined by the equation of constraint $L_1 \cos\theta_1 + L_2 \cos\theta_2 = d$. Any configuration of the bars that violates this equation of constraint is kinematically inadmissible.

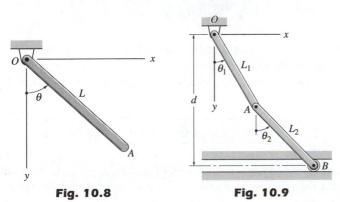

Fig. 10.8   **Fig. 10.9**

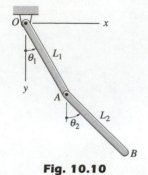

Fig. 10.10

A system with two DOF is shown in Fig. 10.10. It requires two kinematically independent coordinates, such as θ_1 and θ_2, to completely describe its configuration.

c. Implementation of the method of virtual work

When applying the method of virtual work, we must make sure that the virtual displacements of points where the loads are applied (the displacements that contribute to the virtual work) are kinematically admissible. This can be accomplished by the following two steps:

- First, use geometry to relate the coordinates of the points where loads act to the kinematically independent coordinates.
- Then obtain the relationships between the virtual changes of these coordinates (the virtual displacements) by differentiation.

As an example, consider the bar shown in Fig. 10.11. The weight of the bar is W, and its center of gravity is denoted by G. The virtual work done on the bar is

$$\delta U = W \, \delta y_G + F \, \delta x_A$$

Choosing θ as the kinematically independent coordinate, we obtain from geometry

$$y_G = \frac{L}{2}\cos\theta \qquad x_A = L\sin\theta$$

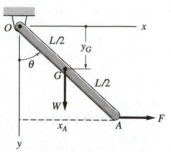

Fig. 10.11

By taking the differentials of the coordinates (recall that virtual changes are identical to differentials), we get

$$\delta y_G = -\frac{L}{2}\sin\theta \, \delta\theta \qquad \delta x_A = L\cos\theta \, \delta\theta$$

Therefore, the virtual work done on the bar is

$$\delta U = \left(-\frac{W}{2}\sin\theta + F\cos\theta\right)L\,\delta\theta$$

If θ is an equilibrium position, then $\delta U = 0$ for any nonzero $\delta\theta$. Consequently, the condition for equilibrium is

$$-\frac{W}{2}\sin\theta + F\cos\theta = 0$$

Consider now a system of bodies with n degrees of freedom with q_1, $q_2, \ldots, q_n$ being the kinematically independent coordinates. If we follow the procedure outlined above, the virtual work done on the system will take the form

$$\delta U = Q_1 \, \delta q_1 + Q_2 \, \delta q_2 + \cdots + Q_n \, \delta q_n \qquad (10.11)$$

where each Q_i is in general a function of $q_1, q_2, \ldots, q_n$. If the system is in equilibrium, then $\delta U = 0$ for any nonzero combination of δq's. This condition can be satisfied only if

$$Q_1 = Q_2 = \cdots = Q_n = 0$$

The Q's are known as the *generalized forces*. If q_i has units of distance, then Q_i has units of force; if q_i is an angle, then Q_i has units of moment of a force.

   When applying the method of virtual work, it is recommended that you begin by drawing an *active-force diagram*, which is a sketch of the body that shows only the forces that do work. Figure 10.11 is an example of an active-force diagram. It displays only the work-producing forces W and F. The pin reactions at O were omitted, because they do no work (point O does not move).

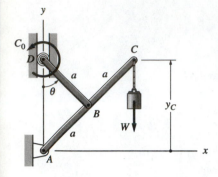

Sample Problem 10.1

Compute the couple C_0 that will support the load W. Neglect the weights of the bars.

Solution

Note that the system possesses one DOF because its configuration can be specified by a single coordinate, such as the angle θ.

The figure is an active-force diagram because only W and C_0 can do virtual work when the system is given a virtual displacement consistent with the constraints. The pin reaction at A is workless because its point of application does not move. The roller reaction at D does no work because it is horizontal, whereas the virtual displacement of end D can only be vertical. The internal forces, including the pin reaction at B, are also workless.

The figure can also be used for the kinematic analysis. We introduce the xy-coordinate system with origin at the fixed point A, and we choose the angle θ as the kinematically independent coordinate. The vertical coordinate of end C is denoted y_C.

Applying the principle of virtual work to the system under consideration, we have

$$\delta U = C_0 \, \delta\theta - W \, \delta y_C = 0 \qquad (a)$$

where $\delta\theta$ is the virtual rotation of the bar BD and δy_C is the vertical virtual displacement of C. The positive directions for $\delta\theta$ and δy_C are, of course, the same as for θ and y_C, respectively. The sign of the first term in Eq. (a) is positive because positive $\delta\theta$ has the same sense as positive C_0. The second term has a negative sign because the positive sense of W is opposite to the positive sense of δy_C.

We now relate y_C to θ using geometry and then obtain δy_C in terms of $\delta\theta$ by differentiation. Referring to the figure, this procedure yields

$$y_C = 2a\cos\theta$$

$$\delta y_C = \frac{dy_C}{d\theta}\, \delta\theta = -2a\sin\theta \, \delta\theta \qquad (b)$$

Substituting Eq. (b) into Eq. (a), we obtain

$$\delta U = C_0 \, \delta\theta - W(-2a\sin\theta \, \delta\theta) = 0$$

or

$$(C_0 + 2Wa\sin\theta)\, \delta\theta = 0 \qquad (c)$$

Equation (c) can be satisfied for a nonzero $\delta\theta$ only if the term in parentheses (which represents the generalized force corresponding to $\delta\theta$) vanishes, which yields

$$C_0 = -2Wa\sin\theta \qquad \textit{Answer}$$

as the condition for equilibrium. The negative sign indicates that the correct sense of C_0 is opposite to that shown in the figure.

Sample Problem 10.2

The mechanism shown in the figure consists of two pin-connected, homogeneous bars of weight W and length L each. The roller at B moves in a horizontal slot, located at the distance $1.5L$ below the pin at O. Determine the force P that will hold the system in equilibrium for $\theta_1 = 30°$.

Solution

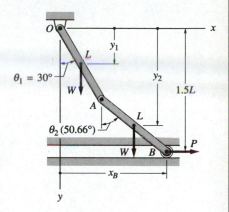

The system has one DOF because only one position coordinate, for example, θ_1 or θ_2, is required to specify its configuration. The figure shown is also an active-force diagram, because it shows only those forces than can perform virtual work on the system. The pin reactions at O and A, and the roller reaction at B, are omitted because their virtual work is zero.

It is convenient to use the same figure for kinematic analysis. We introduce the xy-coordinate system shown, the origin of which is located at the fixed point O. The coordinates y_1 and y_2 locate the centers of gravity of the bars, and x_B is the horizontal coordinate of end B.

If the system is given a virtual displacement consistent with the constraints, the principle of virtual work takes the form

$$\delta U = W\,\delta y_1 + W\,\delta y_2 + P\,\delta x_B = 0 \qquad \text{(a)}$$

All the signs in Eq. (a) are positive because the direction of each force is the same as the positive coordinate direction of its point of application.

The next step is to express δy_1, δy_2, and δx_B as functions of the virtual change in the kinematically independent coordinate. Let us choose θ_1 as the independent coordinate. Because the roller at B is constrained to move in the horizontal slot, the equation of constraint is $L\cos\theta_1 + L\cos\theta_2 = 1.5L$, or

$$\cos\theta_1 + \cos\theta_2 = 1.5 \qquad \text{(b)}$$

which yields $\theta_2 = 50.66°$ when $\theta_1 = 30°$. The evaluation of δy_1, δy_2, and δx_B in terms of the virtual rotation $\delta\theta_1$ now proceeds as follows.

Evaluation of δy_1

From the figure, we see that $y_1 = (L/2)\cos\theta_1$. Forming the differential of both sides and evaluating at $\theta_1 = 30°$, we find that

$$\delta y_1 = \frac{dy_1}{d\theta_1}\,\delta\theta_1 = -\frac{L}{2}\sin\theta_1\,\delta\theta_1 = -0.2500L\,\delta\theta_1 \qquad \text{(c)}$$

Evaluation of δy_2

From the figure, $y_2 = L\cos\theta_1 + (L/2)\cos\theta_2$. Substituting for θ_2 from Eq. (b), and simplifying, gives $y_2 = (L/2)\cos\theta_1 + 0.75L$. Taking the differential of each side, and substituting $\theta_1 = 30°$, yields

$$\delta y_2 = \frac{dy_2}{d\theta_1}\,\delta\theta_1 = -\frac{L}{2}\sin\theta_1\,\delta\theta_1 = -0.2500L\,\delta\theta_1 \qquad \text{(d)}$$

Evaluation of δx_B

From the figure, $x_B = L\sin\theta_1 + L\sin\theta_2$. In principle, θ_2 could be eliminated by using Eq. (b), but this would result in a rather cumbersome expression. It is far easier to form the differentials first and then carry out the substitution. From the chain rule for differentiation, we obtain

$$\delta x_B = \frac{\partial x_B}{\partial\theta_1}\,\delta\theta_1 + \frac{\partial x_B}{\partial\theta_2}\,\delta\theta_2$$

471

which gives

$$\delta x_B = L\cos\theta_1\,\delta\theta_1 + L\cos\theta_2\,\delta\theta_2 \tag{e}$$

From Eq. (b) we obtain the following equation of constraint in terms of $\delta\theta_1$ and $\delta\theta_2$:

$$\frac{\partial}{\partial\theta_1}\cos\theta_1 + \cos\theta_2)\,\delta\theta_1 + \frac{\partial}{\partial\theta_2}\cos\theta_1 + \cos\theta_2)\,\delta\theta_2 = 0$$

or

$$-\sin\theta_1\,\delta\theta_1 - \sin\theta_2\,\delta\theta_2 = 0$$

which gives

$$\delta\theta_2 = -(\sin\theta_1/\sin\theta_2)\,\delta\theta_1 \tag{f}$$

Substituting Eq. (f) into Eq. (e), we find that

$$\delta x_B = L\cos\theta_1\,\delta\theta_1 + L\cos\theta_2\left(-\frac{\sin\theta_1}{\sin\theta_2}\right)\delta\theta_1$$

which, when evaluated at $\theta_1 = 30°$ and $\theta_2 = 50.66°$, yields

$$\delta x_B = 0.4562L\,\delta\theta_1 \tag{g}$$

Virtual Work Equation

Substituting Eqs. (c), (d), and (g) into Eq. (a) gives

$$\delta U = W(-0.2500L\,\delta\theta_1) + W(-0.2500L\,\delta\theta_1) + P(0.4562L\,\delta\theta_1) = 0$$

or

$$(-0.5W + 0.4562P)\,\delta\theta_1 = 0 \tag{h}$$

This equation can be satisfied for a nonzero $\delta\theta_1$ only if the term in parentheses vanishes, which yields

$$P = 1.096W \qquad\qquad\textit{Answer}$$

Because P is positive, its direction is as shown in the figure.

Sample Problem 10.3

The structure shown in Fig. (a) is obtained by fixing point B of the mechanism in Sample Problem 10.2. For $\theta_1 = 30°$, determine the pin reactions B_x and B_y by the following methods: (1) compute B_x and B_y independently; that is, use a separate virtual work equation for each component; and (2) compute B_x and B_y simultaneously; that is, compute both components using one virtual work equation.

Solution

Using $\theta_1 = 30°$ and $\overline{OC} = 1.5L$, we find from geometry that $\theta_2 = 50.66°$ and $\overline{CB} = 1.273L$. (See Sample Problem 10.2.)

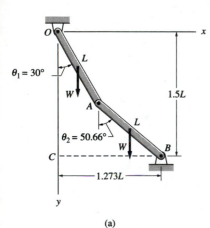

(a)

Part 1

Because the system in Fig. (a) is a structure rather than a mechanism, it has no DOF. To determine a component of the reaction at B by the method of virtual work, we must alter the supports so that the component becomes an active force.

To find B_x, we release the constraint on the horizontal displacement of B, thereby causing B_x to become an active force. The revised system, shown in Fig. (b), possesses one DOF. The value of the force B_x required for equilibrium will be equal to B_x for the original structure, in Fig. (a). Note that Fig. (b) is identical to the mechanism in Sample Problem 10.2 except that P has been relabeled as B_x. Therefore, the solution of Sample Problem 10.2 is applicable for finding B_x:

$$B_x = 1.096W \qquad \qquad \textit{Answer}$$

To find B_y, we release the constraint on the vertical displacement of B in order to convert B_y into an active force, as shown in Fig. (c). The value of B_y required for equilibrium of this revised system will be equal to B_y for the original structure, in Fig. (a). Again we see that the new system has one DOF. If the system is given a virtual displacement consistent with the constraints, the principle of virtual work states that the equation

$$\delta U = W\,\delta y_1 + W\,\delta y_2 + B_y\,\delta y_B = 0 \qquad \qquad \text{(a)}$$

must be satisfied for equilibrium. Each of the signs in Eq. (a) is positive, because each force acts in the same direction as the positive virtual displacement of its point of application. (Recall that the positive directions of y_1, y_2, and y_B in Fig. (c) are also the positive directions of the corresponding virtual displacements.)

We must now express the virtual displacements δy_1, δy_2, and δy_B in Eq. (a) as functions of the virtual change in an independent position coordinate. We choose θ_1 to be the independent position coordinate. Note from Fig. (c) that angles θ_1 and θ_2 are related by the equation of constraint $L\sin\theta_1 + L\sin\theta_2 = 1.273L$, or

$$\sin\theta_1 + \sin\theta_2 = 1.273 \qquad \qquad \text{(b)}$$

The remaining computations are as follows.

Evaluation of δy_1

From Fig. (c) we see that $y_1 = (L/2)\cos\theta_1$. Taking the differential of each side of this equation and evaluating at $\theta_1 = 30°$, we find

$$\delta y_1 = \frac{dy_1}{d\theta_1}\,\delta\theta_1 = -\frac{L}{2}\sin\theta_1\,\delta\theta_1 = -0.2500L\,\delta\theta_1 \qquad \qquad \text{(c)}$$

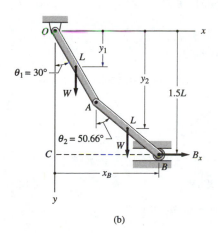

(b)

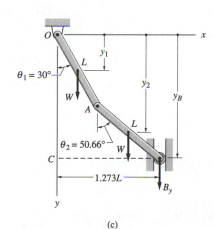

(c)

Evaluation of δy_2

From Fig. (c), $y_2 = L\cos\theta_1 + (L/2)\cos\theta_2$. Because expressing y_2 as a function of θ_1 would involve complicated algebra, it is preferable to obtain the differential of this equation first, and then eliminate θ_2 using Eq. (b). The chain rule for differentiation yields

$$\delta y_2 = \frac{\partial y_2}{\partial\theta_1}\,\delta\theta_1 + \frac{\partial y_2}{\partial\theta_2}\,\delta\theta_2$$

which becomes

$$\delta y_2 = -L\sin\theta_1\,\delta\theta_1 - \frac{L}{2}\sin\theta_2\,\delta\theta_2 \qquad\text{(d)}$$

Taking the differential of Eq. (b), we obtain

$$\frac{\partial}{\partial\theta_1}(\sin\theta_1 + \sin\theta_2)\,\delta\theta_1 + \frac{\partial}{\partial\theta_2}(\sin\theta_1 + \sin\theta_2)\,\delta\theta_2 = 0$$

or

$$\cos\theta_1\,\delta\theta_1 + \cos\theta_2\,\delta\theta_2 = 0$$

which gives

$$\delta\theta_2 = \left(-\frac{\cos\theta_1}{\cos\theta_2}\right)\delta\theta_1 \qquad\text{(e)}$$

Substituting Eq. (e) into Eq. (d), we find that

$$\delta y_2 = -L\,\sin\theta_1\,\delta\theta_1 - \frac{L}{2}\sin\theta_2\left(-\frac{\cos\theta_1}{\cos\theta_2}\right)\delta\theta_1$$

which, evaluated at $\theta_1 = 30°$ and $\theta_2 = 50.66°$, yields

$$\delta y_2 = 0.028\,29L\,\delta\theta_1 \qquad\text{(f)}$$

Evaluation of δy_B

From Fig. (c), $y_B = L\cos\theta_1 + L\cos\theta_2$. The method of relating δy_B and $\delta\theta_1$ is identical to what was used to find δy_2. The result is

$$\delta y_B = 0.5566L\,\delta\theta_1 \qquad\text{(g)}$$

Virtual Work Equation

Substituting Eqs. (c), (f), and (g) into Eq. (a) yields

$$\delta U = W(-0.2500L\,\delta\theta_1) + W(0.028\,29L\,\delta\theta_1) + B_y(0.5566L\,\delta\theta_1) = 0$$

or

$$(-0.2217W + 0.5566B_y)\,\delta\theta_1 = 0$$

For this equation to be satisfied for nonzero $\delta\theta_1$, we must have

$$B_y = 0.398W \qquad\qquad\qquad\textit{Answer}$$

Because B_y is positive, its direction is as shown in Fig. (c).

Part 2

Figure (d) shows the system with both B_x and B_y considered to be active forces, which is achieved by removing all constraints on the displacement of B. In this case, angles θ_1 and θ_2 are independent position coordinates; consequently, the system possesses two DOF. The principle of virtual work states that at equilibrium

$$\delta U = W\,\delta y_1 + W\,\delta y_2 + B_x\,\delta x_B + B_y\,\delta y_B = 0 \tag{h}$$

The virtual displacements in Eq. (h) can be related to $\delta\theta_1$ and $\delta\theta_2$ as follows:

$$y_1 = \frac{L}{2}\cos\theta_1$$

$$\delta y_1 = \frac{dy_1}{d\theta_1}\,\delta\theta_1 = -\frac{L}{2}\sin\theta_1\,\delta\theta_1 \tag{i}$$

$$y_2 = L\cos\theta_1 + \frac{L}{2}\cos\theta_2$$

$$\delta y_2 = \frac{\partial y_2}{\partial\theta_1}\,\delta\theta_1 + \frac{\partial y_2}{\partial\theta_2}\,\delta\theta_2$$

$$\delta y_2 = -L\sin\theta_1\,\delta\theta_1 - \frac{L}{2}\sin\theta_2\,\delta\theta_2 \tag{j}$$

$$x_B = L\sin\theta_1 + L\sin\theta_2$$

$$\delta x_B = \frac{\partial x_B}{\partial\theta_1}\,\delta\theta_1 + \frac{\partial x_B}{\partial\theta_2}\,\delta\theta_2$$

$$\delta x_B = L\cos\theta_1\,\delta\theta_1 + L\cos\theta_2\,\delta\theta_2 \tag{k}$$

$$y_B = L\cos\theta_1 + L\cos\theta_2$$

$$\delta y_B = \frac{\partial y_B}{\partial\theta_1}\,\delta\theta_1 + \frac{\partial y_B}{\partial\theta_2}\,\delta\theta_2$$

$$\delta y_B = -L\sin\theta_1\,\delta\theta_1 - L\sin\theta_2\,\delta\theta_2 \tag{l}$$

Substituting Eqs. (i)–(l) into Eq. (h) and regrouping terms, we obtain

$$\delta U = \left(-\frac{W}{2}\sin\theta_1 - W\sin\theta_1 + B_x\cos\theta_1 - B_y\sin\theta_1\right)L\,\delta\theta_1$$

$$+ \left(-\frac{W}{2}\sin\theta_2 + B_x\cos\theta_2 - B_y\sin\theta_2\right)L\,\delta\theta_2 = 0 \tag{m}$$

Because $\delta\theta_1$ and $\delta\theta_2$ are independent, Eq. (m) will be satisfied only if each of the terms in parentheses is zero; that is,

$$-\frac{3W}{2}\sin\theta_1 + B_x\cos\theta_1 - B_y\sin\theta_1 = 0 \tag{n}$$

and

$$-\frac{W}{2}\sin\theta_2 + B_x\cos\theta_2 - B_y\sin\theta_2 = 0 \tag{o}$$

Substituting $\theta_1 = 30°$ and $\theta_2 = 50.66°$, and solving Eqs. (n) and (o) simultaneously, we obtain

$$B_x = 1.096W \quad \text{and} \quad B_y = 0.398W \qquad \textit{Answer}$$

which agree with the values found in Part 1.

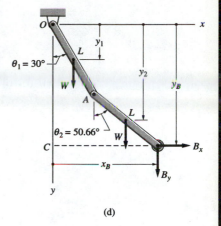

(d)

475

Problems

10.1 Determine the number of DOF for each of the mechanisms shown.

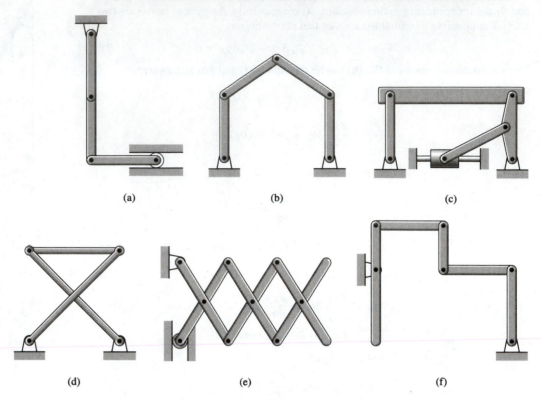

(a) (b) (c)

(d) (e) (f)

Fig. P10.1

Neglect friction in the following problems.

10.2 The uniform bar of weight W is held in equilibrium by the couple C_0. Find C_0 in terms of W, L, and θ.

10.3 Find the ratio W_1/W_2 of the weights for which the scales would be in equilibrium for any value of the angle θ. Neglect the weight of the scale mechanism.

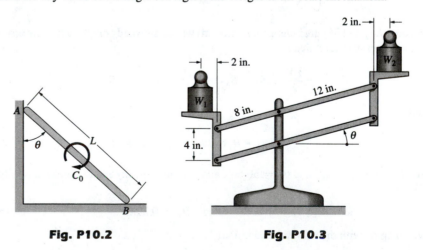

Fig. P10.2 **Fig. P10.3**

10.4 The system is held at rest in the position shown by the couple $C_0 = 610\,\text{N}\cdot\text{m}$ applied to bar AC. The mass of the uniform bar AB is 25 kg, and the masses of the bars AC and BD may be neglected. Determine the angle θ.

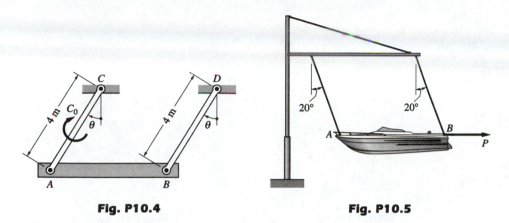

Fig. P10.4   **Fig. P10.5**

10.5 The 1800-kg boat is suspended from two parallel cables of equal length. The location of the center of gravity of the boat is not known. Calculate the force P required to hold the boat in the position shown.

10.6 The 5-lb lamp, with center of gravity located at G, is supported by the parallelogram linkage of negligible weight. Find the tension in the spring AD when the lamp is in equilibrium in the position shown.

10.7 Determine the force P that would hold the mechanism in equilibrium in the position $\theta = 40°$.

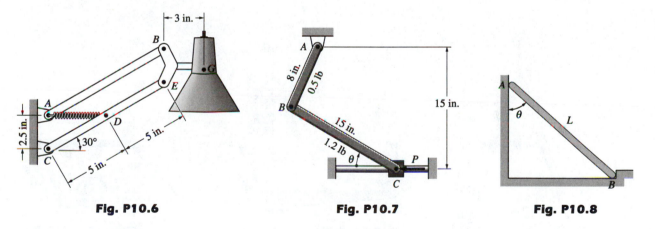

Fig. P10.6   **Fig. P10.7**   **Fig. P10.8**

10.8 Calculate the horizontal reaction at B in terms of the weight W of the homogeneous bar and the angle θ.

10.9 The mass of the uniform top of the folding table is 15 kg. Determine the tension in the rope *AB*. Neglect the masses of the legs.

10.10 The uniform 320-lb bar *AB* is held in the position shown by the cable *AC*. Compute the tension in this cable.

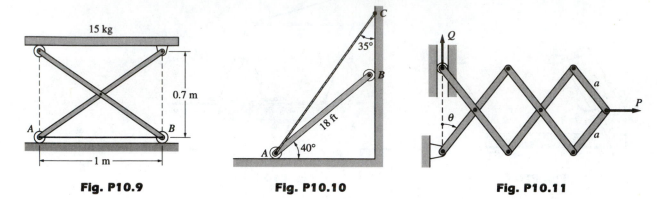

Fig. P10.9   **Fig. P10.10**   **Fig. P10.11**

10.11 Determine the ratio *P/Q* of the forces that are required to maintain equilibrium of the mechanism for an arbitrary angle θ. Neglect the weight of the mechanism.

10.12 Neglecting the weights of the members, determine the force *P* that would keep the mechanism in the position shown. The spring *DE* has a free length of 0.5 m and a stiffness of 1.2 kN/m.

10.13 The linkage of the braking system consists of the pedal arm *DAB*, the connecting rod *BC*, and the hydraulic cylinder *C*. At what angle θ will the force *Q* be four times greater than the force *P* that is applied to the pedal? Neglect the weight of the linkage.

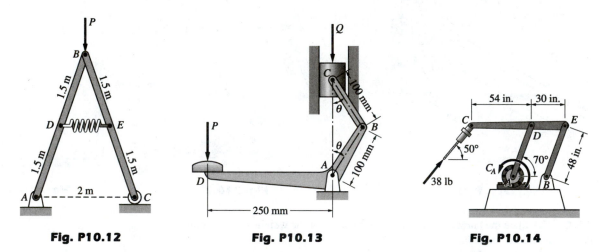

Fig. P10.12   **Fig. P10.13**   **Fig. P10.14**

10.14 The automatic drilling robot must sustain a thrust of 38 lb at the tip of the drill bit. Determine the couple C_A that must be developed by the electric motor at *A* to resist this thrust. Neglect the weights of the members.

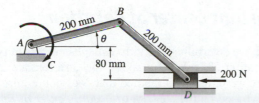

Fig. P10.15

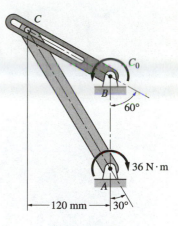

Fig. P10.16

10.15 Determine the couple C for which the mechanism would be in equilibrium in the position $\theta = 25°$. Neglect the weights of the members.

10.16 In the angular motion amplifier, the oscillatory motion of AC is amplified by the oscillatory motion of BC. Neglecting the weights of the members, determine the output torque C_0, given that the input torque is 36 N · m.

10.17 The mechanism shown controls the movement of the fingers of a robotic gripper. The activating force P will cause parallel movement of the fingers, providing the gripping force Q. Determine the relation between P and Q for the position shown. (Hint: Observe that $CDFE$ is a parallelogram linkage.)

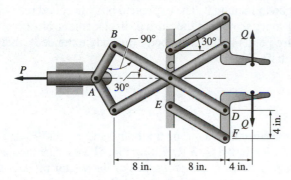

Fig. P10.17

10.18 Calculate the torque C_0 that must be applied to the handle of the screw jack in order to lift the load $P = 3$ kN when $\theta = 30°$. The screw has a pitch of 2.5 mm. Neglect the weight of the linkage.

10.19 Determine the force F and the angle α required to hold the linkage in the position $\theta_1 = 60°$, $\theta_2 = 15°$. Each bar of the linkage is homogeneous and of weight W.

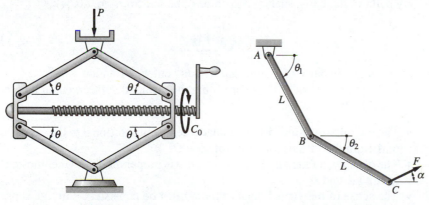

Fig. P10.18   **Fig. P10.19**

*10.5 *Instant Center of Rotation*

In the last article, we determined the virtual displacements of points of interest (points of application of forces) by taking the differentials of their position coordinates. Here we introduce a method that does not require differentiation. This approach is based on the concept of *instant center of rotation*, which is defined as follows:

> *The instant center of rotation of a rigid body is the point in the body that has zero virtual displacement during virtual motion of the body.*

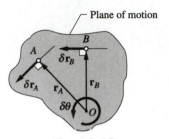

Plane of motion

Fig. 10.12

To find the location of the instant center, consider the virtual motion of the rigid body shown in Fig. 10.12. Let $\delta\mathbf{r}_A$ and $\delta\mathbf{r}_B$ be the virtual displacements of points A and B in the body. Assume, for the time being, that their directions are not parallel. Now draw a line through point A that is perpendicular to $\delta\mathbf{r}_A$, and another line through B that is perpendicular to $\delta\mathbf{r}_B$. The intersection of these two lines, labeled O in the figure, is the instant center of rotation of the body. Sometimes O lies outside the body, in which case we imagine that the body is enlarged to include it. The expanded body is called the *body extended*.

We still must show that the virtual displacement of O is zero. Because A, B, and O are points in the same body (or body extended), their virtual displacements satisfy Eq. (10.5):

$$\delta\mathbf{r}_A = \delta\mathbf{r}_O + \delta\boldsymbol{\theta} \times \mathbf{r}_A \qquad\qquad \text{(a)}$$

$$\delta\mathbf{r}_B = \delta\mathbf{r}_O + \delta\boldsymbol{\theta} \times \mathbf{r}_B \qquad\qquad \text{(b)}$$

where $\delta\boldsymbol{\theta}$ is the virtual rotation of the body, and $\mathbf{r}_A$ and $\mathbf{r}_B$ are the position vectors of A and B relative to O. Recalling that $\delta\boldsymbol{\theta}$ is perpendicular to the plane of motion, we deduce from the properties of the vector product that $\delta\boldsymbol{\theta} \times \mathbf{r}_A$ is parallel to $\delta\mathbf{r}_A$. Therefore, Eq. (a) can hold only if $\delta\mathbf{r}_O$ is also parallel to $\delta\mathbf{r}_A$. In the same way we can argue that $\delta\mathbf{r}_O$ in Eq. (b) must be parallel to $\delta\mathbf{r}_B$. Because a nonzero vector cannot have two different directions, we conclude that $\delta\mathbf{r}_O = \mathbf{0}$.

The term *instant center of rotation* implies that the body appears to be rotating about O during its virtual displacement. Therefore, once the instant center of rotation has been located, the magnitude of the virtual displacement of any point in the body, such as A, can be obtained from Eq. (10.4):

$$\boxed{\delta r_A = r_A \delta\theta} \qquad\qquad \text{(c)}$$

where r_A is the distance of A from the instant center of rotation.

In summary, the rules for determining the virtual displacement of a point A in a rigid body are as follows:

- The magnitude of the virtual displacement is proportional to the distance of A from the instant center of rotation O.
- The direction of the virtual displacement is perpendicular to the line connecting A and O.
- The sense of the virtual displacement must be consistent with the sense of the virtual rotation.

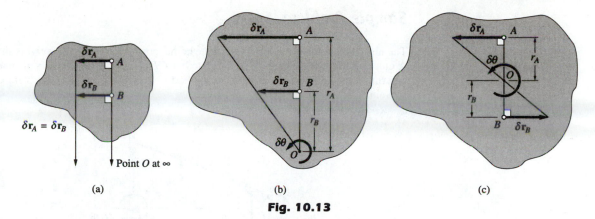

Fig. 10.13

The construction shown in Fig. 10.12 for locating the instant center for a rigid body is valid only if the directions of $\delta\mathbf{r}_A$ and $\delta\mathbf{r}_B$ are not parallel. If the directions are parallel, the instant center can still be located without difficulty, as shown in Fig. 10.13. Figure 10.13(a) shows translation, in which the virtual displacements of all points are equal, and the instant center is located at infinity. Figure 10.13(b) depicts the case where $\delta\mathbf{r}_A$ and $\delta\mathbf{r}_B$ have the same direction but unequal magnitudes, $\delta r_A > \delta r_B$. In Fig. 10.13(c), $\delta\mathbf{r}_A$ and $\delta\mathbf{r}_B$ have parallel but opposite directions. (Whether their magnitudes are equal or not is irrelevant.)

We now derive the formulas for the rectangular components of the virtual displacement of a point. These formulas are frequently used in the computation of the virtual work of a force. Let $\delta\mathbf{r}_A$ be the virtual displacement of a point A that is located at the distance r_A from point O (the instant center of the body containing A). From Fig. 10.14 we see that $\delta x_A = -\delta r_A \sin\theta$ and $\delta y_A = \delta r_A \cos\theta$. Substituting $\delta r_A = r_A\,\delta\theta$, and noting that $r_A \cos\theta = x_A$ and $r_A \sin\theta = y_A$, we obtain

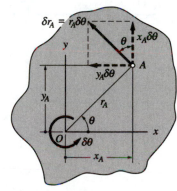

Fig. 10.14

$$\delta x_A = -y_A\,\delta\theta \quad \text{and} \quad \delta y_A = x_A\,\delta\theta \qquad (10.12)$$

The signs in this equation are consistent with the positive directions of x, y, and $\delta\theta$ shown in Fig. 10.14. However, when solving problems, it is easiest to determine the directions of δx_A and δy_A by inspection, rather than by adhering to a rigorous sign convention.

Sample Problem 10.4

The mechanism in Fig. (a) consists of three homogeneous bars with the weights shown. A clockwise 500-lb · ft couple is applied to bar CD. Using instant centers of rotation, determine the couple C_0 that must be applied to bar AB in order to maintain equilibrium.

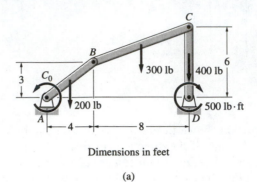

Dimensions in feet

(a)

(b)

Solution

The system possesses one degree of freedom, because only one position coordinate, for example, the angular position of one of the bars, is sufficient to specify its configuration.

Figure (b) shows a virtual displacement of the system that is consistent with the constraints, with $\delta\theta_{AB}$, $\delta\theta_{BC}$, and $\delta\theta_{CD}$ representing the virtual rotations of the bars. Points ①, ②, and ③ indicate the locations of the centers of gravity of the bars. Note that Fig. (a) is the active-force diagram for the system because only the weights and the two couples are capable of doing work on the system. Applying the principle of virtual work, we obtain

$$\delta U = C_0\,\delta\theta_{AB} - 500\,\delta\theta_{CD} - 200\,\delta y_1 - 300\,\delta y_2 - 400\,\delta y_3 = 0 \qquad \text{(a)}$$

where δy_1 is the positive y-component of the virtual displacement of point ①, and so forth. Note the signs in Eq. (a), which follow the rule that virtual work is positive if the force (or couple) has the same direction as the displacement (rotation). If these directions are opposite each other, the virtual work is negative.

The next step in the analysis is to express all of the virtual changes in Eq. (a) in terms of one independent position coordinate, for which we choose θ_{AB}. To utilize instant centers, we must first locate the instant center for each of the three bars. Referring to Fig. (b), we see that A and D, being fixed points, are obviously the instant centers for bars AB and CD, respectively. Therefore, $\delta\mathbf{r}_B$ is perpendicular to AB, and $\delta\mathbf{r}_C$ is perpendicular to CD, each directed as shown in Fig. (c). Because B and C also belong to bar BC, the instant center of BC is point O, where the lines that are perpendicular to $\delta\mathbf{r}_B$ and $\delta\mathbf{r}_C$ intersect. Because point O does not lie on bar BC, it is convenient to think of BC being extended to the triangle BCO, as shown in Fig. (d). From the directions of $\delta\mathbf{r}_B$ and $\delta\mathbf{r}_C$, we see that $\delta\theta_{BC}$ is clockwise.

Figures (c) and (d) show all the dimensions needed for kinematic analysis. We now relate the virtual rotations of BC and CD to $\delta\theta_{AB}$ as follows:

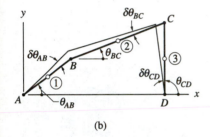

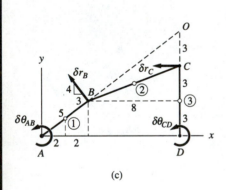

(c)

(d)

B rotates about A: $\delta r_B = 5\,\delta\theta_{AB}$ ft—see Fig. (c)
B rotates about O: $\delta r_B = 10\,\delta\theta_{BC}$ ft—see Fig. (d)

Equating the right-hand sides of these two equations gives

$$\delta\theta_{BC} = \frac{1}{2}\delta\theta_{AB} \qquad\qquad \text{(b)}$$

C rotates about D: $\delta r_C = 6\,\delta\theta_{CD}$ ft—see Fig. (c)
C rotates about O: $\delta r_C = 3\,\delta\theta_{BC}$ ft—see Fig. (d)

Equating the right-hand sides of these two equations, and using Eq. (b), yields

$$\delta\theta_{CD} = \frac{1}{4}\delta\theta_{AB} \qquad\qquad \text{(c)}$$

Next, Eq. (10.12) is used to calculate δy_1, δy_2, and δy_3:

① rotates about A: from Fig. (c) we find

$$\delta y_1 = 2\,\delta\theta_{AB} \qquad\qquad \text{(d)}$$

② rotates about O: from Fig. (d), $\delta y_2 = 4\,\delta\theta_{BC}$. Substituting for δ_{BC} from Eq. (b), we get

$$\delta y_2 = 2\,\delta\theta_{AB} \qquad\qquad \text{(e)}$$

Finally, from Fig. (c) we see that

$$\delta y_3 = 0 \qquad\qquad \text{(f)}$$

Substituting Eqs. (b)–(f) into Eq. (a), we obtain

$$\delta U = C_0\,\delta\theta_{AB} - 500\left(\frac{1}{4}\delta\theta_{AB}\right) - 200(2\,\delta\theta_{AB})$$

$$- 300(2\,\delta\theta_{AB}) - 400(0) = 0$$

which reduces to

$$(C_0 - 1125)\,\delta\theta_{AB} = 0$$

This equation can be satisfied for nonzero $\delta\theta_{AB}$ only if

$$C_0 = 1125\text{ lb} \cdot \text{ft} \qquad\qquad \textit{Answer}$$

Because C_0 is positive, it is directed as shown in Fig. (a)—that is, counterclockwise.

Problems

10.20 Locate the instant center of rotation of bar *AB* for each case shown.

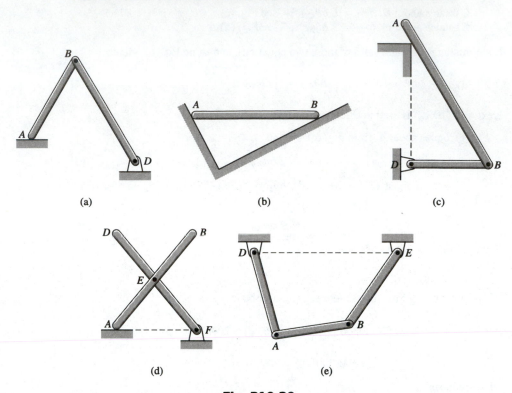

(a)

(b)

(c)

(d)

(e)

Fig. P10.20

The following problems are to be solved using instant centers of rotation. Neglect the weights of the members unless otherwise specified.

10.21 Each of the three uniform bars of the mechanism weighs 12 lb/ft. Determine the couple C_0 that would hold the mechanism in equilibrium in the position shown.

10.22 Compute the force *P* required to maintain equilibrium of the 40-lb uniform bar in the position shown. Neglect friction.

10.23 Determine the tension in the cable attached to the linkage at *B*.

10.24 Find the couple applied by the built-in support at *F* to the linkage.

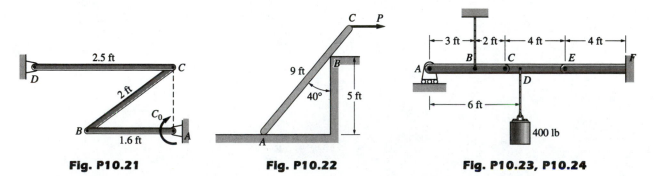

Fig. P10.21

Fig. P10.22

Fig. P10.23, P10.24

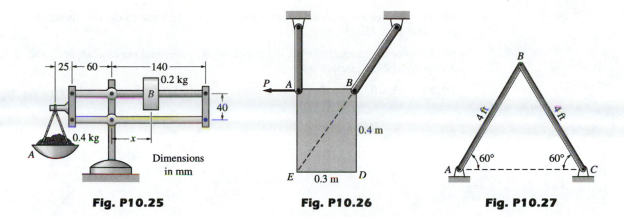

Fig. P10.25   **Fig. P10.26**   **Fig. P10.27**

10.25 Find the position coordinate x of the sliding weight B for which the scales will be balanced.

10.26 Compute the force P that will keep the 15-kg uniform plate $ABDE$ in equilibrium in the position shown.

10.27 Each bar of the structure is uniform and weighs 50 lb. Find the horizontal pin reaction at C.

*10.28 Calculate the couple exerted on the frame by the built-in support at C.

10.29 If the input force to the compound lever is $P = 30$ lb, calculate the output force Q.

10.30 Determine the force P required for equilibrium of the compound lever if $Q = 4200$ N.

10.31 If $Q = 200$ N, determine the couple C_0 required to hold the mechanism in equilibrium in the position $\theta = 25°$. Neglect friction.

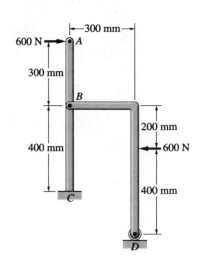

Fig. P10.28

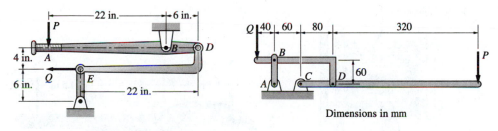

Fig. P10.29   **Fig. P10.30**

Dimensions in mm

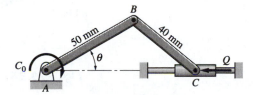

Fig. P10.31

10.32 For the automobile suspension shown, find the force in the coil spring *BE* given that *P* = 2600 N.

10.33 For the structure shown, determine (a) the horizontal pin reaction at *D*; and (b) the vertical pin reaction at *D*.

10.34 What force *P* will produce a tensile force of 25 lb in the cable at *E*?

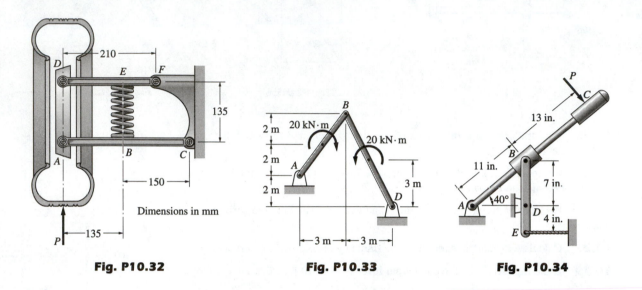

Fig. P10.32   **Fig. P10.33**   **Fig. P10.34**

10.35 If the force *P* acting on the piston in the position shown is equal to 1600 N, compute C_0, the output torque at the crankshaft.

10.36 For the pliers shown, determine the relationship between the magnitude of the applied forces *P* and the magnitude of the gripping forces at *E*. (Hint: Consider *AB* to be fixed.)

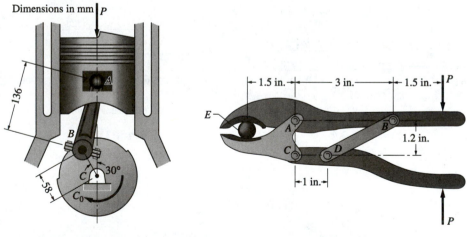

Fig. P10.35   **Fig. P10.36**

10.37 When activated by the force P, the gripper on a robot's arm is able to pick up objects by applying the gripping force F. Given that $P = 120$ N, calculate F in the position shown.

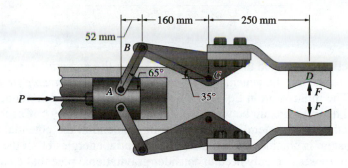

Fig. P10.37

*10.38 (a) Using a scale drawing, locate graphically the instant center of the connecting rod AB in the position shown. (b) Using the results of part (a) and assuming equilibrium, find the couple C_0 that acts on the flywheel if the force acting on the piston is $P = 120$ lb.

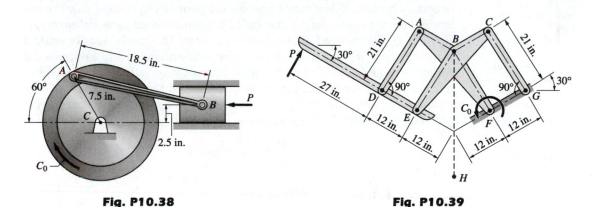

Fig. P10.38   **Fig. P10.39**

*10.39 The hinge is of the type used on some automobiles, in which the door DE appears to rotate about point H. Use a graphical construction, drawn to scale, to locate H. (Hint: $ABED$ and $BCGF$ are parallelogram linkages.)

***10.6** *Equilibrium and Stability of Conservative Systems*

a. *Potential energy*

As explained in Art. 10.4, if a frictionless system is given a kinematically admissible virtual displacement, the virtual work has the form [see Eq. (10.11)]

$$\delta U = Q_1 \, \delta q_1 + Q_2 \, \delta q_2 + \cdots + Q_n \, \delta q_n$$

where the δq's are *virtual changes* in the independent position coordinates and the Q's are called *generalized forces*. The system is classified as *conservative* if there exists a scalar function $V(q_1, q_2, \ldots, q_n)$ such that

$$Q_i = -\frac{\partial V}{\partial q_i} \qquad (i = 1, 2, \ldots, n) \tag{10.13}$$

The function V is called the *potential function*, or the *potential energy*, of the system. Therefore, the generalized forces are said to be *derivable from a potential*. The minus sign in Eq. (10.13), which is part of the definition, has its origin in the relationship between work and potential energy.* If each of the forces acting on a system is derivable from a potential, the potential energy of the system is obtained by summing the potential energies of all the forces. For our purposes, we only need to consider gravitational potential energy and elastic potential energy.

b. Gravitational potential energy

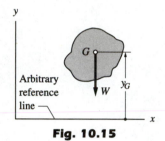

Fig. 10.15

Figure 10.15 shows a body of weight W whose center of gravity G is a distance y_G above an arbitrary reference line that coincides with the x-axis. The generalized force corresponding to the coordinate y_G is $-W$ (the minus sign is necessary because W is directed opposite the positive direction of y_G). The potential function of the weight, also called its *gravitational potential energy*, is $V_g = Wy_G + C$, where C is an arbitrary constant. This result is easily verified by noting that $-dV_g/dy_G = -W$, which agrees with Eq. (10.13). Because the weight W is derivable from a potential, it is a conservative force.† Observe that the value of the additive constant C is irrelevant because it does not contribute to the derivative of V_g. Therefore, C may be taken to be zero, and the gravitational potential energy written as

$$V_g = Wy_G \tag{10.14}$$

c. Elastic potential energy

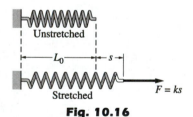

Fig. 10.16

Our discussion of elastic potential energy is limited to ideal springs. An ideal spring has the following properties: (1) the weight of the spring is negligible, and (2) the force exerted by the spring is proportional to the elongation of the spring. (The reader should be aware that our consideration of deformable springs represents a radical departure from the analysis of rigid bodies.) The free (unstretched) length of the spring is denoted by L_0 and the elongation by s, as shown in Fig. 10.16. If the spring is ideal, the force F applied to the spring is related to its elongation s by

$$F = ks \tag{10.15}$$

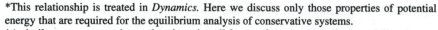

*This relationship is treated in *Dynamics*. Here we discuss only those properties of potential energy that are required for the equilibrium analysis of conservative systems.
†A similar argument can be used to show that all forces of constant magnitude and direction are conservative.

where k is a constant, called the *stiffness* of the spring, or the *spring constant*. The stiffness k has the dimension $[F/L]$; hence the units are lb/ft, N/m, and so forth.

Letting Q_s represent the force exerted *by* the spring, we have $Q_s = -ks$ (the force exerted *by* the spring is opposite the force exerted *on* the spring). Note that this expression for Q_s is valid for both positive s (tension spring) and negative s (compression spring).

The potential energy of an ideal spring is

$$V_e = \frac{1}{2}ks^2 \qquad (10.16)$$

It can be seen that $-dV_e/ds = -ks$, which is indeed the force Q_s exerted by the spring. Because the spring force is derivable from a potential, we conclude that it is a conservative force. The potential energy V_e is called *elastic potential energy*. When using Eq. (10.16), remember that s is the elongation or contraction of the spring, not the length of the spring.

d. Stationary potential energy and stability

In Art. 10.5, Eq. (10.11), we pointed out that a system is in equilibrium only if all the generalized forces vanish—that is, if $Q_1 = Q_2 = \cdots = Q_n = 0$. For a conservative system with potential energy $V(q_1, q_2, \ldots, q_n)$, the equilibrium conditions thus are

$$\frac{\partial V}{\partial q_1} = 0, \quad \frac{\partial V}{\partial q_2} = 0, \ldots, \frac{\partial V}{\partial q_n} = 0 \qquad (10.17)$$

Equation (10.17) represents the *principle of stationary potential energy*:

> *The potential energy of a conservative system is stationary (minimum, maximum, or constant) in an equilibrium position.*

Potential energy can also be used to determine whether an equilibrium position is stable, unstable, or neutral. These three classifications of equilibrium are illustrated in Fig. 10.17: The ball at the bottom of the bowl in (a) is said to be in stable equilibrium—if the ball is displaced a small amount and then released, it will return to the equilibrium position shown. In (b), the ball is in equilibrium at the top of an inverted bowl. Here the equilibrium is unstable—if the ball is displaced a small amount and then released, it will move away from the original equilibrium position. Neutral equilibrium is shown in (c)—if the ball on a flat surface is displaced a small amount to the left or right and then released, the ball will simply remain at rest in the new position.

From the foregoing examples, we can deduce the *principle of minimum potential energy*:

> *The potential energy of a conservative system is at its minimum in a stable equilibrium position.*

As an illustration of this principle, note that when the ball in Fig. 10.17(a) is displaced, its potential energy is increased. When released, the ball returns to its original position of lower potential energy. However, when the ball in

(a) Stable

(b) Unstable

(c) Neutral

Fig. 10.17

Fig. 10.17(b) is displaced, its potential energy decreases. When released, the ball does not return to its original position of higher potential energy. Instead, the ball seeks a position of lower potential energy—it rolls off the bowl.

We restrict our discussion of stability to systems that possess one degree of freedom—that is, systems for which the potential energy $V(q)$ is a function of a single coordinate q. By the principle of stationary potential energy, the equilibrium positions of the system correspond to the roots of the equation $dV/dq = 0$. To determine whether the equilibrium positions are stable or unstable, we must investigate the sign of the second derivative, d^2V/dq^2. If the second derivative is positive at an equilibrium position, the potential energy is a minimum (stable equilibrium); if the second derivative is negative, the potential energy is a maximum (unstable equilibrium).

In summary, if $q = q_0$ is a *stable* equilibrium position, then

$$\left.\frac{dV}{dq}\right|_{q_0} = 0 \qquad \left.\frac{d^2V}{dq^2}\right|_{q_0} > 0 \qquad\qquad (10.18)$$

and if $q = q_0$ is an *unstable* equilibrium position, then

$$\left.\frac{dV}{dq}\right|_{q_0} = 0 \qquad \left.\frac{d^2V}{dq^2}\right|_{q_0} < 0 \qquad\qquad (10.19)$$

Sample Problem 10.5

A light rod is pin-supported at one end and carries a weight W at the other end, as shown in the figure. The ideal spring attached to the rod is capable of resisting both tension and compression, and it is unstretched when the rod is vertical. Find the largest value of W for which the vertical equilibrium position of the rod would be stable.

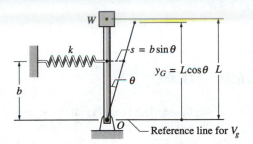

Solution

The potential energy of the system consists of V_g, the gravitational potential energy of the weight, and V_e, the elastic potential energy of the spring. From Eq. (10.14), $V_g = Wy_G$, where y_G is the vertical distance to W measured from an arbitrary reference line. Choosing the horizontal line passing through O as the reference, we find $V_g = WL\cos\theta$.

In order to determine whether a function evaluated at a point is a minimum or maximum, it is sufficient to investigate the function only in a small neighborhood of that point. Therefore, we can confine our attention to small values of θ. Approximating $\cos\theta$ with $(1 - \theta^2/2)$, which is valid for small angles, the gravitational potential energy becomes

$$V_g = WL\left(1 - \frac{1}{2}\theta^2\right) \tag{a}$$

From Eq. (10.16), $V_e = (1/2)ks^2$, where s is the elongation (or contraction) of the spring measured from its unstretched position. From the figure, we see that for sufficiently small θ, $s = b\sin\theta$. Using the approximation $\sin\theta \approx \theta$, the elastic potential energy of the spring becomes

$$V_e = \frac{1}{2}kb^2\theta^2 \tag{b}$$

Combining Eqs. (a) and (b), the potential energy of the system is

$$V = V_g + V_e = WL\left(1 - \frac{1}{2}\theta^2\right) + \frac{1}{2}kb^2\theta^2 \tag{c}$$

which is valid for small values of θ.

For the system to be in stable equilibrium, $d^2V/d\theta^2$ must be positive. After differentiation, we find

$$\frac{dV}{d\theta} = (-WL + kb^2)\theta \tag{d}$$

Observe that $dV/d\theta = 0$ when $\theta = 0$, confirming that the rod is in equilibrium in the vertical position. Differentiating again, we obtain

$$\frac{d^2V}{d\theta^2} = -WL + kb^2 \qquad \text{(e)}$$

We see that $d^2V/d\theta^2$ will be positive; that is, the system will be in stable equilibrium, only if $kb^2 > WL$. Therefore, the maximum value of W for which the system will be in stable equilibrium for $\theta = 0$ is

$$W_{max} = kb^2/L \qquad \qquad \textit{Answer}$$

Sample Problem 10.6

For the system shown in the figure, determine (1) all the values of θ at equilibrium; and (2) the stability of each equilibrium position. The homogeneous rod AB weighs 80 lb, and the ideal spring is unstretched when $\theta = 0$. Neglect friction and the weights of the sliders at A and B.

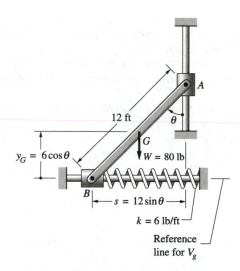

Solution

The potential energy V of the system consists of the potential energy of the weight V_g and the potential energy of the spring V_e. From Eqs. (10.14) and (10.16), we have

$$V = V_g + V_e = Wy_G + \frac{1}{2}ks^2 \qquad \text{(a)}$$

where y_G is the vertical distance of the center of gravity G of bar AB above the chosen reference line, and s is the elongation of the spring. As shown in the figure, we have $y_G = 6\cos\theta$ ft and $s = 12\sin\theta$ ft (recall that the spring is unstretched when the rod is vertical). Therefore, the potential energy in Eq. (a) becomes

$$V = 80(6\cos\theta) + \frac{1}{2}(6)(12\sin\theta)^2$$

$$= 480\cos\theta + 432\sin^2\theta \text{ lb} \cdot \text{ft} \qquad \text{(b)}$$

The first derivative of the potential energy is

$$\frac{dV}{d\theta} = -480 \sin\theta + 864 \sin\theta \cos\theta \text{ lb} \cdot \text{ft} \qquad (c)$$

Using $d(\sin\theta \cos\theta)/d\theta = \cos^2\theta - \sin^2\theta$, the second derivative of the potential energy is

$$\frac{d^2V}{d\theta^2} = -480 \cos\theta + 864(\cos^2\theta - \sin^2\theta) \text{ lb} \cdot \text{ft} \qquad (d)$$

Part 1

According to the principle of minimum potential energy, the values of θ at equilibrium are the roots of the equation $dV/d\theta = 0$. Using Eq. (c), we find that the equilibrium condition is

$$-480 \sin\theta + 864 \sin\theta \cos\theta = 0$$

or

$$\sin\theta(-480 + 864 \cos\theta) = 0$$

The roots of this equation are $\sin\theta = 0$ and $\cos\theta = 480/864$. Consequently, the equilibrium positions are

$$\theta = 0 \qquad \theta = \cos^{-1}\frac{480}{864} = 56.25° \qquad \qquad \textit{Answer}$$

Part 2

Evaluating the second derivative of the potential energy, Eq. (d), at the equilibrium position $\theta = 0$, we find

$$\frac{d^2V}{d\theta^2} = -480 + 864 = 384 \text{ lb} \cdot \text{ft}$$

Because $d^2V/d\theta^2 > 0$, we conclude that $\theta = 0$ is a stable equilibrium position.
For $\theta = 56.25°$, Eq. (d) gives

$$\frac{d^2V}{d\theta^2} = -480 \cos 56.25° + 864(\cos^2 56.25° - \sin^2 56.25°)$$

$$= -597 \text{ lb} \cdot \text{ft}$$

Because $d^2V/d\theta^2 < 0$, we deduce that $\theta = 56.25°$ is an unstable equilibrium position.

Problems

Neglect friction in the following problems, unless otherwise stated.

10.40 Show that $\theta = 0$ represents the only equilibrium position of the uniform bar AB. Is this position stable or unstable?

10.41 The weight W is suspended from end B of the weightless bar that is supported by walls at A and C. Determine the equilibrium value of the angle θ and investigate the stability of equilibrium.

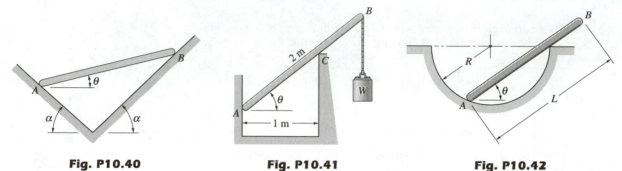

Fig. P10.40   **Fig. P10.41**   **Fig. P10.42**

10.42 The uniform bar of weight W and length $L = 2R$ rests in a hemispherical cavity of radius R. Calculate the angle θ for equilibrium and investigate the stability of equilibrium.

10.43 A slender homogeneous bar is bent into a right angle and placed on a cylindrical surface. Determine the range of b/R for which the equilibrium position shown is stable.

10.44 A uniform rod of length $3R$ and weight W_1 is welded to the bottom of a hemispherical shell of radius R and weight W_2. Determine the range of W_1/W_2 for which the equilibrium position shown is stable.

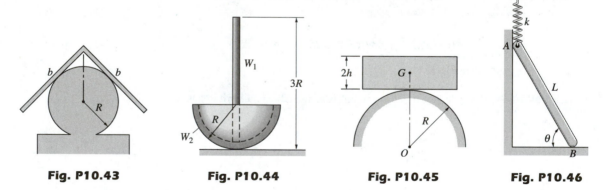

Fig. P10.43   **Fig. P10.44**   **Fig. P10.45**   **Fig. P10.46**

10.45 The uniform block of height $2h$ is balanced on the rough cylindrical surface of radius R. Show that this equilibrium position is stable only if $R > h$. Assume that friction prevents the block from sliding.

10.46 The weight and length of the homogeneous bar AB are $W = 12$ lb and $L = 2$ ft, respectively. The stiffness of the spring at A, which is undeformed when $\theta = \pi/2$, is 6 lb/ft. Find the equilibrium values of the angle θ, $0 < \theta < \pi/2$, and investigate the stability of each equilibrium position.

10.47 Uniform rods of weights W_1 and W_2 are welded to the two pulleys that are connected by a belt. Determine the range of W_1/W_2 for which the equilibrium position shown is stable.

10.48 The torsional spring attached to the base of the uniform rod exerts the couple $C = k\theta$ on the rod, where k is the stiffness of the spring, and θ is the angular displacement of the rod from the vertical (in radians). Determine the range of the force P for which the $\theta = 0$ equilibrium position is stable. Neglect the weight of the rod.

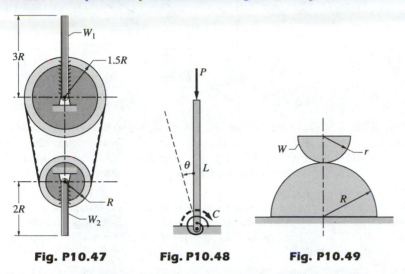

Fig. P10.47   **Fig. P10.48**   **Fig. P10.49**

10.49 The semi-cylinder of radius r is placed on a cylindrical surface of radius R. Assuming no slipping, determine the range of R/r for which the equilibrium position shown is stable.

10.50 The mass per unit length of the bar AC is 150 kg/m. The ideal spring at C, undeformed when the bar is vertical, is capable of carrying tension or compression. The spring stiffness is 6 kN/m. Find the largest dimension L for which the equilibrium position shown remains stable. (Hint: Use $\sin \theta \approx \theta$ and $\cos \theta \approx 1 - \theta^2/2$.)

10.51 Determine the largest weight W for which the hinged bar ABC will be in stable equilibrium in the position shown. The ideal spring of stiffness k is capable of carrying tension and compression and is undeformed in the position shown. (Hint: Use $\sin \theta \approx \theta$ and $\cos \theta \approx 1 - \theta^2/2$.)

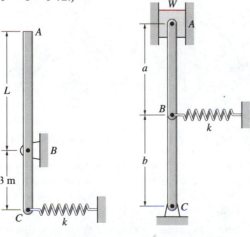

Fig. P10.50   **Fig. P10.51**

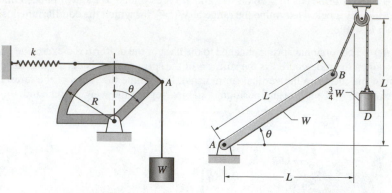

Fig. P10.52   **Fig. P10.53**

10.52   The spring is connected to a rope that passes over the cylindrical surface and is attached to corner A of the rocker. The spring has a stiffness k and is undeformed when $\theta = 0$. When the weight W is suspended from A, the equilibrium position of the rocker is $\theta = 30°$. Determine if this equilibrium position is stable. Neglect the weight of the rocker.

10.53   The homogeneous bar AB of weight W is connected by a cable to the block D that weighs $3W/4$. By plotting the potential energy of the system as a function of the angle θ, $0 \le \theta \le 45°$, determine the approximate value of θ at equilibrium. Is this position stable or unstable?

10.54   The mechanism of negligible weight supports the weight W. Find the value of θ for equilibrium. Is the equilibrium position stable or unstable?

10.55   Solve Prob. 10.54 assuming that A and B are connected by a spring of stiffness $k = 0.3W/b$ and free length b.

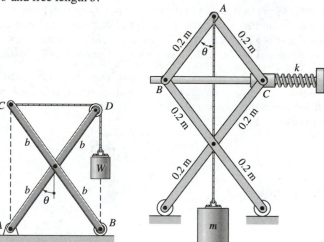

Fig. P10.54, P10.55   **Fig. P10.56, P10.57**

10.56   The stiffness of the ideal spring that is compressed by the slider C is $k = 250$ N/m. The spring is unstretched when $\theta = 20°$. When the mass m is suspended from A, the system is in equilibrium at $\theta = 60°$. Determine the value of m and whether the equilibrium position is stable or unstable.

10.57   Find the stable equilibrium position of the system described in Prob. 10.56 if $m = 2.06$ kg.

10.58 The uniform bar AB of weight $W = kL$ is in equilibrium when $\theta = 65°$. Find the value of θ for which the ideal spring would be unstretched, and investigate the stability of the equilibrium position.

10.59 The weight of the uniform bar AB is W. The stiffness of the ideal spring attached to B is k, and the spring is unstretched when $\theta = 80°$. If $W = kL$, the bar has three equilibrium positions in the range $0 < \theta < \pi$, only one of which is stable. Determine the angle θ at the stable equilibrium position.

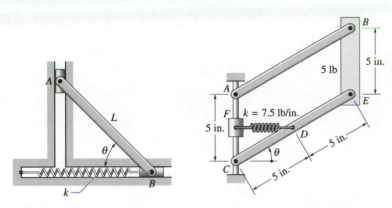

Fig. P10.58, P10.59   **Fig. P10.60**

10.60 The weightless bars AB and CE, together with the 5-lb weight BE, form a parallelogram linkage. The ideal spring attached to D has a free length of 2 in. and a stiffness of 7.5 lb/in. Find the two equilibrium positions that are in the range $0 < \theta < \pi/2$, and determine their stability. Neglect the weight of slider F.

10.61 The three ideal springs supporting the two bars that are pinned together at C are unstretched when $\theta_1 = \theta_2 = 0$. Note that the springs are always vertical because the collars to which they are attached are free to slide on the horizontal rail. Compute the angles θ_1 and θ_2 at equilibrium if $W = kL/10$.

10.62 The bar ABC is supported by three identical, ideal springs. Note that the springs are always vertical because the collars to which they are attached are free to slide on the horizontal rail. Find the angle θ at equilibrium if $W = kL$. Neglect the weight of the bar.

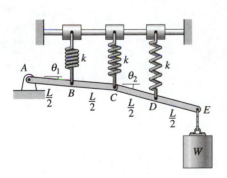

Fig. P10.61

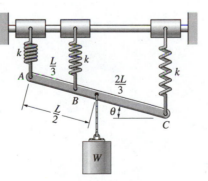

Fig. P10.62

A

Numerical Integration

A.1 Introduction

The purpose of numerical integration, also known as *quadrature,* is to evaluate definite integrals of the type

$$A = \int_a^b f(x)\, dx \qquad (A.1)$$

without using calculus. Quadrature gives only an approximate value for the integral, because calculus is the only method for performing the integration exactly. Numerical integration is useful in the following situations:

- The integration is difficult or tedious to perform analytically.
- The integral cannot be expressed in terms of known functions.
- The function $f(x)$ is unknown, but its values are known at discrete points.

Generally speaking, *integral* is a mathematical term for the sum of an infinite number of infinitesimal quantities. Consequently, the definite integral in Eq. (A.1) represents the summation of all the differential (infinitesimal) areas $dA = f(x)\, dx$ that lie between the limits $x = a$ and $x = b$, as seen in Fig. A.1. In numerical integration, the integral is approximated by adding the areas $A_1, A_2, A_3, \ldots, A_n$, of n (finite) panels, each of width Δx, as shown in Fig. A.2. Because the area of each panel must be estimated (integral calculus would be required to obtain the exact values), quadrature yields only an approximate value of the integral; that is,

$$A \approx \sum_{i=1}^{n} A_i \qquad (A.2)$$

As a rule, a larger number of panels, with correspondingly smaller Δx, yields a more accurate result.

There are several methods available for estimating the areas of the panels. We discuss only the trapezoidal rule and Simpson's rule.

A.2 Trapezoidal Rule

In the trapezoidal rule, each panel is approximated by a trapezoid, as the name suggests. Recalling that the area of a trapezoid equals the base width times the average height, the areas of the panels in Fig. A.2 are $A_1 = [(f_1 + f_2)/2]\Delta x$,

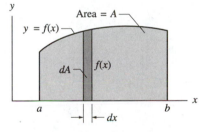

Fig. A.1

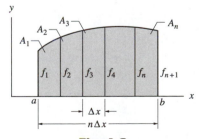

Fig. A.2

$A_2 = [(f_2 + f_3)/2]\Delta x$, $A_3 = [(f_3 + f_4)/2]\Delta x, \ldots, A_n = [(f_n + f_{n+1})/2]\Delta x$. Here, we have introduced the notation $f_1 = f(a)$, $f_2 = f(a + \Delta x)$, $f_3 = f(a + 2\Delta x), \ldots, f_{n+1} = f(a + n\,\Delta x) = f(b)$.

Adding the areas of the panels, we have

$$A \approx \sum_{i=1}^{n} A_i = (f_1 + 2f_2 + 2f_3 + \cdots + 2f_n + f_{n+1})\Delta x/2 \qquad (A.3)$$

Equation (A.3) is known as the *trapezoidal rule*.

The trapezoidal rule is sometimes written in the following form:

$$A \approx \sum_{i=1}^{n+1} W_i f_i \qquad (A.4)$$

where the W_i are known as the *weights* and the expression $\sum_{i=1}^{n+1} W_i f_i$ is called the *weighted summation*. For the trapezoidal rule, the weights are

$$W_1 = W_{n+1} = \frac{\Delta x}{2}$$
$$W_i = \Delta x \quad \text{for } 2 \le i \le n \qquad (A.5)$$

A.3 *Simpson's Rule*

In the trapezoidal rule, the function $f(x)$ is approximated by a straight line within each panel of width Δx; that is, the curvature of $f(x)$ is neglected. This linearization may result in an unacceptably large error in the quadrature, particularly if the curvature of $f(x)$ is large and of the same sign throughout the interval $a \le x \le b$. Simpson's rule overcomes this deficiency by replacing the straight lines with parabolas. Because three points—that is, three values of $f(x)$—are required to define a parabola, Simpson's rule approximates the area of a pair of adjacent panels.

Referring again to Fig. A.2, it can be shown that if a parabola were passed through the three points (a, f_1), $(a + \Delta x, f_2)$, and $(a + 2\Delta x, f_3)$, the sum of the areas A_1 and A_2—that is, the area under the parabola—would be

$$A_1 + A_2 = (f_1 + 4f_2 + f_3)\,\Delta x/3$$

Similarly,

$$A_3 + A_4 = (f_3 + 4f_4 + f_5)\,\Delta x/3$$
$$A_5 + A_6 = (f_5 + 4f_6 + f_7)\,\Delta x/3$$

If these six panels were to represent the entire area, then the quadrature would yield

$$A \approx A_1 + A_2 + A_3 + A_4 + A_5 + A_6$$
$$= (f_1 + 4f_2 + 2f_3 + 4f_4 + 2f_5 + 4f_6 + f_7)\,\Delta x/3$$

Simpson's rule for n panels, where n must be an even number, becomes

$$A \approx \sum_{i=1}^{n} A_i \tag{A.6}$$

$$= (f_1 + 4f_2 + 2f_3 + 4f_4 + \cdots + 2f_{n-1} + 4f_n + f_{n+1})\Delta x/3$$

Introducing the concept of weights W_i, *Simpson's rule* can be written as

$$A \approx \sum_{i=1}^{n+1} W_i f_i \tag{A.7}$$

where the weights are

$$
\begin{aligned}
W_1 &= W_{n+1} = \frac{\Delta x}{3} \\
W_i &= \frac{4\Delta x}{3} \quad i \text{ even} \\
W_i &= \frac{2\Delta x}{3} \quad i \text{ odd}
\end{aligned}
\left.\right\} 2 \leq i \leq n
\tag{A.8}
$$

Because of its greater accuracy, Simpson's rule should be chosen over the trapezoidal rule. If the number of panels is odd, the area of one panel should be calculated using the trapezoidal rule, and then Simpson's rule can be used for the remaining panels.

Sample Problem A.1

Evaluate the integral $A = \int_0^{\pi/2} \sin x \, dx$ (x is measured in radians) with four panels, using (1) the trapezoidal rule; and (2) Simpson's rule.

Solution

Because the range of integration is $0 \leq x \leq \pi/2$ rad and the number of panels is four, we get $\Delta x = (\pi/2)/4 = \pi/8$ rad. The following table is convenient for carrying out the quadrature.

i	x (rad)	$f(x) = \sin x$	Weights W_i Trapezoidal rule	Simpson's rule
1	0	0	$\pi/16$	$\pi/24$
2	$\pi/8$	0.3827	$\pi/8$	$\pi/6$
3	$\pi/4$	0.7071	$\pi/8$	$\pi/12$
4	$3\pi/8$	0.9239	$\pi/8$	$\pi/6$
5	$\pi/2$	1.0000	$\pi/16$	$\pi/24$

Part 1: Trapezoidal Rule

Using the weights from the foregoing table, and Eq. (A.4), we obtain

$$A \approx \sum_{i=1}^{5} W_i f_i$$

$$= \frac{\pi}{16}[0 + 2(0.3827 + 0.7071 + 0.9239) + 1]$$

$$= 0.9871 \qquad \qquad \textit{Answer}$$

Part 2: Simpson's Rule

Substituting the weights from the foregoing table into Eq. (A.7) gives

$$A \approx \sum_{i=1}^{5} W_i f_i$$

$$= \frac{\pi}{24}[0 + 4(0.3827) + 2(0.7071) + 4(0.9239) + 1]$$

$$= 1.0002 \qquad \qquad \textit{Answer}$$

Because the exact value of the integral is 1.0000, you can see that Simpson's rule is considerably more accurate than the trapezoidal rule for this problem. A major source of error in the trapezoidal rule is that the curvature of the function $f(x) = \sin x$ has the same sign throughout the interval $0 \leq x \leq \pi/2$ rad. As mentioned previously, the trapezoidal rule does not perform well in problems of this type.

B

Finding Roots of Functions

B.1 Introduction

The solutions of the equation $f(x) = 0$, where $f(x)$ is a given function, are called the *roots of* $f(x)$. In many practical applications, $f(x)$ is nonlinear in x, in which case it may be difficult or even impossible to find the roots analytically. Examples of such nonlinear functions are $f(x) = e^x \cos x - 1$ and $f(x) = x^4 - 2x^3 + 6x - 5$. Here we introduce two popular numerical methods for root finding: Newton's method (also known as Newton-Raphson iteration) and the secant method. Both methods work by iteration and require a good starting value (initial guess) of the root. If the starting value is not sufficiently close to the root, the procedures may fail. Frequently, the physical principles of a problem suggest a reasonable starting value. Otherwise, a good estimate of the root can be obtained by sketching $f(x)$ versus x. (This involves, of course, evaluating the function at various values of x.) If the initial value is not close enough to the root, two problems may arise:

- The iterative procedure will not converge to a single value of x.
- The procedure will converge to a root that is different from the one being sought. (Recall that nonlinear equations may have multiple roots.)

B.2 Newton's Method

Consider the problem of computing the root x_0 of the function $f(x)$ that is plotted in Fig. B.1. We begin by estimating the value x_1 of the root and computing $f(x_1)$; the corresponding point is denoted A in the figure. The next step is to compute $f'(x_1)$ (the prime indicates differentiation with respect to x), which represents the slope of the straight line that is tangent to $f(x)$ at A. The coordinate of the point where the tangent line crosses the x-axis is denoted x_2. If x_1 is close to x_0, then the tangent line is a good approximation of $f(x)$ in the vicinity of the root. Consequently, x_2 should be a better approximation of the root than x_1.

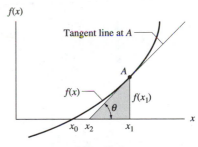

Fig. B.1

The value of x_2 can be computed from the shaded triangle in Fig. B.1: $\tan\theta = f(x_1)/(x_1 - x_2)$. Substituting $\tan\theta = f'(x_1)$ and solving for x_2 yield

$$x_2 = x_1 - \frac{f(x_1)}{f'(x_1)}$$

This completes one cycle of the iteration process. The procedure is then repeated with the output of each iterative step (e.g., x_2) being used as the input for the next step, until the change in x between successive steps is negligible.

The algorithm for Newton's method can thus be summarized as follows:

$$
\begin{aligned}
&\text{estimate } x \\
&\text{do until } |\Delta x| < \epsilon \\
&\qquad \Delta x \leftarrow -\frac{f(x)}{f'(x)} \\
&\qquad x \leftarrow x + \Delta x \\
&\text{end do}
\end{aligned}
\tag{B.1}
$$

where $a \leftarrow b$ means "b replaces a" and ϵ is the convergence parameter (a small number that signals that the desired accuracy has been achieved).

The main drawback of Newton's method is that it requires the derivative of $f(x)$. If $f(x)$ is a simple expression, then deriving $f'(x)$ is only a minor nuisance. However, in cases where $f(x)$ is a complicated function, methods that do not require the derivative are more attractive.

B.3   *Secant Method*

The secant method is based on the same principle as Newton's method. However, instead of requiring the derivative of $f(x)$, it requires two starting values (initial guesses) of the root. These starting values are denoted x_1 and x_2 in Fig. B.2, and the corresponding points on the plot of $f(x)$ are labeled A and B, respectively. (In the figure, we assume that $x_1 > x_2$, but this need not be the case.) The role that was played by the tangent line in Newton's method is now taken over by the chord AB; that is, the chord AB will be a good approximation of $f(x)$ in the vicinity of the root x_0 if the starting values are close enough to the root.

From similar triangles in Fig. B.2 we get

$$
\frac{f(x_2)}{x_2 - x_3} = \frac{f(x_1) - f(x_2)}{x_1 - x_2}
$$

which yields

$$
x_3 = x_2 - f(x_2)\frac{x_1 - x_2}{f(x_1) - f(x_2)}
$$

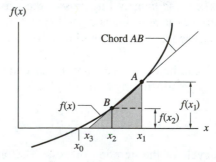

Fig. B.2

Repeating this procedure by using x_2 and x_3 as the new input values (x_1 is discarded) will further improve the estimation of the root. The computations continue until x_3 shows no significant change—that is, until the change in x is insignificant. The summary of the algorithm for the secant method is

$$
\begin{aligned}
&\text{estimate } x_1 \text{ and } x_2 \\
&\text{do until } |\Delta x| < \epsilon \\
&\qquad \Delta x \leftarrow -f(x_2)\,\frac{x_1 - x_2}{f(x_1) - f(x_2)} \\
&\qquad x_3 \leftarrow x_2 + \Delta x \\
&\qquad x_1 \leftarrow x_2 \\
&\qquad x_2 \leftarrow x_3 \\
&\text{end do}
\end{aligned}
\tag{B.2}
$$

Sample Problem B.1

Find the smallest positive, nonzero root of $f(x) = e^x \cos x - 1$ within five significant digits. Use (1) Newton's method; and (2) the secant method.

Solution

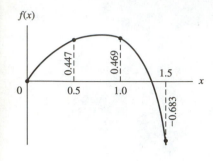

To obtain an approximate value for the desired root, we plot $f(x)$ in increments of $\Delta x = 0.5$, starting at $x = 0$ and ending after $f(x)$ reverses its sign. The results are shown in the figure. By inspection we estimate the root to be about $x = 1.3$.

Part 1: Newton's Method

Newton's method requires the derivative of $f(x)$, which is

$$f'(x) = \frac{d}{dx}(e^x \cos x - 1) = e^x(\cos x - \sin x)$$

The following table shows the computations for the root, based on the algorithm given in Eq. (B.1). The starting value was $x = 1.3$.

x	$f(x)$	$f'(x)$	Δx
1.3	−0.018 47	−2.554 05	−0.007 23
1.292 77	−0.000 18	−2.503 16	−0.000 07
1.292 70	0.000 00	−2.502 64	0.000 00

The final result, $x = 1.2927$, has been obtained in only two iterations because of the accuracy of our initial estimate.

Part 2: Secant Method

We chose $x_1 = 1.2$ and $x_2 = 1.3$ as the starting values for the root. The algorithm in Eq. (B.2) leads to the following sequence of computations.

x_1	x_2	$f(x_1)$	$f(x_2)$	Δx	$x_3 = x_2 + \Delta x$
1.2	1.3	0.203 07	−0.018 47	−0.008 34	1.291 66
1.3	1.291 66	−0.018 47	0.002 58	0.001 03	1.292 69
1.291 66	1.292 69	0.002 58	0.000 03	0.000 01	1.292 70
1.292 69	1.292 70	0.000 03	0.000 00	0.000 00	1.292 70

Referring to the first row in the table, note that x_1 and x_2 are entered first, and then the remaining entries in the row are computed using the algorithm. In the second row, the values for x_1, x_2, and $f(x_1)$ are simply copied from the appropriate columns in the first row. This pattern is repeated in subsequent rows.

Once again, the final result (x_3) is 1.2927, with only three iterations required. In general, the secant method converges a little more slowly than Newton's method.

C

Densities of Common Materials

	ρ kg/m^3	γ lb/ft^3	γ lb/in.^{3}
Aluminum	2 660	166	0.096
Brass	8 300	518	0.300
Brick	2 000	125	0.072
Cast iron	7 200	449	0.260
Concrete	2 400	150	0.087
Copper	8 910	556	0.322
Earth (dry)	1 280	80	0.046
Earth (wet)	1 760	110	0.064
Glass	2 590	162	0.094
Ice	900	56	0.032
Lead	11 370	710	0.411
Oil	900	56	0.032
Steel	7 850	489	0.283
Water (fresh)	1 000	62.4	0.036
Water (ocean)	1 030	64	0.037
Wood, hard (white oak)	800	50	0.029
Wood, soft (Douglas fir)	480	30	0.017

Answers to Even-Numbered Problems

CHAPTER 1

1.2	23.5 lb
1.4	14.75 lb $\cdot$ ft $\cdot$ s^2
1.6	(No answer)
1.8	(a) $[FL]$; (b) $[FT]$; (c) $[F]$
1.10	(a) $[FLT^2]$; (b) $[ML^2]$
1.12	$[c] = [FTL^{-1}]$, $[k] = [FL^{-1}]$, $[P_0] = [F]$, $[\omega] = [T^{-1}]$
1.14	(a) $[FL]$; (b) $[ML^2T^{-2}]$
1.16	7.46×10^{-11} lb
1.18	(No answer)
1.20	2640 km
1.22	(a) 64 mi/h at 21°; (b) 80 mi/h at 42°; (c) 64 mi/h at 99°
1.24	Lift = 6220 lb ↑, Drag = 653 lb →
1.26	(a) $F_x = 173$ lb, $F_y = 100$ lb; (b) $F_{x'} = 184$ lb, $F_{y'} = 163$ lb
1.28	46 kN at 74°
1.30	2210 ft along OB, 2570 ft along OC
1.32	$\alpha = 21.6°$, $\beta = 19.9°$
1.34	(a) 71.8°; (b) 235 lb
1.36	(a) $P_{OA} = 6.83$ tons, $P_{OB} = 3.54$ tons; (b) $P_{OB} = 3.01$ tons, $\theta = 56.2°$
1.38	1907 N, 21.6°
1.40	$-9.64\mathbf{i} + 16.70\mathbf{j} + 22.98\mathbf{k}$ lb
1.42	$7.63\mathbf{i} - 8.90\mathbf{j} + 2.45\mathbf{k}$ ft/s
1.44	(a) $-0.269\mathbf{i} + 0.875\mathbf{j} + 0.404\mathbf{k}$; (b) $-1.61\mathbf{i} + 5.24\mathbf{j} + 2.42\mathbf{k}$ m/s
1.46	$63.0\mathbf{i} - 52.5\mathbf{j} - 87.6\mathbf{k}$ lb
1.48	$\theta_x = 50.8°$, $\theta_y = 63.4°$, $\theta_z = 129.2°$
1.50	$-96.0\mathbf{i} + 123.5\mathbf{j} - 85.8\mathbf{k}$ lb
1.52	$146\mathbf{i} - 48\mathbf{j}$ lb
1.54	$P = 587$ lb, $Q = 455$ lb
1.56	46.4 kN at 74.4°
1.58	(a) $39\mathbf{i} + 63\mathbf{j} - 42\mathbf{k}$ ft^2; (b) $39\mathbf{i} + 26\mathbf{j} + 18\mathbf{k}$ N $\cdot$ m; (c) $-28\mathbf{i} - 25\mathbf{j} + 10\mathbf{k}$ m^2
1.60	$-20\mathbf{i} + 15\mathbf{k}$ in.^{2}
1.62	34.8°
1.64	(c) and (d)
1.66	$-0.464\mathbf{i} - 0.603\mathbf{j} - 0.649\mathbf{k}$
1.68	-17.52 lb
1.70	Parallel: $1.581(0.949\mathbf{i} + 0.316\mathbf{k})$ in., Perpendicular: $6.89(0.218\mathbf{i} + 0.725\mathbf{j} - 0.653\mathbf{k})$ in.
1.72	5 m
1.74	(No answer)
1.76	(No answer)

CHAPTER 2

2.2	413 lb at 53.8°
2.4	$P = 105.8$ kN, $\theta = 19.11°$
2.6	(a) 367 lb; (b) $\cos\theta_x = -0.807$, $\cos\theta_y = 0.530$, $\cos\theta_z = 0.263$; (c) (0, 1.314 ft, 0.651 ft)
2.8	$140.0\mathbf{i} - 169.7\mathbf{j} + 100.8\mathbf{k}$ kN
2.10	$P_1 = 62.3$ kN, $P_2 = 44.6$ kN
2.12	10.0 lb
2.14	$2.30T$
2.16	$Q = 141.7$ N, $\theta = 41.5°$
2.18	$-108.4\mathbf{i} + 121.8\mathbf{j} - 820.4\mathbf{k}$ lb
2.20	(a) 224 lb; (b) 63.4°; (c) $x = 3.5$ ft
2.22	9 kN $\cdot$ m CW
2.24	$\mathbf{M}_A = -291\mathbf{k}$ N $\cdot$ m, $\mathbf{M}_B = 145.5\mathbf{k}$ N $\cdot$ m, $\mathbf{M}_C = \mathbf{0}$, $\mathbf{M}_D = -145.6\mathbf{k}$ N $\cdot$ m
2.26	$-560\mathbf{i} + 200\mathbf{j}$ N
2.28	(a) 400 kN $\cdot$ m CCW; (b) 400 kN $\cdot$ m CW; (c) 0
2.30	10.0 lb
2.32	23.3°
2.34	(a), (d), and (e)
2.36	(a) $-73.0\mathbf{j}$ N $\cdot$ m; (b) $-87.7\mathbf{i} - 121.7\mathbf{k}$ N $\cdot$ m
2.38	(a) $84.5\mathbf{i} + 136.0\mathbf{j} + 181.3\mathbf{k}$ kN $\cdot$ m; (b) $84.5\mathbf{i} + 181.3\mathbf{k}$ kN $\cdot$ m
2.40	$160\mathbf{i} + 299\mathbf{j} - 253\mathbf{k}$ N $\cdot$ m
2.42	$M_O = 233$ lb $\cdot$ in., $\cos\theta_x = 0.487$, $\cos\theta_y = 0.133$, $\cos\theta_z = 0.864$
2.44	$M_O = 7.94$ N $\cdot$ m, $\cos\theta_x = -0.327$, $\cos\theta_y = 0.818$, $\cos\theta_z = -0.473$
2.46	(0, -3 m, 10 m)
2.48	(a) 36 kN $\cdot$ m; (b) 36 kN $\cdot$ m; (c) 32 kN $\cdot$ m; (d) 0; (e) 0
2.50	36 N $\cdot$ m
2.52	415 N $\cdot$ m
2.54	$F = 32.0$ N, $d = 0.450$ m
2.56	$-6.86\mathbf{i} - 9.91\mathbf{j}$ N $\cdot$ m
2.58	$-13.58\mathbf{i} - 18.10\mathbf{k}$ N $\cdot$ m
2.60	120 lb
2.62	528 lb $\cdot$ ft
2.64	0.327 m
2.66	-208 lb $\cdot$ in.
2.68	(b), (c), (d), (f), and (g)
2.70	(a) 40 kN; (b) 43.5 kN
2.72	346 lb $\cdot$ ft
2.74	$-106.2\mathbf{i} + 140.7\mathbf{j} + 280.7\mathbf{k}$ lb $\cdot$ in.
2.76	-208 lb $\cdot$ in.
2.78	339 lb $\cdot$ ft
2.80	$R = 900$ lb, $C_0 = 3120$ lb $\cdot$ in.
2.82	(c) and (e)
2.84	(a) $R = 120$ N ↓, $C^R = 56$ N $\cdot$ m CCW; (b) $F_A = 253$ N ↑, $F_B = 373$ N ↓

2.86 $R = 25.0i - 133.3j$ lb, $C^R = 20.0$ lb·in. CW
2.88 $R = 100$ N ↑, $\mathbf{C}^R = 60i - 50j$ N·m
2.90 $R = -166.4i + 110.9k$ lb, $C^R = 167.5i - 1404j + 1332k$ lb·ft
2.92 $R = 3.15i - 3.75j + 8.49k$ lb, $\mathbf{C}^R = 54.5i + 11.81j + 45.0k$ lb·ft
2.94   97.9 lb
2.96 $T_1 = 79.4$ N, $T_2 = 151.6$ N, $T_3 = 220.3$ N
2.98   65.8 lb
2.100 (a) $1200j + 1200k$ lb·ft; (b) 1477 lb·ft
2.102 (a) Show that $\mathbf{F} \cdot \mathbf{C} = 0$; (b) $(-1.2$ in., -1.6 in., 0)
2.104 $x = L/2$, $R = P$ ↑
2.106 107.3 lb
2.108 20 N

CHAPTER 3

3.2   (a) $\mathbf{R} = -18j$ kN, $C^R = 46$ kN·m CW; (b) $\mathbf{R} = -18j$ kN, $C^R = 80$ kN·m CCW
3.4 $R = 154$ lb ↑, $C^R = 115.2$ lb·in. CW
3.6 $\mathbf{P} = 224i + 129j$ lb, $y = 8.83$ in.
3.8 $R = 120k$ lb, $\mathbf{C}^R = 840i - 815j$ lb·ft
3.10   (a) $\mathbf{R} = 2i + 6j$ lb, $\mathbf{C}^R = -6j$ lb·in.; (b) $\mathbf{R} = 4i + 6j + 3k$ lb, $\mathbf{C}^R = \mathbf{0}$; (c) $\mathbf{R} = 2i + 6j$ lb, $\mathbf{C}^R = -6j$ lb·in.; (d) $\mathbf{R} = \mathbf{0}$, $\mathbf{C}^R = -15i + 8j$ lb·in.; (e) $\mathbf{R} = \mathbf{0}$, $\mathbf{C}^R = -15i + 8j$ lb·in.; (f) $\mathbf{R} = \mathbf{0}$, $\mathbf{C}^R = \mathbf{0}$; (a) and (c) are equivalent, (d) and (e) are equivalent
3.12 $\mathbf{R} = -13.7i - 50.3j + 25.7k$ kN, $\mathbf{C}^R = 102.9i + 150.0j + 54.9k$ kN·m
3.14 $\mathbf{R} = -36i - 30k$ lb, $\mathbf{C}^R = 15i - 162k$ lb·ft
3.16   (1) $\mathbf{R} = 300i$ lb intersecting y-axis at $y = 3$ in.; (2) $\mathbf{R} = 200i - 200j$ N intersecting x-axis at $x = -4$ m; (3) $\mathbf{R} = -600i - 400j$ kN passing through O; (4) $\mathbf{R} = -600i + 800j$ lb intersecting x-axis at $x = -30$ ft
3.18 $P = 348$ lb, $\theta = 33.7°$
3.20   (a) $R = 40$ lb ↑ acting at $x = 5$ ft; (b) $C^R = 240$ lb·ft CW
3.22   (a) $C^R = 3rF$ CCW; (b) $\mathbf{R} = 0.414Fi$ intersecting y-axis at $y = -7.25r$
3.24   (a) $\mathbf{R} = -60i$ N acting through O; (b) $\mathbf{R} = -60i$ N intersecting y-axis at $y = 1.5$ m
3.26 $\mathbf{R} = -17.32i - 50.0j$ N intersecting x-axis at $x = 0.467$ m
3.28 $P_1 = 80.6$ kN, $P_2 = 40.3$ kN, $R = 169.8$ kN
3.30   (a) $\mathbf{R} = -80k$ lb passing through point (4 ft, 5 ft, 0);
(b) $\mathbf{R} = 50k$ kN passing through point (0, -6 m, 0);
(c) $\mathbf{R} = 400k$ lb passing through point (-2.5 in., -3 in., 0);
(d) $\mathbf{R} = 25k$ N passing through point (10 m, 8 m, 0)
3.32 $\mathbf{R} = 82i - 1411j + 406k$ lb at the point of concurrency
3.34 $P_1 = 92.7$ kN, $P_2 = 65.6$ kN, $P_3 = 87.0$ kN
3.36 $\mathbf{C}^R = 18i - 12j$ kN·m
3.38 $P_1 = 28.8$ lb, $P_2 = -16.8$ lb, $C = 540$ lb·in.
3.40 $\mathbf{R} = 580j$ lb passing through point (1.851 ft, 0, 1.301 ft)
3.42 $P = 1063$ lb, $Q = 915$ lb, $\mathbf{R} = -1425j - 3660k$ lb
3.44 $\mathbf{R} = 300i + 400j - 500k$ N, $\mathbf{C}^R = 264i + 352j - 440k$ N·m, passing through point (0.896 m, -2.272 m, 0)
3.46 $R = 13.2$ lb acting at center of sign
3.48 $\mathbf{R} = -4100k$ lb passing through point (3.82 ft, 0)
3.50 $\mathbf{R} = (-w_0L/4)(i + j)$, $x = 0.833L$
3.52 $\mathbf{R} = -76\,970j$ lb passing through point (59.1 ft, 0, 135.6 ft)
3.54 $\mathbf{R} = (-237i - 45.0j) \times 10^3$ lb acting through point (-24.3 ft, 0)

3.56 $T_1 = 361$ lb, $T_2 = 223$ lb
3.58 $P = 450$ N, $Q = 100$ N, $d = 12.0$ m
3.60   (4.67 ft, 5.33 ft, 0)
3.62   (a) $\mathbf{R} = -200i + 300j + 150k$ lb, $\mathbf{C}^R = 1250i + 600j + 1000k$ lb·ft; (b) $\mathbf{R} = -200i + 300j + 150k$ lb, $\mathbf{C}^R = -104.9i + 157.4j + 78.7k$ lb·ft, passing through point (-2.95 ft, 9.03 ft, 0)
3.64 $P = 47.8$ kN, $R = 134.4i$ kN
3.66 $\mathbf{R} = 80i + 240j + 240k$ lb, $\mathbf{C}^R = -960i - 720j - 780k$ N·m
3.68   452 lb
3.70 $T_1 = 654$ N, $T_2 = 425$ N, $R = -1441$ N

CHAPTER 4

4.2   3 unknowns
4.4   3 unknowns
4.6   (a) 3 unknowns; (b) 4 unknowns; (c) 3 unknowns; (d) 3 unknowns
4.8   3 unknowns
4.10   4 unknowns
4.12 $\theta = 33.7°$, $N_A = 0.555W$, $N_B = 0.832W$
4.14   53.6 lb
4.16   38.4 lb·in.
4.18   21.0°
4.20   61.6°
4.22 $T = 387.5$ N, $A_x = 132.5$ N, $A_y = 376.4$ N
4.24   3.35 ft
4.26 $R_A = 192.1$ N at 50° ⟋, $R_B = 192.1$ N at 50° ⟋
4.28 $R_A = 281$ N ↑, $R_B = 254$ N ↑
4.30 $5.36P$
4.32 $R_A = 0$, $R_B = 1.0$ kN ↑
4.34 $T_{AB} = 34.6$ lb, $T_{BC} = 727$ lb
4.36   (b) 32.0 N·m CCW
4.38 $P = 340$ N, $C = 101.9$ N·m
4.40 $R_A = R_B = 960$ N
4.42   6.43 N
4.44 $P = 5W/8$ (θ does not appear in the expression for P)
4.46   (a) $R_A = 647$ N, $N_B = 858$ N; (b) 2.61 m
4.48 $T = 13.66$ N, $R_A = 27.4$ N
4.50 $R_A = 1781$ N, $R_B = 1881$ N
4.52 $R_A = 1150$ lb, $R_B = 920$ lb
4.54   1.975 ft
4.56   7.48 m
4.58   238 N
4.60   50.4 mm
4.62   (a) 6 unknowns, 6 independent eqs.; (b) 8 unknowns, 8 independent eqs.; (c) 8 unknowns, 8 independent eqs.
4.64   6 unknowns, 6 independent eqs.
4.66   6 unknowns, 6 independent eqs.
4.68   9 unknowns, 9 independent eqs.
4.70   (No answer)
4.72   (No answer)
4.74   41.7 lb
4.76   2.70 lb
4.78 $1.5W \cot\theta$
4.80 $R_A = 411$ N, $R_C = 416$ N
4.82 $N_A = 55.6$ kN, $N_B = 10.12$ kN, $N_C = 32.4$ kN
4.84   33.6 lb·in.
4.86 $T = 3530$ N, $N_B = 6530$ N

4.88 280 lb

4.90 7200 lb

4.92 36.7 lb

4.94 24.0 N·m

4.96 (a) $T_A = 68.9$ lb, $T_B = 126.1$ lb; (b) $N_A = 101.1$ lb, $N_B = 63.9$ lb

4.98 2690 N

4.100 $A_x = 55$ kN ←, $A_y = 14$ kN ↓, $E_x = 60$ kN ←, $E_y = 34$ kN ↓, $F_x = 15$ kN →, $F_y = 34$ kN ↑, $P_{CD} = 45$ kN →

4.102 623 lb·in.

4.104 21.2 N

4.106 27.8 kips at A, 31.6 kips at C

4.108 1222 lb at E, 1981 lb at F

4.110 877 N

4.112 $A = 672$ lb, $C = 582$ lb, $E = 336$ lb

4.114 55.6 mm

4.116 $E = 6P$

4.118 10P

4.120 $P = 0.866Q$

4.122 $A = 2400$ lb, $B = 2530$ lb

4.124 29.6 mm

4.126 14.48°

4.128 17.5 mm

4.130 (a) 59.0 lb; (b) 39.2 lb

4.132 $P_{AB} = 0.6P$ (T), $P_{AC} = 0.8P$ (C), $P_{BC} = 0.64P$ (T), $P_{CD} = 0.48P$ (C)

4.134 $P_{AB} = P_{ED} = 1.220P$ (T), $P_{BC} = P_{DC} = 1.829P$ (T), $P_{AC} = P_{EC} = 2.575P$ (C)

4.136 $P_{AD} = 2.24P$ (T), $P_{AB} = P_{BC} = 2.0P$ (C), $P_{BD} = 1.0P$ (T), $P_{CD} = 1.118P$ (C), $P_{DE} = 3.35P$ (T), $P_{CE} = 0.50P$ (T)

4.138 $P_{AB} = 4.24$ kN (C), $P_{AD} = 3.16$ kN (T), $P_{BC} = 4.24$ kN (C), $P_{CD} = 3.0$ kN (T), $P_{BD} = 6.0$ kN (T)

4.140 $P_{AB} = 979$ kN (T), $P_{BC} = 861$ kN (T), $P_{BD} = 171$ kN (T), $P_{CD} = 950$ kN (C)

4.142 1.506 ft

4.144 $P_{HC} = 0.901P$ (T), $P_{HG} = 0.901P$ (C)

4.146 3200 lb (T)

4.148 1.25P (T)

4.150 $P_{BC} = 13\,420$ lb (C), $P_{BG} = 4470$ lb (C), $P_{FG} = 16\,000$ lb (T)

4.152 $P_{EF} = 240$ kN (C), $P_{NF} = 82.0$ kN (T), $P_{NO} = 187.5$ kN (T)

4.154 $P_{BG} = 0.250P$ (T), $P_{CI} = 0.354P$ (T), $P_{CD} = 0.750P$ (C)

4.156 $P = 5.71$ kN, $\alpha = 48.0°$

4.158 $P_{CD} = 12.75$ kN (T), $P_{DF} = 7.39$ kN (C)

4.160 $P = 5170$ lb, $Q = 370$ lb

4.162 259 kN (C) when $P = 200$ kN is located at I, to 760 kN (T) when $P = 0$

4.164 $P_{CD} = 4.0P$ (T), $P_{IJ} = 4.0P$ (C), $P_{NJ} = 0.559P$ (C)

4.166 $P_{BC} = 4.47P$ (C), $P_{BG} = 2.0P$ (C)

4.168 $P_{EF} = 1.828W$ (T), $P_{KL} = 2.83W$ (C)

4.170 $T = 0.909W$, $N_A = 0.466W$, $N_E = 0.943W$

4.172 $T = 23.1(\tan\theta - 1)$ kN, $T = 0$ when $\theta = 45°$

4.174 4.39 kN

4.176 $A = D = 4.22$ kN

4.178 15.12°

4.180 (a) 361 kN (C); (b) 300 kN (T); (c) 631 kN (T)

4.182 20 lb

4.184 (a) 375 lb (T); (b) 4875 lb (T)

4.186 178 lb

4.188 643 N

CHAPTER 5

5.2 5 unknowns

5.4 6 unknowns

5.6 6 unknowns

5.8 6 unknowns

5.10 12 unknowns

5.12 6 unknowns

5.14 424 kN

5.16 7010 lb

5.18 27.3 kN

5.20 4500 lb

5.22 $T_{BC} = 5.82$ kN, $T_{BD} = 14.95$ kN, $R_A = 18.66$ kN

5.24 $A = 557$ N, $B = 375$ N, $C = 510$ N

5.26 14.04°

5.28 $D = 143.6$ N, $E = 82.9$ N, $F = 165.9$ N

5.30 $\mathbf{C} = 400\mathbf{k}$ N, $\mathbf{D} = -400\mathbf{k}$ N

5.32 $P_{EF} = 1000$ lb (C), $C = 318$ lb

5.34 286 lb

5.36 $T_{CD} = 193.5$ N, $\mathbf{A} = 15.7\mathbf{i} + 86.5\mathbf{j} + 83.9\mathbf{k}$ N, $\mathbf{B} = 157.3\mathbf{i} + 6.14\mathbf{k}$ N

5.38 $a = 3.43$ m, $b = 1.93$ m

5.40 $P = 51.4$ lb, $\mathbf{A} = 87.0\mathbf{k}$ lb, $\mathbf{B} = 134.4\mathbf{k}$ lb

5.42 $P_{CD} = 77.9$ lb, $\mathbf{A} = -25.0\mathbf{j} - 58.0\mathbf{k}$ lb, $\mathbf{B} = 40.0\mathbf{j} + 54.5\mathbf{k}$ lb

5.44 $P_{BD} = 1.786P$, $P_{BE} = 1.515P$

5.46 $A = 60.0$ lb, $B = 57.0$ lb, $C = 48.1$ lb

5.48 $P_{BE} = 800$ N (T), $P_{CF} = 640$ N (C)

5.50 1178 lb

5.52 4600 lb

5.54 $T = 25.3$ lb, $A = 78.1$ lb

5.56 $T = 161.3$ lb, $A = 12.0$ lb, $B = 12.0$ lb, $C = 140.5$ lb

5.58 309 lb

5.60 $T = 4750$ N, $\mathbf{A} = 2000\mathbf{j}$ N

5.62 $\mathbf{B} = -341\mathbf{j}$ lb, $\mathbf{D} = 91.4\mathbf{i}$ lb

CHAPTER 6

6.2 $P_1 = 0$, $V_1 = 40$ kN, $M_1 = 60$ kN·m

6.4 $P_1 = 240$ lb (C), $V_1 = 0$, $M_1 = 1200$ lb·ft, $P_2 = V_2 = M_2 = 0$

6.6 $P_1 = V_1 = 0$, $M_1 = 720$ lb·ft, $P_2 = 0$, $V_2 = 72$ lb, $M_2 = 360$ lb·ft

6.8 $P_1 = 28$ lb (T), $V_1 = 0$, $M_1 = 21$ lb·in., $P_2 = 34.6$ lb (T), $V_2 = 8$ lb, $M_2 = 10$ lb·in.

6.10 $P_1 = 644$ N (C), $V_1 = 0$, $M_1 = 162$ N·m, $P_2 = 644$ N (C), $V_2 = 540$ N, $M_2 = 324$ N·m

6.12 $P_3 = 900$ N (C), $V_3 = 400$ N, $M_3 = 300$ N·m, $P_4 = 900$ N (T), $V_4 = 1200$ N, $M_4 = 1800$ N·m

6.14 $P_3 = 255$ N (C), $V_3 = 0$, $M_3 = 29.8$ N·m

6.16 $P_1 = 50$ lb (T), $V_1 = 0$, $M_1 = 600$ lb·in., $P_2 = 50$ lb (C), $V_2 = 37.5$ lb, $M_2 = 150$ lb·in.

6.18 $P_2 = Wx/(\sqrt{5}a)$ (T), $V_2 = Wx/(2\sqrt{5}a)$, $M_2 = 3Wx/4$

6.20 $P_1 = 555$ lb (C), $V_1 = 832$ lb, $M_1 = 4000$ lb·ft

6.22 $V = -w_0x^2/(2L)$, $M = w_0x^3/(6L)$

6.24 For $0 < x < b$: $V = P(L-b)/L$, $M = -P(L-b)x/L$; for $b < x < L$: $V = -Pb/L$, $M = -Pb(L-x)/L$

6.26 For $0 < x < 3$ ft: $V = 600$ lb, $M = 1400 - 600x$ lb·ft; for 3 ft $< x < 4$ ft: $V = 1200 - 200x$ lb, $M = 2300 - 1200x + 100x^2$ lb·ft; for 4 ft $< x < 7$ ft: $V = 800 - 200x$ lb, $M = 700 - 800x + 100x^2$ lb·ft

6.28 For $0 < x < 3$ m: $V = 4 - 0.4x^2$ kN, $M = -4x + 0.1333x^3$ kN·m; for 3 m $< x < 5$ m: $V = 0.4$ kN, $M = -0.4x - 7.2$ kN·m; for 5 m $< x < 7$ m: $V = -4.6$ kN, $M = 4.6x - 32.2$ kN·m

6.30 For $0 < x < 2$ m: $V = -20$ kN, $M = 20x$ kN·m; for 2 m $< x < 4.5$ m: $V = 28$ kN, $M = -28x + 96$ kN·m; for 4.5 m $< x < 7$ m: $V = -12$ kN, $M = 12x - 84$ kN·m

6.32 For $0 < x < 2$ m: $V = -5$ kN, $M = -10 + 5x$ kN·m; for 2 m $< x < 5.5$ m: $V = M = 0$

6.34 For $0 < x < 3$ m: $V = 0.9 - 0.4x$ kN, $M = 0.2x^2 - 0.9x$ kN·m; for 3 m $< x < 6$ m: $V = 0.9 - 0.4x$ kN, $M = 0.2x^2 - 0.9x - 1.8$ kN·m

6.36 (a) $M_A = 60b^2$ lb·ft, $M_C = 960(b - 4)$ lb·ft; (b) $b = 3.31$ ft, $M = 659$ lb·ft

6.38 (a) 2.5 m; (b) for AC: $V = -2.5$ kN, $M = 2.5x$ kN·m; for CD: $V = 0.5$ kN, $M = 7.5 - 0.5x$ kN·m; for DB: $V = 3.5$ kN, $M = 3.5(6 - x)$ kN·m

6.40 (1) For $0 < x < L/2$: $V = -P/2$, $M = Px/2$; for $L/2 < x < L$: $V = P/2$, $M = P(L - x)/2$; (2) For $0 < x < L/2$: $V = Px/L$, $M = Px/2$; for $L/2 < x < L$: $V = -P(L - x)/L$, $M = P(L - x)/2$

6.42 For $0 < x < 2$ ft: $V = 20x^2$ lb, $M = -20x^3/3$ lb·ft; for 2 ft $< x < 6$ ft: $V = 20x^2 - 40(x - 2)^2$ lb, $M = -40(x - 2)^3/3 + 20x^3/3$ lb·ft

6.44–6.62 (No answers)

6.64 $H = 14.70$ ft, $\theta = 15.64°$

6.66 $s = 86.0$ ft, $T = 960$ lb

6.68 $H = 12.76$ ft, $T_B = 2.80$ lb

6.70 (a) 440 N; (b) 33.7 m

6.72 0.338

6.74 36.4 kg

6.76 35.9 ft

6.78 26.7 ft

6.80 $h = 8.13$ ft, $T_{BC} = 72.1$ lb, $T_{CD} = 100$ lb

6.82 $\beta_2 = 27.8°$, $\beta_3 = -3.53°$, $T_{AB} = 5.04$ kips, $T_{BC} = 4.36$ kips, $T_{CD} = 3.87$ kips

6.84 13.03 ft

6.86 $P = 16.62$ kN, $T_{AB} = 18.46$ kN, $T_{BC} = 11.42$ kN, $T_{CD} = 17.73$ kN

6.88 0.420

6.90 1.50 kN

CHAPTER 7

7.2 29.3 lb $\leq P \leq 109.3$ lb

7.4 33.7 lb

7.6 36.9°

7.8 297 lb

7.10 (a) 5.89°; (b) 7.10°

7.12 6.74 N

7.14 (a) $N_A = 94.2$ N, $N_B = 141.2$ N, $N_C = 377$ N; (b) 0.50

7.16 Cylinder cannot be at rest

7.18 0.270

7.20 153.2 N

7.22 Disk is in equilibrium

7.24 (a) Plank will slide at A; (b) Plank will not move

7.26 1.688 m

7.28 Box cannot be in equilibrium

7.30 176.6 N

7.32 115.3 N

7.34 0.686 lb

7.36 47.1°

7.38 Blocks tip

7.40 353 lb

7.42 10.83 kN

7.44 12.68°

7.46 13.14 ft

7.48 8.33 in.

7.50 18.15°

7.52 0.774 lb

7.54 (No answer)

7.56 (a) 1086 N; (b) 0.711 N·m

7.58 (a) 125.3 lb·in.; (b) 0

7.60 $P = 76.2$ lb, $C = 15\,960$ lb·in.

7.62 3.61 lb

7.64 0.190

7.66 40.3°

7.68 (a) $\mu_s/\sin(\beta/2)$

7.70 $(5/8)\mu_k PR$

7.72 (a) 6.01 N·m; (b) 5.89 N·m

7.74 3.77 N·m

7.76 17.25 in.

7.78 $1.461W$

7.80 0.772

7.82 0.203

7.84 Bar is in equilibrium

7.86 $T = \mu_s W/\sqrt{1 + \mu_s^2}$, $\theta = \tan^{-1}\mu_s$

7.88 27.2 N·m

7.90 119.4 N

CHAPTER 8

8.2 $\bar{x} = 2.86$ ft, $\bar{y} = 0.625$ ft

8.4 $\bar{x} = 0.1543$ m, $\bar{y} = 0.281$ m

8.6 $\bar{x} = 2.94$ in., $\bar{y} = 2.03$ in.

8.8 $\bar{x} = 3.33$ in., $\bar{y} = 1.0$ in.

8.10 (a) $\bar{x} = \bar{y} = \dfrac{4}{3\pi}\dfrac{3R^2 + 3Rt + t^2}{2R + t}$

8.12 $\bar{x} = \bar{y} = 2R/\pi$

8.14 $\bar{x} = 2.47$ in., $\bar{y} = 0$

8.16 $\bar{x} = 0$, $\bar{y} = 0.557$ in.

8.18 $\bar{x} = 20.2$ mm, $\bar{y} = 9.81$ mm

8.20 $\bar{x} = 0.872$ in., $\bar{y} = 5.96$ in.

8.22 $\bar{x} = 6.74$ in., $\bar{y} = 2.79$ in.

8.24 $\bar{x} = \bar{y} = \dfrac{b^2 + bt - t^2}{2(2b - t)}$

8.26 2.34 ft

8.28 $\bar{x} = -3.6$ mm, $\bar{y} = 50.9$ mm

8.30 $\bar{x} = 1.0$ ft, $\bar{y} = 1.5$ ft

8.32 $\bar{x} = 200$ mm, $\bar{y} = 115.0$ mm

8.34 $\sqrt{3}$

8.36 $\bar{x} = 6.60$ in., $\bar{y} = 4.32$ in.

8.38 119.7 ft

8.40 $\bar{x} = \bar{y} = 0$, $\bar{z} = R/2$

8.42 $\bar{x} = \bar{z} = 0$, $\bar{y} = 3.125$ in.

8.44 $\bar{x} = \bar{z} = 0$, $\bar{y} = \dfrac{n + 2}{2(2n + 2)}h$

8.46 R/π

8.48 $\bar{x} = \dfrac{2(2h_1 + h_2)}{3(3h_1 + h_2)}a,\ \bar{y} = \dfrac{2(2h_1 + h_2)}{3(3h_1 + h_2)}b$

8.50 $\bar{x} = -(2/\pi)^2 R,\ \bar{y} = 2R/\pi,\ \bar{z} = h/3$

8.52 $\bar{x} = \bar{y} = 3.62$ ft, $\bar{z} = 1.0$ ft

8.54 $\bar{x} = 15.98$ mm, $\bar{y} = 26.5$ mm, $\bar{z} = 20.3$ mm

8.56 $\bar{x} = 28.2$ mm, $\bar{y} = 0$, $\bar{z} = -3.1$ mm

8.58 -29.4 mm

8.60 $\bar{x} = 1.408$ in., $\bar{y} = 1.089$ in., $\bar{z} = 0.957$ in.

8.62 $\bar{x} = 77.8$ mm, $\bar{y} = 161.1$ mm, $\bar{z} = 100$ mm

8.64 $\bar{x} = 18.85$ mm, $\bar{y} = 19.65$ mm, $\bar{z} = -15.72$ mm

8.66 166.7 mm

8.68 $\bar{x} = 0$, $\bar{y} = 1.74$ in., $\bar{z} = 3.21$ in.

8.70 $\bar{x} = 62.7$ mm, $\bar{y} = \bar{z} = 0$

8.72 $\bar{x} = 12$ in., $\bar{y} = 3.83$ in., $\bar{z} = 0$

8.74 $\bar{x} = \bar{z} = 0$, $\bar{y} = 1.640$ m

8.76 $V = 1257$ in.^{3}, $A = 766$ in.^{2}

8.78 $A = 258$ in.^{2}, $V = 366$ in.^{3}

8.80 50.3 in.^{2}

8.82 1923 in.^{3}

8.84 114.2×10^3 mm^3

8.86 7850 m^3

8.88 1.702

8.90 $\bar{x} = -0.056$ m, $\bar{y} = 0$, $\bar{z} = 0.818$ m

8.92 $\bar{x} = 2.5$ in., $\bar{y} = 4.14$ in., $\bar{z} = 0.858$ in.

8.94 $250\tan\alpha$ N

8.96 $\bar{x} = 2.16$ in., $\bar{y} = 4.56$ in., $\bar{z} = 0.71$ in.

8.98 $\bar{x} = 0$, $\bar{y} = -3.12$ mm, $\bar{z} = 5.21$ mm

8.100 3.10 ft

8.102 $27.7°$

8.104 $R = 2780$ lb, $\bar{x} = 7.26$ ft

8.106 $R = 2w_0 a$, $\bar{x} = 0$, $\bar{y} = \pi a/4$

8.108 $R = 0$, $C^R = 9000$ lb · in. CCW

8.110 $R = 149.1$ N, $\bar{x} = 0.0483$ m, $\bar{y} = 0.0644$ m

8.112 (a) $\pi a^2 p_0$; (b) $\pi a^2 p_0$; (c) $2.22 a^2 p_0$; (d) $(ap_0/2)\sqrt{4h^2 + \pi^2 a^2}$

8.114 $1.111 p_0 d^2$

8.116 (a) $\gamma c(h^2 + b^2)/6$; (b) $\gamma c(h^2/6 + 4b^2/15)$; (c) $\gamma c(h^2/6 + b^2/2)$

8.118 Safe against tipping

8.120 $R = 0.622$ N, $\bar{y} = 0$, $\bar{z} = 28.8$ mm

8.122 (No answer)

8.124 $A = 3.22 \times 10^6$ N, $N_B = 3.79 \times 10^6$ N

8.126 $\bar{x} = 8.84$ in., $\bar{y} = 0$, $\bar{z} = 21.45$ in.

8.128 52.1×10^3 mm^2

8.130 $T_A = T_C = 16.94$ lb, $T_B = 13.68$ lb

8.132 $\bar{x} = (11/28)b$, $\bar{y} = (93/280)h$

8.134 44.3 in.

8.136 (a) 276 in.^{2}; (b) 276 in.^{2}

8.138 $\bar{x} = 0$, $\bar{y} = 6.53$ in.

8.140 7.47 lb

CHAPTER 9

9.2 $A = 360$ mm^2, $\bar{I}_x = 24 \times 10^3$ mm^4, $\bar{I}_y = 26 \times 10^3$ mm^4

9.4 826 mm

9.6 $(tL^3 \sin^2 \alpha)/3$

9.8 $R^4 \alpha/2$

9.10 $I_x = 1/28$ m^4, $I_y = 1/20$ m^4

9.12 23.15×10^3 in.^{4}

9.14 $b = 7.35$ in., $h = 12.87$ in.

9.16 $\bar{I}_x = 157.9$ in.^{4}, $\bar{I}_y = 59.1$ in.^{4}

9.18 17.31×10^6 mm^4

9.20 12.96×10^6 mm^4

9.22 $\bar{I}_x = 227$ in.^{4}, $\bar{I}_y = 47.2$ in.^{4}

9.24 16.89 in.^{4}

9.26 $0.433R$

9.28 1.0

9.30 $32.7°$

9.32 $I_x = 63.0 \times 10^6$ mm^4, $I_y = 7.90 \times 10^6$ mm^4

9.34 $h^2 b^2/6$

9.36 $tR^3/2$

9.38 -120×10^3 mm^4

9.40 15.125×10^6 mm^4

9.42 -61.4×10^6 mm^4

9.44 -323×10^3 mm^4

9.46 792 in.^{4}

9.48 -2.76×10^9 mm^4

9.50 (a) $I_1 = 16$ in.^{4}, $I_2 = 9$ in.^{4}, x- and y-axes are principal axes; (b) $I_u = 14.25$ in.^{4}, $I_v = 10.75$ in.^{4}, $I_{uv} = 3.03$ in.^{4}

9.52 $\bar{I}_1 = 103.7$ in.^{4}, $\bar{I}_2 = 29.6$ in.^{4}, $\theta_1 = 29.9°$, $\theta_2 = 119.9°$

9.54 $I_u = 1057$ in.^{4}, $I_v = 3943$ in.^{4}, $I_{uv} = -899$ in.^{4}

9.56 $I_x = 6800$ in.^{4}, $I_y = 5800$ in.^{4}, $I_{xy} = -1200$ in.^{4}

9.58 60×10^6 mm^4

9.60 9.96×10^6 mm^4

9.62 $I_u = I_v = \sqrt{3}a^4/96$, $I_{uv} = 0$ for all θ. Therefore, every axis is a principal axis.

9.64 (a) $I_1 = 0.983 \times 10^6$ mm^4; (b) $\theta_1 = 28.5°$, $\theta_2 = 118.5°$

9.66 7130 in.^{4}

9.68 $I_1 = 143.6 \times 10^6$ mm^4, $I_2 = 19.8 \times 10^6$ mm^4, $\theta_1 = 41.4°$, $\theta_2 = 131.4°$

9.70 $I_u = 33.3 \times 10^6$ mm^4, $I_v = 71.0 \times 10^6$ mm^4, $I_{uv} = 22.5 \times 10^6$ mm^4

9.72 $I_u = 95.7$ in.^{4}, $I_v = 304.3$ in.^{4}, $I_{uv} = 38.5$ in.^{4}

9.74 $I_u = 2 \times 10^6$ mm^4, $I_v = 28 \times 10^6$ mm^4, $I_{uv} = 0$

9.76 (a) $45°$; (b) $I_u = I_v = 5 \times 10^6$ mm^4, $|I_{uv}| = 3 \times 10^6$ mm^4

9.78 (a) 0.983×10^6 mm^4; (b) $\theta_1 = 28.5°$, $\theta_2 = 118.5°$

9.80 (a) $I_x = 56.2 \times 10^6$ mm^4, $I_y = 33.8 \times 10^6$ mm^4; (b) $I_u = 33.2 \times 10^6$ mm^4, $I_v = 56.8 \times 10^6$ mm^4

9.82 $\bar{I}_x = 4.084 \times 10^6$ mm^4, $\bar{I}_y = 0.816 \times 10^6$ mm^4

9.84 $I_1 = 380 \times 10^6$ mm^4, $I_2 = 120 \times 10^6$ mm^4, $\theta_1 = 56.3°$, $\theta_2 = 146.3°$

9.86 3.21%

9.88 (a) $\bar{I}_1 = 73.3$ in.^{4}; (b) $\bar{I}_{xy} = -15.26$ in.^{4}

9.90 $\bar{I}_x = 1.544 \times 10^6$ mm^4, $\bar{I}_y = 0.287 \times 10^6$ mm^4, $\bar{I}_{xy} = 0$

9.92 $I_x = I_y = 21.9$ in.^{4}, $I_{xy} = 21.3$ in.^{4}

9.94 $I_u = 90.5$ in.^{4}, $I_v = 313.5$ in.^{4}, $I_{uv} = -69.7$ in.^{4}

CHAPTER 10

10.2 $(WL\sin\theta)/2$

10.4 $38.5°$

10.6 17.32 lb

10.8 $(W\tan\theta)/2$

10.10 332 lb

10.12 1.697 kN

10.14 624 lb · in.

10.16 24 N · m

10.18 2.07 kN · m

10.20 (No answer)

10.22 71.2 lb

10.24   400 lb · ft
10.26   55.2 N
10.28   180 N · m
10.30   240 N
10.32   15.37 kN
10.34   10.19 lb
10.36   Gripping force = 6.0P
10.38   (a) 27.5 in. directly below B; (b) 700 lb · in.
10.40   Unstable
10.42 $\theta = 32.5°$ is stable

10.44 $W_1/W_2 < 1$
10.46 $\theta = 30°$ is stable, $\theta = 90°$ is unstable
10.48 $P < k/L$
10.50   9.08 m
10.52   Stable
10.54 $\theta = 45°$ is unstable
10.56   2.06 kg, unstable
10.58   79.1°, stable
10.60 $\theta = 29.6°$ is unstable, $\theta = 53.1°$ is stable
10.62   6.15°

Index